Scrivener Publishing
100 Cummings Center, Suite 541J
Beverly, MA 01915-6106

Advance Materials Series

The Advance Materials Series provides recent advancements of the fascinating field of advanced materials science and technology, particularly in the area of structure, synthesis and processing, characterization, advanced-state properties, and applications. The volumes will cover theoretical and experimental approaches of molecular device materials, biomimetic materials, hybrid-type composite materials, functionalized polymers, superamolecular systems, information- and energy-transfer materials, biobased and biodegradable or environmental friendly materials. Each volume will be devoted to one broad subject and the multidisciplinary aspects will be drawn out in full.

Series Editor: Dr. Ashutosh Tiwari
Biosensors and Bioelectronics Centre
Linköping University
SE-581 83 Linköping
Sweden
E-mail: ashutosh.tiwari@liu.se

Publishers at Scrivener
Martin Scrivener(martin@scrivenerpublishing.com)
Phillip Carmical (pcarmical@scrivenerpublishing.com)

Advanced Healthcare Materials

Edited by

Ashutosh Tiwari

Biosensors and Bioelectronics Centre,
Linköping University, Sweden

Scrivener
Publishing

WILEY

Contents

Preface

Advanced healthcare materials are attracting strong interest in fundamental as well as applied medical science and technology. *Advanced Healthcare Materials* summarizes the current state of knowledge in the field of Advanced Materials for functional therapeutics, point-of-care diagnostics, translational materials and up-and-coming bioengineering devices. In this book we have highlighted the key features which enable the design of stimuli-responsive smart nanoparticles, novel biomaterials, and nano/micro devices for either diagnosis or therapy, or both, called theranostics. The latest advancements in healthcare materials and medical technology are also presented. In narrative outline, this volume of the Advanced Materials series includes fourteen chapters divided into four main areas: "Functional Therapeutics," "Point-of-Care Diagnostics," "Translational Materials" and "Up-and-Coming Bioengineering Devices."

The chapter "Stimuli-Responsive Smart Nanoparticles for Biomedical Application," describes the synthesis and engineering of stimuli-responsive polymeric nanosystems and their use in sensors, logic operations, biomedicine, tissue engineering and regenerative medicine, synthetic muscles, "smart" optical or microelectromechanical systems, membranes, electronics and self-cleaning surfaces. The chapter entitled "Diagnosis and Treatment of Cancer – Where We Are and Where We Have to Go!" is an overview of new methods and technology such as functional nanoparticles-based drug delivery and diagnostics systems for overcoming obstacles in cancer diagnosis and treatment. Also, exploratory fundamental and cutting-edge accounts of advanced materials including nanoparticles, nanopolymers, metal-organic frameworks and zeolites in drug delivery and diagnostics are presented in the chapter, "Advanced Materials for Biomedical Application and Drug Delivery." Another chapter, "Nanoparticles for Diagnosis and/or Treatment of Alzheimer's Disease," focuses on the nanotheranostic approach to Alzheimer's treatment.

The chapters "Novel Biomaterials for Human Health: Hemocompatible Polymeric Micro- and Nanoparticles and Their Application in Biosensor" and "The Contribution of Smart Materials and Advanced Clinical Diagnostic Micro-Devices on the Progress and Improvement of Human Health Care," cover the application of advanced healthcare materials for point-of-care diagnostics. The notable advantages and limitations of translational biomaterials are described in the chapters "Hierarchical Modeling of Elastic Behavior of Human Dental Tissue Based on Synchrotron Diffraction Characterization," "Biodegradable Porous Hydrogels," and "Hydrogels: Properties, Preparation, Characterization and Biomedical Applications in Tissue Engineering, Drug Delivery and Wound Care." Up-and-coming bioengineering devices are covered in the chapters entitled "Modified Natural Zeolites – Functional Characterization and Biomedical Application," "Supramolecular Hydrogels Based on Cyclodextrin Poly(Pseudo)Rotaxane for New and Emerging Biomedical Applications," "Polyhydroxyalkanoate-Based Biomaterials for Applications in Biomedical Engineering," "Biomimetic Molecularly Imprinted Polymers as Smart Materials and Future Perspective in Health Care," and "The Role of Immunoassays in Urine Drug Screening."

This book has been written for a large readership including university students and researchers from diverse backgrounds such as chemistry, materials science, physics, pharmacy, medical science, and biomedical engineering. It can be used not only as a textbook for both undergraduate and graduate students, but also as a review and reference book for researchers in materials science, bioengineering, medical, pharmacy, biotechnology and nanotechnology. We hope the chapters of this book will provide readers with valuable insight into the important area of advanced healthcare materials, especially the cutting-edge technology in functional therapeutics, point-of-care diagnostics, translational materials and up-and-coming bioengineering devices. The interdisciplinary nature of the topics in this book will help young researchers and senior academicians. The main credit for this book goes to the contributors who have comprehensively written their updated chapters in the field of Advanced Healthcare Materials.

Ashutosh Tiwari, PhD, DSc
Linköping, Sweden
March 6, 2014

List of Contributors

Debbie P. Anderson works as a researcher in the Bioproducts and Bioprocesses National Science Program, Agriculture and Agri-Food Canada, Government of Canada, and her research interests are centered around applications of biopolymers.

Sophia G. Antimisiaris is Professor in the Department of Pharmacy, University of Patras & FORTH/ICES, Patras), Greece

Peter R. Chang serves as a Research Scientist and Professor in the Bioproducts and Bioprocesses National Science Program, Agriculture and Agri-Food Canada, Government of Canada, and his primary interests reside in functional systems derived from biopolymers.

Q.Z. Chen is an Associate Professor in the Department of Materials Engineering at Monash University, Australia and works in the field of elastomeric biomaterials for applications in soft tissue engineering.

Aleksandra Daković is a Full Research Professor at the Institute for Technology of Nuclear and Other Mineral Raw Materials, Belgrade, Serbia.

Arnab De is in the Department of Microbiology and Immunology, Columbia University where she obtained her PhD and researches in the interface of chemistry and biology.

Khalil Farhadi is a Professor of Chemistry, Department of Chemistry, Urmia University, Iran.

Farnoush Faridbod is an Assistant Professor in analytical chemistry at the University of Tehran, Iran.

Luis P. Fonseca is an Associate Professor in the Department of Bioengineering of Instituto Superior Tecnico at the University of Lisbon,

Portugal and Senior Researcher of Bioengineering Research Group of the Institute for Biotechnology and Bioengineering in Lisbon.

Mehrdad Forough is a PhD Student in the Department of Chemistry, Urmia University, Iran.

Mohammad R. Ganjali is a Professor in the Center of Excellence in Electrochemistry, Faculty of Chemistry at the University of Tehran, Iran.

Rajiv Lochan Gaur is a Research Associate in the School of Medicine, Stanford University, Palo Alto, California, USA.

Jing Hao works as a researcher at the College of Chemical Engineering, Wuhan University of Technology, China and her research interests focus on biomedical materials based on assembly.

Jin Huang is a full professor, College of Chemical Engineering, Wuhan University of Technology, Wuhan, China, and his research interests include, but are not limited to, fabrication and application of assemblies and composites.

Alexander M. Korsunsky is a Professor of Engineering Science whose research group is based at Oxford and Harwell and pursues multi-scale modeling, multi-modal microscopy and "rich" tomography of biomaterials and engineered materials, including metallic alloys, ceramics, polymers, composites, and coatings.

Danina Krajišnik is an Assistant Professor at the Department of Pharmaceutical Technology and Cosmetology of the University of Belgrade, Faculty of Pharmacy, Belgrade, Serbia.

Chun Mao is a Professor at the Jiangsu Key Laboratory of Biofunctional Materials at Nanjing Normal University, China and his research activities focus on surface modification, biomaterials and biological molecule detection.

Eleni Markoutsa is a PhD student in the Department of Pharmacy, University of Patras, Greece.

Jela Milić is a Full Professor at the Department of Pharmaceutical Technology and Cosmetology of the University of Belgrade, Faculty of Pharmacy, Belgrade, Serbia.

Sushil Mishra is a graduate student pursuing his PhD in Dr. Mozumdar's lab at the University of Delhi, India.

Rahim Molaei is a PhD Student in the Department of Chemistry, Urmia University, Iran.

Spyridon Mourtas is a Post-Doc Researcher in the Department of Pharmacy, University of Patras, Greece.

Subho Mozumdar is a pioneer of nanotechnology in India. He obtained his PhD from SUNY, Buffalo and continued his post-doctoral research at Johns Hopkins. In recognition of his discoveries, he recently became the Academic Editor of *Plos One*.

Parviz Norouzi is a Professor in the Center of Excellence in Electrochemistry, Faculty of Chemistry at the University of Tehran, Iran.

Stanley L. Okon is a Resident Physician in the Department of Psychiatry at Advocate Lutheran General Hospital located in Park Ridge, Illinois.

Konstantina Papadia is a PhD student in the Department of Pharmacy, University of Patras, Greece

Niina J. Ronkainen is an Associate Professor of Chemistry at Benedictine University located in Lisle, Illinois, USA.

George E. Rottinghaus is Clinical Professor at the Veterinary Medical Diagnostic Laboratory, College of Veterinary Medicine, University of Missouri, Columbia, USA.

Yasaman Shaabani is a MSc Student at the Faculty of Chemical Engineering, Urmia University of Technology, Iran.

Jian Shen is a Leader at the Jiangsu Key Laboratory of Biofunctional Materials at Nanjing Normal University, China and his current research areas are anticoagulant materials and biomaterials.

Athanassios Skouras, PhD student in the Department of Pharmacy, University of Patras, Greece.

M. Sirousazar is an Assistant Professor of Chemical Engineering , Faculty of Chemical Engineering, Urmia University of Technology, Urmia, Iran.

Richa Srivastava works in the Biotechnology Division, Central Institute of Medicinal and Aromatic Plants, Lucknow, India.

Chong Sun is a doctoral candidate at Nanjing Normal University of Science and Technology, China and her current research interests are nanomaterials and electrochemistry analytical methods.

Tan Sui is a post-doctoral researcher in the Department of Engineering Science, University of Oxford, and specializes in imaging, structural analysis and modeling of mineralized biological tissues.

Fernando Teles works as an Assistant Researcher at the Institute of Hygiene and Tropical Medicine, Lisbon, Portugal with the tasks of Science Manager of the Institute and Microbiology Researcher at its Unit of Medical Microbiology.

Xiaobo Wang is a doctoral candidate at Nanjing Normal University, China and her current research interests are surface modification and biomaterials.

C.H. Zhu is a PhD candidate in the Department of Materials Engineering at Monash University, Australia.

Part 1
FUNCTIONAL THERAPEUTICS

1

Stimuli-Responsive Smart Nanoparticles for Biomedical Application

Arnab De[2], Sushil Mishra and Subho Mozumdar[1],*

[1]Department of Chemistry, University of Delhi, Delhi-110007, India
[2]Department of Microbiology and Immunology, Columbia University, USA

Abstract

Biological systems consist largely of regulation systems; these natural feedback regulation systems are very important to stabilize such non-equilibrium systems like a living organism. One example is release of hormones from secretory cells, which is regulated by physiological cycles or by specific input signals. It is not surprising that regenerative medicine and drug delivery are also utilizing similar responsive strategies in a biomimetic fashion. During the last two decades, scientists have been trying to mimic nature in designing "smart" synthetic materials from various functional molecular building blocks that respond to stimuli such as temperature, pH, ionic strength, light, electric or magnetic field, chemical and biochemical stimuli in order to mediate molecular transport, shape changes, tune adhesion and wettability, or to induce signal transduction of (bio-)chemical or physical stimuli into mechanical, optical or electrical responses. Biomimetic approaches have been employed in the design, synthesis and engineering of stimuli-responsive polymeric systems, which undergo reversible abrupt phase transitions upon variation of a variable around a critical point and their use in a plethora of applications, including sensors, logic operations, biomedicine, tissue engineering and regenerative medicine, synthetic muscles, "smart" optical or microelectromechanical systems, membranes, electronics and self-cleaning surfaces has been explored.

Keywords: Biological systems, nanomedicine, nanoparticles, biomedical applications

---

*Corresponding author: subhoscom@yahoo.co.in

Ashutosh Tiwari (ed.) Advanced Healthcare Materials, (1–36) 2014 © Scrivener Publishing LLC

1.1   A Brief Overview of Nanotechnology

Nanotechnology has emerged in the last decades of the 20th century with the development of new enabling technologies for imaging, manipulating, and simulating matter at the atomic scale. The frontier of nanotechnology research and development encompasses a broad range of science and engineering activities directed toward understanding and creating improved materials, devices and systems that exploit the properties of matter that emerge at the nanoscale. The results promise benefits that will shift paradigms in biomedicine (e.g., imaging, diagnosis, treatment, and prevention); energy (e.g., conversion and storage); electronics (e.g., computing and displays); manufacturing; environmental remediation; and many other categories of products and applications.

Amongst leading scientists, there is growing awareness about the tremendous impact this field will have on society and the economy. It is forecasted to become possibly even more important than, for example, the invention of the steam engine or the discovery of penicillin.

The landmark lecture by eminent Nobel Laureate Richard Feynman in 1959 entitled "There's plenty of room at the bottom," brought life (to) the concept of nanotechnology, which has been influencing all the different fields of research involving hard core science such as chemistry, physics, and other applied fields of science, such as electronics, materials science and biomedical science, agrochemicals, medicine and pharmaceutical sciences etc. [1].

Nanotechnology and nanoscience are widely seen as having a great potential to bring benefits to many areas of research and applications. They are attracting increasing investments from governments and private sector businesses in many parts of the world. Concurrently, the application of nanoscience is raising new challenges in the safety, regulatory, and ethical domains that will require extensive debates on all levels.

The prefix nano is derived from the Greek word dwarf. One nanometer (nm) is equal to one-billionth of a meter, that is, 10^{-9} m. The term "nanotechnology" was first used in 1974, when Norio Taniguchi, a scientist at the University of Tokyo, Japan, referred to materials in nanometers.

At the nanometer scale, the physical, chemical and biological properties of nanomaterials are fundamentally different from those of individual atoms, molecules, and bulk materials. They differ significantly from other materials due to two major principal factors: the increased surface area and quantum effects. A larger surface area usually results in more reactive chemical properties and also affects the mechanical or electrical properties of the materials. At the nanoscale, quantum effects dominate the behaviors of a material, affecting its optical, electrical and magnetic properties. By exploiting these novel properties, the main purpose of research and development in nanotechnology is to understand and create materials, devices and systems with improved characteristics and performances [2].

1.2 Nanoparticulate Delivery Systems

The nanoparticulate system comprises of particles or droplets in the submicron range, i.e., below 1μm, in an aqueous suspension or emulsion, respectively. This small size of the inner phase gives such a system unique properties in terms of appearance and application. The particles are too small for sedimentation; they are held in suspension by the Brownian motion of the water molecules. They have a large overall surface area and their dispersions provide a high solid content at low viscosity.

Historically, the first nanoparticles proposed as carriers for therapeutic applications were made of gelatin and cross-linked albumin [3]. Use of proteins may stimulate the immune system, and to limit the toxicity of the cross-linking agents, nanoparticles made from synthetic polymers were developed. At first, the nanoparticles were made by emulsion polymerization of acrylamide and by dispersion polymerization of methylmethacrylate [4]. These nanoparticles were proposed as adjuvants for vaccines. Couvreur *et al.* [5] proposed to make nanoparticles by polymerization of monomers from the family of alkylcyanoacrylates already used in vivo as surgical glue. During the same period of time, Gurny *et al.* [6] proposed a method for nanoparticle synthesis from another biodegradable polymer consisting of poly(lactic acid) used as surgical sutures in humans. Based on these initial investigations, several groups improved and modified the original processes mainly by reducing the amount of surfactant and organic solvents. A breakthrough in the development of nanoparticles occurred in 1986 with the development of methods allowing the preparation of nanocapsules corresponding to particles displaying a core-shell structure with a liquid core surrounded by a polymer shell [7]. The nanoprecipitation technique was proposed as well as the first method of interfacial polymerization in inverse microemulsion [8]. In the succeeding years, the methods based on salting-out [9], emulsion–diffusion [10], and double emulsion [11] were described. Finally, during the last decade, new approaches were considered to develop nanoparticles made from natural origin such as polysaccharides [12]. These nanoparticles were developed for peptides and nucleic acid delivery. A further development was surface modification of nanoparticles to produce long circulating particles able to avoid the capture by the macrophages of the mononuclear phagocyte system after intravenous administration [13].

1.3   Delivery Systems

The specific delivery of active principles to the target site, organ, tissue, or unhealthy cells by carriers is one of the major challenges in bioactive delivery research. Many of the bioactive compounds have physicochemical characteristics that are not favorable to transit through the biological barriers that separate the administration site from the site of action. Some of the active compounds run up against enzymatic barriers, which lead to their degradation and fast metabolization. Therapeutically, distribution of such active molecules to the diseased target zones can therefore be difficult. Moreover, the accumulation of drugs in healthy tissues can cause unacceptable toxic effects, leading to the abandonment of treatment despite its effectiveness [14].

In order to overcome the above challenges an ideal delivery system must possesses basically two elements: the ability to transport loaded payload to the target site and control its release. The targeting will ensure high efficiency of loaded payload at the site of core interest and reduces any unwanted biological effects. Various delivery devices have been developed and an overview of each type of nanocarrier is given in the following section.

According to the process used for the preparation of nanoparticles, nanospheres or nanocapsules can be obtained. Nanospheres are homogeneous matrix systems in which the drug is dispersed throughout the particles. Nanocapsules are vesicular systems in which the drug is confined to a cavity surrounded by a polymeric membrane [15].

1.3.1   Hydrogels

Hydrogels are three-dimensional networks composed of hydrophilic polymer chains. They have the ability to swell in water without dissolving. The type of cross-linking between the polymer chains can be chemical (covalent bonds) or physical (hydrogen bonds or hydrophobic interactions). The high water content in these materials makes them highly biocompatible. There are natural hydrogels such as DNA, proteins, or synthetic, e.g., poly(2-hydroxyethyl methacrylate), poly(N-isopropylacrylamide) or a biohybrid [16]. The release mechanism can be induced by temperature or pH. Temperature-controlled release is due to the competition between hydrogen bonding and hydrophobic interactions. At lower temperatures, the hydrogen bonding between polar groups of the polymer is predominant, causing the polymer to swell in water. At higher temperatures, the hydrophobic interactions take over, leading to its shrinkage [17]. Glucose-sensitive hydrogels can release insulin in a controlled fashion in response to the demand [18].

1.3.2   Dendrimers

Dendrimers are highly branched cascade molecules that emanate from a central core through a stepwise repetitive reaction sequence. Such a molecule consists of three topologically different regions: a small initiator core of low density and multiple branching units, the density of which increases with increasing distance from core, thus eventually leading to a rather densely packed shell. Finally, outer terminal units for shielding actually amount to an encapsulation that can create a distinct microenvironment around the core moiety and hence affect its properties [19].

Dendrimers can be synthesized in multiple ways. A dendrimer can be synthesized originating form core by repetition of a sequence of reactions, which allows fast growth of the dendrimer in both size and in number of terminal groups [20]. Another method is the convergent method, in which the core is incorporated in the final step of elaboration of the dendrimer [21].

Owing to their large number of surface groups, dendrimers have the ability to create multivalent interactions [22]. Dendritic structures may also be engineered to encapsulate certain hydrophobic drugs like indomethacin [23].

The dendrimeric surface can be tuned for functional groups to induce an electrostatic-type interaction with active molecules. For example, negatively charged DNA chains can be complexed to positively charged dendrimers. Several research groups have demonstrated that dendrimer/DNA complexes, which are very compact, easily penetrate cells by endocytosis and therefore improve transfection [24]. In some cases, the bulkiness of the dendrimer and the density of its structure make the cleavage of the water-soluble and biodegradable bonds of the peripheral layer quite difficult [25]. Delivery of active principles is therefore not so straightforward. In other cases, the encapsulated molecules are not well trapped and may be released prematurely [23]. Nevertheless, the functional groups of dendrimers can be easily tuned and therefore make versatile drug vectors.

1.3.3 Liposomes

Liposomes are vesicles formed by the auto-association of one or several phospholipid bilayers that enclose an aqueous compartment. They have attracted the attention of a number of research groups in various fields, such as physical chemistry, biophysics, and pharmaceutics because of their structure, which is comparable to the phospholipid membranes of living cells [26]. The innocuous nature of phospholipidic components in liposome make them suitable reservoir systems that rapidly became the ideal candidates for drug vectorization in biological media. Liposomes are able to transport both hydrophobic substances anchored into the bilayer, and hydrophilic substances encapsulated in their cavity.

Temperature-sensitive liposomes have also been elaborated using lipids such as 1,2-dipalmitoyl-sn-glycero-3-phosphocholine, which has a phase-transition temperature between 41 and 43 °C. These liposomes could be used in association with hyperthermia treatments, for example, in the delivery of drugs into solid tumors [27]. Ligands can be anchored onto the liposome surface to deliver encapsulated drugs for specific action sites. These ligands can be antibodies, which bind to specific cell receptors, or less-specific ligands, such as folate or selectin [28]. Attachment of PEG to liposomes can also protect them from detection by monocytes and macrophages [29] in the liver and spleen, which allows a prolonged circulation time within the bloodstream. The liposomes utilized in doxil, which is marketed as a chemotherapy drug, are formulated with surface-bound methoxypolyethylene glycol (MPEG). Liposomes are thus versatile reservoir systems. The more they develop, the more sophisticated their compositions become, allowing very specific targeting and completely controlled drug delivery. However, these rather complex systems have to be systematically tuned according to the drug to be encapsulated and the desired application.

The physical and chemical stability of liposomes also limits their use in vectorization. Chemically, their poor stability can be attributed to lipid ester bond hydrolysis, and physically, the aggregation or the fusion of several liposomes can lead to the formation of large-sized objects that are therefore no longer usable in vectorization. Moreover, these objects may be subject to leakage, releasing the encapsulated drugs before they reach their site of action. Their preparation procedure also requires the use of an organic solvent, which can leave toxic residual traces.

1.3.4 Niosomes

Niosomes [14] are made of nonionic surfactants that are organized into spherical bilayers enclosing an aqueous compartment, and have an identical structure to liposomes and polymersomes. Several preparation methods [30] for niosomes have been described in the literature. In most cases, niosome formation requires the addition of molecules such as cholesterol to stabilize the bilayer and molecules that prevent the formation of niosome aggregates by steric or electrostatic repulsion.

In an analogous fashion to liposomes, niosomes are able to vectorize hydrophobic drugs enclosed in their bilayer and hydrophilic substances encapsulated in their aqueous cavity. Unlike phospholipidic liposomes, niosomes, which are made of surfactants, are not sensitive to hydrolysis or oxidation. This is an advantage for their use in biological media. Moreover, surfactants are cheaper and easier to store than phospholipids. A further advantage of niosomes relative to liposomes lies in their formulation, as these vectors can be elaborated from a wide variety of surfactants, the hydrophilic heads of which can be chosen according to the application and the desired site of action [30]. Notably, surfactant niosomes have been obtained with glycerol [31], ethylene oxide [32], crown ethers [33], and polyhydroxylated [34] or sugar-based [35] polar headgroups.

The encapsulation of active substances in niosomes can reduce their toxicity, increase their absorption through cell membranes, and allow them to target organs or specific tissues. Recently, antibody surface-functionalized niosomes [36] were developed in a similar way to virosomes.

Niosomes have been developed to reach the same specific drug delivery objectives as liposomes, thus overcoming the drawbacks of phospholipid use. However, niosome membranes are permeable to low-molecular-weight molecules, and a leakage of drugs encapsulated in the aqueous cavity of niosomes over time has been observed.

1.3.5 Polymersomes

Polymersomes are tank-like systems [37] consisting of a liquid central core enclosed in a thin polymer wall not more than a few nanometres thick [16b]. The polymersome membrane is formed from a block copolymer that is organized in a bilayer, in a similar fashion to those of the liposomes. These polymersomes have an aqueous internal cavity. Polymersomes exhibit versatile transport properties, as hydrophobic drugs can be enclosed in the membrane of the carrier, whereas hydrophilic drugs are encapsulated in their aqueous cavity.

Polymersome systems have been used for delivery of anticancer drugs, such as paclitaxel (hydrophobic) and doxorubicin (hydrophilic). Doxorubicin was encapsulated in the internal cavity of the polymersome, whereas paclitaxel was incorporated into the polymer bilayer during polymer film formation to maximize the anticancer drug efficiency with a cocktail of active substances [38]. Polymersomes were obtained by mixing two block copolymers, namely biodegradable PLA–PEG and inert poly(ethylene glycol)–poly(butadiene) (PEG–PBD). Hydrolysis of PLA–PEG then forms pores in the membrane, which allows the delivery of both drugs to be controlled. Twice as much apoptosis was induced in the tumors by the polymersome–drug cocktail after two days than by the two drugs taken separately.

Despite their efficiency, the major drawback of polymersomes is their instability, leading to leakage of the encapsulated drugs. Moreover, passive encapsulation used in the case of polymersomes requires a high amount of active substances, as the encapsulated concentration is identical to the concentration of the aqueous solution used to rehydrate the polymer film.

1.3.6   Solid Lipid Nanoparticle (SLN)

Nanoparticles [39] composed of lipids, which are solid at room and physiologic temperatures, are referred to as SLNs. These are typically composed of stabilizing surfactants, triglycerides, glyceride mixtures, and waxes. They are usually prepared by various procedures like high-pressure homogenization, microemulsion, and nanoprecipitation. Generally, lipids such as triglycerides are well tolerated by the organism. Moreover, the production of these nanoparticles is much simpler than that of the nanospheres and can be transposed to the industrial scale at lower cost.

The active substance required for the desired application is dissolved or dispersed into the melted lipid phase, and then one of the methods for SLN preparation is applied to obtain the drug-containing nanocarriers. Following fast cooling of the glycerides, an α-crystalline structure [40] is obtained that is unstable and not well ordered. Active molecules then preferentially gather in the amorphous areas of the matrix. However, the α-crystalline structure adopted by the lipids alters during standing to a β-crystalline structure [41], which is more stable and better ordered. During this rearrangement, the increase in the ordering of the lipid phase leads to an expulsion of the active substances into the amorphous regions [42]. Control of the lipid matrix transformation from the α form to the β form (for example, by temperature control) should therefore allow an on-command [39] release of the drug. However, to date, these SLN with controlled crystalline transformation have not been fully mastered.

As the drug loading capacity of the particles relies essentially on the structure and the polymorphism of the lipid forming the nanoparticles, some new types of lipid particles exhibiting amorphous zones have been developed [43]. These lipid particles, which are partially crystalline, can be composed of a mixture of glycerides with different fatty acids possessing various chain lengths and degrees of unsaturation, leading to an imperfect material, and therefore offering a better drug-loading rate. A second type of lipid particle, called multiple lipid particles, is obtained by mixing liquid lipids with solid lipids when preparing the nanoparticles. The active substances become localized in the oily compartments contained in the solid lipid particles. Finally, an amorphous system can be obtained with a particular mixture of lipids. The incorporation of active molecules into this kind of solid nanoparticles is one of the most efficient.

The use of these solid nanoparticles in drug vectorization is now under development, as both in vitro and in vivo studies have proved that these carriers are well tolerated. However, the polymorphism of these lipid matrixes and possible crystal rearrangements has to be controlled to avoid stability problems in these structures (gelification problems) [44]. Moreover, the release of the active molecules incorporated into these solid nanoparticles is not always well controlled, which limits their applications in vectorization.

1.3.7   Micro- and Nanoemulsions

Emulsions are heterogenous dispersions of two immiscible liquids such as oil in water (O/W) or water in oil (W/O). Without surfactant molecules, they are susceptible for rapid degradation by coalescence or flocculation leading to phase separation [45]. The use of micro- and nanoemulsions are becoming increasingly common in drug delivery systems. Microemulsion is used to denote a thermodynamically stable, fluid, transparent (or translucent) dispersion of oil and water, stabilized by an interfacial film of amphiphilic molecules [46]. The striking difference between a conventional emulsion (1–10 μm) and the microemulsion (200 nm–1 μm) is that the latter does not need any mechanical input for its formation as it is thermodynamically more stable. On the other hand, nanoemulsions (20–200 nm) are at best kinetically more stable.

Nanoemulsions are of great interest as pharmaceutical, cosmetic, etc., formulations [47]. Nanoemulsions are used as drug delivery systems for administration through various systemic routes. Parenteral administration [48] of nanoemulsions is employed for a variety of purposes, controlled drug delivery of vaccines or as gene carriers [49]. The benefit of nanoemulsions in the oral [50], ocular [49a, 51] administration of drugs has been also reported. Cationic nanoemulsions have been evaluated as DNA vaccine carriers to be administrated by the pulmonary route [52]. They are also interesting candidates for the delivery of drugs or DNA plasmids through the skin after topical administration [53]. The drawback in emulsion systems is the use of high concentration of surfactant, which leads to toxicity and embolism.

1.3.8 Micelles

Micelles are aggregates of amphiphilic molecules in which the polar headgroups are in contact with water and the hydrophobic moieties are gathered in the core to minimize their contact with water. The main driving force in the auto-association process of these surfactants is their hydrophobicity. The micelles form above a certain concentration, known as the critical micelle concentration (CMC). The mean size of these objects usually varies from 1 nm to 100 nm. The micellar systems are dynamic in nature, as the surfactants can exchange freely and rapidly between the micellar structure and the aqueous solution.

In addition to surfactants, block copolymers (having both a hydrophilic and a hydrophobic part) or triblock copolymers (with one hydrophobic and two hydrophilic parts or one hydrophilic and two hydrophobic parts) can also self-assemble to form polymeric micelles. These polymeric micelles have a mean diameter of 20 to 50 nm and are practically monodisperse. Polymeric micelles are generally more stable than surfactant micelles, and form at markedly lower CMCs. These objects are also much less dynamic than those formed from surfactants.

Polymeric micelles are more frequently used in vectorization than surfactant micelles. The slow degradation kinetics of polymeric micelles has contributed to their success in vectorization applications, usually for anticancer hydrophobic drug delivery (such as paclitaxel) to tumors.

Polymeric micelles also have the advantage of being able to deliver an active principle to its specific site of action if the polymer structure is tuned to make them sensitive to the medium in which they are found. An example is the development of pH-sensitive copolymers by inclusion of amine [54] or acid functional groups [55] into the copolymer skeleton, which changes the solubility of the polymer and therefore the stability of the vectors according to the pH. The active principles can then be delivered by micelle destabilization at a site of action possessing a specific pH.

The major drawback of micellar vectors, and in particular surfactant vectors, is their tendency to break up upon dilution. This is not the case for polymeric micelles, but their synthesis can sometimes prove difficult for use in biological applications, which have specific requirements, such as nontoxicity, biocompatibility, degradability, and accurate molecular weight.

1.3.9 Carbon Nanomaterials

Carbon nanomaterials for drug delivery applications mainly include fullerenes and carbon nanotubes (single and multiwalled). Considerable amount of work has been done to utilize them as nanocarriers for drug delivery [56]. The inert surface of these materials has posed challenges in terms of surface modifications to make them water soluble, biocompatible, and fluorescent. But despite all these, a number of recent reports establish that carbon nanotubes are toxic [57]. More recently glucose-derived functionalized carbon spheres [58] seem to present hopes as efficient nanocarriers. They have been shown to be nuclear targeting and nontoxic. But more detailed studies on their mechanism of entry and other possible applications are awaited.

1.4 Polymers for Nanoparticle Synthesis

The polymers that can degrade into biologically compatible components [59] under physiologic conditions present a far more attractive alternative for the preparation of delivery systems. Degradation may take place by a variety of mechanisms, although it generally relies on either erosion or chemical changes to the polymer. Degradation by erosion normally takes place in devices that are prepared from soluble polymers. In such instances, the device erodes as water is absorbed into the system causing the polymer chains to hydrate, swell, disentangle, and, ultimately, dissolve away from the dosage form. Alternatively, degradation could result from chemical changes to the polymer including, for example, cleavage of covalent bonds or ionization/protonation either along the polymer backbone or on pendant side-chains. As a necessity due to this process, biodegradable polymers and their degradation products must be biologically compatible and non-toxic. Consequently, the monomers typically used in the preparation of biodegradable polymers are often molecules that are endogenous to biological systems. Few biocompatible and biodegradable polymers used for nanoparticle synthesis for delivery purposes are discussed below.

1.4.1 Polyesters

A variety of hydrolytically labile polyesters have been evaluated in delivery applications [60]. Among these, however, poly(glycolide), poly(lactide), and various copolymers of poly(lactide-co-glycolide) are the ubiquitous choice because of their proven safety and lack of toxicity, their wide range of physicochemical properties, and their flexibility to be processed into a variety of physical dosage forms. These polymers remain popular for a variety of reasons including the fact that both of these materials have properties that allow hydrolytic degradation. Once degraded, natural pathways remove the degradation products, namely, the monomeric components of each polymer, glycolic acid that can be converted to other metabolites or eliminated by other mechanisms, and lactic acid that can be eliminated through the tricarboxylic acid (TCA) cycle [61].

Homopolymers of poly(D-lactide) and poly(L-lactide) tend to be semi-crystalline. As a result, water transport into these polymers is slow. Because of the slow uptake of water and the structural integrity introduced by crystallites, degradation rates of these polymers tend to be relatively slow (i.e., 18–24 months). In contrast, poly(D,L-lactide) (PLA) is amorphous and is observed to degrade somewhat faster (i.e., 12–16 months). Adding increasing proportions of glycolide into PLA lowers Tg and generally increases polymer hydrophilicity. In contrast, poly(L-lactide-co-glycolide) (PLGA) is amorphous when the glycolide content is 25–70 mole%. The most rapid degradation rate (i.e., 2 months) is observed in PLGA copolymers containing 50% glycolide. Poly(glycolide), (PGA) despite being semi-crystalline, is found to degrade relatively fast (i.e., 2–4 months) even compared to the amorphous PLA. This is attributed to the much greater hydrophilicity of the glycolide over the lactide. Actual degradation times, though, will depend on environmental conditions, polymer molecular weight, system geometry and morphology, and processing conditions [59].

PLGA-loaded nanoparticles have been developed for oral delivery of active molecule such as ellagic acid [62], streptomycin [63], estradiol [64] and cyclosporine [65]. PLGA nanoparticles showed an initial burst release and then sustained release phase for adriamycin [66], an anticancer drug. The cisplatin-loaded PLGA-mPEG nanoparticles appeared to be effective in delaying tumor growth in HT 29 tumor-bearing SCID mice. The group of mice treated with intravenous injection of cisplatin-loaded nanoparticles [67] exhibited a higher survival rate compared with the free cisplatin group. PLGA microspheres with an incorporated antigen [68] represent a good antigen delivery system for both cellular and humoral responses.

1.4.2   Poly-ε-caprolactone

Poly-ε-caprolactone (PCL) is derived by the ring opening polymerization of ε-caprolactone [59]. PCL is a biodegradable, biocompatible [69] and semicrystalline polymer having a melting point in the range of 59–64°C and very low glass transition temperature (Tg) of −60°C. PCL was developed as synthetic plastic material to be used in biodegradable packaging designed to reduce environmental pollution, like container for aerial planting of conifer seedlings [59]. Slow degradation of PCL has led it to the application in the preparation of different delivery systems in the form of microspheres, nanospheres and implants [69].

PLA degrade in two phases. In the first phase, a random hydrolytic chain scission occurs, which results in a reduction of the polymer molecular weight. In the second phase, the low molecular fragments and the small polymer particles are carried away from the site of implantation by solubilization in the body fluids or by phagocytosis [70], which results in a weight loss. Complete degradation and elimination of PCL homopolymers may last for 2 to 4 years. The degradation rate of PCL is still slower than other biodegradable polymers, thus making it suitable for long-term biological implantable systems. U.S. FDA approved Levonorgestrel containing an implantable contraceptive [71], Capronor®, has been fabricated by PCL matrix.

Indomethacin loaded submicron system of PCL developed by Calvo *et al.* [72] showed 300% ocular bioavailability in comparison to commercial solution. PCL has been used to develop other anti-inflammatory agents like flubiprofen [73] and diclofenac [74] containing nanospheres. Isradipine [75], a antihypertensive agent, was encapsulated by PCL as delivery system for oral administration, to reduce the initial hypotensive peak and to prolong the antihypertensive effect of the drug.

1.4.3   Poly(alkyl cyanoacrylates)

Poly(alkyl cyanoacrylates) (PACA) are synthesized from cyanoacrylates. They have excellent adhesive properties as a result of the strong bonds that can form with polar substrates including living tissues and skin [76]. They are widely used as surgical adhesives [77]. Alkylcyanoacrylates [78], commonly known as Superglue® (Super Glue Corporation, USA/Henkel Loctite, Germany), have been used as suture materials for more than four decades.

PACA are used in several biomedical applications and more recently with increasing interest in the field of nanotechnology [79] for targeting of bioactives, including low molecular weight drugs, peptides, proteins, and nucleic acids. As polymerized nanoparticles they were introduced to the area of drug delivery by Couvreur *et al.* in the 1970s [78]. The extensive interest in PACA nanoparticles as drug carriers is due to the biocompatibility and biodegradability of the polymer, the ease of preparation of the particles, and their ability to entrap bioactives, including subunit antigens.

In 2006 BioAlliancePharma [80] announced clinical phase II/III trials for Transdrug®, doxorubicin (DOX)-loaded poly(isohexyl cyanoacrylate) (PiHCA) nanoparticles suitable for intra-arterial, intravenous (IV), or oral administration. However, because of acute pulmonary damage, phase II trials of Transdrug® were suspended in July 2008.

The major in vivo degradation mechanisms consist of the hydrolysis of the ester bond of the alkyl side chain of the polymer. Degradation products consist of an alkylalcohol and poly(cyanoacrylic acid), which are soluble in water and can be eliminated in vivo via kidney filtration. This degradation has been shown to be catalyzed by esterases from serum, lysosomes and pancreatic juice. The degradation rate of PACA nanoparticles [79], and therefore the drug release, depends on the alkyl side chain length, and an increase in length leads to a decrease in the degradation rate.

PACA nanoparticles have been used for oral peptide delivery of insulin [81] and calcitonin [82]. Pilocarpine-loaded PACA nanoparticles [83] administered in a Pluronic® gel were more promising and significantly increased the bioavailability of the drug. Polysorbate 80-coated PACA nanoparticles were shown to enable the transport of anti-tumor antibiotic doxorubicin [84] across the blood–brain barrier (BBB) to the brain after intravenous administration and to considerably reduce the growth of brain tumors in rats.

1.4.4   Polyethylene Glycol

Polyethylene oxide (PEO) and polyethylene glycol (PEG) are essentially identical polymers. PEO has the repeat structural unit $-CH_2CH_2O-$ and PEG has general structure of $HO-(CH_2CH_2O)_n-CH_2CH_2-OH$, possesses [85] a similar repeating unit of PEO, and has hydroxyl groups at each end of the molecule. PEO and PEG are highly biocompatible [61].

PEG and PEO employed modification has emerged as a common strategy to ensure such stealthshielding and long-circulation of therapeutics or delivery devices. PEG-modification is often referred to as PEGylation [85], a term that implies the covalent binding or non-covalent entrapment or adsorption of PEG onto an object. PEG coating of nanospheres provides protection against interaction with the blood components, which induces removal of the foreign particles from the blood [86]. PEG-coated nanospheres may function as circulation depots of the administered drugs [87].

The terminal hydroxyl groups of PEG can be activated for conjugation to different types of polymers and drugs. Amphiphilic block co-polymers, such as poloxamers (commercially available as Pluronics) and poloxamines (Tetronics), consisting of blocks of hydrophilic PEG (or PEO) and hydrophobic poly(propylene oxide) (PPO) are additional forms of PEG derivatives, often employed for modification by surface adsorption or entrapment [85].

PEG has little limitation in its biological use as these are usually excreted in urine or feces but at high molecular weights they can accumulate in the liver, leading tomacromolecular syndrome. Apart from limitations, still U.S. FDA [88] has approved some of PEG conjugates for marketing namely, PEG–asparaginase (Oncaspar®) for acute lymphoblastic leukemia, PEG–adenosine deaminase (Adagen®) for severe combined immunodeficiency disease, PEG–interferon α2a (Pegasys®) for Hepatitis C, PEG–G-CSF (peg-filgrastim, Neulasta®) for treating of neutropenia during chemotherapy and PEG–growth hormone receptor antagonist (Pegvisomant, Somavert®) for curing Acromegaly.

Peracchia *et al.* [89] showed the polymeric nanoparticle coated with PEG can reduce either protein adsorption and complement consumption as a function of the PEG density. The effect of PEO surface density on long-circulating PLA-PEO nanoparticles synthesized by Vittaz *et al.* [90] has shown some advantages in preventing opsonization and thereby avoiding the mononuclear phagocytes system (MPS) uptake. Jaeghere *et al.* [91] studied the freeze-dried PEO-surface modified NPs as a function of PEO chain length and surface density to avoid the MPS uptake.

1.5   Synthesis of Nanovehicles

The approaches for synthesis of nanomaterials are commonly categorized into top-down approach, bottom-up approach, and hybrid approach.

1.5.1   Top-Down Approach

This approach starts with a block of material and reduces the starting material down to the desired shape in nanoscale by controlled etching, elimination, and layering of the material. For example, a nanowire fabricated by lithography impurities and structural defects on the surface. One problem with the top-down approach is the imperfections of the surface structure, which may significantly affect the physical properties and surface chemistry of the nanomaterials. Further, some uncontrollable defects may also be introduced even during the etching steps. Regardless of the surface imperfections and other defects, the top-down approach [92] is still important for synthesizing nanomaterials usually contains. This technique employs two very common high-energy shear force methods [59] viz., milling and high-pressure homogenization. Milling yields nanoparticle in dry state and high-pressure homogenization of suspension form.

1.5.2 Bottom-Up Approach

In a bottom-up approach, materials are fabricated by efficiently and effectively controlling the arrangement of atoms, molecules, macromolecules or supramolecules. The synthesis of large polymer molecules is a typical example of the bottom-up approach, where individual building blocks, monomers, are assembled into a large molecule or polymerized into bulk material. The main challenge for the bottom-up approach is how to fabricate structures that are of sufficient size and amount to be used as materials in practical applications. Nevertheless, the nanostructures fabricated in the bottom-up approach usually have fewer defects, a more homogeneous chemical composition and better short and long range ordering [92]. In bottom-up approach precipitation, crystallization and single droplet evaporation processes [93] are used produce nanoparticles. Few techniques used for fabrication of nanoparticles for bottom-up approach are detailed in further sections of the same chapter.

1.5.3 Hybrid Approach

Though both the top-down and bottom-up approaches play important roles in the synthesis of nanomaterials, some technical problems exist with these two approaches. It is found that, in many cases, combining top-down and bottom-up method into an unified approach that transcends the limitations of both is the optimal solution [92]. A thin film device, such as a magnetic sensor, is usually developed in a hybrid approach, since the thin film is grown in a bottom-up approach, whereas it is etched into the sensing circuit in a top-down approach.

1.6   Dispersion of Preformed Polymers

1.6.1   Emulsification-Solvent Evaporation

A hydrophobic polymer organic solution is dispersed into nanodroplets, using a dispersing agent and high-energy homogenization [94], in a non-solvent or suspension medium such as chloroform, dichloromethane (ICH, class 2) or ethyl acetate (ICH, class 3) [95]. The polymer precipitates in the form of nanospheres in which the drug is finely dispersed in the polymer matrix network. The solvent is subsequently evaporated by increasing the temperature under pressure or by continuous stirring. The size can be controlled by adjusting the stir rate, type and amount of dispersing agent, viscosity of organic and aqueous phases, and temperature [96]. In the conventional methods, two main strategies are being used for the formation of emulsions: the preparation of single-emulsions, e.g., oil-in-water (o/w) or double-emulsions, e.g., (water-in-oil)-in-water, (w/o)/w [97]. Even though different types of emulsions may be used, oil/water emulsions are of interest because they use water as the nonsolvent; this simplifies and thus improves process economics, because it eliminates the need for recycling, facilitating the washing step and minimizing agglomeration. However, this method can only be applied to liposoluble drugs, and limitations are imposed by the scale-up of the high energy requirements in homogenization. Frequently used polymers are PLA [98], PLGA [99], PCL [100], and poly(h-hydroxybutyrate) [101]. Few drugs encapsulated were texanus toxoid [102], loperamide [98] and cyclosporin A [103] by this technique.

1.6.2   Solvent-Displacement, -Diffusion, or Nanoprecipitation

A solution of polymer, drug and lipophilic stabilizer (surfactant) in a semi-polar solvent miscible with water is injected into an aqueous solution (being a non-solvent or anti-solvent for drug and polymer) containing another stabilizer under moderate stirring. Nanoparticles are formed instantaneously by rapid solvent diffusion and the organic solvent is removed under reduced pressure. The velocity of solvent removal and thus nuclei formation is the key to obtain particles in the nanometer range instead of larger lumps or agglomerates [15]. As an alternative to liquid organic or aqueous solvents, supercritical fluids can be applied. Fessi *et al.* [104] proposed a simple and mild method yielding nanoscale and monodisperse polymeric particles without the use of any preliminary emulsification for encapsulation of indomethacin. Both, solvent and non-solvent must have low viscosity and high mixing capacity in all proportions, e.g., acetone (ICH, class3) [95] and water. Another delicate parameter is the composition of the solvent/polymer/water mixture limiting the feasibility of nanoparticle formation. The only complementary operation following the mixing of the two phases is to remove the volatile solvent by evaporation under reduced pressure. One of the most interesting and practical aspects of this method is its capacity to be scaled up from laboratory to industrial amounts, since they can be run with conventional equipment.

This method has been applied to various polymeric materials, such as PLA [105] and PCL [106]. Barichello *et al.* [107] demonstrated application of this method for entrapment of valproic acid, ketoprofen, vancomycin, phenobarbital, and insulin by using PLGA polymer.

1.6.3   Emulsification-Solvent Diffusion (ESD)

The encapsulating polymer is dissolved in a partially water-soluble solvent such as propylene carbonate, and saturated with water to ensure the initial thermodynamic equilibrium of both liquids. To produce the precipitation, it is necessary to promote the diffusion of the solvent of the dispersed phase by dilution with an excess of water when the organic solvent is partly miscible with water or with another organic solvent in the opposite case. Subsequently, the polymer-water saturated solvent phase is emulsified in an aqueous solution containing stabilizer, leading to solvent diffusion to the external phase and the formation of nanospheres or nanocapsules, according to the oil-to-polymer ratio. Finally, the solvent is eliminated [96]. Several drug-loaded nanoparticles were produced by the ESD technique, including doxorubicin-PLGA conjugate nanoparticles [108], plasmid DNA-loaded PLA-PEG nanoparticles [109], cyclosporin (Cy-A)-loaded gelatin and cyclosporin (Cy-A)-loaded sodium glycolate nanoparticles [110].

1.6.4   Salting-Out

Salting-out is based on the separation of a water-miscible solvent from aqueous solution via a salting-out effect. The salting-out [96–97] procedure can be considered as a modification of the emulsification/solvent diffusion. Polymer and drug are initially dissolved in a solvent such as acetone (ICH, Class3), which is subsequently emulsified into an aqueous gel containing the salting-out agent (electrolytes, such as magnesium chloride, calcium chloride, and magnesium acetate, or non-electrolytes such as sucrose) and a colloidal stabilizer such as polyvinylpyrrolidone or hydroxyethylcellulose. This oil/water emulsion is diluted with a sufficient volume of water or aqueous solution to enhance the diffusion of acetone into the aqueous phase, thus inducing the formation of nanospheres. The selection of the salting-out agent is important, because it can play an important role in the encapsulation efficiency of the drug. Both the solvent and the salting-out agent are then eliminated by cross-flow filtration.

In a work carried out by Song *et al.* [111], PLGA nanoparticles were prepared by employing NaCl as the salting-out agent instead of $MgCl_2$ or $CaCl_2$.

1.6.5   Dialysis

Polymer is dissolved in an organic solvent and placed inside a dialysis tube with proper molecular weight cutoff. Dialysis is performed against a non-solvent miscible with the former miscible. The displacement of the solvent inside the membrane is followed by the progressive aggregation of polymer due to a loss of solubility and the formation of homogeneous suspensions of nanoparticles. Dialysis method was used for synthesizing PLGA [112], PLA [113] and dextran ester [114] nanoparticle. Poly(ε-caprolactone) grafted poly(vinyl alcohol)copolymer nanoparticles [115] were investigated as drug carrier models for hydrophobic and hydrophilic anti-cancer drugs; paclitaxel and doxorubicin. In vitro drug release experiments were conducted; the loaded NPs reveal continuous and sustained release form for both drugs, up to 20 and 15 days for paclitaxel and doxorubicin, respectively.

1.6.6   Supercritical Fluid Technology

Conventional methods, such as in situ polymerization and solvent evaporation, often require the use of toxic solvents and surfactants. Supercritical fluids allow attractive alternatives for the nanoencapsulation process because these are environmentally friendly solvents. The commonly used methods of supercritical fluid technology [97, 116] are the rapid expansion of supercritical solution (RESS) and the supercritical anti-solvent (SAS) method. A supercritical fluid is a substance that is used in a state above the critical temperature and pressure where gases and liquids can coexist. It is able to penetrate materials such as gas, and to dissolve materials such as liquid. For example, use of carbon dioxide or water in the form of a supercritical fluid allows substitution for an organic solvent.

In the RESS method, a polymer is solubilized in a supercritical fluid and the solution is expanded through a nozzle. Thus, the solvent power of supercritical fluid dramatically decreases and the solute eventually precipitates. A uniform distribution of drug inside the polymer matrix, e.g., PLA nanospheres, can be achieved only for low-molecular-mass (<10,000) polymers because of the limited solubility of high-molecular-mass polymers in supercritical fluids. Chernyak et al. [117] produced droplets of poly(perfluoropolyetherdiamide) from the rapid expansion of CO_2 solutions. Sane and Thies [118] presented method for developing poly(l-lactide) nanoparticle by using CO_2+THF solution.

In the SAS method, the solution is charged with the supercritical fluid in the precipitation vessel containing a polymer in an organic solvent. At high pressure, enough anti-solvent will enter into the liquid phase so that the solvent power will be lowered and the polymer precipitates. After precipitation, the anti-solvent flows through the vessel to strip the residual solvent. When the solvent content has been reduced to the desired level, the vessel is depressed and the solid nanoparticles are collected. Meziani *et al.* [119] reported the preparation of poly(heptadecafluorodecylacrylate) by nanoparticles by this technique.

1.7   Emulsion Polymerization

Emulsion polymerization is the most common method used for the production of a wide range of specialty polymers. The use of water as the dispersion medium is environmentally friendly and also allows excellent heat dissipation during the course of the polymerization. Based on the utilization of surfactant, it can be classified as conventional and surfactant-free emulsion polymerization [97].

1.7.1   Conventional Emulsion Polymerization

In conventional emulsion polymerization [97], initiation occurs when a monomer molecule dissolved in the continuous phase collides with an initiator molecule that may be an ion or a free radical. Alternatively, the monomer molecule can be transformed into an initiating radical by high-energy radiation, including γ-radiation, ultraviolet or strong visible light. Phase separation and formation of solid particles can take place before or after the termination of the polymerization reaction. Brush-type amphiphilic block copolymers of polystyrene-b-poly[poly(ethylene glycol) methyl ether methacrylate] [120] was synthesized by conventional emulsion polymerization.

1.7.2   Surfactant-Free Emulsion Polymerization

This technique has received considerable attention for use as a simple, green process for nanoparticle production without the addition and subsequent removal of the stabilizing surfactants [97]. The reagents used in an emulsifier free system include deionized water, a water-soluble initiator (potassium persulfate) and monomers, more commonly vinyl or acryl monomers. In such polymerization systems, stabilization of nanoparticle occurs through the use of ionizable initiators or ionic co-monomers. The emulsifier-free monodisperse poly(methyl methacrylate) (PMMA) microspheres [121] was synthesized with microwave irradiation. The emulsifier-free core–shell polyacrylate latex [122] nanoparticles containing fluorine and silicon in shell were successfully synthesized by emulsifier-free seeded emulsion polymerization with water as the reaction medium.

1.7.3 Mini-Emulsion Polymerization

Mini-emulsion polymerization formulation consists of water, monomer mixture, co-stabilizer, surfactant, and initiator [97]. The key difference between emulsion polymerization and mini-emulsion polymerization is the utilization of a low molecular mass compound as the co-stabilizer and also the use of a high-shear device (ultrasound, etc.). Mini-emulsions are critically stabilized, require a high-shear to reach a steady state and have an interfacial tension much greater than zero. Polymethylmethacrylate [123] and poly(n-butylacrylate) [124] nanoparticles were produced by employing sodium lauryl sulfate/dodecyl mercaptan and sodium lauryl sulfate/hexadecane as surfactant/co-stabilizer systems, respectively.

1.7.4 Micro-Emulsion Polymerization

In micro-emulsion polymerization, an initiator, typically water-soluble, is added to the aqueous phase of a thermodynamically stable micro-emulsion containing swollen micelles. The polymerization starts from this thermodynamically stable, spontaneously formed state and relies on high quantities of surfactant systems, which possess an interfacial tension at the oil/water interface close to zero. Furthermore, the particles are completely covered with surfactant because of the utilization of a high amount of surfactant [97]. Initially, polymer chains are formed only in some droplets, as the initiation cannot be attained simultaneously in all microdroplets. Later, the osmotic and elastic influence of the chains destabilize the fragile microemulsions and typically lead to an increase in the particle size, the formation of empty micelles, and secondary nucleation. Synthesis of a functional copolymer of methyl methacrylate and N-methylolacrylamide (NMA) [125] and polymerization of vinyl acetate [126] in microemulsions was prepared with Aerosol OT.

1.7.5 Interfacial Polymerization

Interfacial polymerization involves step polymerization of two reactive monomers or agents, which are dissolved respectively in two phases (i.e., continuous- and dispersed-phase), and the reaction takes place at the interface of the two liquids [127]. The relative ease of obtaining interfacial polymerization has made it a preferred technique in many fields, ranging from encapsulation of pharmaceutical products to preparation of conducting polymers [97]. α-tocopherol-loaded polyurethane and poly(ether urethane)-based nanocapsules [128] were reported by Bouchemal el al. Core-shell biocompatible polyurethane [129] nanocapsules encapsulating ibuprofen was obtained by interfacial polymerization.

1.8   Purification of Nanoparticle

Purification of nanoparticle is needed to remove impurities and excess of reagents involved during manufacture. Depending on the method of preparation, impurities include organic solvents, oil, surfactants, residual monomers, polymerization initiators, salts, excess of surfactants or stabilizing agents, and large polymer aggregates. It becomes essential to obtain highly purified nanoparticle suspensions for dosages form synthesized for administered by specifically in vivo route.

There are several suitable methods that can be applied to purify nanoparticle dispersions. They include evaporation under reduced pressure, centrifugation, ultracentrifugation techniques filtration through mesh or filters, dialysis, gel filtration, ultrafiltration, diafiltration and cross-flow microfiltration.

1.8.1   Evaporation

Evaporation under reduced pressure is the most common approach to remove large quantities of volatile organic solvents and a part of water. This process is usually used after the obtaining of nanoparticle suspensions by nanoprecipitation [130], emulsification-reverse salting-out, emulsification-solvent diffusion [131] and interfacial poly-condensation [128] combined with spontaneous emulsification.

1.8.2   Filtrations Through Mesh or Filters

Filtrations through mesh or filters are applied to remove large particles or polymer aggregates which formed during preparations [132]. Such purification is systematically applied on nanoparticle suspensions designed for intravenous injections.

1.8.3   Centrifugation

A centrifugation at low gravity force can also be applied to remove aggregates and large particles on most of the polymer nanoparticle suspensions. It does not warranty the elimination of all particles with a diameter above a very define size as filtration on calibrated membrane does. Moreover, it is not suitable to purify nanoparticles having a high density because they will sediment with aggregates. For instance, this restriction applies in the case of metal colloids containing nanoparticles, which are designed for applications in diagnosis by imaging techniques or in techniques based on thermal treatments applied in cancer therapy.

1.8.4   Ultracentrifugation

Ultracentrifugation methods consist in very high speed centrifugations. For example, ultracentrifugations are performed at 100,000–110,000×g for 30 to 45 min. The nanospheres, those having a slightly higher density than water, can sediment and concentrate in a pellet form. The main problem of this technique is that nanospheres are not always easy to re-disperse after ultracentrifugation. Aggregates may remain and the uses of vortex or ultrasounds are often mentioned as methods used to redispersed pellets after ultracentrifugation. Nanocapsules are more difficult to separate from the dispersing medium because the cream remains semi-liquid. In addition, they are fragile and the application of several cycles of ultracentrifugation is hazardous because they can break easily. Despite these drawbacks, ultracentrifugation appeared as a method of choice to facilitate the transfer nanoparticle from one dispersion medium to another, and nanoparticle washing can also be applied. Ultracentifugation was used for separation and purification of PEG-coated poly(isobutyl 2-cyanoacrylate) (PIBCA) nano-particles [133]. Budesonide loaded poly(lactic acid) [134] and doxorubicin-loaded human serum albumin [135] nanoparticle were purified by this method.

1.8.5   Dialysis

Purification by dialysis can be performed using different kinds of cellulose membranes of various molecular weight cut off, allowing substances having low or high molecular weight to diffuse toward the counter dialysing medium. While purifying nanoparticles, premature release of nanoparticle payload can occur during the long purification period it requires, and because large volume of counter dialysing medium are required to make the purification efficient. Furthermore, the application of dialysis in a large-scale is disputable from an economical point of view and from the high risk of microbial contamination of the final product due to the long duration of the process [136].

1.8.6   Gel Filtration

It is much faster than methods based on simple dialysis but it is greatly limited by the relatively small volume of sample that can be processed at a time. In addition, irreversible adsorption of actives onto the column stationary phase and the poor resolution between large impurities and small nanoparticles can restrict the use of this technique for purification of drug-loaded nanoparticulate formulations. Beck el al. [136] presented a purification method by gel filtration [137] method in Sephadex® G 50 medium.

1.9   Drying of Nanoparticles

Storage of nanoparticles as suspensions presents many disadvantages. The major obstacle that limits the use of these nanoparticles is due to the physical instability (aggregation/particle fusion) and/or to the chemical instability (hydrolysis of polymer materials forming the nanoparticles, drug leakage of nanoparticles and chemical reactivity of medicine during the storage) which are frequently noticed when these nanoparticle aqueous suspensions are stored for an extended period. Other risk includes microbiological contamination, premature polymer degradation by hydrolysis, physicochemical instability due to particle aggregation and sedimentation and loss of the biological activity of the drug. To circumvent such problems, pharmaceutical preparations are stored under a dry form. In general, the transformation of a liquid preparation into a dry product can be achieved using freeze-drying or spray-drying processes.

1.9.1   Freeze Drying

Freeze drying, also known as lyophilization, is a very common technique of conservation used to ensure long-term stability of pharmaceutical and biological products preserving their original properties. The basic principle of this process consists of removing water content of a frozen sample by sublimation and desorption under vacuum. In general, freeze-drying processes can be divided in three steps:

- Freezing of the sample (solidification);
- Primary drying corresponding to the ice sublimation; and
- Secondary drying corresponding to desorption of unfrozen water.

During the freeze drying process, several problems may arise, which can lead to a loss of integrity of the nanoparticle characteristics. For instance, crystallization of ice may exert a mechanical stress on nanoparticles leading to their destabilization. This effect is more critical during the lyophilization of nanocapsules, which are very fragile upon lyophilization. The high concentration in nanoparticles in the final dried product may favor aggregation and even in some cases irreversible coalescence of nanoparticles [138].

The addition of cryoprotectants can improve the resistance of nanoparticles toward freezing and drying stresses and also increase stability during long-term storage. Sugars including trehalose, mannose, sucrose, glucose, lactose, maltose and mannitol are often used.

1.9.2   Spray-Drying

The spray-drying technique transforms liquids into dried particules under a continuous process. Spray-drying process includes four important steps: atomization of the feed, i.e., nanoparticle suspension, into a spray, spray-air contact, drying of the spray and separation of the dried product from the drying gas [139]. Nanoparticle formulations submitted to spray-drying are generally aqueous suspensions and contain one soluble compound added as drying auxiliary. Examples of drying auxiliaries are lactose, mannitol, trehalose and PVP [140].

1.10   Drug Loading

A successful nanodelivery system should have a high drug-loading capacity, thereby reducing the quantity of matrix materials for administration. Drug loading can be accomplished by two methods. The incorporation method requires the drug to be incorporated at the time of nanoparticle formulation. The adsorption/absorption methods call for absorption of the drug after nanoparticle formation; this is achieved by incubating the nano-carrier with a concentrated drug solution. Drug loading and entrapment efficiency depend on drug solubility in the excipient matrix material (solid polymer or liquid dispersion agent), which is related to the matrix composition, molecular weight, drug–polymer interactions, and the presence of end functional groups (i.e., ester or carboxyl) in either the drug or matrix [132b, 141]. A polymer of choice for some nanoparticle formulations is PEG, which has little or no effect on drug-loading and interactions [142]. In addition, the macromolecules, drugs, or protein encapsulated in nanoparticles [143] show the greatest loading efficiency when they are loaded at or near their isoelectric point (pI) [144].

1.11 Drug Release

It is important to consider both drug release and polymer biodegradation when developing a nanoparticulate delivery system. In general, the drug release rate depends on:

1. Drug solubility
2. Desorption of the surface-bound or adsorbed drug
3. Drug diffusion through the nanoparticle matrix
4. Nanoparticle matrix erosion or degradation
5. Combination of erosion and diffusion processes.

Hence, solubility, diffusion, and biodegradation of the particle matrix govern the release process. In the case of nanospheres, where the drug is uniformly distributed, drug release occurs by diffusion or erosion of the matrix. If the diffusion of the drug is faster than matrix erosion, then the mechanism of release is largely controlled by a diffusion process. The rapid, initial release, or "burst," is mainly attributed to weakly bound or adsorbed drug to the relatively large surface of nanoparticles. It is evident that the method of incorporation has an effect on the release profile. If the drug is loaded by the incorporation method, then the system has a relatively small burst effect and sustained release characteristics. If the nanoparticle is coated by polymer, the release is then controlled by diffusion of the drug from the polymeric membrane [145].

Membrane coating acts as a drug release barrier; therefore, drug solubility and diffusion in or across the polymer membrane becomes a determining factor in drug release. Furthermore, the release rate also can be affected by ionic interactions between the drug and auxiliary ingredients. When the entrapped drug interacts with auxiliary ingredients, a less water-soluble complex can form, which can slow the drug release—having almost no burst release effect. Whereas if the addition of auxiliary ingredients, e.g., ethylene oxide–propylene oxide block copolymer (PEO-PPO) to chitosan, reduces the interaction of the drug with the matrix material due to competitive electrostatic interaction of PEO-PPO with chitosan, then an increase in drug release could be achieved. Various methods can be used to study the release of drug from the nanoparticle: (1) side-by-side diffusion cells with artificial or biological membranes; (2) dialysis bag diffusion; (3) reverse dialysis bag diffusion; (4) agitation followed by ultracentrifugation/centrifugation; or (5) ultra-filtration. Usually the release study is carried out by controlled agitation followed by centrifugation. Due to the time-consuming nature and technical difficulties encountered in the separation of nanoparticles from release media, the dialysis technique is generally preferred [144].

1.12   Conclusion

The mimic nature in designing "smart" synthetic materials from various functional molecular building blocks can respond to stimuli including temperature, pH, ionic strength, light, electric or magnetic field, chemical and biochemical stimuli. The biomimetic approaches employed in the design, synthesis and engineering of stimuli-responsive polymeric systems, which undergo reversible abrupt phase transitions upon variation of a variable around a critical point and their use in sensors, logic operations, biomedicine, tissue engineering, regenerative medicine, synthetic muscles, "smart" optical or microelectromechanical systems, membranes, electronics and self-cleaning surfaces.

References

1. R.P. Feynman, *Engineering and* Science, Vol. 23, pp. 22–36, 1960.
2. D. Thassu, M. Deleers, and Y. Pathak, *Nanoparticulate Drug Delivery Systems*, Vol. 166, 1 ed., Informa Healthcare, New York, 2007.
3. a) U. Scheffel, B. A. Rhodes, T.K. Natarajan, and H. N. Wagner, Jr., *J Nucl Med*, Vol. 13, pp. 498–503, 1972; b) J. J. Marty, R. C. Oppenheim, and P. Speiser, *Pharm Acta Helv*, 53, pp. 17–23, 1978.
4. a) G. Birrenbach, P.P. Speiser, *J Pharm Sci*, Vol. 65, pp. 1763–1766, 1976; b) J. Kreuter and P. P. Speiser, *Infection and Immunity*, Vol. 13, pp. 204–210, 1976.
5. P. Couvreur, B. Kante, M. Roland, P. Guiot, P. Bauduin, and P. Speiser, *J Pharm Pharmacol*, Vol. 31, pp. 331–332, 1979.
6. R. Gurny, N.A. Peppas, D.D. Harrington, and G.S. Banker, *Drug Development and Industrial Pharmacy*, Vol. 7, pp. 1–25, 1981.
7. a) N. Al Khouri Fallouh, L. Roblot-Treupel, H. Fessi, J. P. Devissaguet, and F. Puisieux, *International Journal of Pharmaceutics*, Vol. 28, pp. 125–132, 1986; b) P. Legrand, G. Barratt, V. Mosqueira, H. Fessi, and J. P. Devissaguet, *STP Pharma Sciences*, Vol. 9, pp. 411–418, 1999.

8. M. R. Gasco and M. Trotta, *International Journal of Pharmaceutics*, Vol. 29, pp. 267–268, 1986.

9. E. Allémann, R. Gurny, and E. Doelker, *International Journal of Pharmaceutics*, Vol. 87, pp. 247–253, 1992.

10. a) D. Quintanar-Guerrero, E. Allemann, E. Doelker, and H. Fessi, *Pharm Res*, 15, pp. 1056–1062, 1998; b) D. Quintanar-Guerrero, E. Allemann, H. Fessi, and E. Doelker, *Int J Pharm*, Vol. 188, pp. 155–164, 1999.

11. M. F. Zambaux, F. Bonneaux, R. Gref, P. Maincent, E. Dellacherie, M. J. Alonso, P. Labrude, and C. Vigneron, *J Control Release*, Vol. 50, pp. 31–40, 1998.

12. a) K. A. Janes, P. Calvo, and M. J. Alonso, *Adv Drug Deliv Rev*, Vol. 47, pp. 83–97, 2001; b) M. Prabaharan and J. F. Mano, *Drug Deliv*, Vol. 12, pp. 41–57, 2005; c) Z. Liu, Y. Jiao, Y. Wang, C. Zhou, and Z. Zhang, *Adv Drug Deliv Rev*, Vol. 60, pp. 1650–1662, 2008.

13. R. Gref, Y. Minamitake, M.T. Peracchia, A. Domb, V. Trubetskoy, V. Torchilin, and R. Langer, *Pharm Biotechnol*, Vol. 10, pp. 167–168, 1997.

14. E. Soussan, S. Cassel, M. Blanzat, and I. Rico-Lattes, *Angew Chem Int Ed Engl*, Vol. 48, pp. 274–288, 2009.

15. A. Lamprecht, *Nanotherapeutics: Drug Delivery Concepts in Nanoscience*, Pan Stanford Pub, 2009.

16. a) N. A. Peppas, J.Z. Hilt, A. Khademhosseini, and R. Langer, *Advanced Materials*, Vol. 18, pp. 1345–1360, 2006; b) K. Letchford, H. Burt, *European Journal of Pharmaceutics and Biopharmaceutics*, 65, 259–269, 2007.

17. a) Y. H. Bae, T. Okano, and S. W. Kim, *Pharmaceutical Research*, Vol. 8, pp. 531–537, 1991; b) Y. H. Bae, T. Okano, and S. W. Kirn, *Pharmaceutical Research*, Vol. 8, pp. 624–628, 1991.

18. S. H. Yuk, S. H. Cho, and S. H. Lee, *Macromolecules*, Vol. 30, pp. 6856–6859, 1997.

19. S. Hecht and J. M. J. Fréchet, *Angewandte Chemie International Edition*, Vol. 40, pp. 74–91, 2001.

20. D.A. Tomalia, H. Baker, J. Dewald, M. Hall, G. Kallos, S. Martin, J. Roeck, J. Ryder, and P. Smith, *Polymer Journal*, Vol. 17, pp. 117–132, 1985.

21. C.J. Hawker and J.M.J. Fréchet, *Journal of the American Chemical Society*, 112, pp. 7638–7647, 1990.

22. M. Mammen, S.-K. Choi, and G. M. Whitesides, *Angewandte Chemie International Edition*, Vol. 37, 2754–2794, 1998.

23. M. Liu, K. Kono, and J.M.J. Fréchet, *Journal of Controlled Release*, Vol. 65, pp. 121–131, 2000.

24. a) M. X. Tang, C. T. Redemann, F. C. Szoka, *Bioconjugate Chemistry*, Vol. 7, pp. 703–714, 1996; b) B. H. Zinselmeyer, S. P. Mackay, A. G. Schatzlein, and I. F. Uchegbu, *Pharm Res*, Vol. 19, pp. 960–967, 2002; c) C. Loup, M.-A. Zanta, A.-M. Caminade, J.-P. Majoral, and B. Meunier, *Chemistry – A European Journal*, Vol. 5, pp. 3644–3650, 1999; d) A.-M. Caminade, C.-O. Turrin, and J.-P. Majoral, *Chemistry – A European Journal*, Vol. 14, pp. 7422–7432, 2008; e) M. X. Tang and F. C. Szoka, *Gene Ther*, Vol. 4, pp. 823–832, 1997.

25. J.F.G.A. Jansen, E.W. Meijer, and E. M. M. de Brabander-van den Berg, *Journal of the American Chemical Society*, 117, pp. 4417–4418, 1995.
26. A. Samad, Y. Sultana, amd M. Aqil, *Current Drug Delivery*, Vol. 4, pp. 297–305, 2007.
27. D. Needham and M.W. Dewhirst, *Advanced Drug Delivery Reviews*, Vol. 53, pp. 285–305, 2001.
28. E. Forssen and M. Willis, *Advanced Drug Delivery Reviews*, Vol. 29, **1998**, *29*, 249–271.
29. A.A. Gabizon, *Cancer Research*, Vol. 52, pp. 891–896, 1992.
30. I.F. Uchegbu and S.P. Vyas, *International Journal of Pharmaceutics*, Vol. 172, pp. 33–70, 1998.
31. S. Lesieur, C. Grabielle-Madelmont, M.-T. Paternostre, J.-M. Moreau, R.-M. Handjani-Vila, and M. Ollivon, *Chemistry and Physics of Lipids*, Vol. 56, pp. 109–121, 1990.
32. E. Gianasi, F. Cociancich, I.F. Uchegbu, A.T. Florence, and R. Duncan, *International Journal of Pharmaceutics*, Vol. 148, pp. 139–148, 1997.
33. I. A. Darwish and I.F. Uchegbu, *International Journal of Pharmaceutics*, Vol. 159, pp. 207–213, 1997.
34. T.P. Assadullahi, R.C. Hider, and A.J. McAuley, *Biochimica et Biophysica Acta (BBA) - Lipids and Lipid Metabolism*, Vol. 1083, pp. 271–276, 1991.
35. A. Polidori, B. Pucci, J. G. Riess, L. Zarif, A.A. Pavia, *Tetrahedron Letters*, Vol. 35, pp. 2899–2902, 1994.
36. E. Hood, M. Gonzalez, A. Plaas, J. Strom, and M. VanAuker, *International Journal of Pharmaceutics*, Vol. 339, pp. 222–230.
37. a) C.J.F. Rijcken, O. Soga, W. E. Hennink, and C. F. v. Nostrum, *Journal of Controlled Release*, Vol. 120, pp. 131–148, 2007; b) F. Meng, G.H.M. Engbers, and J. Feijen, *Journal of Controlled Release*, Vol. 101, pp. 187–198, 2005.
38. F. Ahmed, R. I. Pakunlu, A. Brannan, F. Bates, T. Minko, and D. E. Discher, *Journal of Controlled Release*, Vol. 116, pp. 150–158, 2006.
39. R. H. Muller, K. Mader, and S. Gohla, *Eur J Pharm Biopharm*, Vol. 50, pp. 161–177, 2000.
40. H. Bunjes, K. Westesen, and M. H. J. Koch, *International Journal of Pharmaceutics*, Vol. 129, 159–173, 1996.
41. K. Westesen, B. Siekmann, and M. H. J. Koch, *International Journal of Pharmaceutics*, Vol. 93, 189–199, 1993.
42. J. Pietkiewicz, M. Sznitowska, and M. Placzek, *International Journal of Pharmaceutics*, Vol. 310, pp. 64–71, 2006.
43. a) R.H. Müller, M. Radtke, and S.A. Wissing, *International Journal of Pharmaceutics*, Vol. 242, pp. 121–128, 2002; b) R.H. Müller, M. Radtke, and S.A. Wissing, *Advanced Drug Delivery Reviews*, Vol. 54, Supplement, pp. S131-S155, 2002; c) S.A. Wissing, O. Kayser, and R.H. Müller, *Advanced Drug Delivery Reviews*, Vol. 56, pp. 1257–1272, 2004.
44. W. Mehnert and K. Mäder, *Advanced Drug Delivery Reviews*, Vol. 47, pp. 165–196, 2001.

45. S. Fukushima, S. Kishimoto, Y. Takeuchi, and M. Fukushima, *Advanced Drug Delivery Reviews*, Vol. 45, pp. 65–75, 2000.

46. I. Danielsson, B. Lindman, *Colloids and Surfaces*, Vol. 3, pp. 391–392, 1981.

47. C. Solans, P. Izquierdo, J. Nolla, N. Azemar, and M. J. Garcia-Celma, *Current Opinion in Colloid & Interface Science*, Vol. 10, pp. 102–110, 2005.

48. S. Tamilvanan, S. Schmidt, R.H. Müller, and S. Benita, *European Journal of Pharmaceutics and Biopharmaceutics*, Vol. 59, pp. 1–7, 2005.

49. a) S. Tamilvanan, *Progress in Lipid Research*, Vol. 43, pp. 489–533, 2004; b) G. Pan, M. Shawer, S. Øie, and D. R. Lu, *Pharmaceutical Research*, Vol. 20, pp. 738–744, 2003.

50. G. Nicolaos, S. Crauste-Manciet, R. Farinotti, and D. Brossard, *International Journal of Pharmaceutics*, Vol. 263, pp. 165–171, 2003.

51. L. Rabinovich-Guilatt, P. Couvreur, G. Lambert, and C. Dubernet, *Journal of Drug Targeting*, Vol. 12, pp. 623–633, 2004.

52. M. Bivas-Benita, M. Oudshoorn, S. Romeijn, K. van Meijgaarden, H. Koerten, H. van der Meulen, G. Lambert, T. Ottenhoff, S. Benita, H. Junginger, and G. Borchard, *Journal of Controlled Release*, Vol. 100, pp. 145–155, 2004.

53. a) J.-Y. Fang, Y.-L. Leu, C.-C. Chang, C.-H. Lin, and Y.-H. Tsai, *Drug Delivery*, Vol. 11, pp. 97–105, 2004; b) H. Wu, C. Ramachandran, A. U. Bielinska, K. Kingzett, R. Sun, N.D. Weiner, B.J. Roessler, *International journal of pharmaceutics* **2001**, *221*, 23–34.

54. T. J. Martin, K. Procházka, P. Munk, and S. E. Webber, *Macromolecules*, Vol. 29, pp. 6071–6073, 1996.

55. Y. Mitsukami, M.S. Donovan, A.B. Lowe, and C.L. McCormick, *Macromolecules*, 34, pp. 2248–2256, 2001.

56. a) A. Bianco and M. Prato, *Advanced Materials*, Vol. 15, pp. 1765–1768, 2003; b) A. Bianco, K. Kostarelos, and M. Prato, *Current Opinion in Chemical Biology*, Vol. 9, 674–679, 2005; c) N.W.S. Kam and H. Dai, *Journal of the American Chemical Society*, Vol. 127, pp. 6021–6026, 2005.

57. G. Jia, H. Wang, L. Yan, X. Wang, R. Pei, T. Yan, Y. Zhao, and X. Guo, *Environmental Science & Technology*, Vol. 39, pp. 1378–1383, 2005.

58. B. R. Selvi, D. Jagadeesan, B. S. Suma, G. Nagashankar, M. Arif, K. Balasubramanyam, M. Eswaramoorthy, and T. K. Kundu, *Nano Letters*, Vol. 8, pp. 3182–3188, 2008.

59. J. Swarbrick, *Encyclopedia of Pharmaceutical Technology*, Informa Healthcare, New York, 2006.

60. M. S. Muthu, *Asian Journal of Pharmaceutics*, Vol. 3, p. 266, 2009.

61. P. K. Chu and X. Liu, *Biomaterials Fabrication and Processing Handbook*, 1 ed., CRC Press, Florida, 2008.

62. I. Bala, V. Bhardwaj, S. Hariharan, S. V. Kharade, N. Roy, and M. N. Ravi Kumar, *J Drug Target*, Vol. 14, pp. 27–34, 2006.

63. R. Pandey and G. K. Khuller, *Chemotherapy*, Vol. 53, pp. 437–441, 2007.

64. G. Mittal, D. K. Sahana, V. Bhardwaj, and M. N. Ravi Kumar, *J Control Release*, Vol. 119, pp. 77–85, 2007.

65. J. L. Italia, D. K. Bhatt, V. Bhardwaj, K. Tikoo, and M. N. Kumar, *J Control Release*, Vol. 119, pp. 197–206, 2007.

66. S. Davaran, M. R. Rashidi, B. Pourabbas, M. Dadashzadeh, and N. M. Haghshenas, *Int J Nanomedicine*, Vol. 1, pp. 535–539, 2006.

67. G. Mattheolabakis, E. Taoufik, S. Haralambous, M. L. Roberts, and K. Avgoustakis, *Eur J Pharm Biopharm*, Vol. 71, pp. 190–195, 2009.

68. K. M. Lima and J. M. Rodrigues Junior, *Braz J Med Biol Res*, Vol. 32, pp. 171–180, 1999.

69. V.R. Sinha, K. Bansal, R. Kaushik, R. Kumria, and A. Trehan, *International Journal of Pharmaceutics*, Vol. 278, pp. 1–23, 2004.

70. a) G.G. Pitt, M.M. Gratzl, G.L. Kimmel, J. Surles, and A. Sohindler, *Biomaterials*, Vol. 2, pp. 215–220, 1981; b) S. C. Woodward, P. S. Brewer, F. Moatamed, A. Schindler, and C.G. Pitt, *The Intracellular Degradation of Poly(Epsilon-Caprolactone)*, Vol. 19, 1985.

71. a) S.J. Ory, C.B. Hammond, S.G. Yancy, R.W. Hendren, C.G. Pitt, *Am J Obstet Gynecol*, Vol. 145, pp. 600–605, 1983; b) P.D. Darney, S.E. Monroe, C.M. Klaisle, A. Alvarado, *Am J Obstet Gynecol*, 160, pp. 1292–1295, 1989.

72. P. Calvo, M.J. Alonso, J.L. Vila Jato, and J. R. Robinson, *Journal of Pharmacy and Pharmacology*, Vol. 48, pp. 1147–1152, 1996.

73. F. Gamisans, F. Lacoulonche, A. Chauvet, M. Espina, M. Garcia, and M. Egea, *International Journal of Pharmaceutics*, Vol. 179, pp. 37–48, 1999.

74. C. Müller, S. Schaffazick, A. Pohlmann, L. De Lucca Freitas, and D. A. S. Pesce, *Pharmazie*, Vol. 56, pp. 864–867, 2001.

75. M. Leroueil-Le Verger, L. Fluckiger, Y.-I. Kim, M. Hoffman, and P. Maincent, *European Journal of Pharmaceutics and Biopharmaceutics*, Vol. 46, pp. 137–143, 1998.

76. a) A. J. Singer, J. E. Hollander, S. M. Valentine, T. W. Turque, C. F. McCuskey, and J. V. Quinn, *Acad Emerg Med*, Vol. 5, pp. 94–99, 1998; b) A. J. Singer and H. C. Thode, Jr., *Am J Surg*, Vol. 187, pp. 238–248, 2004; c) M. E. King and A. Y. Kinney, *Nurse Pract*, Vol. 24, pp. 66, 69–70, 73–64, 1999.

77. a) M. Ryou and C. Thompson, *Techniques in Gastrointestinal Endoscopy*, Vol. 8, pp. 33–37, 2006 b) T. B. Reece, T. S. Maxey, and I. L. Kron, *Am J Surg*, Vol. 182, pp. 40S-44S, 2001.

78. A. Graf, A. McDowell, and T. Rades, *Expert Opinion on Drug Delivery*, Vol. 6, pp. 371–387, 2009.

79. C. Vauthier, C. Dubernet, E. Fattal, H. Pinto-Alphandary, and P. Couvreur, *Advanced Drug Delivery Reviews*, Vol. 55, pp. 519–548, 2003.

80. M. Fanun, *Colloids in Drug Delivery*, Vol. 148, 1 ed., CRC Press/Taylor & Francis, Florida, 2010.

81. Saı, amp, x, P. C. Damgé, A. S. Rivereau, A. Hoeltzel, and E. Gouin, *Journal of Autoimmunity*, Vol. 9, 713–721, 1996.

82. P. J. Lowe and C. S. Temple, *Journal of Pharmacy and Pharmacology*, Vol. 46, pp. 547–552, 1994.

83. S. D. Desai and J. Blanchard, *Drug Delivery*, Vol. 7, pp. 201–207, 2000.

84. S. E. Gelperina, A. S. Khalansky, I. N. Skidan, Z. S. Smirnova, A. I. Bobruskin, S. E. Severin, B. Turowski, F. E. Zanella, and J. Kreuter, *Toxicology Letters*, Vol. 126, pp. 131–141, 2002.

85. L. E. Vlerken, T. K. Vyas, and M. M. Amiji, *Pharmaceutical Research*, Vol. 24, pp. 1405–1414, 2007.

86. R. Gref, Y. Minamitake, M. T. Peracchia, V. Trubetskoy, V. Torchilin, and R. Langer, *Science*, Vol. 263, pp. 1600–1603, 1994.

87. R. Gref, A. Domb, P. Quellec, T. Blunk, R. H. Müller, J. M. Verbavatz, and R. Langer, *Advanced Drug Delivery Reviews*, Vol. 16, pp. 215–233, 1995.

88. F. M. Veronese and G. Pasut, *Drug Discovery Today*, Vol. 10, pp. 1451–1458, 2005.

89. M. T. Peracchia, C. Vauthier, C. Passirani, P. Couvreur, and D. Labarre, *Life Sciences*, Vol. 61, pp. 749–761, 1997.

90. M. Vittaz, D. Bazile, G. Spenlehauer, T. Verrecchia, M. Veillard, F. Puisieux, and D. Labarre, *Biomaterials*, Vol. 17, pp. 1575–1581, 1996.

91. D. F. Jaeghere, E. Allémann, J.-C. Leroux, W. Stevels, J. Feijen, E. Doelker, and R. Gurny, *Pharmaceutical Research*, Vol. 16, pp. 859–866, 1999.

92. K. M. Teli, S. Mutalik, and G. K. Rajanikant, *Current Pharmaceutical Design*, Vol. 16, pp. 1882–1892, 2010.

93. H.-K. Chan and P. C. L. Kwok, *Advanced Drug Delivery Reviews*, Vol. 63, pp. 406–416, 2011.

94. T. R. Tice, R. M. Gilley, *Journal of Controlled Release*, Vol. 2, pp. 343–352, 1985.

95. G. ICH, in *ICH Vol. Q3C(R5)* (Ed.: I. E. W. Group), ICH 2011.

96. C. Pinto Reis, R. J. Neufeld, A. J. Ribeiro, and F. Veiga, *Nanomedicine: Nanotechnology, Biology and Medicine*, Vol. 2, pp. 8–21, 2006.

97. J. P. Rao and K. E. Geckeler, *Progress in Polymer Science*, Vol. 36, pp. 887–913, 2011.

98. M. Ueda and J. Kreuter, *Journal of Microencapsulation*, Vol. 14, pp. 593–605, 1997.

99. Y. Tabata and Y. Ikada, *Pharmaceutical Research*, Vol. 6, pp. 296–301, 1989.

100. R. Gref, Y. Minamitake, M. Peracchia, V. Trubetskoy, V. Torchilin, and R. Langer, *Science*, Vol. 263, pp. 1600–1603, 1994.

101. F. Koosha, R. H. Muller, S. S. Davis, and M. C. Davies, *Journal of Controlled Release*, Vol. 9, pp. 149–157, 1989.

102. M. Tobío, R. Gref, A. Sánchez, R. Langer, and M. J. Alonso, *Pharmaceutical Research*, Vol. 15, pp. 270–275, 1998.

103. J. Jaiswal, S. Kumar Gupta, and J. Kreuter, *Journal of Controlled Release*, Vol. 96, pp. 169–178, 2004.

104. H. Fessi, F. Puisieux, J. P. Devissaguet, N. Ammoury, and S. Benita, *International Journal of Pharmaceutics*, Vol. 55, pp. R1–R4, 1989.

105. F. Némati, C. Dubernet, H. Fessi, A. Colin de Verdière, M. F. Poupon, F. Puisieux, and P. Couvreur, *International Journal of Pharmaceutics*, Vol. 138, pp. 237–246, 1996.

106. J. Molpeceres, M. Guzman, M. R. Aberturas, M. Chacon, and L. Berges, *Journal of Pharmaceutical Sciences*, Vol. 85, pp. 206–213, 1996.

107. J. M. Barichello, M. Morishita, K. Takayama, and T. Nagai, *Drug Development and Industrial Pharmacy*, Vol. 25, pp. 471–476, 1999.

108. H. S. Yoo, J. E. Oh, K. H. Lee, and T. G. Park, *Pharmaceutical Research*, Vol. 16, pp. 1114–1118, 1999.

109. C. Perez, A. Sanchez, D. Putnam, D. Ting, R. Langer, and M. J. Alonso, *Journal of Controlled Release*, Vol. 75, pp. 211–224, 2001.

110. E.-S. M.H, *International Journal of Pharmaceutics*, Vol. 249, pp. 101–108, 2002.

111. X. Song, Y. Zhao, W. Wu, Y. Bi, Z. Cai, Q. Chen, Y. Li, and S. Hou, *Int J Pharm*, 350, pp, 320–329, 2008.

112. S.-W. Choi, J.-H. Kim, *Journal of Controlled Release*, Vol. 122, pp. 24–30, 2007.

113. M. Liu, Z. Zhou, X. Wang, J. Xu, K. Yang, Q. Cui, X. Chen, M. Cao, J. Weng, and Q. Zhang, *Polymer*, Vol. 48, 5767–5779, 2007.

114. S. Hornig and T. Heinze, *Carbohydrate Polymers*, Vol. 68, 280–286, 2007.

115. F. Sheikh, N. Barakat, M. Kanjwal, S. Aryal, M. Khil, and H.-Y. Kim, *Journal of Materials Science: Materials in Medicine*, Vol. 20, pp. 821–831, 2009.

116. C. Cristian and P. Karol, *Encyclopedia of Nanoscience and Nanotechnology*, 2 ed., CRC Press Florida, 2009.

117. Y. Chernyak, F. Henon, R. B. Harris, R. D. Gould, R. K. Franklin, J. R. Edwards, J. M. DeSimone, and R. G. Carbonell, *Industrial & Engineering Chemistry Research*, Vol. 40, pp. 6118–6128, 2001.

118. A. Sane, M. C. Thies, *The Journal of Supercritical Fluids,* Vol. 40, 134–143, 2007.

119. M. J. Meziani, P. Pathak, R. Hurezeanu, M. C. Thies, R. M. Enick, and Y.-P. Sun, *Angewandte Chemie International Edition*, Vol. 43, pp. 704–707, 2004.

120. A. Muñoz-Bonilla, A. M. van Herk, and J. P. A. Heuts, *Macromolecules*, Vol. 43, pp. 2721–2731, 2010.

121. J. Bao and A. Zhang, *Journal of Applied Polymer Science*, Vol. 93, pp. 2815–2820, 2004.

122. X. Cui, S. Zhong, and H. Wang, *Polymer*, Vol. 48, pp. 7241–7248, 2007.

123. D. Mouran, J. Reimers, and F. J. Schork, *Journal of Polymer Science Part A: Polymer Chemistry*, Vol. 34, pp. 1073–1081, 1996.

124. J. R. Leiza, E. D. Sudol, and M. S. El-Aasser, *Journal of Applied Polymer Science*, Vol. 64, pp. 1797–1809, 1997.

125. E. R. Macías, L. A. Rodríguez-Guadarrama, B. A. Cisneros, A. Castañeda, E. Mendizábal, and J. E. Puig, *Colloids and Surfaces A: Physicochemical and Engineering Aspects* Vol. 103, pp. 119–126, 1995.

126. N. Sosa, E. A. Zaragoza, R. G. López, R. D. Peralta, I. Katime, F. Becerra, E. Mendizábal, and J. E. Puig, *Langmuir*, Vol. 16, pp. 3612–3619, 2000.

127. S. K. Karode, S. S. Kulkarni, A. K. Suresh, and R. A. Mashelkar, *Chemical Engineering Science*, Vol. 53, pp. 2649–2663, 1998.

128. K. Bouchemal, S. Briançon, E. Perrier, H. Fessi, I. Bonnet, and N. Zydowicz, *International Journal of Pharmaceutics*, Vol. 269, pp. 89–100, 2004.

129. F. Gaudin and N. Sintes-Zydowicz, *Colloids and Surfaces A: Physicochemical and Engineering Aspects*, Vol. 331, pp. 133–142, 2008.

130. a) P. Legrand, S. Lesieur, A. Bochot, R. Gref, W. Raatjes, G. Barratt, and C. Vauthier, *International Journal of Pharmaceutics*, Vol. 344, pp. 33–43, 2007; b) C. Duclairoir, E. Nakache, H. Marchais, and A. M. Orecchioni, *Colloid & Polymer Science*, Vol. 276, pp. 321–327, 1998; c) C. Giannavola, C. Bucolo, A. Maltese, D. Paolino, M. A. Vandelli, G. Puglisi, V. H. L. Lee, and M. Fresta, *Pharmaceutical Research*, Vol. 20, pp. 584–590, 2003.

131. D. Moinard-Chécot, Y. Chevalier, S. Briançon, L. Beney, and H. Fessi, *Journal of Colloid and Interface Science*, Vol. 317, pp. 458–468, 2008.

132. a) H. Murakami, M. Kobayashi, H. Takeuchi, and Y. Kawashima, *International Journal of Pharmaceutics*, Vol. 187, pp. 143–152, 1999; b) T. Govender, S. Stolnik, M. C. Garnett, L. Illum, and S. S. Davis, *Journal of Controlled Release*, Vol. 57, pp. 171–185, 1999.

133. M. T. Peracchia, C. Vauthier, F. Puisieux, and P. Couvreur, *Journal of Biomedical Materials Research*, Vol. 34, pp. 317–326, 1997.

134. U. B. Kompella, B. Nagesh, and A. S. P., *Drug Development and Delivery*, Vol. 1, pp. 1–7, 2001.

135. S. Dreis, F. Rothweiler, M. Michaelis, J. Cinatl Jr., J. Kreuter, and K. Langer, *International Journal of Pharmaceutics*, Vol. 341, pp. 207–214, 2007.

136. I. Limayem, C. Charcosset, and H. Fessi, *Separation and Purification Technology*, Vol. 38, pp. 1–9, 2004.

137. P. Beck, D. Scherer, J. Kreuter, *Journal of Microencapsulation*, Vol. 7, pp. 491–496, 1990.

138. a) W. Abdelwahed, G. Degobert, and H. Fessi, *International Journal of Pharmaceutics*, Vol. 324, pp. 74–82, 2006; b) W. Abdelwahed, G. Degobert, S. Stainmesse, and H. Fessi, *Advanced Drug Delivery Reviews*, Vol. 58, pp. 1688–1713, 2006.

139. C. Vauthier and K. Bouchemal, *Pharmaceutical Research*, Vol. 26, pp. 1025–1058, 2009.

140. a) R. Vehring, *Pharmaceutical Research*, Vol. 25, pp. 999–1022, 2008; b) P. Tewa-Tagne, S. Briançon, and H. Fessi, *European Journal of Pharmaceutical Sciences*, Vol. 30, pp. 124–135, 2007.

141. T. Govender, T. Riley, T. Ehtezazi, M. C. Garnett, S. Stolnik, L. Illum, and S. S. Davis, *International Journal of Pharmaceutics*, Vol. 199, pp. 95–110, 2000.

142. M. T. Peracchia, R. Gref, Y. Minamitake, A. Domb, N. Lotan, and R. Langer, *Journal of Controlled Release*, Vol. 46, pp. 223–231, 1997.

143. P. Calvo, C. Remuñan-López, J. L. Vila-Jato, and M. J. Alonso, *Pharmaceutical Research*, Vol. 14, pp. 1431–1436, 1997.

144. R. Singh, and J. W. Lillard Jr., *Experimental and Molecular Pathology*, 86, pp. 215–223, 2009.

145. a) B. Magenheim, M. Y. Levy, and S. Benita, *International Journal of Pharmaceutics*, Vol. 94, pp. 115–123, 1993; b) M. Fresta, G. Puglisi, G. Giammona, G. Cavallaro, N. Micali, and P. M. Furneri, *Journal of Pharmaceutical Sciences*, Vol. 84, pp. 895–902, 1995.

2

Diagnosis and Treatment of Cancer—Where We are and Where We have to Go!

Rajiv Lochan Gaur[1],* and **Richa Srivastava[2]**

[1]*Department of Pathology, Stanford School of Medicine, Stanford University, Palo Alto, CA , USA*
[2]*Biotechnology Division, Central Institute of Medicinal and Aromatic Plants, Lucknow , India*

Abstract

Cancer is a group of diseases that can be managed if diagnosed early. In the past few decades, we have made considerable advances in diagnosis, and opened new avenues for early identification of cancer. Even with tremendous efforts, it appears that we have a long way to go in terms of making a full-proof diagnosis at the early stage of cancer development. The key hurdle in cancer diagnosis is the availability of cancer cells and markers in body fluids. In the contemporary era of new materials in pace with advance technology, we can look forward for better diagnosis. Treatment is another revolting aspect of cancer. Present diagnostics can pick up cancer cells after plentiful development; by this time metastasis might have already been started. Decimation of cancer stem cells deep into tissue provides a safe niche to hide and facilitate cancer cells to proliferate to new tumors. Metastasis enables a good amount of cancerous cells to remain away from drugs. Although the search for new anticancer drugs is always in demand, at the same time, it is necessary to find out the mode of delivering the drug to every single niche which can hide and nourish the cancer cells. In the present age of treatment, we have to look forward for using new methods and technology like nanoparticle drug delivery system to overcome this obstacle. There is a lot of scope to ameliorate the existing technology with advance materials for making substantial progress in cancer diagnosis and treatment.

Keywords: Cancer, diagnosis, tumor, surface marker, nanoparticle

Corresponding author: rlgaur@stanford.edu

Ashutosh Tiwari (ed.) Advanced Healthcare Materials, (37–48) 2014 © Scrivener Publishing LLC

2.1   Cancer Pathology

Cancer is a wide range term used for the group of diseases. The hallmark events of cancer are: uncontrolled growth, immortality, invasion, and metastasis. The principal cause claimed for cancer formation is a variety of lifestyle-related stresses, which can range from the environment to diet, and eventual alteration of genes [3]. Once genes modify, the cells start independent replication to form numerous replica and mass of cells known as tumors (in the case of solid tumors). At this point, the tumor may be benign, which rarely spread, but if it is malignant, it will grow faster. A benign tumor has the cancer cells in the center surrounded by normal cells. In contrast, malignant tumor has mixed cancer and normal cells (Fig 1).

For a certain size of tumor, core cells can survive, as the requirement of oxygen can be accomplished through blood from surrounding blood vessels. Beyond threshold limit, the core cells can die in the absence of oxygen and nutrients, as all living cells remain 100–200 μm vicinity of blood vessel [3]. At the stage of intermediate size, a tumor may remain dormant, as the supply of oxygen and nutrients are just enough to keep the cancer cells alive. More nutrients and oxygen are needed for cancer cells to grow and divide, which can be accomplished by angiogenesis (formation of blood vessels) [3]. After completion of angiogenesis, cancer cells multiply under no control as nutrients and oxygen are in abundance. Researchers have proved that high angiogenic potential leads to highly malignant tumors [25]. Solid tumors are graded according to their severity; increases in grade show advanced stage of disease [9]. Once formed, cancer cells move to the distal part of the body to form secondary tumors by a process know as metastasis [6]. It is still not known that what are the influences which

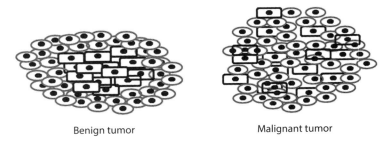

Benign tumor   Malignant tumor

Figure 1 Basic composition of cells in the tumor. In benign tumor, cancer cells ▭ restrict to the core, surrounded by non-cancerous cells ◉. There is no zonal distribution in a malignant tumor as cancerous cells are intermingled with non-cancerous cells.

remove the cells from tumors to go to settle to distal organs and develop new tumors [19]. More than 100 years ago, Stephen Paget (1889) proposed the "seed and soil" hypothesis, which is still convincing, but at the same time, it is intriguing that we know very little about the "soil" which attracts the "seed." [7]. In light of the facts given above, it is explicit that there is more to know about cancer. Early and accurate diagnosis could be the way to treat cancer and halt it at the initial stage.

In this chapter, we will dissect the possibility of cancer detection, the need for advance diagnostic and the technological advantage to do so. In the next part, we discuss the possibility of cancer treatment with newer technology and how it is better in comparison to old and existing treatment.

2.2   Cancer Diagnosis

According to the World Health Organization, 7.6 million people died of cancer in 2008 [28] and this number can go up in the near future. Most of the cancers are detected in late metastatic stage with secondary tumors. Researchers have shown that several advanced techniques can be useful in quick cancer detection, for example, polymerase chain reaction can detect cancer from a very small fraction of a tissue sample [14]. Although this technique has great success in detection of cancer cells, even thought it needs lots of proficiency, mutation in the gene can provide discordant results. The PCR based detection comes with other choices of RT-PCR (reverse transcription), quantitative PCR etc., with the cost of increased expense and equipment sophistication [20]. Another possibility is the blood test [8], which is being utilized with confidence in a different part of the globe. Most of the blood base diagnosis kits are measuring the cancer markers as surrogates for cancer cells. Several of these tests are already in use, like the prostate-specific antigen (PSA) blood test for prostate [16]. PSA provides a good preliminary screening for prostate cancer, but new markers are needed for detection [23].

New technology is not beneficial for cancer detection at every single time. X-rays are used to detect lung cancer, and recently the computed tomography (CT) scan has been introduced as better test. In year 2002, the National Cancer Institute started a big study for comparison of the X-ray and CT scan, and the study concluded that CT scans are doing more harm when used for detection [18]. Mammographic detection techniques are in use for breast cancer worldwide, through results are not very satisfactory [31]. Magnetic Resonance Imaging (MRI) techniques are used for breast cancer. Marker diagnosis is another way to diagnose cancer. Breast cancer type susceptibility proteins (BRCA1, 2) are markers used for cancer detection, with pros and cons [32]. Likewise, for ovarian cancer, pelvic examination, blood tests (for markers like CA-125), and ultrasounds are routine diagnostics [15]. Biopsies are used for cancer detection and confirmation for all solid tumors [29]. Although current tests offer variety and make the backbone of treatment, detection limit and ease need more refinement. Better tests are in need for more sensitive, rapid, easy, and accurate diagnosis.

Nanomaterials (NMs) with promising newer intervention are receiving acclaim for cancer diagnosis [27]. Variety, ease, size, manipulation, and surface charge are manageable properties of NMs that are used for diagnosis of cancer [5]. NMs can be manipulated for their surface charge, hydrophobicity, and surface structure. For example, large NMs can be easily cleared by the reticuloendothelial system or mononuclear phagocytic system. Various components of the reticuloendothelial system (e.g., macrophage and monocytes) recognize the antediluvian cells and remove them from circulation and tissue. The basic steps involved in the removal of these cells are recognition, tagging (opsonization) and removal. Large NMs are falling in the category of antediluvian cells and clear in the same fashion [13]. Once cancer cells filter out from active circulation to tissue, the destination would be either liver or spleen. Thus, liver and spleen are good targets for identification of cancer cells. Smaller NMs can escape to filter and remain in circulation or in tissue for a long time. Base material of NMs is an essential aspect to consider. Soon after phagocytosis, enzyme acts on NMs to break apart for disposal. There are at least three mechanisms known for phagocytosis of opsonized NMs: i) non-specific association of phagocytes with opsonized NMs; ii) activation of opsonins followed by recognition via phagocytic receptor; and iii) by activation of complementary systems [21]. Once encapsulated, they can be cleared by an oxidative burst or enzymatic mechanisms (Fig 2). During degradation, base material plays a crucial role as if it is non-biodegradable, will retain in the reticuloendothelial system or can be cleared by renal system if biodegradable.

Surface charge and hydrophobicity are aspects of consideration for design of NMs. From adherence to opsonins, followed by clearance through phagocytosis, surface charge and hydrophobicity play a crucial role [1]. The opsonization process is expedited for more hydrophobic particles as they attract and absorb more protein on their surface to be recognized by phagocytic cells. Likewise, charges on nanoparticles (NPs) equally influence its uptake for clearance. As a rule of thumb, low or neutral charges are easy to accommodate. Negatively charged particles can be cleared up quickly; on the contrary, positive charges enable NPs to stick to cells non-specifically. In the preparation of NPs, proper consideration should be given to zeta

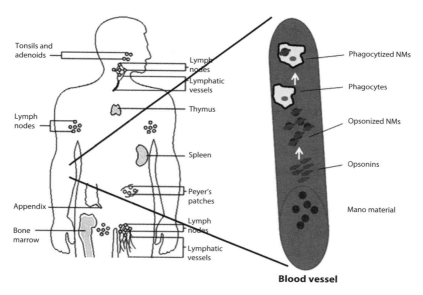

Figure 2 Nanomaterial removal by immune cells from blood circulation.

potential (ζ-potential). One other consideration is self aggregation, which can be prevented by surface charge balance. ζ-potential equal to or higher than ± 30 mv is considered as stable, and prevents aggregation of NPs.

The basic NP removal activity of the immune system is widely exploited in identification of cancerous cells or tissue. The versatile manipulation of size, material, charge, and surface structure provide a great possibility to utilize NPs for identification of cancer. For example, to save NPs from opsonization and eventual clearing by the immune system, polyethylene glycols (PEG) are widely used [12]. PEG containing NPs do not allow opsonin to attach and escape from being phagocytosed. The ability to cross-different compartments of the human body are explored in the design and utilization of NPs.

Most of the diagnosis aims to identify the cancer cells via imaging. Imaging provides an advantage to identify the location and maybe the size, if not number of cancer cells [35]. The other advantage of imaging could be targeting the metastatic cells. It is debatable to detect micro-metastasis by imaging as signal cells or few cells may have a weak signal to read.

Cancer cells differ in physiology from normal cells, and thus the difference can be targeted by NPs [30]. In healthy cells, aerobic metabolism is coupled with gaseous exchange (removal of CO_2 and absorption of O_2) and the process maintains the pH homeostasis at intracellular pH ≈ 7.2 and extracellular pH ≈ 7.4. To the contrary, in tumors (solid), because of high metabolic rate and poor vascular supply, aerobic and anaerobic breakdown of glucose combined with less O_2 intake and less CO_2 removal make an imbalance in pH. Because of biochemical reactions, intracellular pH remains neutral (pH≈ 7.2) but extracellular pH drops to acidic. This physiological imbalance has been exploiting to make functional NPs, which respond to pH.

Cell surface markers are unique signatures of cancer cells. These markers are a lucrative target for diagnosis. Some of these markers are also present on the surface of normal cell but will differ in number, and these markers may be specific to cancer cells. The specificity of cancer cells by several physical and physiological means make it identifiable. There is a long list of markers for different cancers, like human epidermal growth factor receptor 2 (HER2) for breast cancer [27], cancer antigen 125 (CA-125) for ovarian cancer [30], prostate specific antigen (PSA) for prostate cancer [29] and so on. A common strategy for identification of cancer cells is to target these cell surface markers. NPs can be conjugated with active recognition moiety, to recognize cancer cells. The other strategy is to make NPs as ligand to target to cancer cells.

Inorganic NPs are also exploited for identifying the cancer cells. Metal NPs can be used as colloidal solution because of their size and light scattering character [26]. Magnetic resonance imaging (MRI) has already been in use for decades. Oxides of iron ($Fe_3O_4^{--}$) are well studied and the first choice for making magnetic NPs. The concept of supermagnetism is also being tested to develop newer diagnostics [33]. The concept of supermagnetism has been tested to develop newer diagnostics as some compounds have intrinsic property to show strong magnetism in the presence of magnetic field, and lost it after removal of magnetic field. In conjunction of elements, biological materials are also used for cancer diagnostics. Protein-based fluorophores are radially used for diagnostics and can be organic fluorophores or recombinant fluorescent protein. In recent years, semiconductor nanocrystals and Quantum Dots (QDs) were tested as a newer strategy for cancer diagnosis [34]. Single Nucleotide Polymorphisms (SNPs) provides a fair method for identification of cancer cells [22]. The specificity, sensitivity and its use is in the development phase, but initial results indicate it as a bright candidate for future cancer diagnosis. Some approaches like gold conjugated NPs attached to DNA were tested.

Thus, with increasing cancer incidence, early diagnosis is the key to control. The ultimate lead would be to identify cancer quickly. The key to cancer control is detection of metastasis, which is still far from diagnostics. NPs provide seamless possibilities to be explored for diagnosis. More research will needed with collaboration of various disciplines to make a full proof diagnosis for cancer.

2.3   Treatment

Radiotherapy [24], and chemotherapy [4] form the backbone of cancer treatment with associated side effects. Radiotherapies lack specificity and generate equal harm to normal cells [17]. Radiation meant to harm DNA of cancer cells, even after best-targeted radiation, generate toxic radicals from water and oxygen present in surrounding areas to damage non cancerous cells [11]. Chemotherapies are designed for rapidly dividing cells and routed via bloodstreams. Before reaching the real target, drugs can harm the normal dividing cells falling in their path [2]. The magnitude of the adverse effects depends on the site of drug introduction (more distance from cancer cells will lead the drug to remain in circulation for longer before reaching cancer cells) and the drug. A better approach to treat cancer can be achieved by enhanced permeability and retention (EPR) effect for cancer drugs with NPs. EPR results in delivery and localization of NPs inside cancer cells. Various approaches have been tested to pack NPs inside the delivery vehicle. Once packed, NPs can remain in blood circulation for long extents of time. The release of the drug can be achieved by degradation (chemical or enzymatic) of the vehicle. NPs containing active cancer targeting molecules (on the surface) with anticancer drugs (in the core) are well taken by cancer cells. Once internalized in lysosome, the drug and NP get separated and thus anticancer drugs deliver to their target. In the present era of advanced nanotechnology, complex organic molecules can be attached to NPs; one example of this kind of NPs is gold-NP. One working example for this class is NP protected by PEG and functionally activated by tumor necrosis factor alpha (TNF-α) on the surface [10]. One approach to deliver anticancer drugs is the magnetic NPs under the influence of external magnetic field. Various NPs are under trial like magnetolippsomes. Carbon nanotubes containing magnetic nickel NPs tend to get in the cells with great success. Hydrophobicity can be compensated by NPs. Certain hydrophobic drugs can be delivered to cancer cells by attaching to the NPs and can be further protected by packing in vehicle. Toxicological consideration is one of the criteria to be considered before making NPs. Though most of the strategies utilize NPs in very small quantity, its proper disposal out of the host should be the first criteria to consider.

NPs have been greatly exploited for radiation-based therapy. Radiations are tough to control and target; photons generated near cancer cells harm the surrounding non-cancerous cells and pose toxic effects. The ionizing radiations generate free radicals, which are equally harmful for non-cancerous cells. Thus, radiation may be a non-targeted therapy. It is hard to decide whether radiation therapy does more harm than benefit. One way to overcome the side effects is to manage the radiation at a lower dose.

Elements of different atomic numbers were studied to target cancer tissue with increased dose of radiation [17]. By now, scientists aim to find out non toxic elements to explore radiation therapy and recently, in vivo testing show up some promising results [11]. The NPs present in the tumor facilitate the radiation therapy to regress the tumor. Inspired by these studies, several researchers reported their work in finding the utility of elements at the interface of cancer cells and radiation to get more out of radiation in the cure and treatment of cancer.

Conclusion

Nanomaterials are becoming a tool to combat a variety of diseases, including cancer. Certain newer NPs are gaining popularity in battle against cancer, like carbon nanoparticles. Multiplex NPs make diagnosis more promising; though in the juvenile stage, it is promising as it can measure more markers at a time to confirm cancer. More research is needed for accurate, easy, quick, and cheap diagnostics of cancer. Accurate and early diagnoses of cancer followed by targeted treatment are demand of time, which can be explored by NPs. In the current scenario, single drugs are not significant in killing cancer cells. Alternate medicine and newer drugs are available more frequently than ever, and drug delivery methods are in the process of improving. In the present paradigm, we lack early diagnosis and can start treatment after advanced stage of cancer in most cases. Early diagnosis and treatment of cancer will certainly enhance the efficacy of anticancer drugs to a certain extent. NP-based cancer marker tests could be the best possibility of cancer diagnosis by now. But there is a need of more advanced, sophisticated, early, and accurate diagnosis followed by equally efficient drugs and their delivery to the accurate target. It would be relevant to have some monitoring system for emerging or relapsing tumors. Conclusively, having a foolproof strategy against cancer means identification at a very early stage with precisely targeted treatment. Nanomaterials have proved to be a contribution against cancer and will certainly help us to come up with advanced and useful ways to eliminate the disease we call cancer.

References

1. F. Alexis, E. Pridgen, L.K. Molnar, and O.C. Farokhzad, "Factors affecting the clearance and biodistribution of polymeric nanoparticles." *Molecular Pharmaceutics*, Vol. 5 pp. 505–515, 2008.

2. E.E. Calle and R. Kaaks, "Overweight, obesity and cancer: epidemiological evidence and proposed mechanisms." *Nature reviews Cancer*, Vol. 4, pp. 579–591, 2004.

3. P. Carmeliet and R.K. Jain, "Angiogenesis in cancer and other diseases." *Nature*, Vol. 407, pp. 249–257, 2000.

4. B.A. Chabner and T.G. Roberts, Jr., "Timeline: Chemotherapy and the war on cancer." *Nature reviews Cancer*, Vol. 5, pp. 65–72, 2005.

5. O.C. Farokhzad and R. Langer, "Impact of nanotechnology on drug delivery." *ACS nano*, Vol. 3, pp. 16–20, 2009.

6. I.J. Fidler, "The pathogenesis of cancer metastasis: the 'seed and soil' hypothesis revisited." *Nature reviews Cancer*, Vol. 3, pp. 453–458, 2003.

7. I.J. Fidler and G. Poste, "The 'seed and soil' hypothesis revisited." *The Lancet Oncology*, Vol. 9, p.808, 2008.

8. T.R. Fleming, "Surrogate markers in AIDS and cancer trials." *Statistics in Medicine* Vol. 13, pp. 1423–1435; discussion 1437–1440, 1994.

9. D.F. Gleason and G.T. Mellinger, 1974. "Prediction of prognosis for prostatic adenocarcinoma by combined histological grading and clinical staging." *The Journal of Urology*, Vol. 111, pp. 58–64, 1974.

10. R. Goel, N. Shah, R. Visaria, G.F. Paciotti, and J.C. Bischof, "Biodistribution of TNF-alpha-coated gold nanoparticles in an in vivo model system." *Nanomedicine*, Vol. 4, pp. 401–410, 2009.

11. J. F. Hainfeld, D.N. Slatkin, and H.M. Smilowitz, "The use of gold nanoparticles to enhance radiotherapy in mice." *Physics in Medicine and Biology*, Vol. 49, pp. N309–315, 2004.

12. S. Hak, E. Helgesen, H.H. Hektoen, E.M. Huuse, P.A. Jarzyna, W.J. Mulder, O. Haraldseth, and L. de C. Davies, "The effect of nanoparticle polyethylene glycol surface density on ligand-directed tumor targeting studied in vivo by dual modality imaging." *ACS Nano*, Vol. 6, pp. 5648–5658, 2012.

13. K. Knop, R. Hoogenboom, D. Fischer, U.S. Schubert, "Poly(ethylene glycol) in drug delivery: pros and cons as well as potential alternatives." *Angewandte Chemie*, Vol. 49, pp. 6288–6308, 2010.

14. E.S. Lianidou and A Markou, "Molecular assays for the detection and characterization of CTCs." *Recent Results in Cancer Research, Fortschritte Der Krebsforschung Progrès, Dans Les Recherches Sur Le Cancer*, Vol. 195, pp. 111–123, 2012.

15. T. Maggino, F. Sopracordevole, M. Matarese, C. Di Pasquale, and G. Tambuscio, "CA-125 serum level in the diagnosis of pelvic masses: comparison with other methods." *European Journal of Gynaecological Oncology*, Vol. 8, pp. 590–595, 1987.

16. S.K. Martin, T.B. Vaughan, T. Atkinson, H. Zhu, and N. Kyprianou, "Emerging biomarkers of prostate cancer." (Review). *Oncology Reports*, Vol. 28, pp. 409–417, 2012.

17. H. Matsudaira, A.M. Ueno, I. Furuno, "Iodine contrast medium sensitizes cultured mammalian cells to X rays but not to gamma rays." *Radiation Research*, Vol. 84, pp. 144–148, 1980.

18. C.H. McCollough, A.N. Primak, N. Braun, J. Kofler, L. Yu, J. Christner, "Strategies for reducing radiation dose in CT." *Radiologic Clinics of North America*, Vol. 47, pp. 27–40, 2009.

19. P. Mehlen and A. Puisieux, "Metastasis: a question of life or death." *Nature reviews Cancer*, Vol. 6, pp. 449–458, 2006.

20. M. Mitas, K. Mikhitarian, C. Walters, P.L. Baron, B.M. Elliott, T.E. Brothers, J.G. Robison, J.S. Metcalf, Y.Y. Palesch, Z. Zhang, W.E. Gillanders, and D.J. Cole, "Quantitative real-time RT-PCR detection of breast cancer micrometastasis using a multigene marker panel." *International Journal of Cancer, Journal international du cancer*, Vol. 93, pp. 162–171, 2001.

21. S.M. Moghimi, A.C. Hunter, J.C. Murray, "Long-circulating and target-specific nanoparticles: theory to practice." *Pharmacological Reviews*, Vol. 53, pp. 283–318, 2001.

22. J.L. Murray, P. Thompson, S.Y. Yoo, K.A. Do, M. Pande, R. Zhou, Y. Liu, A.A. Sahin, M.L. Bondy, and A.M. Brewster. "Prognostic value of single nucleotide polymorphisms of candidate genes associated with inflammation in early stage breast cancer." *Breast Cancer Research and Treatment*, Vol. 138, pp. 917–924, 2013.

23. L. Ng, N. Karunasinghe, C.S. Benjamin, L.R. Ferguson. "Beyond PSA: are new prostate cancer biomarkers of potential value to New Zealand doctors?" *The New Zealand Medical Journal*. Vol. 125, pp. 59–86, 2012.

24. K. Ogawa, Y. Yoshioka, F. Isohashi-, Y. Seo, K. Yoshida, and H. Yamazaki. "Radiotherapy targeting cancer stem cells: current views and future perspectives." *Anticancer Research*, Vol. 33, pp. 747–754, 2013.

25. M. J. Plank and B.D. Sleeman. "A reinforced random walk model of tumour angiogenesis and anti-angiogenic strategies." *Mathematical Medicine And Biology : A Journal of the IMA*, Vol. 20, pp. 135–181, 2003.

26. I. Romer, T.A. White, M. Baalousha, K. Chipman, M.R. Viant, and J. R. Lead, "Aggregation and dispersion of silver nanoparticles in exposure media for aquatic toxicity tests." *Journal of Chromatography A*, Vol. 1218, 4226–4233, 2011.

27. A. Schroeder, D.A. Heller, M.M. Winslow, J. E. Dahlman, G. W. Pratt, R. Langer, T. Jacks, and D.G. Anderson. "Treating metastatic cancer with nanotechnology." *Nature reviews Cancer*, Vol. 12, pp. 39–50, 2012.

28. sheet" Cf. http://www.who.int/mediacentre/factsheets/fs297/en/. 2013.

29. D. Sidransky, "Emerging molecular markers of cancer." *Nature reviews Cancer*, Vol. 2, pp. 210–219, 2002.

30. R. Sinha, G.J. Kim, S. Nie, D.M. Shin. "Nanotechnology in cancer therapeutics: bioconjugated nanoparticles for drug delivery." *Molecular Cancer Therapeutics*, Vol. 5, pp. 1909–1917, 2006.

31. D.H. Smetherman. "Screening, imaging, and image-guided biopsy techniques for breast cancer." *The Surgical Clinics of North America*, Vol. 93, pp. 309–327, 2013.

32. M. M. Tilanus-Linthorst, H.F. Lingsma, D.G. Evans, D. Thompson, R. Kaas, P. Manders, C.J. van Asperen, M. Adank, M.J. Hooning, G. E. Kwan Lim,

R. Eeles, J. C. Oosterwijk, M. O. Leach, E.W. Steyerberg, "Optimal age to start preventive measures in women with BRCA1/2 mutations or high familial breast cancer risk." *International Journal of Cancer, Journal International du Cancer*, Vol. 133, pp. 156–163, 2013.

33. M.V. Yezhelyev, X. Gao, Y. Xing, A. Al-Hajj, S. Nie, and R.M. O'Regan , "Emerging use of nanoparticles in diagnosis and treatment of breast cancer." *The Lancet Oncology*, Vol. 7, pp. 657–667, 2006.

34. M.Z. Zhang, R.N. Yu, J. Chen, Z.Y. Ma, and Y.D. Zhao, "Targeted quantum dots fluorescence probes functionalized with aptamer and peptide for transferrin receptor on tumor cells." *Nanotechnology* Vol. 23, pp. 485104, 2012.

35. Z. Zhou, L. Wang, X. Chi, J. Bao, L. Yang, W. Zhao, Z. Chen, X. Wang, X. Chen, J. Gao, "Engineered iron-oxide-based nanoparticles as enhanced t1 contrast agents for efficient tumor imaging." *ACS Nano*, Vol. 7, 3287–3296, 2013.

3

Advanced Materials for Biomedical Application and Drug Delivery

Salam J.J. Titinchi[1,*], Mayank P. Singh[2,3], Hanna S. Abbo[1] and Ivan R. Green[1]

[1]Department of Chemistry, University of the Western Cape, Bellville, 7535, Cape Town, South Africa
[2]Department of Chemical & Biological Sciences, Polytechnic Institute of New York University Brooklyn, New York USA 11201
[3]ICL-IP America R&D, Ardsley, New York USA, 10502

Abstract

This chapter includes the following topics:

- Anticancer drug entrapped zeolite structures as drug delivery systems
- Mesoporous silica nanoparticles and multifunctional magnetic nanoparticles in biomedical applications
- BioMOFs: Metal-Organic Frameworks for biological and medical applications

Advanced materials (nanoparticles, nanopolymers, metal-organic frameworks and zeolites) provide a very attractive insight into materials science and thereby open new unexplored horizons for the application of these materials in numerous exciting ways and in this manner giving access to unconventional functions in every material class. Nanostructural composites are known to exhibit nanostructures and properties that show promise for different fields of medicinal usage, particularly in drug delivery. Most significant and efficient progress has been made over the past decade employing several advanced materials and nanoparticles in various important biological applications *viz.*, drug delivery and diagnostic application.

Keywords: Advanced material, nanoparticle, drug delivery, zeolite, mesoporous silica, biomedical application, metal organic framework

---

*Corresponding author: stitinchi@uwc.ac.za

Ashutosh Tiwari (ed.) Advanced Healthcare Materials, (49–86) 2014 © Scrivener Publishing LLC

3.1 Introduction

For the last two decades, intense interest has been developed around the nanotechnology field of research. Nanoparticles have been widely used in diverse areas such as advanced materials, catalysis, electronics, environmental remedies as well as pharmaceuticals [1–4].

On the other hand, advanced materials have also been widely used in biomedical applications due to their specific physical and chemical properties, which are totally different from the bulk materials, and thus the former have a great effect on activities such as catalytic biological and biomedical. As a result, these smart nanomaterials showed many promising applications in various diverse areas such as advanced material, industry, biology, electronics, biosensing, environmental detection and pharmaceuticals, protein purification, drug delivery, and medical imaging [5–7].

The design of materials with specific functional and effective properties is of great interest, and there is enormous potential in the application of biomedical sciences and drug delivery. With the remarkable development of innovative methods for the synthesis of new advanced materials in last decade, new approaches based on the state-of-the-art nanotechnology have been developed and are still receiving significant attention.

The main focus on the development of technologies in the field of drug delivery has been the delivery of the medicine directly to the specific disease sites in order to maximize therapeutic outcomes by promoting medication adherence and persistence, to minimize the risk of side effects, and most importantly, to enhance the patients' quality of life.

This chapter focuses on a number of novel advanced materials for drug delivery and addresses the problems related to their synthesis and their side effects.

3.2 Anticancer Drug Entrapped Zeolite Structures as Drug Delivery Systems

Zeolites have special features, such as their three-dimensional stable microstructure, presence of variable channels and cavity systems, and pore sizes that can be tailored to suit the need, high surface areas and above all, lack of cytotoxicity. All of these above features make zeolites an excellent candidate for drug release. The cavities formed by the framework units have diameters ranging from about 2 to 50 angstroms, which permits relatively easy movement of drugs inside the cavities. This in turn will facilitate the release of guest molecules through the escape from the cavities that form the openings through the channels.

Zeolite molecular sieves are crystalline, hydrated alumino silicates of group I and II elements and are involved with infinitely extending a three-dimensional aluminosilicate network of tetrahedra links. Zeolites, in general, may be represented by the empirical formula $M_{x/y}^{n+}[(AlO_2)_x(SiO_2)_y] \cdot mH_2O$ having pores of uniform size (3–10 Å), which are uniquely determined by the unit structure of the crystal. The framework of zeolites is formed between AlO_2 and SiO_4 by the sharing of all the oxygen atoms. This framework contains channels and interconnected voids, which are occupied by exchangeable cations and water molecules [8–10].

Zeolites are being explored in a variety of applications, which include the catalysis of various reactions, chemical engineering process technology, removal of sulfur and carbon dioxide from natural gas, and the recovery of radioactive ions from waste solutions [8].

There are several examples of biomedical applications of zeolites reported in the literature [11–24].

Zeolite nanocrystals have been also used in the immobilization of enzymes for biosensing and magnetic resonance imaging [25].

In the early 1980s, Unger and co-workers [26] visualized the use of silica as a drug support. Nowadays, various inorganic silica porous materials are being developed as promising drug hosting systems. In the last decade, numerous microspheres and porous materials, such as zeolites, silica, xerogels, titania, lamellar clays, and ceramics have been explored for their use as carriers for a variety of drugs.

On the other hand, zeolites have been investigated for encapsulation of drugs in their cavities that make them suitable for the storage and release of drugs and may thus also be used as carriers in a variety of drug delivery systems [27–33].

Different zeolites can be exchanged with cations, functionalized and loaded or encapsulated with drug molecules and used for specific biomedical applications. In this regard, zeolite-based compositions loaded with zinc and erythromycin have been used in the treatment of acne [34]. The ability of zeolite-Y to act as a slow release agent for various anthelmintic drugs has been published earlier [21, 23, 35, 36].

Microencapsulation is another potential route for a slow drug release carrier. Commercial zeolite-Y was used as a carrier for some anthelmintic drugs and was demonstrated to be a slow-release carrier. The results indicate that zeolite-Y is therefore a most suitable slow-release carrier and has led to an improvement of the drug's efficiency [21, 37].

Doxorubicin and paraquat molecules have been successfully loaded into a trimethylsilyl functionalized version of zeolite-Y. The controlled release of sulfonamide antibiotics adsorbed on zeolite-Y has been investigated by changing the pore size, modifying the external surface, and the nature of the grafted functional groups [19, 20]. The controlled release of ibuprofen encapsulated in dealuminated zeolite-Y is another example of an efficient drug delivery [38, 39].

The zeolite structure with the lowest framework density (ITQ-40), which contains large pore openings *viz.*, 15- and 16-membered ring channels, have been synthesized. The results indicated that zeolites have a high potential for use in the controlled delivery of drugs to specific sites in addition to then being available for electronics and catalysis [40].

One of the most promising applications for nano-engineered drug carriers is gene silencing, in which small bits of RNA are deployed to shut down the activity of crucial cancer genes through a process known as RNA interference.

Various zeolites with different Si/Al ratios were studied to explore their ability to encapsulate and to release drugs [41]. Changing the Si/Al ration often changes their adsorption properties [42]. Zeolite materials are used in the pharmacological field for skin-related conditions due to the negligible dermal uptake of the zeolite on the undamaged skin upon long term exposure [43].

New horizons in clinical medicine have opened, using synthetic nanozeolite biomaterials in various medical applications, in particular the multifunctional therapeutic delivery systems. [41, 44–48]. These non-toxic zeolitic materials, together with the physical–chemical properties, offer considerable advantages in their use compared to the polymeric materials utilized at present [49].

Immobilization of famotidine (sulphonamide derivative) on two zeolite structures (ZSM-5 and MOR) revealed that the amount of famotidine adsorbed was generally greater in ion-exchanged-zeolites, suggesting that these cations act as binding sites for the famotidine molecules. The equilibrium and kinetic characteristics of the drug on these materials were studied by varying the incubation time, the famotidine concentration and the pH of the solution [50].

Mixed matrix membrane (MMM) loaded with NaX zeolite was used as devices for the transdermal controlled release of the model drug, tramadol. The release of tramadol from the MMMs was done to evaluate the effects that the zeolite has on the release kinetics of the drug [51–53].

Algieri *et al.* revealed that the most promising combination for an effective transdermal device was achieved when the Polydimethylsiloxane (PDMS) membrane containing 17 wt% of zeolite and 0.2 wt% of drug was used [53].

Natural enzymes frequently suffer limitations in various applications due to denaturation. They are easily deactivated under extreme temperatures and/or pH. For these reasons, zeolites were also used as drug carriers and delivery systems for a number of enzymes such as glucose oxidase and antibodies [54, 55].

New drug delivery and controlled-release systems were also designed by surface solubilization or adsolubilization of drugs by zeolite–surfactant complexes [56]. Surfactant-modified zeolites incorporate the hydrophobic surfactant micelles to solubilize water-insoluble compounds in the hydrophobic core. The efficient adsolubilization of the drug chloroquin by zeolite–surfactant complexes has been described by Hayakawa et al. [45].

Zeolite-X and a zeolitic product obtained from a co-crystallization of zeolite-X and zeolite-A were examined in order to investigate their ability to encapsulate and to release drugs viz., Ketoprofen. Thus ketoprofen (800 mg) was encapsulated in 2 g of a zeolite matrix by a soaking procedure. The absence of drug release in acid conditions suggests that after activation, these materials offer good potential for a modified release delivery system of ketoprofen [41].

Zeoliote encapsulated drugs (ZED) have a profound impact in the area of drug delivery with several advantages as delivery systems. For example, one of the key aspects in drug delivery is the controlled release of the drugs. These zeolites were studied in order to investigate their ability to encapsulate and to release drugs. In particular, a zeolite-X and a zeolitic product obtained from a co-crystallization of zeolite-X, Y and zeolite-A were examined.

Neves and Baltazar et al. [23, 57] studied the encapsulation of an anticancer drug α-cyano-4-hydroxycinnamic acid (CHC) into faujasite (FAU) and Linde type A (LTA) zeolite cavities (Figure 3.1). They revealed unequivocal

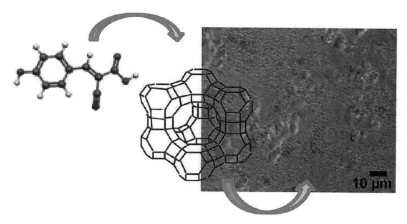

Figure 3.1 Encapsulation of an anticancer drug α-cyano-4-hydroxycinnamic acid (CHC) into faujasite (FAU) zeolite cavities.

evidence for the encapsulation of the drug in the framework structure of the zeolite. These ZEDs were used as a model for colon carcinoma treatment. Their results indicate the potential for drug loading of CHC@NaY and delivering it to the cancer cells leading to an inhibition of cell viability up to 110-fold compared to the nonencapsulated drug.

Various model drugs, such as aspirin [57], and ketoprofen [41] have been chosen for the design of controlled-release dosage systems. Due to the undesirable side effects of aspirin taken by mouth, especially in higher doses, these drugs were encapsulated into the zeolite's cavities by various methodologies in which the drug is released slowly over a period of time. The amount of drug entrapped in the zeolite matrices was estimated by thermogravimetric behavior.

In the case of ketoprofen, it was found that the total amount of 800 mg was encapsulated in 2 g of activated zeolite matrix. The amount of drug released was measured by HPLC at different pHs to mimic gastrointestinal fluids. The result was promising and the zeolite materials used offer good potential for a modified release delivery system for ketoprofen. This is due to the absence of drug release under acidic conditions.

The principal aspects in drug delivery are: (i) to transport the drug molecules to the target organ without the loss of its pharamaceutical and therapeutic activity; (ii) controlled release of drugs, i.e., concentration (in the dosage needed). The amount of drug released should be in the range of the needed dosage. Increasing the amount of drug released would cause toxic side effects, while if released below this limit, the drug would be ineffective.

3.3   Mesoporous Silica Nanoparticles and Multifunctional Magnetic Nanoparticles in Biomedical Applications

Mesoporous silica nanoparticles (MSNs) are another form of silica materials which were synthesized in 1990. Originally, mesoporous silica was first found between 1968 and 1971 [58–60].

These types of Mesoporous silica nanoparticles, such as MCMs and SBA-15, have various applications in the biomedical field, such as in medicine, imaging and drug delivery. Due to their high surface area, these mesoporous silica nanoparticles can accommodate a drug in their cavities and may thus be considered to be used as a Trojan Horse. There are several examples of biomedical applications of zeolites reported for drug delivery [61].

Another type of silica *viz.*, UVM-7, with different pore systems was used as support for ibuprofen storage and delivery [62–65].

The drug-storage and delivery study based on the use of nanoparticulate bimodal mesoporous silicas as supports was investigated by Amors *et al.* [62–65]. In the drug-charge a high ibuprofen loading was achieved owing to the decrease in pore-blocking effects in comparison to unimodal mesoporous materials *viz.*, MCM-41 as well as the availability of intra-nanoparticle mesopores and large textural voids.

The drug-delivery processes for UVM-7/ibuprofen nanocomposites were monitored by spectrometric techniques. The bimodal porosity results in two-stage drug-delivery processes, which were analysed through kinetic models.

Figure 3.2 shows a schematic diagram of the drug-delivery mechanism for a two-stage profile from UVM-7 silica. The first stage involves the ibuprofen molecules being released from the external surfaces of the pores (or macropores), while the second stage is correlated to the release of the drug molecules located inside the intraparticle mesopores.

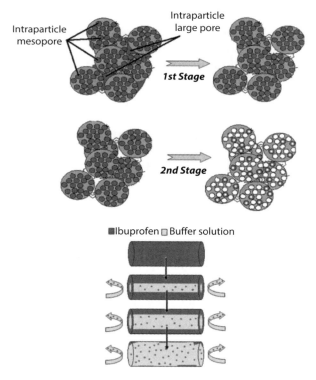

Figure 3.2 Schematic diagram of the drug-delivery mechanism of ibuprofen from UVM-7 silica.

The first article on sorption of an anticancer drug "taxol" into ordered mesoporous silicas was published by Kuroda *et al.* [66] followed by Vallet-Regi *et al.* who described a comprehensive study on the sorption and release of ibuprofen supported on MCM-41 silicas [67, 68]. Since then, various mesoporous M41s- materials have been explored as supports for hosting varieties of drugs and screened their potential capability for drug storage and release [69–74]. These drug carrier systems were studied for both in vitro and in vivo stages as well as in targeted cancer therapy [75]. Several drugs were also loaded on hollow porous spheres, which involve mesoporous shells and large micrometric core voids [76–77]. Recent review articles on the state of art in this emerging topic have been reported [54, 55, 78–82].

Very recently, an article on the synthesis of mesoporous hollow silica nanospheres using polymeric micelles as template was reported [83]. These mesoporous hollow silica nanospheres were applied as a drug-delivery carrier by an in vitro method using ibuprofen as a model drug. The study revealed that mesoporous hollow silica nanospheres exhibited a higher storage capacity than the normal mesoporous materials. They additionally found that the amine functionalized hollow nanospheres showed more sustained drug-release behavior than the unfunctionalized counterparts. This led to a suggestion that a huge potential of hollow silica nanospheres was involved in the controlled delivery of small drug molecules.

Vallet-Regi's research group pioneered the synthesis of the first silica mesoporous-based system for the controlled release of a chemical species. In their study, mesoporous materials loaded with drug systems *viz.*, the MCM-41 scaffold loaded with ibuprofen, was among the very first to be evaluated as a potential drug delivery system. Later several other successful drug releasing systems were designed by the group by selecting the appropriate matrix and loading materials and concentrations [67].

Some years later, the same group reported on the functionalization of the silica surface with certain organic moieties to provide for supramolecular interactions with the guest, which could both accommodate the molecules and control the delivery rate [54, 71].

A further advance in the field of control delivery applications was accomplished by Fujiwara *et al.* [84, 85] who developed the first gate-like hybrid material in which the functional group on the external surface was capable of being opened or closed. This innovative open-closing system was designed to regulate the storage and release of guest molecules in the pore void of coumarin-modified mesoporous materials by irradiation with the appropriate light wavelength. A conceptual scheme of photo-switched storage-release controlled release by coumarin-modified MCM-41 is shown in Figure 3.3. Since then, several photo-triggered based systems

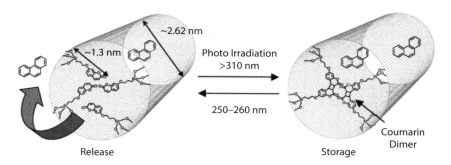

Figure 3.3 Conceptual scheme of photo-switched storage-release controlled release by coumarin-modified MCM-41.

have been developed to be utilized in the photo-switched controlled release of included compounds [86–90].

In the last decade, many drugs, proteins, and other biogenic molecules have been encapsulated in mesoporous materials [91–93].

Berlier and co-workers recently reported on the immobilization of the unstable guest molecule "Trolox" inside the hexagonal MCM-41 porous structure. Trolox, a water-soluble derivative of vitamin E, is used as a standard antioxidant in biochemical studies. The mesoporous support acts as a vehicle and facilitates the reductive photo-degradation of the active ingredient.

Analysis of the results showed that Trolox was mainly located within the mesopores of the carrier. The resulting complexes were tested for the influence of the host on the release of Trolox. In vitro diffusion tests showed a slower release of Trolox after insertion and with an increase in the photo-stability of the complex [94].

Different approaches were employed for the absorption of (-)-menthol on nanoporous silica materials for isothermal release of (-)-menthol once again. Four parameters were found to play a role in the isothermal release of (-)-menthol *viz.*, pore size, structure, wall thickness and surface functionality of the nanoporous adsorbents. By tuning these parameters, controlled release of (-)-menthol was achieved [95].

A broad review on various ordered mesoporous materials used as drug delivery systems has shown that these mesoporous materials offer promising potential for controlling drug release. Modification of mesoporous silicas by functionalization resulted in higher drug loading and provided for an improved controlled drug release. Table 3.1 summarizes some drugs loaded on various mesoporous materials having different surface areas and pore diameters [79].

Table 3.1 Comparison of the surface area for various mesoporous solid-drug delivery systems.

Mesoporous solid	SBET (m²/g)	Pore diameter (Å)	Drug Loading (wt%)	Reference
MCM-41	1157	36	Ibuprofen 34	[67]
AlSi-MCM41	1124	43	Diflunisal 8.7	[69]
AlSi-MCM41	1124	43	Naproxen 7.3	[69]
AlSi-MCM41	1124	43	Ibuprofen 6.4	[69]
AlSi-MCM41	1124	43	Ibuprofen Na salt 6.9	[69]
Si-MCM41	1210	27.9	Captopril 32.5	[96]
Si-MCM41-A	1157	25	Ibuprofen 2.9	[71]
Si-MCM41-A	1024	35.9	Aspirin 3.88	[97]
Si-SBA-15	787	61	Gentamicin 20.0	[68]
Si-SBA-15	787	88	Erythromycin 34	[98]
Si-SBA-15-C8T	559	82	Erythromycin 13	[98]
Si-SBA-15-C18ACE	71	54	Erythromycin 15	[98]
Si-SBA-15	602	86	Ibuprofen 14.6	[99]
Si-SBA-15-APTMS-O	571	86	Ibuprofen 16.9	[99]
Si-SBA-15-APTMS-P	473	78	Ibuprofen 20.6	[99]
Si-SBA-15	602	86	Bovine serum albumin 9.9	[99]
Si-SBA-15-APTMS-O	571	86	Bovine serum albumin 28.5	[99]
Si-SBA-15-APTMS-P	473	78	Bovine serum albumin 1.1	[99]
Si-SBA-15	787	49	Amoxicillin 24	[100]
HMS	1152	–	Ibuprofen 35.9	[101]

Mesoporous solid	SBET (m²/g)	Pore diameter (Å)	Drug Loading (wt%)	Reference
MCM-41	1210	26.7	Ibuprofen 74.4	[101]
HMS	1244	27.1	Ibuprofen 96.9	[76]
HMS-N-TES	1083	25.2	Ibuprofen 76.8	[76]
HMS-NN-TES	1036	24.6	Ibuprofen 74.2	[76]
HMS-NNNTES	990	24.7	Ibuprofen 70.9	[76]
Si-MSU	1200	42	Pentapeptide –	[102]
MCM-41	1200	33	Ibuprofen 41	[103]
SBA-3	1000	26	Ibuprofen 33	[103]
SBA-1	1000	18	Ibuprofen 25	[103]
SBA-16	490	85	ZnNIA 14.3	[104]
SBA-16	490	85	ZnPCB 18.3	[104]
MCM-48	1166	36	Ibuprofen 28.7	[105]
LP-Ia3d	857	57	Ibuprofen 20.1	[105]
MCM-48	1166	36	Erythromycin 28.0	[105]

Other drugs that have been loaded on mesoporous silica other than those mentioned in Table 3.1 to be used as a drug delivery system were vancomycin [106], famotidine [107] and itraconazole [108].

An envelope-type mesoporous silica nanoparticle was designed for tumor-triggered targeting drug delivery to cancerous cells. The antineoplastic drug, doxorubicin hydrochloride, was loaded on the surface of the mesoporous silica core linked with β-cyclodextrin via disulfide bonding. Thereafter, the surface of the nanoparticles was allowed to interact with different peptide sequences via a host-guest interaction. The resultant nanoparticles were further decorated with poly(aspartic acid) to obtain the desired material. The purpose of the later step was to protect the targeting ligand and preventing the nanoparticles from being taken up by normal cells. In vitro studies revealed that the envelope- mesoporous silica nanoparticles were shielded against normal cells and enhanced cell growth inhibition efficiency in cancerous cells [109].

This new type of mesoporous silica nanoparticle was first introduced by a group of Japanese scientists [110].

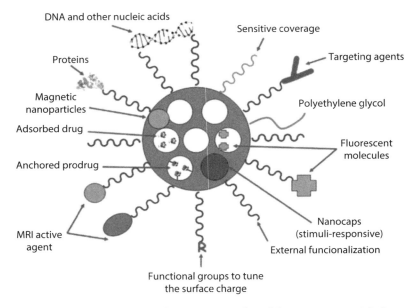

Figure 3.4 Various possibilities in designing smart drug delivery nanomaterials for numerous therapeutic applications.

Very recently, Regi *et al.* reviewed the use of the mesoporous silica nanoparticles for the design of smart delivery nanodevices [111]. This review pictured the multifunctionality and drug loading possibilities for the design of smart delivery nanomaterials (Figure 3.4).

Loading of small drug molecules into the mesopores occurred via the adsorption approach. Whereas the prodrugs could be attached to the functional groups present in the inner part of mesoporous silica walls, the surface of the silica nanoparticles can be modified by using different functional groups to further enhance controlling drug release. The review describes the use of different nanocaps to design stimuli-responsive drug delivery systems and incorporate magnetic nanoparticles into the silica matrix to provide the mesoporous silica nanoparticles with magnetic properties suitable for various applications such as magnetic guidance, hyperthermia and magnetic resonance imaging (MRI).

Mesoporous silica nanoparticles (MSNPs) are attracting an increasing interest in the nanomedicine field and are consequently receiving enormous attention due to the possibility for high drug loading into the mesopore cavities and covering the pore entrances with different nanogates. These features and functionality can be tailored to serve specific clinical needs.

A summary of various mesoporous materials employed for encapsulation of enzymes was reported and is summarized in Table 3.2 [112].

Table 3.2 Various mesoporous silica materials used to encapsulate different enzymes.

Mesoporous materials	Silica source	Template	Description	Pore diameter (nm)	Refs.
MCM-41	TEOS, sodium Silicate	C_nTMA^+ ($n = 12-18$)	2D hex. channels	2-10	[58]
MCM-48	TEOS, sodium silicate	CTAB, $C_{16}H_{33}(CH_3)_2N(CH_2)(C_6H_5)$, Gemini C_{m-12-m}	Bicontinuous	2-4	[113]
FSM-16	Polysilicate kanemite	C_nTMA+ ($n = 12-18$)	2D hex. channels	~4	[114]
SBA-1	TEOS	$C_nH_{2n+1}N(C_2H_5)_3X$ ($n = 12-18$), $18B_{4-3-1}$, C_{n-s-1} ($n = 12-18$)	3D cubic mesostructure	2-3	[115]
SBA-15	TEOS, sodium silicate	P_{123}, P_{85}, P_{65}, B50—1500 ($B0_{10}EO_{10}$), Brij 97($C_{18}H_{35}EO_{10}$)	2D hex. channels	5-30	[116]
SBA-16	TEOS, TMOS	F127, F108, or F98	Spherical cages	5-30	[117]
MCF	TEOS	F127 ($EO_{106}PO_{70}EO_{106}$) with TMB	Cellular foam	10-50	[118]
HMS	TEOS	$C_mH_{2m+1}NH_2$ ($m= 8-22$)	Disordered mesostructure	2-10	[119]
MSU-X	TEOS, TMOS	C_mEO_n ($m= 11-15$), C_8PhEO_n, $EO_{13}PO_{30}EO_{13}$	Disordered mesostructure	2-15	[120]
IBN-X	TEOS	F108, F127, P65, P123 with FC-4 and TMB	Nanoparticle	5-20	[121]
PMOs	$(RO)_3Si-R-Si(OR)_3$	CTAB, OTAB, CPB, P123, F127, Brij 56, Brij 76	2D or 3D hex.	2-20	[122]

Many researchers revealed that mesoporous structures are selective for the size of the drug and also are able to control the level of loading over the target molecule [123–126].

The concept of controlling the mesoporous silica nanoparticles pore size was demonstrated in a study by Min *et al.* [126]. The study revealed that loading of large amounts of plasmid DNA could be achieved using mesoporous silica NPs with pores more than 15 nm.

Other researchers prepared carboxylic acid modified mesoporous silica nanoparticle carriers by surface modification. Surface modification was achieved by functionalization of the surface via a two-step process: (i) 3-amino-propyltriethoxysilane; and (ii) reaction with succinic anhydride. The carboxylic-modified mesoporous silica materials showed high adsorption capacity (~50 wt.%) for sulfadiazine. In vitro release studies showed that the functionalized mesoporous materials with carboxylic groups are appropriate carriers for prolonged release of sulfadiazine. Also the drug-loaded silica materials demonstrated no cytotoxicity to the Caco-2 cell line [127].

Another family of modified silica mesoporous supports with various alkyl groups anchored on the pore outlets of mesoporous MCM-41 has been prepared [128]. These supports could be modified to display a unique architecture to design gated systems able to achieve zero release. The hybrid mesoporous silica surface provides adequate controlled drug release features through advanced controlled release from the pore voids by control of the open-close gate operation (Figure 3.5). The rate of delivery

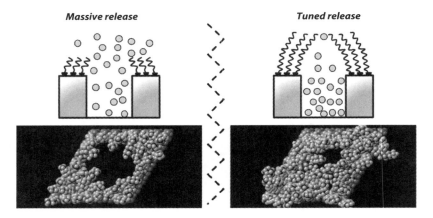

Figure 3.5 Schematic representation on guest release rate from different alkyl modified surface support with the surface perspective views of obtained models for alkyl chains containing (left) 8 and (right) 16 carbon atoms.

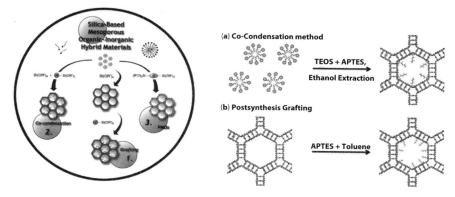

Figure 3.6   Schematic diagrams illustrating the functionalization of mesoporous materials by:

1. Grafting (subsequent attachment of organic components onto a pure silica matrix; **2.** co-condensation (one-pot synthesis) simultaneous reaction of condensable inorganic silica species and silylated organic compounds; **3.** periodic mesoporous organosilicas (PMOs), hydrolysis and condensation reactions of bridged organosilica precursors, i.e., use of bissilylated organic precursors [130].

(A) Co-condensation: organosilane with the desired functional group is added during the synthesis so that the organic functionality is incorporated into mesoporous silica directly during the synthesis.
In postsynthesis grafting, and **(B)** postsynthesis grafting: calcined mesoporous silica is treated with the organosilane that reacts with surface silanol groups [129].

was found to be dependent on the length of the alkyl chain anchored on the pore outlets.

Amine functionalized MCM-41 with organic functional groups was employed for aspirin loading and release studies. The modification methods used for functionalization were co-condensation and post synthesis grafting routes as illustrated in Figure 3.6.

The carboxylic acid group of aspirin interacts with surface silanol groups or amino groups modified with aminopropyltriethoxysilane (APTES) functional groups on the pore walls which make these materials useful for controlled drug release. The loading and release of aspirin from the mesoporous silica was studied to provide insight into the molecular level interactions of drug with the mesoporous host and how these interactions influence loading and release properties [129].

The release profiles for the amine functionalized silica were studied at pH = 7.4 and 37.4 °C and the aliquots were analysed for aspirin with

UV–Vis spectroscopy at $\lambda = 296$ nm. The amount of released aspirin was calculated using the literature reported relation [97].

Drug release data indicated that the distribution and loading of the amine groups and the method of loading (co-condensation or post synthesis) are important factors and also affect the release of aspirin. The difference in release models was explained by the relatively weaker interaction of aspirin with the parent MCM-41 relative to the amine functionalized samples. On the other hand, it was found that materials prepared by co-condensation demonstrated that the maximum release of the drug decreased significantly (from 71 to 50%) due to increased pore blocking by APTES.

The same group also successfully loaded aspirin into the pores of various (SiO_2/Al_2O_3) ratios of zeolite HY [107]. The study revealed that the amount of aspirin loaded was found to increase with decreasing of SiO_2/Al_2O_3 ratios. Nitrogen adsorption–desorption experiments revealed that aspirin was loaded into the internal pore surface of these materials.

The nature of intermolecular aspirin-zeolite interactions were obtained from the spectroscopic methods viz., FT-IR, ^{27}Al and ^{13}C magic angle spinning nuclear magnetic resonance spectra. The extent of loading has been determined by analytical techniques viz., N_2 adsorption/desorption isometry.

Aspirin was bonded to silanol sites via hydrogen bonding on higher SiO_2/Al_2O_3 ratio of HY zeolites. The theoretical quantum calculations provided aspirin/host interactions and the zeolite properties that control the loading and release. The drug release study from the zeolite matrix at pH = 7.4 in aqueous solution revealed nearly complete release of aspirin HY-5, while HY-30 and 60 exhibited partial release of aspirin (Figure 3.7). The partial release of the drug was attributed to the increased hydrophobicity in HY-30 and HY-60 materials due to an increase of van der Waals interactions. The aim of the study was to control the delivery of the drug, i.e, the optimal release of the drug in effective amounts while minimizing the side effects that can occur depending on the dosage forms [131].

Several articles on loading drugs, genes and proteins (i.e., cytochrome c) on mesoporous silicas have been reported [124, 132–135]. The release of drug/genes/proteins from the mesoporous silica was studied in vitro within human cells when showing integral activity.

Wheatley and co-workers in 2006 published the synthesis of systems that involve coordination of NO to ion-exchanged sites within the zeolite framework [136, 137].

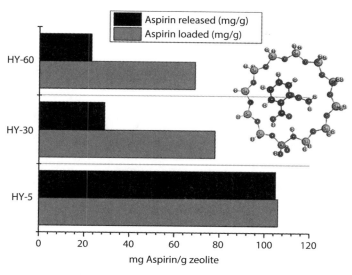

Figure 3.7 Drug release study from HY-5, while HY-30 and 60 zeolite matrix at pH = 7.4 in aqueous [131].

This system was releasing nearly 100% of the incorporated NO over ~40 min in moisture-rich air, which is very difficult to control and cannot avoid cellular damage due to exposure to excess NO.

Recently, Mascharak and his team developed a photoactive system that releases bactericidal amounts of nitric oxide (NO) under the control of low-power (10–100 mW) levels of visible light [138]. The system consists of photoactive manganese nitrosyl, $[Mn(PaPy_3)(NO)](ClO_4)$, which was entrapped within the extended pores of the mesoporous material MCM-41 as well as aluminosilicate-based material that has a negatively charged host structure.

In this study, the loading efficiency and leaching and the NO-releasing capacities from Si- and Al-MCM-41 systems have been determined (Scheme 3.1). The loading efficiency was improved using Al-MCM-41. Leaching of the nitrosyl from the host was found to be minimal in physiological saline.

The NO-releasing capacities from these two hybrid materials by visible light exposure results in rapid release of NO from the entrapped complex while the photoproducts are retained in the host structure. Therefore these hybrid NO-donors could be employed to deliver NO under the control of light with very little toxicity from the biocompatible host that also retains the photoproducts. Thus these NO-porous materials could be

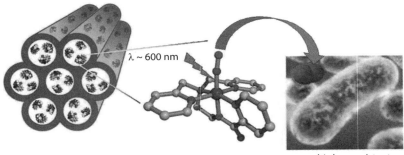

multi-drug resistant
Acinetobacter baumannii

Scheme 3.1 A photoactive and multi drug resistance manganese nitrosyl loaded into the columnar pores of MCM-41 host as NO delivery for treatment for Acinetobacter baumannii infections.

applied to infected wounds for treatment for *A. baumannii* infections in battlefield wounds.

3.4   BioMOFs: Metal-Organic Frameworks for Biological and Medical Applications

3.4.1   Introduction

Metal-organic frameworks (MOFs) are some of the most exciting and high-profile areas that have emerged as an important field of innovative research over the past decade. Metal-organic frameworks (MOFs), composed of metal ions or metal ion clusters as "nodes" and organic ligands as "linkers," has been the subject of current research due to an upsurge of scientific interest in the creation of new porous materials and their enormous potential in applications such as photonics, heterogeneous catalysis and separation [139–153]. Recently MOFs have been extensively reviewed in the field of gas storage and are therefore often highlighted for their gas-storage properties.

The coordination compounds with infinite one-, two-, three-dimensional network structures with backbones constructed by metal ions and ligands are called "coordination polymers" [154]. Metal organic frameworks (MOFs) can be defined as having an extended array of metal ions or complexes that are linked by multifunctional bridging organic ligands. MOFs are specially defined as strong bonding (such as covalent bonding) three dimensional networks with geometrically well-defined structure [155]. In this chapter, we will discuss the potential applications of MOFs in biological systems, an area that is only recently being developed. In order to evaluate any in vivo biological application, there is a need for a strict regulatory approval which in general is not the case for other commercial activities. The regulatory approval for human and animal application requires capital investments, but good signs are that early studies on MOFs have demonstrated some excellent results which serve to facilitate continuation in the research on their therapeutic and diagnostic applications.

3.4.2 Synthesis, Properties and Structures of MOFs

The synthesis of MOFs can be considered to be aligned to a building-block approach by connecting metal ions with organic linkers illustrated in Figure 3.8.

Typically, MOFs have low density (0.2–1 g/cm^3), high surface area (500–4500 m^2/g), very high porosity, and moderate thermal and mechanical stability. The greatest advantage of MOFs over other well-known nanoporous materials such as zeolites is the ability researchers have to tune the structure, functionality and properties of MOFs directly during synthesis by changing the shape of building blocks. This type of MOFs design approach is well known as reticular synthesis [141, 155]. The broad ranges of metal and linkers offers a theoretically large number of materials with a wide range of structural and different physical properties in addition to chemical, optical, magnetic, and electrical properties. The wide variety and numbers of MOFs available afford both a challenge and an opportunity for their implementation in practical applications. The wide range of available

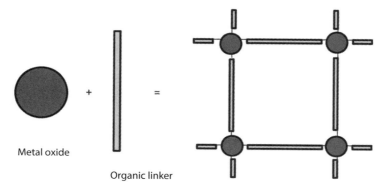

Metal oxide

Organic linker

Figure 3.8 Building block approach of synthesis of MOFs.

pore sizes, topologies, and functionalities suggests that existing MOFs will have useful properties in several fields of biomedical applications. Despite the fact that MOFs are a rapidly expanding area of research, it still remains a challenge to predict which frameworks are best suited for a specific biomedical application.

MOFs are generally synthesized using hydrothermal or solvothermal techniques in which the crystals are grown slowly from the mother liquor of metal precursors and the organic moiety [156–158]. There are several organic linkers which have been used for the synthesis of bio-MOFs and a few are listed in Scheme 3.2.

MOFs are usually characterized by X-ray crystallography due to their high crystallinity. Since MOFs contain large pores, they become filled by the solvent used in the synthesis. Removal of the solvent molecules from the pores is referred to as the activation of MOFs. Generally, the activation collapses the frameworks and results in decomposition of MOFs. Most of the MOFs have unsaturated metal sites (open metal sites) on the walls of pores. These open sites are further responsible for binding the guest molecules through various interactions [159–161].

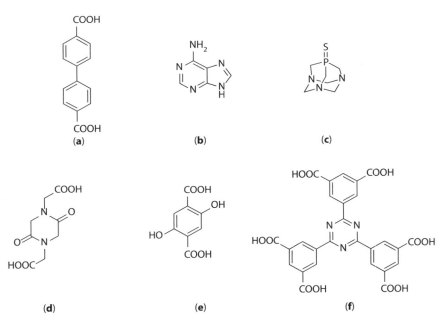

Scheme 3.2 Building blocks for BioMOFs. (a) Biphenyl dicarboxylate. (b) Adenine. (c) 2,5-Piperazinedione-1,4-diacetic acid. (d). (e) 2,5-Dihydroxyterephthalic acid (f) 5,5′,5″-(1,3,5-triazine-2,4,6-triyl)tris(azanediyl)triisophthalate.

One very important aspect to be considered while considering using MOFs in biomedical application is their stability. For example, IRMOF-1 is hydrolytically unstable. Theoretical simulations and experiments have shown that it collapse in 3.9% of water. This is mainly due to the replacement of oxygen of MOFs which is coordinated to Zn^{2+} by the oxygen of water [162, 163].

Toxicology of the materials is another key factor when considering health, biomedical or biological application. Another very important issue which has to be considered is the toxicology of the metal ions present in the MOFs. More research work is needed to explore the behavior of carboxylate-based MOFs as these are the better known members of the MOFs family.

3.4.3 MOFs as Drug Delivery Agents

One of the major challenges faced in drug delivery using MOFs is the efficient delivery of the drug using non-toxic nano-carriers. In order to deliver the drug into the cell, the carriers should have the following features: (1) efficiently drug trapping within the pores of MOFs; (2) controlled release; (3) controlled degradation; and (4) be detectable by imaging techniques (Figure 3.9). Recently Patricia Horcajada's research group demonstrated the ibuprofen release [164], R. Morris *et al.* [165, 166] demonstrated delivery of NO gas for antithrombosis and vasodilatation and Lin *et al.* [167–169] demonstrated the imaging application using MOFs.

A series of biologically and environmentally favourable non-toxic carboxylate MOFs have been reported by Ferey and coworkers [170]. The MIL family, synthesized from Cr^{3+} centers and benzene dicarboxylic acid have large pore sizes (25–34 Å) and outstanding surface areas (3100–5900 m^2/g), and are the ideal systems for drug delivery.

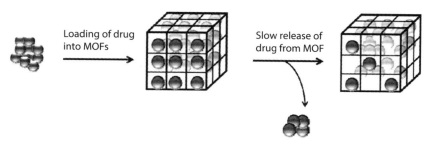

Figure 3.9 Absorption and release of drug molecule through MOFs pores.

The storage and release of ibuprofen with chromium-based MIL-101 and MIL-100 materials showed high ibuprofen loading, with 0.347 g ibuprofen/g MOF for MIL-100 and 1.376 g ibuprofen/g MOF for MIL-101. The difference in drug loading between the two materials is due to the pore sizes of the materials; MIL-101 has larger pore volumes of 12700 and 20600 Å^3 (8200 and 12 700 Å^3 for MIL-100). The kinetics of ibuprofen release was investigated by suspending the ibuprofen-loaded materials in simulated body fluid (SBF) at 37 °C.

These MOFs contain toxic chromium, and thus, the use of these materials for drug delivery is very limited. A less-toxic analog, MIL-101(Fe) has been developed and reported by Horcajada as a biocompatible alternative, and should be a much more appropriate drug carrier [164]. These MOFs have been modified into nanoparticles for efficient delivery of anti HIV and anticancer drugs (busulfan, azidothymidine triphosphate, doxorubicin or cidofovir) (Figure 3.10) [171].

The physical properties and the structures of the drugs are summarized in Table 3.3 and Scheme 3.3.

Recently, Zhong-Min Su and coworkers reported on chiral MOFs synthesized from 5,5′,5″-(1,3,5-triazine-2,4,6-triyl)tris(azanediyl)triisophthalate and zinc nitrate [172]. Single crystal X-ray diffraction analysis revealed that enantiomers exhibit homochiral 3D structures with nanoscale porous and helical channels (Figure 3.11a). The large pore size of MOFs (27579.3 Å^3) facilitate drug storage and subsequent drug (anticancer 5-FU D) release was observed with no burst effect. The delivery of 5-FU occurred within a week and 86.5% of the loaded drug was released (Figure 3.11b). The interaction between pore walls and the guest drug are mainly hydrogen bonding and π-π interaction.

Another interesting BioMOFs has been reported by Rosi and coworkers employing adinine, diphenyl dicarboxylate and zinc acetate [173]. The

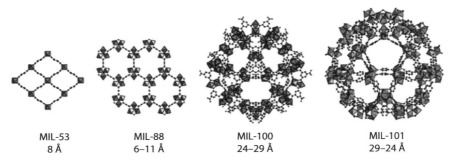

| MIL-53 | MIL-88 | MIL-100 | MIL-101 |
| 8 Å | 6–11 Å | 24–29 Å | 29–24 Å |

Figure 3.10 Porous iron MOFs for drug delivery applications.

Table 3.3 The physical properties of and the structures of the drugs.

	MIL89	MIL88A	MIL100	MIL53
Organic linker	Muconic acid	Fumaric acid	Trimesic acid	Terphthalic acid
Structure				
Flexiblity	Yes	Yes	No	Yes
Pore size (Å)	11	6	25(5.6)	8.6
Particle size (nm)	50–100	150	200	350
Bu loading(%)	9.8(4.2)	8 (3.3)	25.5(31.9)	14.3(17.9)
AZT-TP loading (%)	–	0.6(6.4)	21.2(85.5)	0.24(2.8)
CDV loading (%)	14(81)	2.6(12)	16.1(46.2)	–
Doxorubicin loading (%)	–	–	9.1(11.2	–
Ibuprofen loading (%)	–	–	33(11.0)	22 (7.3)
Caffeine- loading (%)	–	–	24.2(16.5)	23.1(15.7)
Urea loading (%)	–	–	69.2(2.1)	63.5(1.9)
Benzophenone 4 loading (%)	–	–	15.2(22.8)	5(7.5)
Benzophenone 3 loading (%)	–	–	1.5(74.0)	–
Doxorubicin loading (%)	–	–	9.1(11.2)	–

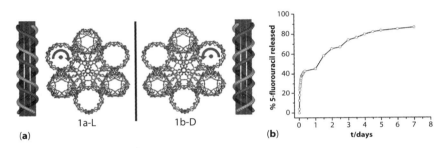

Busulfan (BU)

Azidothymidine triphosphate (AZT-TP)

Cidofovir (CDV)

Doxorubicin

Ibuprofen

Caffeine

Benzophenone 4

Benzophenone 3

Scheme 3.3 Structures of various anticancer drugs.

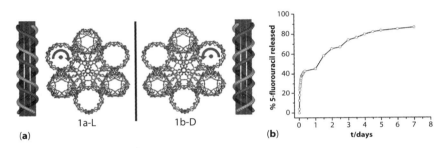

1a-L

1b-D

(a)

(b)

Figure 3.11 (a) The enantiomeric nature of MOFs viewed down the [−111] direction. Side view of left-handed and right-handed double-stranded 3_1 helical chains in isomeric MOFs, respectively, C grey, O red, N blue, Zn green. (b) The release process of 5-FU from the drug-loaded 1 (% 5-FU vs. time).

MOFs have a large pore volume and surface area which greatly facilitates loading of the drug (procainamide) efficiently. Since the drug has a very short in vivo half-life, the patient is required to be dosed every 3–4 hours. Complete loading (0.22 g/g material) was achieved after 15 days, and it has been estimated that ~2.5 procainamide molecules per formula unit remains in the pores while the rest adhere to the surface. Due to the ionic interactions between host and guest, cations can be used to trigger procainamide release from the framework (Figure 3.12).

Lin's research group designed and synthesized a nano-MOF (NCP-1) from Tb^{3+} ions and c,c,t-(diamminedichlorodisuccinato)Pt(IV) (disuccinatocisplatin, DSCP), a cisplatin prodrug for cancer treatment [174]. The nanoparticles of NCP-1 (58.3-11.3 nm) were encapsulated with silica to enhance the half-life time in HEPES buffer at 37 °C. These

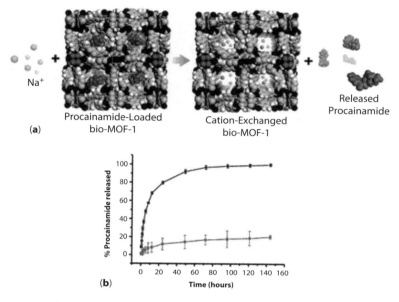

Figure 3.12 Scheme depicting cation-triggered procainamide release from bio-MOF-1. (B) Procainamide release profiles from bio-MOF-1 (blue, PBS buffer; red, deionized nanopure water).

silica coated nanoparticle were further functionalized with c(RGDfk), a cycllic peptide which targets the αnβ3 integrin, which is over expressed in many cancers. These particles displayed a lower IC50 (inhibitory concentration, 50%) than that of cisplatin (9.7 mM versus 13.0 mM for cisplatin), while the untargeted particle did not exhibit significant cell death. The improved cytotoxicity of the functionalized particles suggests that these particles are taken up by receptor-mediated endocytosis, followed by reduction to the active cis platinPt(II) species inside the cell (Figure 3.13).

3.4.4   Applications of MOFs as NO storage

Metal organic frameworks have a great capability of storing gas since they have large pore volumes and large surface areas incorporated into their structures. NO storage is one of the most important but less studied fields of MOFs. NO is a gaseous radical which is involved in most of the biological processes such as neural, immune and vascular systems. In vivo and in vitro delivery of NO through MOFs is thus considered to be a most attractive and challenging task. There are several materials and polymers such as zeolites

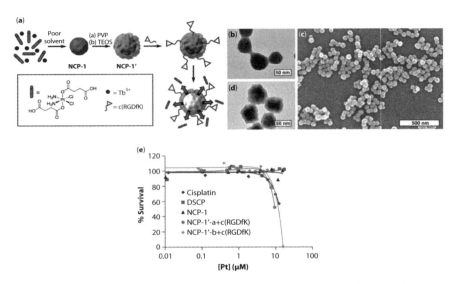

Figure 3.13 (a) Schematic showing the synthesis of Tb-DSCP NMOF (designated as NCP-1) and its subsequent coating with silica shell (NCP-1′) and conjugation with cyclic peptide (PVP, polyvinylpyrollidone; TEOS, tetraethylorthosilicate); (b) TEM micrograph for as-synthesized NCP-1; (c) TEM; and (d) SEM micrographs for NCP-1′; (e) In vitro cytotoxicity assay curves for HT-29 cells obtained by plotting the % cell viability against the Pt concentration of various samples and cisplatin control.

which are available in the literature but have a disadvantage in that they are responsible for also delivering the carcinogenic by-products inside the living cells. Therefore development of target oriented NO delivering materials would be of immense interest and this would ensure that no side products would remain inside the living cells. An ideal material should be able to load NO efficiently and be able to release the NO the specific target sites efficiently.

Xiao and co-workers reported a MOF which was derived from benzene tricarboxylic acid and copper acetate (also known as HKUST-1) [175]. This MOF has open metal sites available for NO adsorption after activation. They measured NO adsorption gravimetrically and observed that HKUST adsorbs NO with a large adsorption capacity of 9 mmol/g at 1 bar and 196K, which is significantly greater than any other adsorption capacity reported for a poro us solid. However, at 1 bar and 298K it only absorbs 3 mmol/g. Infra-red experiments further confirms that the NO binds to the empty copper metal sites in HKUST-1. They confirmed NO release from HKUST-1 on contact with water vapor. However, the total amount of NO released was around 1 µmol NO/g MOF.

Morris and co-workers reported two MOFs *viz.*, $[M_2(C_8H_2O_6)(H_2O)_2] \cdot 8H_2O$ where (M) = Co, Ni (CPO-27-Co and CPO-27-Ni) which perform exceptionally well for the adsorption, storage and water-triggered delivery of NO [176]. They further confirmed by powder X-ray diffraction studies, that each of the unsaturated metal ion sites coordinates to one NO molecule. The high adsorption of NO by these MOFs is as a result of chemisorption which was later confirmed by IR studies. The stored NO in these MOFs could be maintained for several months. The capability of storing NO for long term and high adsorption capacity (~ 7 mmol NO/g of MOF which is ~ 7000 times more than the HKUST-1) inspired researchers to study these types of MOFs further. It was found that since Cr-MIL53 and AL-MIL53 do not possess open metal sites they adsorb very little NO.

3.4.5   Applications of Bio-MOFs as Sensors

Apart from the fascinating property for gas storage, MOFs have also demonstrated a capacity to behave as sensors. MOFs possessing a luminescent property have a great advantage as biomedical diagnostic tools that are apparent from a few reports in the literature on luminescence [177, 178].

Rosi and coworkers first reported that bio-MOF-1 derived from adenine and biphenyldicarboxylic acid is an efficient sensor for certain visible and NIR-emitting lanthanides (Tb^{3+}, Sm^{3+}, Eu^{3+} and Yb^{3+}) [179]. The lanthanide ions were inserted into the pores of bio-MOF-1 as nitrate salts in DMF solution. The lanthanide loading doesn't disturb the crystallinity order, which was further confirmed by X-ray powder diffraction pattern (Figure 3.14). Similar experiments were performed in water as solvent. Despite the fact that water is a strong luminescent quenching solvent, the bio-MOF-1 showed a strong luminescent property because the material protects the lanthanide within its pores enabling the use of NIR-emission. The encapsulation of large lanthanide cations within the defined space is the main reason for the bio-MOF-1 having such a high luminescence intensity.

In another report, Chen et al. mentioned a new application of MOFs as a temperature dependent luminescent thermometer [180]. They synthesized isostructural MOFs, $[Tb_2(dmbdc)_3]_n$ or $[Eu_2(dmbdc)_3]_n$, based on (2,5-dimethoxy-1,4-benzenedicarboxylate) (Figure 3.15). The organic linker acted as a sensor which upon excitation at 381 nm underwent a p-p* electron transition between the linker and the metal centres occurred. By increasing the temperature from 10K to 300K the emission intensity gradually decreased due to the thermal non-radiative decay. Doping of $[Tb_2(dmbdc)_3]_n$ with Eu^{3+} ions using a one-pot synthesis led to the formation of $[(Eu_{0.0069}Tb_{0.9931})_2(dmbdc)_3]_n$. Interestingly, the mixed-MOF shows a novel temperature-dependent luminescence. At 300 K, the Eu^{3+} emission dominates the whole spectrum, which is most likely due to the temperature-dependent photon-assisted Förster transfer mechanism.

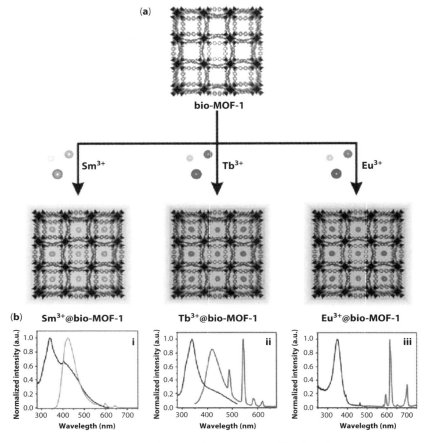

Figure 3.14 Bio-MOF-1 encapsulation and sensitization of lanthanide cations. (a) Schematic illustration of Ln3þ incorporation into bio-MOF-1 and subsequent Ln^{3+}sensitization by the framework. (b) Excitation and emission spectra of Sm^{3+}@bio-MOF-1 (i), Tb^{3+}@bio-MOF-1 (ii), and Eu^{3+}@bio-MOF-1 (iii).

In another example, Nenoff *et al.* reported the Indium based white light emitter MOF $[In_3(btb)_2(oa)_3]_n)$ (btb = (btb=1,3,5-tris(4-carboxyphenyl) benzene and oa=oxalic acid) [181]. The emission is based on a LMCT (ligand to metal charge transfer) mechanism. After doping with 9% Eu^{3+} ions onto the $[In_3(btb)_2(oa)_3]_n)$, the doped-MOF $[(Eu_{0.09}In_{0.91})_3(btb)_2(oa)_3]_n$ behaved as a red light emitter. The objective of the doping in this case was to enhance the properties such as color rendering index (CRI) and thus correlated color temperature (CCT), and chromaticity to approach the requirements for solid-state lighting (SSL).

(a)

(b)

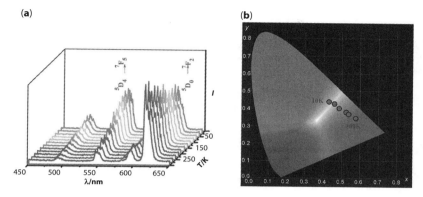

Figure 3.15. a) Temperature-dependent luminescent spectra of the mixed Tb/Eu framework between 10 to 300 K; and b) the corresponding CIE chromaticity diagram showing the change of luminescent color by increasing the temperature.

3.5   Conclusions

Most significant and efficient progress has been made over the past decade employing MOFs in various important biological applications *viz.*, drug delivery and diagnostic application. The distinguishing properties of bio-MOFs made them an important class of new biologically applicable materials. MOFs were previously known for their impressive gas storage application. Their latest emerging biological application has, however, made them extraordinary new candidates for further exploration in this newly discovered research enterprise. A few issues still remain to be addressed *viz.*, toxicity, instability, biocompatibility and biodegradability all of which hinder the biological applications of MOFs. However, with the ever-increasing numbers of groups working in this fascinating area there is no doubt that all these problems will be solved in the near future.

References

1. R.A. Petros and J. M. Desimone, *Nature Reviews Drug Discovery*, Vol. 9 (8), p. 615, 2010.
2. M. Ferrari, *Nature Nanotechnology*, Vol. 3 (3), p. 131, 2008.
3. K.E. Sapsford, W. R. Algar, L. Berti, K. B. Gemmill, B. J. Casey, E. Oh, M. H. Stewart and I.L. Medintz, *Chemical Reviews,* Vol. 113, p. 1904, 2013.
4. A. Korkin, P. S. Krstić and J. Carter Wells, *Nanotechnology For Electronics, Photonics and Renewable Energy.* Springer 2010.

5. G.Z. Cao, *Nanostructures and Nanomaterials: Synthesis, Properties, and Applications,* Imperial College Press, 2004.

6. P. Knauth, J. Schoonman, Nanostructured Materials: Selected Synthesis Methods, Properties and Applications, Kluwer Academic Publishers, 2002.

7. C.N.R. Rao, A. Müller and A. K. Cheetham, *The Chemistry of Nanomaterials: Synthesis, Properties and Applications,* John Wiley & Sons, 2004.

8. D.W. Breck, *Zeolite Molecular Sieves: Structure, Chemistry and Use,* New York, Wiley, 1974.

9. R.M. Barrer, *Zeolites and Clays as Sorbents and Molecular Sieves,* London, Academic Press, 1978.

10. H. van Bekkum, E. M. Flanigen and J. C. Jansen, *Introduction to Zeolite Science and Practice,* Amsterdam, Elsevier, 1991.

11. S.W. Young, F. Qing, D. Rubin, K. J. Balkus Jr., J. S. Engel, J. Lang, W. C. Dow, J. D. Mutch and R. A. Miller, *Journal of Magnetic Resonance Imaging,* Vol. 5, p. 499, 1995.

12. K.J. Balkus Jr. and J. Shi, *The Journal of Physical Chemistry,* Vol. 100, p. 16429, 1996.

13. É. Csajbók, I. Bányai, L. V. Elst, R. N. Muller, W. Zhou and J. A. Peters, *Chemistry - A European Journal,* Vol. 11, p. 4799, 2005.

14. M. M. Tsotsalas, K. Kopka, G. Luppi, S., Wagner, M. P. Law, M. Schäfers and L. De Cola, *ACS Nano,* Vol. 4, p. 342, 2010.

15. N. Ndiege, R. Raidoo, M. K. Schultz and S. Larsen, Langmuir, Vol. 27, p. 2904, 2011.

16. W.-S. Li, Z.-F. Li, F.-Y. Jing, X.-G. Yang, X.-J. Li, F.-K. Pei, X.-X. Wang and H. Lei, *Acta Chimica Sinica,* Vol. 65, p .2029, 2007.

17. I. Bresinska and K. J.Balkus Jr., *Journal of Physical Chemistry,* Vol. 98, p. 12989, 1994.

18. H. Zhang, Y. Kim and P. K. Dutta, *Microporous Mesoporous Matterails,* Vol. 88, p. 312, 2006.

19. I. Braschi, G. Gatti, G. Paul, C. E. Gessa, M. Cossi and L. Marchese, *Langmuir,* Vol. 26, p. 9524, 2010.

20. I. Braschi, S. Blasioli, L. Gigli, C. E. Gessa, A. Alberti and A. Martucci, *Journal of Hazardous Materials,* Vol. 178, p. 218, 2010.

21. A. Dyer, S. Morgan, P. Wells and C. Williams, *J. Helminthol,* Vol. 74, p. 137, 2000.

22. A. Martucci, L. Pasti, N. Marchetti, A. Cavazzini, F. Dondi and A. Alberti, *Microporous Mesoporous Mater,* Vol. 148, p. 174, 2012.

23. N. Vilaca, R. Amorim, O. Martinho, R. M. Reis, F. Baltazar, A. M. Fonseca and I. C. Neves, *J. Mater. Sci.,* Vol. 46, p. 7511, 2011.

24. M. Arruebo, R. Fernández-Pacheco, S. Irusta, J. Arbiol and M. R. Ibarra, J. Santamaria, *Nanotechnology,* Vol. 17, p. 4057, 2006

25. P.M. Danilczuk, K. Dugopolska, T. Ruman and D. Pogocki, *Mini Rev Med Chem.,* Vol. 8, p. 1407, 2008.

26. K. Unger, H. Rupprecht, B. Valentin and W. Kircher, *Drug. Dev. Ind. Pharm.,* Vol. 9, p. 69, 1983.

27. M.E. Davis, *Nature,* Vol.417, p. 813, 2002.

28. M. Ahola, P. Kortesuo, I. Kangasniemi, J. Kiesvaara and A. Yli-Urpo, *Int. J. Pharm.,* Vol. 195, p. 219 2000.

29. V. Ambrogi, G. Fardella, G. Grandolini and L. Perioli, *Int. J. Pharm.,* Vol. 220, p. 23, 2001.

30. P. Kortesuo, M. Ahola, S. Karlsson, I. Kangasniemi, A. Yli-Urpo, J. Kiesvaara, *Biomaterials,* Vol. 21, p. 193, 2000.

31. P. Kortesuo, M. Ahola, M. Kangas, M. Jokinen, T. Leino, L. Vuorilehto, S. Laakso, J. Kiesvaara, A. Yli-Urpo and M. Marvola, *Biomaterials,* Vol. 23, p. 2795, 2002.

32. T. Farías, A.R. Ruiz-Salvador and A. Rivera, *Microporous and Mesoporous Materials.* Vol. 61, p. 117, 2003.

33. A. Kulak, S. R. Hall and S. Mann, *Chem. Commun.,* p. 576, 2004.

34. M. De Gennaro and G. Cerri, Pharmaceutical zeolite-based compositions containing zinc and erythromycin, to be used in the treatment of acne. Patent: WO 02/100420, 2002.

35. T. Ukmar, T. Cendak, M. Mazaj, V. Kaucic and G. Mali, *J. Phys. Chem. C,* Vol. 116, p. 2662, 2012.

36. M. Betsiou, G. Bantsis, I. Zoi and C. Sikalidis, *Clay Miner.* Vol. 46, p. 613, 2011.

37. M.V. Cattaneo and T. M. Chang. *ASAIO Trans,* Vol. 37, p. 80, 1991.

38. P. Horcajada, C. Márquez-Alvarez, A. Rámila, J. Pérez-Pariente and M. Vallet-Regí, *Solid State Sciences,* Vol. 8, p. 1459, 2006.

39. P. Horcajada, A. Rámila, G. Férey and M. Vallet-Regí, *Solid State Sciences,* Vol. 8, p. 1243, 2006.

40. A. Corma, M.J. Díaz-Cabañas, J. Jiang, M. Afeworki, D. L. Dorset, S. L. Soled and K. G. Strohmaier, *PNAS,* Vol. 107, p. 13997, 2010.

41. M. G. Rimoli, M.R. Rabaioli, D. Melisi, A. Curcio, S. Mondello, R. Mirabelli and E. Abignente, *J Biomed Mater Res A,* Vol. 87A, p. 156, 2007.

42. C. Martinez and A. Corma. *Coord Chem Rev.,* Vol. 255, p. 1558, 2011.

43. C. Fruijtier-Pölloth, *Arch Toxicol,* Vol. 83, p. 23, 2009.

44. F.A. Mumpton, *Proc. Nat. Acad. Sci. USA* Vol. 96, p. 3463, 1999.

45. K. Hayakawa, Y. Mouri, T. Maeda, I. Satake and M. Sato, *Coll. Polym. Sci.* Vol. 278, p. 553, 2000.

46. Z. Popovic, M. Otter, G. Calzaferri and L. De Cola, *Angew. Chem. Int. Ed.* Vol. 46, p. 6188, 2007.

47. G. Calzaferri and K. Lutkouskaya, *Photochem. Photobiol. Sci.* Vol. 7, p. 879, 2008.

48. P. Tavolaro, A. Tavolaro and G. Martino, *Colloids Surf., B,* Vol. 70, p. 98, 2009.

49. C.A. Strassert, M. Otter, R.Q. Albuquerque, A. Hone, Y. Vida, B. Maier and L. De Cola, *Angew. Chem. Int. Ed.* Vol. 48, p. 7928, 2009.

50. A. Tavolaro, I.I. Riccio and P. Tavolaro, *Microporous and Mesoporous Materials,* Vol. 167, p. 62, 2013.

51. T. Salsa, F. Veiga and M.E. Pina, *Drug Dev Ind Pharm,* Vol. 23, p. 929, 1997.

52. M. Mulder, *Basic Principles of Membrane Technology,* London, Kluwer Academic Publishers, p. 59, 1991.

53. L. Donato, G. Barbaro, E. Drioli and C. Algieri, *Journal of Membrane and Separation Technology*, Vol. 1, p. 137, 2012.
54. M. Vallet-Regí, F. Balas and D. Arcos, *Angew. Chem., Int. Ed.* Vol. 46, p. 7548, 2007.
55. I. I. Slowing, J.L. Vivero-Escoto, C.W. Wu and V.S. Lin, *Adv. Drug. Del. Rev.* Vol. 80, p. 1278, 2008.
56. R. Sharma, "Surfactant Adsorption and Surface Solubilization." ACS Symposium Series 615, Washington, DC, 1995.
57. R. Amorim, N. Vilaca , O. Martinho, R.M. Reis, M. Sardo, J. Rocha, A.M. Fonseca, F. Baltazar and I.C. Neves, *J. Phys. Chem. C* Vol. 116, p. 25642, 2012.
58. C.T. Kresge, M.E. Leonowicz, W.J. Roth, J.C. Vartuli and J.S. Beck, *Nature*, Vol. 359, p. 710, 1992.
59. J.S. Beck, J.C. Vartuli, W.J. Roth, M.E. Leonowicz, C.T. Kresge, K.D. Schmitt, C.T.W. Chu, D.H. Olson, E.W. Sheppard, S.B. McCullen, J.B. Higgins and J.L. Schlenker, *J. Am. Chem. Soc.* Vol. 114, p. 10834, 1992.
60. T.J. Barton, L.M. Bull, W.G. Klemperer, D.A. Loy, B. McEnaney, M. Misono, P.A. Monson, G. Pez, G.W. Scherer, J.C. Vartuli and O.M. Yaghi, *Chem. Mater.* Vol. 11, p. 2633, 1999.
61. K. Pavelic, M. Hadzija, *Medical Applications of Zeolites, in Handbook of Zeolite Science and Technology* BY S.M. Auerbach, K.A. Carrado, P.K. Dutta, (Eds), NY., Marcel Dekker Inc., 2003.
62. P. Burguete, A. Beltrn, C. Guillem, J. Latorre, F.Prez-Pla, D. Beltrn and P. Amors, *ChemPlusChem*, Vol. 77, p. 817, 2012.
63. J. El Haskouri, D. Ortiz de Z_rate, C. Guillem, J. Latorre, M. Cald_s, A. Beltr_n, D. Beltr_n, A. B. Descalzo, G. Rodr_guez-L_pez, R. Mart_nez- M_Çez, M. D. Marcos and P. Amors, *Chem. Commun.* p. 330, 2002.
64. L. Huerta, C. Guillem, J. Latorre, A. Beltrn, R. Martnez-MaÇez, M. D. Marcos, D. Beltrn and P. Amors, *Solid State Sci.*, Vol. 8, p. 940, 2006.
65. J. El Haskouri, J. M. Morales, D. Ortiz de Zrate, L. Fernndez, J. Latorre, C. Guillem, A. Beltrn, D. Beltrn and P. Amors, *Inorg. Chem.*, Vol. 47, p. 8267, 2008.
66. H. Hata, S. Saeki, T. Kimura, Y. Sugahara and K. Kuroda, *Chem. Mater.* Vol.11, p. 1110, 1999.
67. M. Vallet-Regi, A. Ramila, R.P. del Real and J. Perez-Pariente, *Chem. Mater.* Vol. 13, p. 308, 2001.
68. A. L. Doadrio, E.M.B. Sousa, J.C. Doadrio, J. Perez Pariente, I. Izquierdo-Barba and M. Vallet-Regi, *J. Controlled Release*, Vol. 97, p. 125, 2004.
69. G. Cavallaro, P. Pierro, F.S. Palumbo, F. Testa, L. Pasqua and R. Aiello, *Drug Delivery*, Vol. 11, p. 41, 2004.
70. C. Charnay, S. Bégu, C. Tourné-Péteilh, L. Nicole, D. A. Lerner and J.M. Devoisselle, *Eur. J. Pharm. Biopharm.* Vol. 57, p. 533,2004.
71. B. MuÇoz, A. Ramila, J. Perez-Pariente, I. Diaz and M. Vallet-Regi, *Chem. Mater.*, Vol. 15, p. 500, 2003.
72. C. Tourné-Péteilh, D. Brunel, S. Bégu, B. Chiche, F. Fajula, D. A. Lerner and J.- M. Devoisselle, *New J. Chem.*, Vol. 27, p. 1415, 2003.

73. P. Horcajada, A. Ramila, J. Perez-Pariente and M. Vallet-Regi, *Microporous Mesoporous Mater.* Vol. 68, p. 105, 2004.
74. A. Nieto, M. Colilla, F. Balas and M. Vallet-Regi, *Langmuir*, Vol. 26, p. 5038, 2010.
75. J.M. Rosenholm, V. Mamaeva, C. Sahlgren and M. Linden, *Nanomedicine* , Vol. 7, p. 111, 2012.
76. Y. Zhu, J. Shi, W. Shen, H. Chen, X. Dong and M. Ruan, *Nanotechnology*, Vol. 16, p. 2633, 2005.
77. W. Zhao, H. Chen, Y. Li, L. Li, M. Lang and J. Shi, *Adv. Funct. Mater.* Vol. 18, p. 2780, 2008.
78. M. Vallet-Regi, F. Balas and D. Arcos, *Angew. Chem. Int. Ed.*, Vol. 119, p. 7692, 2007.
79. S. Wang, *Microporous Mesoporous Mater.*, Vol. 117, p. 1, 2009.
80. M. Manzano and M. Vallet-Regi, *J. Mater. Chem.* Vol. 20, p. 5593, 2010.
81. Q.J. He, J.L. Shi, *J. Mater. Chem.* Vol. 21, p. 5845, 2011.
82. M. Vallet-Regi and E. Ruiz-Hernández, *Adv. Mater.*, Vol. 23, pp. 5177 –5218, 2011.
83. M. Sasidharan, H. Zenibana, M. Nandi, A. Bhaumik and K. Nakashima, *Dalton Trans.*, Vol. 42, p. 13381, 2013.
84. N.K. Mal, M. Fujiwara and Y. Tanaka, *Nature*, Vol. 421, p. 350, 2003.
85. N.K. Mal, M. Fujiwara, Y. Tanaka, T. Taguchi and M. Matsukata, *Chem. Mater.*, Vol. 15, p. 3385, 2003.
86. Y. Zhu and M. Fujiwara, *Angew. Chem. Int. Ed.*, Vol. 46, p. 2241, 2007.
87. H. Yan, C. Teh, S. Sreejith, L. Zhu, A. Kwok,W. Fang, X. Ma, K. T. Nguyen, V. Korzh and Y. Zhao, *Angew. Chem. Int. Ed.*, Vol. 51, p. 8373, 2012.
88. J.L. Vivero-Escoto, I.I. Slowing, C.-W. Wu and V. S.-Y. Lin, *J. Am. Chem. Soc.*, Vol. 131, p. 3462, 2009.
89. H.S. Qian, H.C. Guo, P.C. Ho, R. Mahendran and Y. Zhang, *Small*, Vol. 20, p. 2285, 2009.
90. J. Lu, M. Liong, Z. Li, J.I. Zink and F. Tamanoi, *Small*, Vol. 6, p. 1794, 2010.
91. M. Hartmann, *Chem. Mater.*, Vol. 17, p. 4577, 2005.
92. H.H.P. Yiu and P.A. Wright, *J. Mater.Chem.*, Vol. 15, p. 3690, 2005.
93. Y. Wang, A.D. Price and F. Caruso, *J. Mater. Chem.*, Vol. 19, p. 6451, 2009.
94. L. Gastaldi, E. Ugazio, S. Sapino, P. Iliade, I. Milettob and G. Berlier, *Phys. Chem. Chem. Phys.*, Vol. 14, p. 11318, 2012.
95. J. Zhang, M. Y.-P. Yuan, G. Lu and C. Yu, *J Incl Phenom Macrocycl Chem.*, Vol. 71, p. 593, 2011.
96. F.Y. Qu, G.S. Zhu, S.Y. Huang, S.G. Li and S.L. Qiu, *Chemphyschem.*, Vol. 7, p. 400 , 2006.
97. W. Zeng, X.F. Qian, Y.B. Zhang, J. Yin and Z.K. Zhu, *Mater. Res. Bull.* Vol. 40, p. 766, 2005.
98. J.C. Doadrio, E.M.B. Sousa, I. Izquierdo-Barba, A.L. Doadrio, J. Perez-Pariente and M. Vallet-Regi, *J. Mater. Chem.*, Vol. 16, p. 462, 2006466.
99. S.W. Song, K. Hidajat and S. Kawi, *Langmuir*, Vol. 21, p. 9568, 2005.

100. M. Vallet-Regi, J.C. Doadrio, A.L. Doadrio, I. Izquierdo-Barba and J. Perez-Pariente, *Solid State Ionics*, Vol. 172, p. 435, 2004.
101. Y.F. Zhu, J.L. Shi, H.R. Chen, W.H. Shen and X.P. Dong, *Turret Mat.* Vol. 84, p. 218, 2005.
102. C. Tourné-Péteilh, D.A. Lerner, C. Charnay, L. Nicole, S. Bégu and J.-M. Devoisselle, *Chemphyschem,* Vol. 4, p. 281, 2003.
103. J. Andersson, J. Rosenholm, S. Areva and M. Linden, *Chem. Mater.*, Vol. 16, p. 4160, 2004.
104. V. Zelenak, V. Hornebecq and P. Llewellyn, *Micropor. Mesopor. Mat.*, Vol. 83, p. 125, 2005.
105. I. Izquierdo-Barba, A. Martinez, A.L. Doadrio, J. Perez-Pariente, M. Vallet-Regi, *Eur. J. Pharm. Sci.*, Vol. 26, p. 365, 2005.
106. C.Y. Lai, B.G. Trewyn, D.M. Jeftinija, K. Jeftinija, S. Xu, S. Jeftinija and V.S. Lin, *J Am Chem Soc.*, Vol. 125, p. 4451, 2003.
107. Q. Tang, Y. Xu, D. Wu and Y. Sun, *J Solid State Chem.* Vol. 179, p. 1513, 2006.
108. R. Mellaerts, R. Mols, J.A. Jammaer, C.A. Aerts, P. Annaert, J. van Humbeeck, G. Van Den Mooter, P. Augustijns and J.A. Martens, *Eur J Pharm Biopharm.* Vol. 69, p. 223, 2008.
109. J. Zhang, Z-F. Yuan, Y. Wang, W.-H. Chen, G.-F. Luo, S.-X. Cheng, R.-X Zhuo and X.-Z. Zhang, *J. Am. Chem. Soc.*, Vol. 135, p. 5068, 2013.
110. K. Kogure, R. Moriguchi, K. Sasaki, M. Ueno, S. Futaki and H. Harashima, *Journal of Controlled Release*, Vol. 98, p. 317, 2004.
111. M. Colilla, B. Gonzáleza and M. Vallet-Regi, *Biomater. Sci.*, Vol. 1, p. 114, 2013.
112. C.-H. Leea, T.-S. Linb and C.-Y. Mou, *Nano Today,* Vol. 4, p. 165, 2009.
113. A. Monnier, F. Schuth, Q. Huo, D. Kumar, D. Margolese, R.S. Maxwell, G.D. Stucky, M. Krishnamurty, P. Petroff, A. Firouzi, M. Janicke and B.F. Chmelka, *Science*, Vol. 261, p. 1299, 1993.
114. S. Inagaki, Y. Fukushima and K. Kuroda, *J. Chem. Soc. Chem. Commun.*, Vol. 680, 1993.
115. Q. Huo, D.I. Margolese, U. Ciesla, P. Feng, T.E. Gier, P. Sieger, R. Leon, P.M. Petroff, F. Schüth and G.D. Stucky, *Nature*, Vol. 368, p. 317, 1994.
116. D. Zhao, J. Feng, Q. Huo, N. Melosh, G.H. Fredrickson, B.F. Chmelka and G.D.Stucky, *Science*, Vol. 279, p. 548, 1998.
117. D. Zhao, Q. Huo, J. Feng, B.F. Chmelka and G.D. Stucky, *J. Am. Chem. Soc,* Vol. 120, p. 6024, 1998.
118. P. Schmidt-Winkel, W.W.J. Lukens, D. Zhao, P. Yang, B.F. Chmelka and G.D. Stucky, *J. Am. Chem. Soc.* Vol. 121, p. 254, 1999.
119. P.T. Tanev and T.J. Pinnavaia, *Science*, Vol. 267, p. 865, 1995.
120. C.Y. Chen, H.X. Li and M.E. Davis, *Micro-Mesoporous Mater.*, Vol. 2, p. 17, 1993.
121. Y. Han and J.Y. Ying, *Angew. Chem., Int. Ed.*, Vol. 44, p. 288, 2005.
122. F. Hoffmann, M. Cornellius, J. Morell and M. Froba, *Angew. Chem., Int. Ed.*, Vol. 45, p. 3216, 2006.

123. I.I. Slowing, J.L. Vivero-Escoto, C.W. Wu and V.S.Y. Lin, *Adv. Drug Delivery Rev.*, Vol. 60, p. 1278, 2008.

124. I.I. Slowing, B.G. Trewyn, S. Giri and V.S.Y. Lin, *Adv. Funct. Mater.*, Vol. 17, p. 1225, 2007.

125. D.D. Li, J.H. Yu, R.R. Xu, *Chem. Commun.* Vol. 47, p. 11077, 2011.

126. M.H. Kim, H.K. Na, Y.K. Kim, S.R. Ryoo, H.S. Cho, K.E. Lee, H. Jeon, R. Ryoo and D. H. Min, *ACS Nano*, Vol. 5, p. 3568, 2011.

127. M.D. Popova, A. Szegedi, I.N. Kolev, J. Mihaly, B.S. Tzankov, G.Tz. Momekov, N.G. Lambov and K.P. Yoncheva, *International Journal of Pharmaceutics*, Vol. 436, p. 778, 2012.

128. E. Aznar, F. Sancenoon, M. D Marcos, R. Martínez-Mannez, P. Stroeve, J. Cano and P. Amoros, *Langmuir*, Vol. 28, p. 2986, 2012.

129. A. Datt, I. El-Maazawi and S.C. Larsen, *J. Phys. Chem. C,* Vol. 116, p. 18358, 2012.

130. F. Hoffmann, M. Cornelius, J. Morell and M. Froba, *Angew. Chem., Int. Ed.* Vol. 45, p. 3216, 2006.

131. A. Datt, D. Fields and S.C. Larsen, *J. Phys. Chem. C.,* Vol. 116, p. 21382, 2012.

132. J. Lu, M. Liong, J.I. Zink and F. Tamanoi. *Small,* Vol. 3, p. 1341, 2007.

133. Q. Lin, Q. Huang, C. Li, C. Bao, Z. Liu, F. Li and L. Zhu, *J Am Chem Soc,* Vol. 132, p. 10645, 2010.

134. I.S. Carino, L. Pascua, F. Testa, R. Aiello, F. Puoci, F. Iemma and N. Picci, *Drug Deliv.,* Vol. 14, p. 491, 2007.

135. N.W. Fadnavis, V. Bhaskar, M.L. Kantam and B.M. Choudary. *Biotechnol Prog.,* Vol. 19, p. 346, 2003.

136. P.S. Wheatley, A.R. Butler, M.S. Crane, S. Fox, B. Xiao, A.G. Rossi, I.L. Megson and R.E. Morris, *J. Am. Chem. Soc.*, Vol. 128, p. 502, 2006.

137. S. Fox, T.S. Wilkinson, P.S. Wheatley, B. Xiao, R.E. Morris, A. Sutherland, A.J. Simpson, P.G. Barlow, A.R. Butler, I.L. Megson and A.G. Rossi, *Acta Biomater.*, Vol. 6, p. 1515, 2010.

138. B.J. Heilman, J. St. John, S.R.J. Oliver and P.K. Mascharak, *J. Am. Chem. Soc.* Vol. 134, p. 11573, 2012.

139. S. Kitagawa, R. Kitaura and I.S. Noro, *Angew. Chem. Int. Ed.,* Vol. 43, p. 2334, 2004.

140. S. Kitagawa and R. Matsuda, *Coord. Chem. Rev.* Vol. 251, p. 2490, 2007.

141. O.M. Yaghi, M. O'Keeffe, N.W. Ockwig, H. Chae, M. Eddaoudi and J. Kim, *Nature*, Vol. 423, p. 705, 2003.

142. G. Ferey, *Chem. Soc. Rev.* Vol. 37, p. 191, 2008.

143. G.S. Papaefstathiou and L.R. MacGillivary, *Coord. Chem. Rev.* Vol. 246, p. 169, 2003.

144. O.R. Evans and W. Lin, *Acc. Chem. Res.* Vol. 35, p. 511, 2002.

145. B.F. Abrahams, B.F. Hoskins, D.M. Michall and R. Robson, *Nature*, Vol. 369, p. 727, 1994.

146. C.N.R. Rao, S. Natarajan and R. Vaidhyanaathan, *Angew. Chem. Int. Ed.* Vol. 43, p. 1466, 2004,

147. C.J. Kepert, *Chem. Commun.* p. 695, 2006.

148. D. Bradshaw, J.B. Claridhe, E.J. Cussen, T.J. Prior and M.J. Rosseinsky, *Acc. Chem. Res.*, Vol. 38, p. 273, 2005.

149. Y. Liu, J.F. Eubank, A.J. Cairns, J. Eckert, V.C. Kravtsov, R. Luebke and M. Eddaoudi, *Angew. Chem. Int. Ed.* Vol. 46, p. 3278, 2007.

150. B.Q. Ma, K.L. Mulfort and J.T. Hupp, *Inorg. Chem.*, Vol. 44, p. 4912, 2005.

151. M. Dinca, A.F. Yu and J.R. Long, *J. Am. Chem. Soc.*, Vol. 128, p. 8904, 2006.

152. X. Lin, A.J. Blake, C. Wilson, X.Z. Sun, N.R. Champness, M.W. George, P. Hubberstey, R. Mokaya and M. Schroder, *J. Am. Chem. Soc.* Vol. 128, p. 10745, 2006.

153. E.Y. Lee and M.P.A Suh, *Angew. Chem. Int. Ed.*, Vol. 43, p. 2798, 2004.

154. K. Uemura, R. Matsuda and S. Kitagawa, *J. Solid State Chem.* Vol. 178, p. 2420, 2005.

155. J.L.C. Rowsell and O.M. Yaghi, *Microporous Mesoporous Mater.* Vol. 73, p. 3, 2004.

156. H. Li, M. Eddaoudi, M. O'Keeffe and O.M. Yaghi, *Nature*, Vol. 402, p. 276, 1999.

157. A. Pichon, A. Lazuen-Garay and S.L. James, *Cryst. Eng. Comm.*, Vol. 8, p. 211, 2006.

158. O.M. Yaghi and H.L. Li, *J. Am. Chem. Soc.*, Vol. 117, p. 10401, 1995.

159. A.O. Yazaydin, A.I. Benin, S.A. Faheem, P. Jakubcza, J.J. Low, R.R. Willis and R.Q. Snurr, *Chem. Mater.*, Vol. 21, p. 1425, 2009.

160. J.R. Karra and K.S. Walton, *Langmuir*, Vol. 24, p. 8620, 2008.

161. Q. Yang and C. Zhong, *J. Phys. Chem. B*, Vol. 110, p. 655, 2006.

162. J.A. Greathouse and M.D. Allendorf, *J. Am. Chem. Soc.*, Vol. 128, p. 10678, 2006.

163. S.S. Kaye, A. Dailly, O.M. Yaghi and J.R. Long, *J. Am. Chem. Soc.* Vol. 129, p. 14176, 2007.

164. P. Horcajada, C. Serre, M. Vallet-Regi, M. Sebban, F. Taulelle and G. Ferey, *Angew. Chem. Int. Ed.*, Vol. 45, p. 5974, 2006.

165. N.J. Hinks, A.C. McKinlay, B. Xiao, P.S. Wheatley and R.E. Morris, *Microporous Mesoporous Mater.* Vol. 129, p. 330, 2010.

166. A.C. McKinlay, R.E. Morris, P. Horcajada, G. Ferey, R. Gref, P. Couvreur and C. Serre, *Angew. Chem., Int. Ed.*, Vol. 49, p. 6260, 2010.

167. W.J. Rieter, K.M. Pott, K.M.L. Taylor and W.B. Lin, *J. Am. Chem. Soc.* Vol. 130, p. 11584, 2008.

168. W.J. Rieter, K.M.L. Taylor, H.Y. An, W.L. Lin and W.B. Lin, *J. Am. Chem. Soc.* Vol. 128, p. 9024, 2006.

169. W.J. Rieter, K.M.L. Taylor and W.B. Lin, *J. Am. Chem. Soc.* Vol.129, p. 9852, 2007.

170. G. Ferey, C. Mellot-Draznieks, C. Serre, F. Millange, J. Dutour, S. Surble and I.A. Margiolaki, *Science*, Vol. 309, p. 2040, 2005.

171. P. Horcajada, T. Chalati, C. Serre, B. Gillet, C. Sebrie, T. Baati, J.F. Eubank, D. Heurtaux, P. Clayette, C. Kreuz, J.S. Chang, Y.K. Hwang, V. Marsaud, P.N.

Bories, L. Cynober, S. Gil, G. Ferey, P. Couvreur and R. Gref, *Nat. Mater.*, Vol. 9, p. 172, 2010.

172. C-Y. Sun, C. Qin, C-G. Wang, Z-M. Su, S. Wang , X-L. Wang, G-S. Yang, K.-Z. Shao, Y-Q. Lan and E-B. Wang. *Adv. Mater.*, Vol. 23, p. 5629, 2011.

173. J. An, S.J. Geib and N.L. Rosi, *J. Am. Chem. Soc.*, Vol. 131, p. 8376, 2009.

174. R.C., Huxford, J. D. Joseph Della Rocca and W. Lin, *Current Opinion in Chemical Biology* Vol. 14, p. 262, 2010.

175. B. Xiao, P.S. Wheatley, X.B. Zhao, A.J. Fletcher, S. Fox, A.G. Rossi, I.L. Megson, S. Bordiga, L. Regli, K.M. Thomas and R.E. Morris, *J. Am. Chem. Soc.* Vol. 129, p. 1203, 2007.

176. A.C. McKinlay, B. Xiao, D.S. Wragg, P.S. Wheatley, I.L. Megson and R.E. Morris, *J. Am. Chem. Soc.*, Vol. 130, p. 10440, 2008.

177. B.L. Chen, L.B. Wang, F. Zapata, G.D. Qian and E.B. Lobkovsky, *J. Am. Chem. Soc.* Vol. 30, p. 6718, 2008.

178. B.L. Chen, L.B. Wang, Y.Q. Xiao, F.R. Fronczek, M. Xue, Y.J. Cui and G.D.A Qian, *Angew. Chem., Int. Ed.*, Vol. 48, p. 500, 2009.

179. J. An, C.M. Shade, D.A. Chengelis-Czegan, S. Petoud and N.L. Rosi, *J. Am. Chem. Soc. Vol.*. 133, p. 1220, 2011.

180. Y. Cui, H. Xu, Y. Yue, Z. Guo, J. Yu, Z. Chen, J. Gao, Y. Yang, G. Qian and B. Chen, *J. Am. Chem. Soc.* Vol. 134, p.3979, 2012.

181. D. F. Sava, L. E. S. Rohwer, M. A. Rodriguez and T. M. Nenoff, *J. Am. Chem. Soc.* Vol. 134, p.3983, 2012.

4

Nanoparticles for Diagnosis and/or Treatment of Alzheimer's Disease

S.G. Antimisiaris[1,2,*], S. Mourtas[1], E. Markoutsa[1], A. Skouras[1], and K. Papadia[1]

[1]*Laboratory of Pharmaceutical Technology, Dept. of Pharmacy, University of Patras, Rio, Greece*
[2]*Institute of Chemical Engineering Sciences (ICES), Foundation for Research and Technology Hellas (FORTH), Platani, Greece*

Abstract

Recent attempts to develop nanoparticulate systems for diagnosis and/or therapy of neurodegenerative diseases, with main focus on Alzheimer's Disease (AD) are presented. Since the main obstacle to treat brain-located pathologies is the blood-brain barrier, a brief description of its physiology, and methodologies used for studying transport of drugs across the BBB, are mentioned. All types of nanoparticulates which have been employed to-date to target AD (and deliver drugs or imaging substances to AD-related pathological features) are described, and the results accomplished so far together with advantages/disadvantages of each specific category of nanoparticles are mentioned. The philosophy and main physicochemical characteristic prerequisites of nanoparticles used as systems to target diseases in general, and brain-located pathologies particularly, are analyzed. Finally, the main current accomplishments and challenges for the future are summarized.

Keywords: Nanoparticulate systems, diagnosis and therapy, neurodegenerative diseases, Alzheimer's disease

4.1 Introduction

Neurological disorders (ND) (such as Alzheimer's disease [AD], Parkinson's disease [PD], multiple sclerosis [MS], and primary brain tumors) occurrence

---

Corresponding author: santimis@upatras.gr

Ashutosh Tiwari (ed.) Advanced Healthcare Materials, (87–180) 2014 © Scrivener Publishing LLC

is constantly increasing in the last years [1], mainly as a result of the continuously increasing aging population and life expectancy. Thereby, strategies for ND early detection and treatment are currently among the most urgent and challenging areas in therapeutics. Since NDs are "located" in the central nervous system (CNS), the blood-brain barrier (BBB) presents a serious impediment for their diagnosis and treatment. The ability of nanoparticulate systems or nanoparticles (NPs) to traverse the BBB provides alternate means for targeted drug delivery to the CNS and novel therapeutic applications for ND [2], as well as increased opportunities for their early diagnosis. "Theranosis" (a word derived by combining the two greek-origin words "therapy" and "diagnosis") has been proposed recently, as a term to describe systems which deliver imaging and therapeutic agents simultaneously. This chapter will focus on recent nanotechnological approaches (mainly development of nanoparticulate systems, or else, nanoparticles) for detection and/or therapy (theranosis) of AD. Therapy and diagnosis are two major categories in the clinical treatment of disease. AD is the most common form of dementia in people over the age of 65. Today, millions of people are affected by AD, resulting in a heavy social and economic burden, which is projected to seriously increase and profoundly impact the health care systems, if no efficient early detection and therapies become available. Various types of nanoparticles are currently under investigation for their potential to target AD-related pathologies and thus be used for development of diagnostic and/or therapeutic systems.

4.2 Nanoparticles

NPs have been considered as alternative and ideal nano-diagnostics/nano-therapeutics (compared to traditional diagnostics and therapeutics) due to their advantages [3], the most important of which are: (1) That they can be easily loaded (or they can integrate) with more than one kind of imaging or therapeutic agents, which makes them potential multifunctional nanoplatforms for theranosis; (2) That they can accommodate large amounts of imaging agents or drugs (by simple loading or chemical conjugation) as a result of their large surface area and/or interior volume (depending on the specific NP type); (3) That they can target disease sites, after decoration with specific targeting moieties or after optimization of their physicochemical properties (size, surface charge and hydrophobicity, etc). In fact it has been confirmed that decoration with more than one targeting molecule can highly increase NP target binding capability and specificity (compared to single molecule [or targeting ligand] decoration) due to multivalent effects [4]; (4) That their circulation time in the blood can be enhanced by modification of their size and/or surface characteristics, due to reduced degree of opsonization and subsequent uptake by the reticuloendothelial system (RES) [5]. Indeed, in most cases NPs can be engineered to have specific blood circulation times, depending on the specific application requirements.

Various types of nanoparticles, composed of organic (polymers, proteins, lipids, polysaccharides, etc.) or inorganic materials (iron, gold, silica, etc.) have been developed, and are currently under clinical or preclinical evaluation, or even in the market, for delivery of imaging agents and/or drugs (Table 4.1) [6, 7]. The structural characteristics of most NP types is presented in Figure 4.1 [6].

Table 4.1 Categories of Nanoparticles (NPs) used in theranosis.

TYPE	ADVANTAGES	DISADVANTAGES
Organic Nanoparticles		
-**Polymeric NPs or micelles** -**Biological NPs**	-Good loading for lipophilic drugs/agents; Homogeneity; biological synthesis; biodegradable/ biocompatible;	-Bad loading of hydrophilic molecules; some polymers or stabilizers have possible: toxicity, slow biodegradability, etc; -Difficult modification; handling;
Lipidic NPs: -**Solid Lipid NP (SLN)** -**Liposomes**	-Small uniform size; Good easy loading of lipophilic molecules; easy scale-up of production; -FDA approved carriers; good/easy loading of any type of molecule; easy surface modification (many chemistries available; commercially available building blocks);	-High content of surfactants; possible toxicity/biocompatibility problems; problem to load hydrophilic molecules; -Poor in vivo stability; limited sustained release potential;
Dendrimers	-High loading capacity; controllable & uniform size/chemical composition;	Limited synthetic possibilities; toxicity issues;

(Continued)

Table 4.1 (*Cont.*)

TYPE	ADVANTAGES	DISADVANTAGES
Inorganic NPs		
Iron NPs	Clinically available contrast agent (MRI); biocompatible; controllable size/shape/properties;	poor aqueous stability;
Gold NPs	Optical quenching capability; controllable size/shape; easy surface modification; biocompatible;	poor aqueous stability;
Quantum Dots	Very good imaging properties; high S/N; in vivo longevity;	Cytotoxicity; questionable biocompatibility and clearance;
Carbon-based NPs	Structural rigidity; mechanical properties; good electrical properties;	Potential toxicity;
Silica NPs	Biocompatible; high loading capacity;	Mechanical stability;

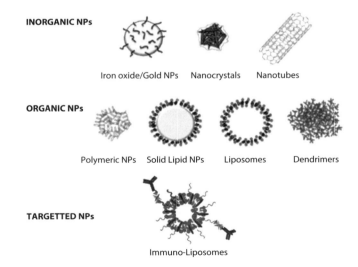

Figure 4.1 Structure of the main types of NPs currently used or being under investigation for therapeutic and/or diagnostic applications. Part of the figure was extracted from Ref [6] after permission.

4.2.1   Types of NPs Used for Therapy and/or Diagnosis

As stated above, NPs can be categorized in two main groups: those consisting of organic components (or building blocks); i.e., **Organic NPs**, and those consisting of inorganic building blocks, i.e., **Inorganic NPs**. In Table 4.1 a brief presentation of the advantages and disadvantages of the basic types of NPs, in regards to their applicability for theranostic applications, is given.

Over the past decade, several kinds of organic NPs, such as polymeric NPs or micelles (consisting of various synthetic or naturally occurring polymers), Dendrimers, Biological NPs (mainly consisting of proteins as albumin etc.), and lipidic NPs having sphere (Solid Lipid NPs-SLNs) or capsule geometry (liposomes), have been used in various applications for diagnosis and therapy [8, 9]. These organic NPs can extend the blood circulation time of therapeutic agents and selectively accumulate in the pathological sites via the enhanced permeability and retention (EPR) effect [10]. NPs carrying imaging agents such as radionuclides, dyes, MRI contrast agents etc., have shown potential for early diagnosis of diseases, while various organic or inorganic NPs have also shown great potential as theranostic systems. Furthermore, stimuli-responsive organic NPs can release their payloads only in response to an appropriate stimulus such as pH, temperature, light, or ultrasound. Controlled release upon these stimuli results in a more specific delivery of the payload to the target sites, and this strategy is currently under intensive preclinical evaluation for targeted drug delivery and biomedical imaging [11, 12]. A brief description for each NP-type follows:

Polymeric NPs and Micelles have attracted great attention as carriers of hydrophobic molecules because they are able to provide a narrow size distribution, incorporate multiple functionalities into one system, increase the solubility of hydrophobic drugs, control their release over extended time periods, and exhibit slow elimination through the RES [13, 14]. Polymeric micelles are composed of a hydrophobic core, and the hydrophilic part forms the shell. The core of the micelles can contain hydrophobic imaging or therapeutic agents, while the shell provides stability under aqueous conditions and binding locations for targeting moieties.

Dendrimers are spherical, branched or hyper-branched macromolecules with a tree-like structure. Imaging and therapeutic agents can be encapsulated in the internal cavity or on the surface of the dendrimers with high loading capacity [15]. By monomer selection and control of polymerization degree, dendrimer size, molecular weight and chemical composition can be modulated. Due to their high loading and capability to control the polymer structure, dendrimers are favorable platforms for multimodal imaging and theranosis. Polyamidoamine (PAMAM) dendrimers derived from polypropyleneimine and a 1,2-diaminoethane core have shown great promise in diagnostic and therapeutic applications because of their bio-compatibility, small (< 5nm diameter) and uniform size and shape, mono-disperse molecular weight distribution, good blood clearance through renal excretion and availability of numerous amino groups to conjugate targeting moieties and/or imaging/therapeutic agents [16].

Other types of polymers, which may be considered as **Biological polymers**, such as proteins, antibodies, enzymes, lipoproteins, and viruses, also possess the required properties to make significant contributions for theranosis. A few of the various advantages they offer, compared to synthetic polymer-based nanoparticles, are good biocompatibility, biodegradability, quick and easy synthetic steps, and homogeneity [17], which make them desirable drug carriers. Albumin, a blood-plasma protein that transports various molecules through the circulatory system, has been used as a carrier because of its innate transport role, and excellent biocompatibility / biodegradability. Natural source derived polymers, such as chitosan, alginate, collagen, and hyaluronic acid, are also useful biomaterials for delivery of imaging or therapeutic agents due to their easy modification (by simple chemical methodologies) as well as their minimal inflammation induction and good biodegradability.

Lipidic NPs: Such NPs are constructed of lipids as building blocks and are mainly solid lipid NPs (SLNs) or nanocapsules or liposomes (LIPs).

SLNs are generally constituted by a matrix of lipids that remain solid at room and body temperature. These solid lipids are mostly physiological lipids such as fatty acids, mono-, di- or triglycerides, glycerin mixtures and waxes [18]. A related class of nanocarriers is called lipid nanocapsules, and typically consist of a mixture of triglycerides and phospholipids [19]. Both types of the nanocarriers mentioned above require surfactants (non-ionic or ionic) for stabilization and may facilitate sustained or controlled release of therapeutic agents. In addition, lipophilic particulate matters are known to cross the BBB well [20] and enter the CNS compartment even without surface modification. They therefore demonstrate strong potential for CNS drug delivery. Some examples of SLNs known to bind apolipoproteins and target brain tissues will be described below [21].

LIPs have been extensively studied as carriers for both imaging and therapeutic agents. They are small artificial lipid-based vesicles of spherical shape usually produced from natural non-toxic phospholipids and cholesterol. They were initially invented by Alec Bangham as a model for cell membranes and from the 70s and on, they are extensively investigated for drug delivery applications. Because of their diverse physicochemical character (both hydrophilic and hydrophobic regions exists in one single vesicle allowing the incorporation of hydrophobic and encapsulation of hydrophilic molecules), their versatile structure that can be easily tailored (allowing adjustment of size, lamellarity and surface properties) and their biocompatible and non-toxic components, liposomes are excellent systems for the delivery and targeting of drugs and/or imaging agents. Depending on their lipid composition, size, surface charge and method of preparation, their physichochemical properties and, as a consequence, their in vivo distribution, may be modulated. Liposomes can range between 25–50 nm up to several micrometers in diameter, and may consist of one (unilamellar) or more (multilamellar) homocentric bilayers of amphipathic lipids (usually phospholipids). They can be classified according to their lamellarity (number of lamellae) (and may thus be characterized as unilamellar vesicles [UV] or multilamellar vesicles [MLV]) and according to their size (and may thus be characterized as small [SUV, if they are also unilamellar] or large [LUV]). Liposomes may also be named (frequently found in the relevant literature) by the technique used for their preparation (as DRV, REV). The main advantages of liposomal drug delivery systems are that: (i) they are structurally versatile and can be easily tailored to bear the specific properties required for each application; (ii) they can accommodate both hydrophilic drugs (in their aqueous compartments) and lipophilic or ampliplilic drugs (in their lipid bilayers); and (iii) they are non-toxic, non-immunogenic and fully biodegradable, since they consist of naturally occurring phospholipids. The choice of lipidic components determines the "rigidity" (or "fluidity") and the charge of the liposome bilayer. The introduction of positively or negatively charged lipids in their bilayer provides a surface charge to liposomes. Liposome surfaces can be readily modified by attaching polyethylene glycol (PEG)-units to the bilayer (producing what is known as stealth liposomes) to prolong their circulation time in the bloodstream. Furthermore, antibodies or ligands can be conjugated to the surface of liposomes, to enhance target-specific drug therapy. Today, several different types of liposomes are available, mainly conventional, long circulating (or "Stealth," sterically stabilized, or Pegylated), targeted (or ligand bearing or immunoliposomes [when antibodies are used as targeting ligands]), cationic (for genetic material delivery) and deformable or elastic (or "transferosomes") with applications in transdermal delivery [22, 23].

The second main category of NPs is **Inorganic NPs**. Several types of NPs consisting of inorganic building blocks have been identified as candidates for potential applications in theranosis, such as iron oxide NPs, gold NPs, Q-dots, silica NPs or carbon-based NPs (Table 4.1). A brief description of each one of these NP-types is given below.

Iron Oxide NPs. Considerable interest in iron oxide nanoparticles (IONPs) as multifunctional nanoplatforms for imaging and therapy has recently emerged, due to their unique properties such as the intrinsic ability to enhance MR contrast, easy surface tailoring, and biocompatibility [24]. IONPs are synthesized by co-precipitation of Fe^{2+} and Fe^{3+} ions in basic aqueous media or by the thermal decomposition method (for more uniform and highly crystalline structures). They provide large T2 relaxation effects, and have been used as T2-weighted MRI contrast agents. Generally, for these applications, IONPs should be coated with biocompatible polymers, such as dextran, dextran derivatives, or PEG, to confer stability in the aqueous environments of biological systems. These surface functionalities can also bind different therapeutic agents, imaging agents, or targeting moieties for the development of IONP platforms for multimodal imaging. There has also been increasing interest in the combination of IONPs with PET or SPECT probes for MRI/PET or MRI/SPECT dual-modality imaging. Such probes can be developed by conjugation of chelates (e.g., DOTA and DTPA) to the polymer surface coating of IONPs, for radionuclide complexation (^{64}Cu, ^{111}In and ^{124}I). IONPs have received great attention towards the development of theranostic nanomedicines, since they are not only used as contrast enhancement agents for MRI, but can also deliver therapeutic agents, such as anticancer drugs and siRNA, to disease sites [25]. IONPs can also emit heat upon exposure to an alternating external magnetic field (by converting electromagnetic energy into heat) and thus be used for hyperthermia (therapy) in addition to imaging [26]. Since MRI is not sufficiently sensitive to monitor drug delivery or target-specific accumulation of drug carriers in a diseased site, a variety of IONP-based drug carriers have been combined with imaging probes, including optical imaging and nuclear imaging (SPECT or PET) probes, to provide real-time, noninvasive imaging of drug delivery or monitoring of therapy [27].

Gold NPs, which can be spherical, rod, or nanoshell-shaped, have been designated as multi-functional probes due to their unique properties such as optical quenching [28], X-ray absorption [29], and surface enhanced Raman scattering (SERS) [30]. These sensing properties are applied to single diagnostic imaging techniques, but other gold nanomaterial properties like size controllability, good biocompatibility, easy surface modification, and photothermal effects can be further utilized towards the development of multi-modal theranostic agents [31]. Therapeutic agents and photothermal properties can be easily integrated with gold nanomaterials through surface modification or shape control, while nucleic acids can be easily integrated onto them through thiol groups and desorbed through intense local heat induced by their photothermal properties [32].

Quantum Dots (QDs) are NPs composed of ZnS or Cd/Se. Traditional methods for imaging, such as fluorescent or organic dyes are prone to photobleaching and pH sensitivity, highlighting the need for more stable contrast agents. QDs are resistant to chemical degradation and photo-bleaching, and have tuneable emission spectra, making them excellent candidates for molecular imaging and biosensing [33]. Furthermore, QDs are being employed as probes given their brighter fluorescence than traditional dyes, which is due to a higher molar extinction coefficient (which means that they can absorb more light energy). Recent applications of QDs for in vivo imaging include peptide-conjugated QDs, which target tumor tissue and allow optical detection [33]. This may have applications for in vivo imaging of brain tumors. However, their biodegradability and toxicity towards cells is questionable.

Silica NPs. Silica is a natural component of sand, glass, and quartz, and exists as the chemical compound silicon dioxide (SiO_2). For a long time, it has been widely applied in glass and ceramic industries, and now, many researchers use it as a nanomedicine component because of its biocompatibility, availability to chemical or physical modification, and mechanical properties. Bradbury and colleagues showed that silica NPs could be used for in vivo multimodal imaging of tumor tissue [34]. Mesoporous silica NPs (MSNs) have been recently considered as attractive tools for imaging and drug delivery. The high porosity of MSNs results in a very large surface area to carry drugs or imaging agents. The location of molecules inside the pores can provide superior loading efficiency and stable storage/protection from the external microenvironment. Furthermore, stimuli-responsive systems based on MSNs, can regulate the release of loaded molecules, which is a useful property for target-specific delivery of active agents [35]. Recently multifunctional hollow mesoporous silica nanocages (FITC labeled to conduct intracellular tracking through fluorescence microscopy) were developed as a single system for imaging, drug delivery and photodynamic therapy [36].

Carbon-based NPs. Carbon-based materials (mainly carbon nanotubes and graphene), have been recently considered for theranostic applications. Carbon nanotubes (CNTs) are hollow graphitic nanomaterials, structured as rolled sheets of benzene ring carbon atoms in the form of cylindrical tubes. Depending on the structure, CNTs can be categorized into two groups: single-walled carbon nanotubes (SWCNTs) and multi-walled carbon nanotubes (MWCNTs). SWCNTs consist of a one-layer sheet with diameter of 1–2 nm and a length ranging from several 10s to 100s of nanometres, while MWCNTs consist of multiple concentric layers of SWCNTs. In nanomedicine, SWCNTs have been used more frequently than MWCNTs. Due to their unique properties such as electrical conductance, piezoresistance, and electrochemical bond expansion, CNTs have been considered more for applications in electrochemistry and sensors. Other CNT properties, such as high absorption in the NIR region, strong Raman shift, photoacoustic properties and photothermal ability, are more interesting for biomedical applications (imaging and therapy) [37]. PEG chain conjugated SWCNTs coated with cyclic RGD peptides for efficient tumor targeting (labeled with ^{64}Cu for radioimaging) have been developed [38]. Graphene is a one-atom-thick single sheet of sp2-hybridized carbon atoms. Geim and Novoselov were awarded the Nobel Prize in Physics for the study of two-dimensional graphene properties [39]. Because of its common benzene carbon sheet with CNTs, graphene also demonstrates useful properties such as NIR light absorption and photothermal effects. Graphene is relatively amenable, compared to CNTs, to micro- and nanofabrication of complex structures of high aspect ratios; however, the structural rigidity of CNTs provides more stability under intracellular conditions. Therefore these two carbon-based materials should be carefully selected for specific applications.

4.2.2   Physicochemical Properties and their Effect on the in vivo Fate of Nanoparticle Formulations

A number of requirements concerning the physicochemical properties of NPs exist, in order to be used for drug or imaging agent delivery. First of all, they should be able to load sufficient amounts of drug or imaging agents, and retain the required quantity while in the blood, at least until they reach their target. For successful theranosis, the efficient delivery of imaging agents and drugs is critical, to provide sufficient signal (for imaging) or drug concentration (for therapy) at the targeted disease site. A second important feature is the blood circulation time of NPs. Intravenously administered NPs' major clearance is through renal and hepatic routes. The physicochemical properties of NPs, such as size (i.e., hydrodynamic diameter, HD), hydrophobicity, flexibility, and surface charge, have a profound effect on their in vivo biodistribution and clearance [40, 41]. Rigid NPs with HD<6 nm can be excreted through the renal route, whereas larger NPs are rapidly removed from the blood by the RES, leading to rapid uptake in the liver and spleen and subsequent excretion through the hepatobiliary route. Surface charge also plays a crucial role in the rapid elimination of NPs from the blood by the RES uptake because NP HD may increase by non-specific adsorption of plasma proteins, which increases opsonization. In general, appropriate surface modification of NPs using biocompatible, hydrophilic, and neutral polymers, such as polyethylene glycol (PEG) or polysialic acids, will reduce opsonization and uptake into the RES and facilitate efficient clearance of NPs from the body. The enhanced permeation and retention (EPR) effect in fenestrated vessels can enable high accumulation of nanoparticles at angiogenic disease sites, as tumors. This strategy to target disease sites is known as *passive targeting*; by modulating nanoparticle physicochemical properties (size, zeta potential and surface properties) it is possible to prolong blood circulation time and target such diseased sites passively. NP accumulation and cellular uptake could be further enhanced by *active targeting* with target-cell-specific antibodies or peptides conjugated on the surface of nanoparticle [Figure 4.1].

Biocompatibility, biodegradability and *low-toxicity* are additional important required features for in vivo administration of NPs. Some NPs composed of biocompatible and biodegradable materials can be metabolized into clearable components, although their elimination from the body is extremely slow in many cases. Many inorganic NPs consist of metal elements, such as gadolinium, cadmium (e.g., quantum dots), which are known to be toxic. Furthermore, some studies showed that carbon-based NPs have potential biological toxicity [42]. Therefore, before practical application of NPs and administration into the human body for diagnostic or therapeutic purposes, comprehensive assessments of potential toxicity should be carefully considered. One has to keep in mind that toxicity/biocompaticility issues may be more pronounced when therapeutic (or theranostic) NP-systems are to be designed, compared to diagnostics, since the amount of drug required for therapy is usually much higher compared to required levels of imaging agents for detection; thereby, the carrier material is used in much larger amounts. Additionally, it is very important to carry out such toxicity assessment studies at the nano range (i.e, evaluate the toxicity of the NPs and not the NP building blocks as pure materials), since at this size dimension, the surface area, and thus also the contact are a between material and biological milieu, is significantly increased, leading to enhanced interactions [43].

A few diagnostic or therapeutic materials formulated in NPs have been approved by the Food and Drug Administration (FDA) for clinical use, and many more are under clinical investigation. The currently marketed products are mainly simple formulations with no specificity (e.g., Doxil^s, Abraxane^s, or Feridex^s) [22, 23]; such formulations are considered as "first generation nanomedicines." Today, various NPs with high target specificity are being actively investigated, some being multifunctional nanoparticles with more than one clinical function. To summarize, in order to successfully translate NPs into clinical treatment, several issues have to be considered, such as: (1) Reasonable blood half-life, retaining their nano-size and content; (2) Favorable physiological behavior with minimal off-target distribution; (3) Effective clearance of NPs from the human body or possible metabolism to clearable components; and (4) Potential toxicity.

4.3   Physiological Factors Related with Brain-Located Pathologies: Focus on AD

4.3.1   Neurodegenerative Diseases; AD and Related Pathologies

Neurodegenerative diseases (NDs) are one of the main classes of CNS-located disorders. In NDs, such as Alzheimer's diseases (AD), Parkinson's diseases (PD) and multiple sclerosis, patients experience symptoms related to movement, memory, and dementia due to gradual loss of neurons or else neurodegeneration. The causes of NDs are complex and associated with many factors such as advancing age, environmental issues, and disordered immunity, and less with the host genetics [44, 45]. The unique and complicated environment and restricted anatomical access of the CNS due to the blood–brain barrier (BBB), makes any diagnosis or surgery-based therapy here more difficult than in any other diseased site. [46]. Furthermore, clinical neuroscience faces great challenges due to the extremely heterogeneous cellular and molecular environment and the complexities of brains anatomical and functional "wiring" and associated information processing. As mentioned in the introduction, nanotechnology provides hope that it will revolutionize diagnosis and treatment of CNS disorders.

Since this chapter is focused on AD, a description of the disease and related pathologies follows:

AD is a ND which currently affects > 24 million people worldwide. There is no available therapy, but only symptomatic treatment [2]. It is characterized by cognitive dysfunction and progressive memory loss. The clinical course of AD is generally characterized by dementia as well as learning and memory impairment, accompanied by behavior and cognition changes [47]. AD diagnosis is thus based on detection of behavioral changes (as mood swings, confusion, and irritability). Two main pathological hallmarks are connected with AD, β-**amyloid (Aβ) plaques** and **neurofibrillary tangles**. Aβ peptides consist of 39–42 amino acids and are formed by abnormal processing—or better cutting—of the larger amyloid precursor protein (APP) [48]. The deposits of Aβ produced by this abnormal processing are insoluble and thereby accumulate, resulting in a cascade of destructive effects on neurons, and increased oxidative stress. This abnormal accumulation of Aβ seems to be the cause of AD. Metal ions such as copper (Cu) and zinc (Zn) tend to further promote deposition of these extracellular plaques [49]. **Neurofibrillary tangles** are the other major findings in AD; These are hyperphosphorylated, paired helical filaments of microtubule-associated-protein called tau (τ). In healthy cells they are associated with growth and development, but their intracellular accumulation eventually leads to cell death by disturbing the normal cytoskeleton. Other manifestations of AD, such as neuronal loss (particularly of the cholinergic type that plays an important role in memory cognition) have also been observed, while other findings of clinical importance also exist as deposition of amyloid plaques in blood vessels, granulovacuolar degeneration, chromosomal mutations, oxidative stress, etc. [50, 51]. More details about the two main pathological hallmarks of AD will be given below in the section focused on past and current approaches to target AD for diagnosis and therapy. For more details on AD pathophysiology many recent review articles are available [52–54].

4.3.2 The Blood Brain Barrier (BBB)

4.3.2.1 BBB Physiology

The BBB's main function is to maintain the chemical composition of the neuronal milieu, protecting the functions of neurons. It is readily permeable only to lipophilic molecules or those of a molecular weight below 400–600 Da [55], and thus the options for potential therapeutic and diagnostic tools are limited. The BBB is localized at the interface between the blood and the cerebral tissue [56, 57] formed by endothelial cells (ECs) of cerebral blood vessels, which display a unique phenotype characterized by the presence of intercellular tight junctions and expression of numerous transport systems. The anatomical and functional site of the BBB is the brain endothelium. Under physiological conditions, extracellular base membrane, adjoining pericytes, astrocytes, and microglia are, in addition to brain capillary endothelial cells, parts of the BBB supporting system. Together with surrounding neurons, these components form a complex and functional "neurovascular unit" [58] (Figure 4.2). Apical junctional complexes, notably tight junctions (**TJs**), are present in brain endothelium forming the blood-brain-barrier (BBB), which controls cerebral homeostasis and provides the central nervous system (CNS) with a unique protection against the toxicity of many xenobiotics and pathogens.

TJs are located on the apical side of endothelia cells and are structurally formed by an intricate complex network made of a series of parallel, interconnected, transmembrane and cytoplasmatic strands of proteins

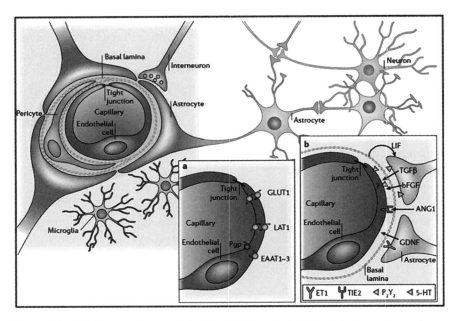

Figure 4.2 Cellular constituents of the BB. a) Brain endothelial cell features observed in cell culture. The cells express a number of transporters and receptors, some of which are shown. Excitatory amino acid transporters 1–3(EAAT1–3); glucose transporter 1(GLUT1); system for large neutral amino acids (LAT1); P-glycoprotein (P-gp). b) Examples of bidirectional astroglial–endothelial induction necessary to establish and maintain the BBB. Figure is obtained from Ref [59] after permission.

[57, 59, 60]. The high level of integrity of TJs is reflected by the high electrical resistance of the BBB (1500–2000 Ω cm^2), which depends on extracellular Ca^{2+} ion concentration. Extensive reviews on TJ characteristics are available for those interested in more details [57, 60, 61]. Adherens junctions (**AJs**) are located below the TJs in the basal region of the lateral plasma membrane. They are composed of transmembrane glycoproteins (represented by the large family of cadherins) and are linked to the cytoskeleton (by cytoplasmatic proteins), thus providing additional tightening structure between the adjacent endothelial cells at the BBB. Another blood-brain interface is localized at the choroid plexus epithelium, which controls the exchanges between the blood and the cerebro-spinal-fluid (so-called the blood-CSF barrier); these specialized epithelial cells also express TJs and a number of transporters. The problem of drug transport to the brain or else the unique biological characteristics of the BBB are mainly the following: 1) Lack of fenestrations with very few pinocytotic vesicles, and a relatively large number and volume of mitochondria in endothelial cells [62, 63]; 2) The presence of TJs between adjacent endothelial cells, formed by an intricate complex of transmembrane proteins (junctional adhesion molecule-1, occludin, and claudins) with cytoplasmic accessory proteins. They are linked to the actin cytoskeleton, [64], forming the most intimate cell-to-cell connection. TJs are further strengthened and maintained by interaction/communication between astrocytes or pericytes and endothelial cells [56]; 3) The expression of various transporters including GLUT1 glucose carrier, amino acid carrier LAT1, transferrin receptors, insulin receptors, lipoprotein receptors and ATP family of efflux transporters such as p-glycoprotein (P-gp) and multidrug resistance-related proteins MRPs [59]. Some of these aid the transport into the brain while others prevent the entry of many molecules; 4) The synergistic inductive functions and upregulating of BBB features by astrocytes, astrocytic perivascular endfeet, pericytes, perivascular macrophages and neurons, as suggested by the strong evidence from cell culture studies [65–67]; and: 5) The lack of lymphatic drainage, absence of major histocompatibility complex (MHC) antigens and on-demand inducible immune reactivity for maximum protection to neuronal function [68]. The BBB has a strict limit for the passage of immune cells, especially lymphocytes, [69] and its immune barrier is made by the association between BBB endothelial cells, perivascular macrophages and mast cells [70]. Additionally, this immune barrier is reinforced by local microglial cells [71].

As seen in Figure 4.3 [72], the routes for penetration across the BBB can be regulated by different types of receptors or transporters, depending on the specific molecule. Thereby: (1) Small hydrophilic molecules such as amino acids, glucose, and other molecules necessary for the survival of brain cells use *transporters* expressed at the luminal (blood) and basolateral (brain) side of the endothelial cells. (2) Large and/or hydrophilic essential macro molecules, such as hormones, transferrin for iron, insulin, and lipoproteins use specific receptors that are highly expressed on the luminal side of the endothelial cells. These receptors function for the endocytosis and transcytosis of compounds across the BBB. (3) Small lipophilic molecules can diffuse passively across the BBB into the brain but will be exposed to efflux pumps (P-glycoprotein [P-gp], some Multidrug Resistance Proteins [MRP], Breast cancer Resistance Protein [BCRP] and others) expressed on the luminal side of the BBB and exposed to degrading enzymes (ecto- and endo-enzymes) localized in the cytoplasm of endothelial cells before brain penetration.

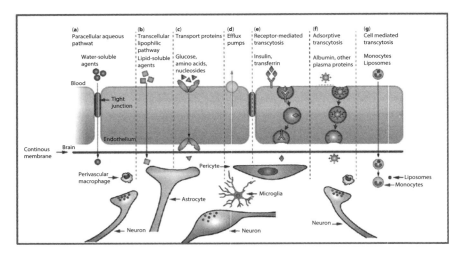

Figure 4.3 Mechanisms for transport of molecules across the BBB. Pathways a–f apply for solutes and pathway g for cells or NP-drug/imaging agent-carriers. From Ref [72], after permission.

4.3.2.2   Methods to Overcome the BBB

In addition to the mechanisms available for solutes to pass the BBB, there are specific methodologies that may be applied **to overcome the BBB**. These can be divided into invasive and non-invasive (when specific transport systems available on the BBB cell components, are utilized) approaches.

Invasive approaches to overcome the BBB include transport by intra-cerebro-ventricular infusion (ICV) or direct injection in the CNS. **Intra-cerebro-ventricular infusion** (ICV) or insertion of a small-caliber catheter into the brain parenchyma (**Convection-enhanced delivery** (CED)) have been used as methods to overcome the BBB but the limitations of these methods were that the diffusion of the drug in the brain parenchyma is very low and proper drug delivery depends on the placement of catheters [73]. Intra-cerebral injection or use of implants is limited by the fact that diffusion in the brain decreases exponentially with distance. Thereby, the injection site has to be very precisely mapped to get efficacy and overcome the problem associated with diffusion of drugs in the brain parenchyma. Polymer wafers, currently used in the clinic, are considered as the gold standard of intra-cerebral drug therapy [74]. Another type of invasive approach to pass the BBB is via tight junction opening (enhanced by biological chemical and physical stimuli). Disruption of the BBB can open access of the brain to components in the blood by making the tight junction between the endothelial cells more permeable (by osmotic disruption, by

MRI-guided focused ultrasound BBB disruption technique, by application of bradykinin-analogue). All these approaches are relatively costly, require anesthesia and hospitalization, and are non-patient friendly. Neurons may be damaged permanently from unwanted blood components entering the brain. Drug transport across the blood–brain barrier for drugs which cannot pass the BBB normally (as free molecules) can be also achieved by *non-invasive approaches*, which can be: a) physiological methods; b) pharmacological methods; or c) other methods. These types of BBB transport methods are briefly presented below:

a. *Physiological methods*: The brain requires essential substances for metabolism and survival, such as glucose, insulin, growth hormone, low density lipoprotein (LDL), which are recognized by specific receptors or transport mechanisms, resulting in specific transport mechanisms into the brain. Thereby molecules can overcome the BBB *via transport vectors*, *adsorptive mediated transcytosis*, and *receptor-mediated transcytosis*. Furthermore, a *cell-mediated transcytosis* mechanism is available (Figure 4.3). Indeed, peptides and small molecules may use specific transporters expressed on the luminal and basolateral side of the endothelial cells forming the BBB to cross into the brain. At least eight different nutrient transport systems have been identified, each one transporting a group of nutrients of similar structure. Drugs may be modified such that their transport is increased by using a carrier-mediated transporter expressed on the endothelial cells forming the BBB. Use of small molecules that directly target transporters to overcome BBB restrictions eliminate the need for the drug to be transformed, for example by conjugation to antibodies [75] a method used to deliver the metabolic precursor of dopamine. Using these methods to overcome the BBB, multiple factors must be considered, such as the kinetics available to transport physiologic molecules, the structural binding requirements of the transporter and the therapeutic compound manipulation so that the compound binds but also remains active in vivo. Adsorptive-mediated transcytosis (AMT), also known as the pinocytosis route, is triggered by an electrostatic interaction between a positively charged substance, usually the charged moiety of a peptide, and the negatively charged plasma membrane surface (i.e., heparin sulphate proteoglycans). The development of many new drug delivery technologies focusing on

AMT has similarity to receptor-mediated mechanisms, but are not so specific. Peptides and proteins with a basic isoelectric point ("cationic" proteins) bind initially to the luminal plasma membrane (mediated by electrostatic interactions with anionic sites), which triggers adsorptive endocytosis. Receptor-mediated transcytosis can be achieved via the different receptors, which are known to be overexpressed on the brain endothelial cells, as: (1) The transferrin receptor (TfR), the function of which is to provide iron to cells. Drug targeting to the TfR can be achieved by using the endogenous ligand transferrin, or by using antibodies directed against the TfR. For transferrin (Tf) the in vivo application is limited due to high endogenous concentrations of Tf in plasma. Transferrin is an essential protein needed for iron delivery to cells and is found at mg/ml amounts in plasma. Using the antibody approach against TfR, the receptor specific monoclonal antibody (mAb) binds to the receptor on the endothelial cells, and allows the associated therapeutic agent to cross the BBB via receptor-mediated transcytosis [76, 77]. For antibodies against the TfR, proof of concept studies in rats have demonstrated that a mAb that binds to a distinct epitope from TfR (OX26) can be used as a brain delivery agent [78]; (2) The low-density lipoprotein receptor related proteins 1 and 2 (LRP1 and LRP2 receptors). LRP mediates the internalization and degradation of multiple ligands involved in diverse metabolic pathways [79, 80]. LRP is a multiligand lipoprotein receptor which interacts with a broad range of secreted proteins and resident cell surface molecules (eq. apoE (apolipoprotein E), α2 M (α2 macroglobulin), tPA (tissue Plasminogen Activator), PAI-1 (Plasminogen Activator Inhibitor 1), APP (Amyloid Precursor Protein), Factor VIII, Lactoferrin, etc., mediating their endocytosis or activating signaling. Nanoparticles functionalized with specific to LRP ligands; have been used for drug delivery to the brain [81–83]; (3) The insulin receptor (IR). Pardridge *et al.* have extensively documented the use of the insulin receptor for the targeted delivery of drugs to the brain using specific antibodies directed against the IR [84]. For example, using the 83–14 mouse mAb against the human insulin receptor (HIR) in rhesus monkeys shows that total uptake of the mAb is 4%, which corresponds to 0.04%/g brain tissue 3 h after IV injection [85]. Both the chimeric antibody and a fully humanized form of the 83–14 antibody against HIR have been developed [86] and shown to be able to transport an associated/conjugated molecule across the BBB; (4) The diphteria toxin receptor. The utility of CRM197, a non-toxic mutant of diphtheria toxin, as a targeting vector for drug delivery to the brain has been demonstrated [87]. CRM197 has been shown to endocytose after binding to the membrane precursor of heparin binding epidermal growth factor (HB-EGF) [88], also known as the diphtheria toxin receptor (DTR).

Finally, cell-mediated drug transport employs specific cells that take up drug-loaded nano, or microcarriers traffic them through the BBB and deliver the drugs to their target sites inside the brain. In this case, cells act as Trojan horses. Compared to other transport pathways, cell-mediated drug delivery has attracted far less attention for brain drug transport but there have been some very promising results reported. Most noteworthy is the work carried out [89] on RGD-anchored magnetic-liposomes for monocyte/neutrophil-mediated brain targeting using an IL-1β induced brain inflammation rat model.

b. *Pharmacological methods*: This approach consists of modifying, through medicinal chemistry, a molecule that is known to be active against a CNS target to enable it to penetrate the BBB. Modification of drugs through a reduction in the relative number of polar groups increases the transfer of a drug across the BBB. Lipid carriers have been used for transport for many years, and there are successful early examples of both these approaches [90]. The limitations include that the modifications necessary to cross the BBB often result in loss of the drug activity. Additionally increasing the lipophilicity of a molecule to improve transport can also result into making it a good substrate for the P-glycoprotein-efflux pump (P-gp).

c. *Other routes*: Intranasal administration has emerged as a compelling method for delivering drugs or diagnostic agents to the brain as it bypasses the BBB non-invasively [91]. Mechanisms of intranasal drug delivery to the CNS remain to be elucidated. Evidence suggests that nerves connecting the nasal passages to the brain and spinal cord as well as the vasculature, cerebrospinal fluid, and lymphatic system contribute to transportation of molecules to the CNS following adsorption from the nasal mucosa [93]. The major route of intranasal delivery is the olfactory nerve pathway. Drugs travel via the olfactory nerve axons, accumulate in olfactory bulbs (OB), and diffuse into the brain [92]. This administration route has been exploited for pain management, particularly post-operative pain and moderate-to-severe cancer pain due to ease of delivery and high brain uptake [93, 94].

4.3.3   In vitro and in vivo Models for BBB Permeability and AD Diagnostic/Therapeutic Approach Assesment

Methods used to investigate BBB transport or for screening of potential delivery systems for their capability to facilitate drug or imaging agent transport across the BBB, (in vitro systems, cellular models, and animals [wild type (WT) or transgenic (TR)]), as well as animal models which have been developed up-to-date for assessment of AD diagnostic and/or therapeutic approaches, are mentioned and briefly described.

4.3.3.1   In vitro Methods

The in vitro methods can be categorized in two main groups: non-cell and cell-based permeability assays. **Non-Cell-Based Permeability Assays**: About 30 years ago, efforts were initiated toward the development of methodologies for molecule membrane permeability screening. Various techniques, such as high-performance liquid chromatography (HPLC) (by development of affinity columns), "immobilized artificial membranes" (IAMs) that mimicked the properties of biological membranes, etc., have been identified. Some of the later were moderately successful and capable of ranking compounds according to BBB permeability [95], but they were not found suitable for medium- to high-throughput operation. A more promising technology is the parallel artificial membrane permeability assay (PAMPA), which is suitable for the study of passively permeating compounds, showing moderately good correlation with data derived from the human colonic epithelial cell line (Caco-2) and from in vivo permeability studies. By modifying the lipid composition of the artificial membrane, which is the basic component of the PAMPA system, it appears to be capable of predicting CNS permeability with reasonable accuracy [96]. **Cell-Based Permeability Assays**: Several cell culture models (summarized in Table 4.2) have been developed and used for BBB permeability prediction studies. Most in vitro BBB cellular models proposed so far are based on co-cultures of brain endothelial cells (ECs) and astrocytes (or glial cells) [97],

Table 4.2 Cellular models of the BBB.

BRAIN ENDOTHELIAL CELLS	
PRIMARY ENDOTHELIAL CELLS	
Origin	*Ref*
Bovine	[98]
Porcine	[102]
Rat	[103]
Mouse	[101]
Human	[104]
IMMORTALIZED BRAIN ENDOTHELIAL CELL LINES	
Bovine SV-BEC	[111]
Porcine PBMEC	[113]
Rat RBE4	[108]
Mouse bEND5	[112]
Human hCMEC/D3	[116]
CO-CULTURES	
Glial Cells/Astrocytes	[98]
Pericytes	[106]
Astrocytes and Pericytes	[97]
Neurons	[105]

in two-chamber cell culture systems. In particular, whereas bovine brain endothelial cells alone only partly recapitulate BBB properties, a co-culture system with rat glial cells has been extensively validated as a reference BBB model [98]. Because these cells express tight junctions as well as numerous membrane transporters, they constitute a valuable alternative or complement to the epithelial cell lines Caco-2 and MDCK, currently used for drug screening by pharmaceutical industries because of their very high permeability restriction [99]. Alternative BBB models are also available, using pig, mouse, rat or human brain ECs [100–106]. In addition, stable immortalized rat EC lines were produced and validated as in vitro models of brain endo-thelium: first the RBE4 cell line [107–109], followed by a number of other cell lines [110–113], some of which have been widely used for biochemical, immu-nological and toxicological studies. More recently, the human hCMEC/D3 brain EC line, which retains most of the morphological and functional char-acteristics of brain endothelial cells, was developed. This cell line expresses TJs and multiple active transporters, receptors and adhesion molecules, even without co-culture with glial cells, thus appearing as a unique and easy to maintain in vitro model of the human BBB [114–116]. Recently drug per-meability across hCMEC/D3 monolayers was significantly decreased when the system was used under three-dimensional flow conditions [117, 118] (compared to the non-flow setup), improving the predictability of the sys-tem. All together, these in vitro BBB models are helpful for screening of new drugs as well as for unraveling the molecular mechanisms of BBB-control under physiological conditions and in various CNS diseases.

For measurement of *BBB-permeability*, cell monolayers are generated by growing the cells on permeable filters, mounted in a device that separates the apical (luminal) compartment and the basal (abluminal) compartment (Ussing chamber, Transwell system, etc.). The drug is added to one com-partment (the donor) and the amount appearing in the other compartment (receiver) is determined over time. For measurement of *Uptake by Cells*: the study typically involves incubation of the cells with a tracer molecule, stop-ping the uptake process at set time-points (e.g., by washing with cold buffer solution, or using a stopping solution containing specific inhibitors to block transport), lysing the cells and analyzing the contents for presence of the tracer (by radioactive counting, quantification of fluorescence, HPLC or liq-uid chromatography-mass spectrometry [LC/MS]), as well as total protein content. In order to better mimic the in vivo situation, some *three-dimen-sional "dynamic" in vitro BBB models* (DIV–BBB) have been developed. In these, *intraluminal flow* is incorporated, providing an optimal combination of conditions, which encourage BBB differentiation. For this, endothelial cells are seeded inside porous tubes with astrocytes on the outside, and medium flow is maintained via the lumen. Such systems provide high TEER values while numerous transporters and receptors (characteristic of the BBB) are also upregulated [119, 120]. Although difficult to use for kinetic studies, such models have been used to test drug permeability.

4.3.3.2   In vivo (and in situ) Methods

For BBB permeability: There are two methods that are used for assess-ing BBB permeability in vivo [121]; these include determinations of brain:plasma ratio (log BB); and measurement of the permeability surface area product (PS or log PS), from which permeability (P) can be derived provided that the vessel surface area (S) can be estimated. Most pharmaceutical companies generate log BB data in animals (often rat) as part of standard pharmacokinetic profiling of compounds [122]. These measurements are generally made over several hours, with a number of animals required per data point, making the studies costly and labor intensive. Moreover, a number of factors, including metabolism and binding, affect the brain distribution and therefore log BB may not be an accurate measure of BBB permeability. Determining the unidirectional influx coefficient (Kin) by using the in situ saline-based perfusion method more accurately reflects the BBB permeation step, effectively isolating this "kinetic" element of drug penetration [121, 122]. Because of the accurate quantitation, the Kin (or PS) measurements are considered as "gold standard" references for other methods [123]. For drugs acting on the basic CNS target sites (membrane receptors, transporters), the critical concentration is the free concentration in brain interstitial fluid (ISF). Brain/plasma ratio measured especially at longer times and for more lipophilic agents will be affected by drug distribution into brain lipids and nonspecific binding. Measuring free concentration with a microdialysis probe is possible but technically difficult, and there is a particular problem of recovery of more lipophilic agents [124, 125].

For assessment of proposed AD diagnostic and therapeutic approaches— AD mouse models: Transgenic mice models that mimic a range of Alzheimer's disease–related pathologies are currently available (summarized in Table 4.3). They have been widely used in the preclinical testing of potential therapeutic modalities and have played a pivotal role in the development of immunotherapies for Alzheimer's disease, which are currently in clinical trials. The most common approach to create genetically manipulated mice is to microinject a complementary DNA (cDNA) transgene (usually with a pathogenic mutation). However, some transgenic animals have been created with a whole, wild-type human genomic fragment that includes promoter, introns and flanking sequence (the whole APP gene [126], PS-1 gene [127], or tau gene [128]). Other approaches target the endogenous mouse gene with pathogenic mutations or human gene sequence (the knock-in approach) [129]. There have been problems thus far with the uniformity of results due to the different mouse strains and methods used.

Amyloid-forming APP mice: There have been many attempts to create models of amyloidogenesis, some of which replicate the amyloid pathology rather well. The first mouse model with extensive AD-like neuropathology

Table 4.3 Transgenic Mice Models that Mimic Alzheimer's Disease or Related Pathologies.

Trans/Line	Pathology	Refs
PD-APP	Aβ deposits, neuritic plaques, synaptic loss, astrocytosis and microgliosis	[130]
Tg2576	Hamster Prp Behavioral, biochemical and pathological abnormalities	[131]
TgAPP23	Congophilic senile plaques that are immunoreactive for hyperphosphorylated tau	[132]
TgCRND-8	Hamster PrP Aβ amyloid deposits, dense-cored plaques and neuritic pathology, early impairment in acquisition and learning reversal	[125]
TgR1.40YAC	Aβ deposits, pathology accelerated inhomozygotes, or when bred to mutant PS-1YAC transgenic mice	[126]
PS/APP	Fibrillar Aβ deposits in the cerebral cortex and hippocampus, reduced spontaneous alternation performance in a "Y" maze	[133]
PS-1	Elevated levels of the highly amyloidogenic 42- or 43-amino acid peptide Aβ42(43)	[134]
Alz27	Tau present in nerve cell bodies, axons and dendrites was phosphorylated at sites that are hyper-phosphorylated in paired helical filaments	[135]
JNPL3	Murine PrP Motor and behavioral deficits, with age- and gene-dose-dependent development of neurofibrillary tangles	[136]
hTau40	Axonal degeneration in brain and spinal cord, axonal dilations with accumulation of neurofilaments, mitochondria and vesicles	[137]
hTau	Hyperphosphorylated tau accumulating as aggregated paired helical filaments in the cell bodies and dendrites of neurons in a spatiotemporally relevant distribution	[138]
PS1/APP/Tau	Development of plaques and tangles; synaptic dysfunction, including long term potentiation (LTP) deficits, manifesting in an age-related manner (before plaque and tangle pathology)	[139]

was created by Exemplar_Athena Neuroscience in 1995 (the PD_APP line). This mouse line developed fibrillar and diffuse amyloid pathology in the cortex and hippocampus [140]. A subsequent model (Tg2576) created by Hsiao and colleagues showed age-related cognitive impairment in addition to Aβ plaques [131]. Surprisingly no model showed overt cell loss [141] and amyloid associated cell loss has only been reported in one (APP-23), although it was very marginal [134]. It is likely however, that neurodegeneration is occurring in these models, as several of the mutant APP mice have now been shown to be cognitively impaired [125, 131, 133].

Presenilin mice: Both wild-type and mutant PS-1 mice have been created to determine the effects of PS-1 on APP processing. [128]. Studies on these mice indicated that PS-1 mutations had the effect of increasing Aβ 1–42. The level of Aβ 42 generated in these mice was not sufficient for amyloid to be formed. When mutant PS-1 mice were crossed to mutant APP mice, Aβ aggregation into plaques was greatly accelerated in the mouse line studied, indicating either that Aβ levels are critical for amyloid formation or that these AD genes were interacting synergistically [133]. PS-1 knockout (KO) mice have demonstrated decreased Aβ levels, further emphasizing the APP processing role of PS-1 [142], and PS-1 is now thought to be synonymous with the APP-cleaving enzyme, γ-secretase [143].

ApoE mice: ApoE KO, knock-in or cDNA mice have been created as tools to understand the role of this protein in AD [143]. When ApoE KO mice were crossed with PD_APP mice, the mice had significantly reduced Aβ deposition. When ApoE4 mice were crossed with PD_APP mice, higher levels of amyloid were seen compared with an ApoE2/PD_APP cross, supporting the pathogenic role of ApoE4 in AD [144]. The mechanism by which ApoE exerts its effect is unknown, although some theories have been proposed.

Tau mice: The first transgenic tau mouse was created in 1995 [145], but further neurofibrillary pathology was not apparent. Use of pathogenic mutations in tau cDNA transgenes led to the development of mice that form robust neurofibrillary (tangle) pathology of relevance to FTD-17 and AD [136]. Recent developments using a genomic, wild-type tau transgene has led to a mouse model that forms more AD-like tangles [138]. More recently created mouse models have cortical/hippocampal pathology and associated cell loss; these should be more informative for the study of pathogenic tau formation (including the contribution of hyperphosphorylated or aggregated tau) and neurodegeneration/cell death. Unfortunately tangle formation in these models is not associated with Aβ aggregation, and the tau transgenic mice do not form amyloid plaques. Interestingly, exposing mutant tau mice to elevated Aβ (either through crossing to an

APP-overexpressing mouse or by injecting Aβ into the brain [145]) leads to increased tau pathology, suggesting that there is an interplay between Aβ and tau. Recently, a triple transgenic model (expressing mutant APP, PS-1 and tau) has been created, in which the APP and tau transgenes were co-injected (leading to co-integration) into a PS-1 knock-in line [139]. Progeny express all three transgenes in the same background strain. They develop both plaques and tangles (plaques come before tangles in this model) and have shown synaptic dysfunction, although behavioral studies have not been reported.

4.4   Current Methodologies to Target AD-Related Pathologies

Methodologies to target AD-related pathologies that are currently available or (in the majority of cases) under preclinical or clinical research are mainly focusing on the two basic—hallmark—pathologies; i.e., Tau neurofilaments and Amyloid plaques. In most cases a prerequisite to target these pathologies is the passage of the BBB, however, some research approaches are designed in a way that passing the BBB is not required (as the "sink" theory approach) [146, 147].

A basic description of the various approaches currently under clinical or pre-clinical investigation will be given in this section, together with detailed lists of potential ligands that may be used as targeting ligands to direct nanotechnologies to the specific AD-related pathological features. For presentational reasons the ligands are divided in three main categories: small molecule ligands (Table 4.4), peptide (or peptide-type) ligands (Table 4.5) and antibody ligands (Table 4.6).

4.4.1   Tau-targeted Strategies—Available Ligands

Tau, one of the main microtubule-associated proteins, will be hyperphosphorylated and lose the ability to bind microtubules when the homeostasis of phosphorylation and dephosphorylation is disturbed in neurons. Tau hyperphosphorylation seems to be required, but is not sufficient, to induce tau aggregation alone. Due to the better understanding of the mechanisms of tau phosphorylation and its quantitative importance (85 phosphorylation sites), the approach of targeting tau phosphorylation by inhibiting tau kinases seems to be a feasible strategy to prevent tau aggregation and associated pathological effects. However, tau protein, regardless of its

Table 4.4 Small Molecule Ligands for Aβ or Tau Targeting.

Molecule	Aβ aggregation (IC50 μM)	Tau formation (IC50 μM)	Reference
A] MOLECULES ATTACHED TO NPs			
Curcuminoids	SPR, Kd: 1-5 nM		[201, 200–204]
Phosphatidic acid (PA) (incorporated)	SPR, Kd: 22-60nm		[203, 205, 206]
Cardiolipins (incorporated)	SPR, Kd: 22-60nm		[203, 206]
Monosialogangliosides (GM1) (incorporated)	SPR, Kd: 0.2 μM		[203, 205–209]
2-methyl-N-(2'-aminoethyl)-3-hydroxyl-4-pyridinone			[210]
Sialic acid			[211, 212]
Maltose			[213]
Cationic lipids (incorporated)			[214]
B] POTENTIAL LIGANDS FOR ATTACHMENT TO NPs			
Anthracyclines: Daunomycin			[194]
Benzothiazoles: 2-(4-Aminophenyl)-6-Methylbenzothiazole; Basic blue 41; 2-[4-(Dimethylamino)phenyl]-6-methylbenzothiazole; 3,3'-Dipropyl thiodicarbocyanine iodide; Thioflavin T	Aβ$_{1-40}$ 0.30 – 2.4 ; Aβ$_{1-42}$ 0.12 - 122*	>200	[156, 187, 197]

(Continued)

Table 4.4 (*Cont.*)

Molecule	Aβ aggregation (IC50 μM)	Tau formation (IC50 μM)	Reference
Conco Red derivatives/Sulfonated Dyes: BSB; Chlorazol black E; Congo Red; FSB; Ponceau SS; Chicago Sky Blue 6B; Phenol red; Direct Red 80; Orange G	$Aβ_{1-40}$ 6.4; $Aβ_{1-40}$ 0.3; $Aβ_{1-40}$ 0.9; $Aβ_{1-42}$ 10; $Aβ_{1-40}$ 1.9; $Aβ_{1-40}$ 1.2; $Aβ_{1-42}$ 2.58; $Aβ_{1-42}$ 426	18.2; >200; 2.2; 35.7; >200	[156, 187, 197]
Isoflavones: Daidzein	$Aβ_{1-40}$ >40	>200	[197]
Polyene macrolides: Amphotericin B; Filipin III; Mycostatin	$Aβ_{1-40}$ 2.2 ; 14.6; 9.3	>200	[156, 197]
Polyphenols: Apigenin; Baicalein; Catechin; (+)-catechin ; (-)-cathechin gallate; Chlorogenic acid; Cyanidin; Curcumin; 2,2′-Dihydroxybenzophenone; Dopamine chloride; Delphinidin; (-)-Epicatechin and derivatives; Exifone; (-)-gallocatechin and derivatives; Gingerol; Gossypetin; Hinokiflavone; Hypericin; Kaempferol; Luteolin; Myricetin; Morin; Naringenin; Nordihydroguaiaretic acid (NDGA); 2,3,4,2′,4′-Pentahydroxybenzophenone; Procyanidin B1; Procyanidin B2; Pseudohyperici; Purpurogallin; Quercetin Resveratrol; Rosmarinic Acid *Rutin*; (+)-α-tocopherol; 2,3,4′-Trihydroxybenzophenone; (+)-Taxifolin; 2,2′,4,4′-Tetrahydroxybenzophenone; Teaflavin; Tannic acid	$Aβ_{1-40}$ 0.14 - >40 $Aβ_{1-42}$ 0.63 - 5.3 [369]*;	1.8 - >200	[187, 190, 193, 195, 197, 186, 187, 215, 216]
Porphyrins: Ferric dehydroporphyrin IX; Hematin; Hemin; Hemin chloride; Phthalocyanine tetrasulfonate; Protoporphyrin IX; Ferulic acid	$Aβ_{1-40}$ 0.1 - 65.0 $Aβ_{1-42}$ 0.4-15.9	1.4 - 67	[156, 187, 197]

Molecule	Aβ aggregation (IC50 μM)	Tau formation (IC50 μM)	Reference
Rifamycins: Rifampicin; Rifamycin B; Rifamycin SV; Rodamine B	$A\beta_{1-40}$ 3.1 – 9.7 $A\beta_{1-42}$ 9.1 -309	>200	[156, 187, 189, 196, 197]
Steroids: Taurochenodeoxycholic acid; Taurohydroxycholic acid; Taurolithocholic acid; Taurolithocholic acid 3-sulfate; Tauroursodeoxycholic acid	$A\beta_{1-40}$ > 40	>200	[197]
Terpenoids: Asiatic acid; β-Carotene; Ginkgolide A (B and C)	$A\beta_{1-40}$ 2.4 - >40 $A\beta_{1-42}$ 5.2	>200	[197, 217]
Tetracycline derivaives: Rolitetracycline; Tetracycline and derivatives	$A\beta_{1-42}$ 1.87 - 10 $A\beta_{1-42}$ 10		[187, 189, 196, 218]
Polyene macrolides: Amphotericin B; Filipin III; Mycostatin	$A\beta_{1-40}$ 2.2 - 14.6	>200	[156, 197]
OTHER MOLECULES: Apomorphine; Bromocriptine; Coenzyme CoQ$_{10}$; 4,5-Dianilinophthalimide; Dihydrolipoic acid; 2, 4-Dinitrophenol; L-Dopa; Dopamine; Eosin Y; Ferulic acid; Hexadecyl trimethyl ammonium bromide; Indomethacin; Juglone; α-Lipoic acid; Magnolol; Meclocycline sulfosalicylate; Methyl yellow; 1,2-Napthoquinone; Neocuproine; R(-) Norapomorphine hydrobromide; 3-Nitrophenol; Pergolide; Retinol; Retinoic acid; Sesamin; Tetradecyl trimethyl ammonium bromide; o-Vanilin; Zinc protoporphyrin IX	$A\beta_{1-40}$ 0.01 - 80 $A\beta_{1-42}$ 0.04 – 67.7 [958]*		[156, 187, 189, 190, 196, 197, 219]

* Higher IC50 values were measured when different techniques were utilized (Elisa)

Table 4.5 Peptide Ligands for Aβ or Tau Targeting.

Peptide type	Results	Reference
A) PEPTIDES ATTACHED TO NPs		
-ApoE3 (sequence based on) -KLVFF peptide (sequence based on) -CLPFFD peptide (seq:THRPPMWSPVWP) -TGN peptide (seq:TGNYKALHPHNG) -Based on Aβ1-42 -Based on Aβ1-40	-Reversal of Aβ-induced toxicity (neuroblastoma cell); binding to Aβ1–42 peptides (in vitro); -Inhibition of Aβ1-42 aggregation; -Inhibition/redirection of Aβ fibrillization; -Improvement of spatial learning (Morris water maze experiment); -Targeting of amyloid plaques (in vivo); -Effect on Aβ peptide aggregation	[236, 257, 258]
B] POTENTIAL LIGANDS FOR ATTACHMENT TO NPs		
Based on Aβ- or KLVFF sequences	Prevention/inhibition of Aβ fibrilization/aggregation (in vitro); fibril disassembling, inhibition of Aβ-neurotoxicity; SPR studies	[233-246]
Based on LPFFD sequence (Proline)	Inhibition/de-fibrillization of Aβ; *reduction of toxicity*; Protection of neurons (in vitro and in vivo)	[220-222, 224, 225]
Based on Methyl-Amino Acids	Prevention of Aβ aggregation in vitro; fibril disassembling in vitro; cell toxicity inhibition; **phase I & II clinical trials**	[247-250]
Selected from Combinatorial Libraries:	Inhibit Aβ aggregation and toxicity; modulates Aβ oligomerization; in vitro and in vivo results (tg mice); reduce toxicity in neuroblastoma cells	[227, 228, 230, 251-253]

Peptide type	Results	Reference
Other peptides	-Aβ specific; inhibition of Aβ aggregation/fibril formation; reduction of Aβ cell toxicity; inhibited fibril formation (and/or reduced toxicity); staining of Aβ deposits (post mortem and ex vivo)	[226, 227, 229, 231, 232, 245]

Table 4.6 Antibodies for Aβ or Tau Targeting (or Passive immunotherapy).

Antibody	Epitope	Phase	Company/ reference
A) mAbs attached to NPs			
Therapy			
Anti-Aβ1-42 antibody	Not disclosed	preclinical	(Stab Vida) / [267, 268]
Anti-Aβ antibody	Fibrils	preclinical	Mayo Clinic/ [269]
Diagnosis			
anti-AβPP		preclinical	[270]
Anti-Aβ(1-42)		preclinical	[271, 272]
Aβ(1–40) antibody		preclinical	[271-274]
mAbs specific to ADDL		clinical	[275]
l anti-tau mAb		clinical	[276]
B) Potential Candidates attachment to NPs			
3D6(anti-Aβ1-6)10D5(anti-Aβ1-36)	N-terminal (anti-Aβ1-6)	preclinical	[277]
m266(anti-Aβ1-6)	N-terminal (anti-Aβ1-6)	preclinical	[260]
Bapineuzumab (AAB-001, ELN15727 humanized version of clone 3D6)	N-terminal (1–5)	III	Janssen-Pfizer-Elan

(Continued)

Table 4.6 (*Cont.*)

Antibody	Epitope	Phase	Company/ reference
ACC-001	N-terminus Aβ 1-6	II	Janssen
AFFITOPE / ADO2	N-terminus Aβ 1-6	II	Affiris AG
CAD 106	N-terminus Aβ 1-6	II	Novartis
12A11	N-terminal region		
WO2	N-terminus region		
Gantenerumab (RO4909832)	N- terminus and central regions	I	La Roche I
M266(anti-Aβ13-28)	Central region (anti-Aβ13-28)	preclinical	[278]
Solenazumab (LY2062430)	Central region	III	Eli Lilly III
4G8	Central region (Aβ17-28)		A.H. Tammer
(anti-Aβ1-28)	(anti-Aβ1-28)	preclinical	[279]
Ponezumab (humanized version of 2H6) (PF-04360365)	C- terminus (Aβ28-42)	II	Pfizer-Rinat/ [280]
1A10	C- terminus (Aβ28-42)		[281]
22C4	C-terminus (Aβ 28-42)		[282]
AAB-002	Oligomers and protofibrils	preclinical	
None	Oligomers and protofibrils	preclinical	Genenteche-ACImmune
A887755	Oligomers		[283]
mAb158	Aβ protofibrils		[284]
WO1	Aβ fibrils		[285]
BAN2401	Aβ protofibrils	I	Eisai
Crenezumab (MBAT5102A)	Not disclosed	II	Genentech

Antibody	Epitope	Phase	Company/reference
GSK933776A	Not disclosed	I	GlaxoSmith Kline
Gammaguard	IVIg mix	III	Baxter
Nanobodies	Aβ peptides	Preclinical	Boehringer-Ablynx

post-translational modifications, can also be toxic [148]. Furthermore, the suppression of tau protein, blocks Aβ-induced apoptosis and reduces memory deficit [149]. Due to the presence of Neurofibrillary tangles (NFTs) in AD brains, tau protein level is 8-fold higher than in control brains [150]. These data suggest that reduction of the overall tau levels may constitute a neuroprotective strategy to combat tauopathies [151]. Therefore, studying tau regulation at the transcriptional and translational levels is of high interest for understanding the physiological role of tau and its involvement in human pathologies.

Strategies for Targeting Tau Oligomers: Different strategies exist today to develop therapeutics directly or indirectly targeting tau oligomers. These strategies are categorized as: (1) Prevention of phosphorylation of tau: phosphorylation of tau is controlled by different kinases and phosphatases. Tau kinase inhibitors are used as AD treatments. As an example, protein phosphatase (PP)-2A may increase dephosphorylation of tau. (2) Prevention of tau misfolding: Activation of molecular chaperones might prevent the misfolding of tau; this would subsequently reduce the development of NFTs. Heat shock proteins have been shown to activate chaperones that prevent misfolding and even promote tau binding with microtubules. (3) Tau immunotherapy: there has been growing interest in immunotherapies targeted to tau, with the objective to reduce tau levels, as a method to treat AD. One of the possible immunotherapeutic approaches currently under consideration is the selective reduction of pathological forms of tau.

4.4.1.1 Ligands Available for Tau Targeting

The different types of ligands that have been found to exert tau-targeting capabilities are seen in Tables 4.4–4.6, as categorized in small molecules, peptides, and antibodies.

Small-molecule Ligands: Recent studies using cell models have demonstrated that certain **small-molecule** inhibitors are able to prevent tau protein aggregation and even dissolve the developed aggregates. Such small molecule ligands include anthraquinones, polyphenols, thiacarbocyanine dyes, N-phenylamines, thiazolyl-hydrazides, rhodanines (thioxothiazolid-inones), quinoxalines, aminothienopyridazines, phenylthiazolyl-hydra-zide, N-phenylamines, phenothiazines benzothiazoles and others [152].

The rhodanine core for tau aggregation inhibition has been investigated by synthesis of an appropriate small-molecule library. Rhodanines (R1 = S and R2 = S, Figure 4.4), thiohydantoin (R1 = S and R2 = N), thioxooxa-zolidine (R1 = S and R2 = O), oxazolidinedione (R1 = O and R2 = O), and hydantoin (R1 = O and R2 = N) were screened for activity on tau aggregation inhibition (IC50) and disaggregation of preformed tau aggregates (DC50). The following results were found: rhodanine (IC50/DC50 (mM); 0.8/0.1) > thiohydantoin (6.1/0.4) >> oxazolidinedione (3.5/2.2) = thioxooxazolidi-none (3.1/2.4) >> hydantoin (22.6/54.3). From this experiment besides the importance of thioxo group, the importance of R1 and R3 groups, the total length of the molecule and the importance of aromatic side-chains were also proved [153]. After optimization, two interesting compounds were obtained; one with IC50 = 170nM; DC50 = 130nM (K19Tau) and a second with IC50 = 82 nM and DC50 = 100nM (K19 Tau) (Figure 4.5).

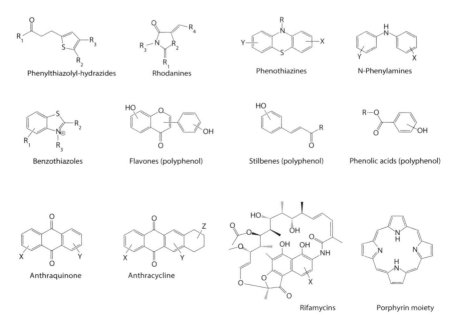

Figure 4.4 Structure of some examples of small molecule ligands against Tau proteins.

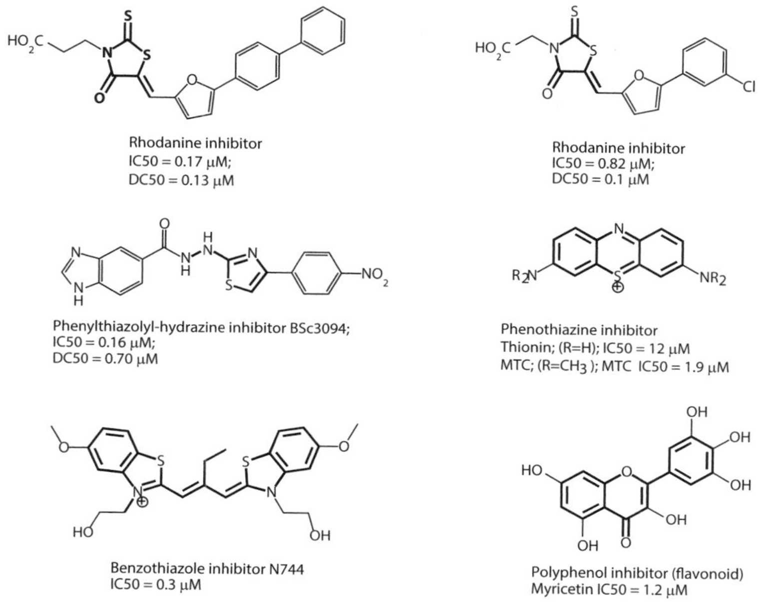

Rhodanine inhibitor
IC50 = 0.17 µM;
DC50 = 0.13 µM

Rhodanine inhibitor
IC50 = 0.82 µM;
DC50 = 0.1 µM

Phenylthiazolyl-hydrazine inhibitor BSc3094;
IC50 = 0.16 µM;
DC50 = 0.70 µM

Phenothiazine inhibitor
Thionin; (R=H); IC50 = 12 µM
MTC; (R=CH3); MTC IC50 = 1.9 µM

Benzothiazole inhibitor N744
IC50 = 0.3 µM

Polyphenol inhibitor (flavonoid)
Myricetin IC50 = 1.2 µM

Figure 4.5 Examples of derivatives with high inhibition potency of Tau (the characteristic category of each derivative is marked in bold).

In case of *Phenylthiazolyl-hydrazine inhibitors*[154], (Figure 4.5), a novel derivative BSc3094 was reported with IC50 = 160 nM; DC50 = 70 nM. STD-NMR experiments showed that the binding of this compound with Tau (construct K19) is specific (KD = 62µM). *N-phenylamines* displayed lower potencies in vitro and in vivo, which have been attributed to the non-planar conformation of the molecule [155]. Significant features distinguish phenothiazines and benzothiazoles from the N-phenylamines, mainly their cationic charge, planarity and a high level of aromatic conjugation. In case of phenothiazine derivatives it was registered that quinoxaline has an IC50 = 2.4µM; Thionin IC50 = 12µM; MTC IC50 = 1.9µM; Azure A/B/C = 2–7µM (Figure 4.5). On the other hand, perphenazine showed IC50>200µM. This high IC50 value for perphenazine was attributed to the fact that this compound was non-planar, non-conjugated [156]. *Benzothiazoles* (like phenothiazines) have the characteristic positive charge and extensive aromatic conjugation. A representative example of benzothiazole based inhibitor, N744, with high inhibitory potency, has been reported (IC50 = 0.3µM)[157].

IC50 values of a wide range of derivatives are presented in Table 4.4. It is worth mentioning that, despite extensive efforts, the lack of structural similarity of selected tau inhibitors makes inferring conclusions from structure/activity relationships, and hence rational design of new tau inhibitors, a rather difficult task. For example in case of Anthracyclines, Daunorubicin hydrochloride presents a low inhibition of tau (IC50 >200μM). On the other hand, Emodin, Daunorubicin, Adriamycin, and PHF016 were able to inhibit the aggregation of the K19 tau construct and also induced the disaggregation of preformed aggregates, with IC50 values between 1.1 μM and 2.4 μM. In a second example, benzo-thiazole tau inhibitors: 2-(4-Aminophenyl)-6-Methylbenzothiazole; Basic blue 41; 2-[4-(Dimethylamino)phenyl]-6-methylbenzothiazole and 3,3'-Dipropyl thiodicarbocyanine iodide, present low tau inhibition ability (IC50>200μM), while N744, displays high inhibitory potency (IC50=0.3 μM) [157].

Other derivatives with excellent inhibitory potencies are: Congo Red (IC50 = 2.2μM), Polyphenols: Baicalein (IC50=2.7μM); (-)-gallocatechin-gallate (IC50=1.0μM); Gossypetin (IC50=2.0μM); Myricetin (IC50=1.2μM) (Figure 4.5); Pentahydroxybenzophenone (IC50=2.4μM), Porphyrins: Ferric dehydroporphyrin IX (IC50=1.4μM). These derivatives along with the corresponding bibliography are presented in Table 4.4.

Peptide Ligands: Although the design and preparation of peptides that inhibit tau aggregation is underestimated in comparison to the preparation of small molecules that inhibit Aβ aggregation, there have been some efforts in this area [158]; Researchers have chosen, as a target for inhibitor design, the hexapeptide VQIVYK (residues 306–311) of the tau protein, which is known to form intracellular amyloid fibrils in AD [159]. This segment has been shown to aggregate in solution through β-sheet interactions to form straight and twisted filaments similar to those formed by tau protein in Alzheimer's neurofibrillary tangles [160]. In order to design a D-amino-acid hexapeptide sequence that interacts favorably with the VQIVYK, and prevents further addition of tau molecules to the fibril, computer modeling resulted to four D-amino acid peptides: D-TLKIVW, D-TWKLVL, D-DYYFEF and D-YVIIER, in which the prefix signifies that all a-carbon atoms are in the D configuration (Figure 4.6) [158]. More analysis on the molecular interactions can be found in the original manuscript. In another approach, the preparation of macrocyclic β-sheet peptides that inhibit the aggregation of Tau-protein-derived peptide Ac-VQIVYK-NH$_2$ (AcPHF6) was attempted. It was found that macrocycles containing the pentapeptide VQIVY in the "upper" strand, delay and suppress the onset of aggregation of the AcPHF6 peptide [161]. Inhibition was particularly pronounced for the peptide macrocycles presented in Figure 4.6.

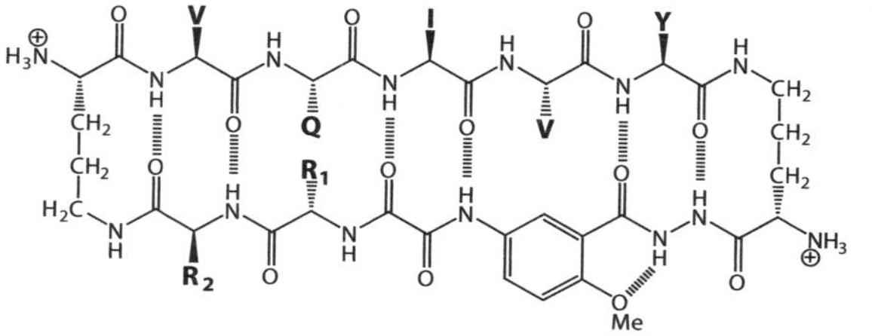

VQIVY-R$_1$R$_2$ macrocycles

1. VQIVY-KL macrocycles

2. VQIVY-KV macrocycles

3. VQIVY-RL macrocycles

Figure 4.6 Macrocyclic β-sheet peptides that inhibit the aggregation of Tau-protein.

Lately, Cheng *et al.* described a family of robust β-sheet macrocycles with inhibitory properties against amyloid-β peptide and tau protein through an interesting mechanism [162].

Antibody Ligands: After searching in the relevant literature, it was concluded that the development of antibody ligands against tau pathologies has not been extended up to now. In one study, based on in vivo investigations in a mouse model, immunotherapies targeting tau appear to provide a viable potential for treatment of AD and other tauopathies. Although preliminary evidence on the safety and efficacy of this approach appears promising, several questions still remain. Key information on the ideal adjuvant for active immunization, the mechanism of passive immunization antibody entry, and epitope specificity of antibody-mediated tau clearance will need to be provided in the future, in order to have a better feeling about the potential of such therapeutic approaches [163].

4.4.2 Amyloid Plaque or Aβ- species Targeted Strategies

4.4.2.1 Aβ Peptide Formation

As mentioned above, amyloid plaques are deposits of fibrous amyloid β (Aβ) composed most usually of 40–42 amino acid long peptides. These peptides are proteolytic fragments obtained by the action of β and γ secretases (membrane enzymes) on amyloid-precursor protein (APP).

APP can undergo a variety of proteolytic cleavages carried out by enzymes or enzyme complexes with α-, β- and γ-secretase activity leading to the formation of large soluble secreted fragments and membrane-associated C-terminal fragments (CTF). The enzymes with α-secretase activity belong to the ADAM family (a disintegrin and metalloproteinase enzyme family) [164], β-secretase activity has been identified in the β-site of APP-cleaving enzyme 1 (BACE1, a type I integral membrane protein belonging to the pepsin family of aspartyl proteases) and γ-secretase is an enzyme complex composed of presenilin 1 or 2, nicastrin, anterior phar-ynx defective and presenilin enhancer 2. APP can be processed through the prevalent non amyloidogenic pathway in which the α-secretase cleaves the APP 83 amino acids from the C-terminus, producing a membrane-retained CTF of 83 residues and a large N-terminal soluble ectodomain fragment (sAPPα) released into the extracellular space. The CT fragment is subsequently cleaved by the γ-secretase with the production of a short fragment called p3. In the non-amyloidogenic pathway of APP process-ing, **α-secretase** cleavage occurs in the Aβ sequence and thus prevents formation of Aβ peptides. Oppositely, in the amyloidogenic pathway, the first cleavage of APP is carried out by β-**secretase** at the 99th amino acids from the C-terminus producing a soluble ectodomain fragment (sAPPβ) released into the extracellular space and an alternative CTF of 99 amino acids (C99) retained in the membrane. The C99 fragment begins at residue 1 of the Aβ region. The following cleavage of C99 by γ-**secretase** leads to the release of the Aβ peptide. Most of the Aβ produced is the variant of 40 residues (Aβ40), even though a longer form of 42 residues (Aβ42) can also be produced. This last variant is more hydrophobic than the shorter one and is the basic constituent of Aβ plaques [165]. It has also been shown that APP is a substrate for caspase cleavage. However, the effective role of cas-pases in Aβ accumulation is not clear. In normal conditions, the levels of all peptides are a direct consequence of the balance between their produc-tion and catabolism; several molecular and cellular studies in transgenic mouse models and in AD patients have demonstrated that the levels of Aβ-degrading enzymes decrease during disease progression [166].

4.4.2.2   *Aβ Transport Across the BBB-Strategies for Therapy*

Aβ is transported bidirectionally across the BBB; that is, both in the brain-to-blood (efflux) and the blood-to-brain (influx) directions. Separate transport-ers are responsible with blood-to-brain transport being primarily mediated by receptor for advanced glycation end products (RAGE) and the brain-to-blood transport being primarily meditated by low-density lipoprotein receptor-related proteins (LRP)[167] and P-glycoprotein transporter (P-gp) (which seems to have an effect on the brain-to-blood transport of Aβ) [168] while other related transporters have also been proposed [169]. Increased blood-to-brain transport by RAGE and decreased efflux by LRP-1 and P-gp, act synergistically to enhance the uptake or retention of Aβ in AD [167]. Convincing evidence exists that efflux is an important determinant of Aβ accumulation in the brain and cognitive impairment. Indeed, knockdown of LRP-1 with antisense results in decreased Aβ efflux, increased brain levels of Aβ1-42, and cognitive impairment [170]. Mutations in LRP-1 result in decreased Aβ efflux [167] and P-gp knockout mice have Aβ accumulations in their brain and cognitive deficits. It is unclear how P-gp interacts with LRP-1 in the efflux of Aβ, but several theories have been proposed. However, Aβ has the ability to promote memory at lower concentrations [171] than those at which it impairs it. It may be postulated that efflux/influx transporters act in tandem to maintain the CNS levels of Aβ at their most optimal levels; the transporters being differentially regulated for this purpose. For example, Aβ influx is altered by its binding to apolipoproteins [172]. Transport in both directions is likely influenced by the primary structure of Aβ as exemplified by the fact that Aβ mutations are transported at lower amounts. Finally, inflammation induces an increased influx and decreased efflux of Aβ across the BBB [170], changes which may in part be mediated by inhibition of P-gp [171]. Several therapeutic options are directed towards alterations of Aβ transport; some of which have been tested and found to have beneficial effects. Knocking down APP expression results in recovery of Aβ efflux, suggesting that antisense directed against APP could be used therapeutically to correct efflux [173]. In an AD mouse model with impaired Aβ efflux, treatment with APP-directed antisense reduces oxidative stress and treatment with antioxidants leads to improved cognition [174]. *Small molecule effects on Aβ transport*: Treatment with indomethacin, a nonsteroidal anti-inflammatory drug (NSAID) restores the inflammation-induced inhibition of efflux, but not the enhancement of influx [170]. Indomethacin, but not necessarily other NSAIDs, has been associated epidemiologically with protection against AD. Vitamin D also enhances Aβ efflux [175]; the calcium channel blocker nilvadipine increases Aβ efflux, reduces Aβ brain levels and improves cognition in an AD mouse model, while calcium channel blockers as amlodipine and nifedipine do not affect efflux. Thus, evidence exists that treatment with APP antisense, antioxidants, vitamin D, NSAIDs, and calcium channel blockers can restore the Aβ clearance from brain to normal levels.

4.4.2.3   Aβ Peptide Species

The presence of Aβ in various organs and body fluids, and the fact that the body has evolved sophisticated mechanisms for its metabolism, back-up the theory that Aβ has a physiological role [176], and that with old age or more specifically with AD onset, Aβ either losses its physiological function or gains a pathological one [177]. Although this transformation of Aβ from a physiological to a pathological agent has not been elucidated, it is a fact that Aβ exerts neurotoxic and synaptotoxic affects both in vitro and in vivo [178].

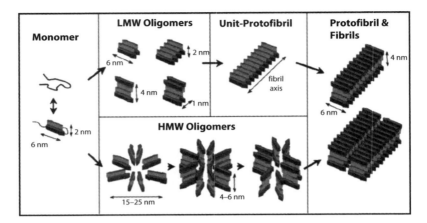

Figure 4.7 A model showing the sequence of assembly and possible structures of Ab42 monomers, low molecular weight (LMM) oligomers, high molecular weight (HMM) oligomers, unit-protofibrils, and protofibrils/fibrils. From Ref [183] after permission.

Researchers have recently focused on the effect of Aβ structure on its toxicity. It was found that soluble Aβ monomers assume a random coil or α-helix conformation; however, in AD they undergo a structural change into pleated β-sheets [179]. This induces the peptide to form low molecular weight oligomers, higher molecular weight complexes (protofibrils and amyloid-β derived diffusible ligands or ADDLs), mature fibrils and amyloid plaques (APs) in the neuropil and the vasculature [180] (Figure 4.7). In vitro studies have shown that amyloidogenesis and fibrillogenesis can be affected not only by the type of Aβ produced and its conformation, but also by factors such as time, concentration, temperature, pH and metal ion concentration [181]. For many years it was believed that the toxic effects of Aβ were mainly the results of mature Aβ fibrils; however, later studies suggest that low molecular weight, soluble, oligomeric forms of Aβ1–42 rather than Aβ1–40 are more neurotoxic than the mature Aβ fibrils [182]. Indeed, the severity of AD correlates better with cerebral concentrations of soluble Aβ rather than insoluble Aβ load.

4.4.2.4 Ligands Available to Target Aβ

Ligands to target Aβ are again categorized in three groups: small molecules, peptides and antibodies.

Small-molecule Ligands: Since oligomeric forms of Aβ are considered as the most toxic species [184], efforts have been focused on the synthesis of **small molecules** that may block early stages of amyloid self-assembly [185]. Several molecules were tested as Aβ ligands (peptides, proteins, small molecules). Small molecules with aromatic moieties insert within β-sheets created grooves in both, the soluble oligomeric forms as well as in the fibrils [186]. Several types of molecules were found to be active inhibitors of Aβ aggregation, including Benzothiazole derivatives, Congo Red derivatives, Isoflavones, Polyene macrolides, Polyphenols, Porphyrins, Rifamycins, Steroids, Terpenoids, Tetracycline and its derivatives and other (Table 4.4). Different types of molecules were found to be active in the inhibition of Aβ aggregation, with a wide range of activity, from < 100 nM to > 100 μM, as analytically determined. Some specific molecules are presented below:

Congo Red is a hydrophilic symmetrical sulfonated azodye that binds specifically to amyloid fibrils. Thioflavin T (ThT) and Thioflavin S (ThS) are also known to stain amyloid deposits. Both Congo Red and ThT have also been shown to inhibit fibril formation at higher concentrations [187]. Many small Congo Red and ThT derivatives were also reported. Especially those which preserve their planarity and aromaticity were found to be potent inhibitors of Aβ species.

Nordihydroguaiaretic acid (NDGA), a small molecule with two ortho-dihydroxyphenyl rings symmetrically bound by a short carbohydrate chain, was found to be an effective inhibitor of Aβ aggregation [188]. The activity of NDGA was ascribed to its compact and symmetric structure that might be suitable to specifically bind soluble Aβ, thus inhibiting its aggregation. Moreover, this molecule might specifically bind to Aβ fibrils and lead to destabilization of their β-sheet conformation.

Dopamine exhibited potent anti-amyloidogenic and fibril-destabilizing effects and the effective concentration able to inhibit the formation or extension of Aβ fibrils was in the order of 0.01–0.10 μM. When dopamine and L–DOPA were used in a solution of Aβ (1–40), it was found that these molecules were able to inhibit the formation of Aβ fibrils in a range from 10 to 100 μM. In a comparative study, molecules able to inhibit the formation of Aβ fibrils and also lead to their destabilization were ranked in the order: Dopamine > Selegiline ~ nordihydroguaiaretic acid (NDGA) > L-Dopa ~ Pergolide > Bromocriptine ~ Rifampicin (RIF) [189].

Other molecules that present very high activity against amyloid fibrils and plaque formation are *polyphenols* [190]. These compounds were found to be potent inhibitors of Aβ fibril formation [Aβ (1–40) and Aβ (1–42)] and also destabilized preformed Aβ1–40 and Aβ1–42 fibrils (dose–dependent effect). Within this class of compounds, *tannic acid* showed the best activity. The activity of tannic acid as well as the anti–amyloidogenic and fibril-destabilizing effects of *myricetin, rifampicin, tetracycline*, and *NDGA* were tested. *Tannic acid* exhibited potent anti–amyloidogenic and fibril–destabilizing effects and the EC50 for the formation or extension of Aβ fibrils or for destabilization of Aβ fibrils were in the order of 0.012–0.065 μM for both Aβ1–40 and Aβ1–42. Anti–amyloidogenic and fibril-destabilizing activity of *NDGA* and *wine-derived polyphenols* can be classified in the following order: Tannic acid > NDGA = Myricetin = morin = quercetin > kaempferol > (+)–catechin = (–)–epicatechin > tetracycline. The activity of these molecules is mainly attributed to their propensity to bind to Aβ [191].

Curcumin (diferulomethane) is another naturally occurring phytochemical with antioxidant and anti-inflammatory activity and a favorable toxicity profile [192] that protects the brain from lipid peroxidation. Curcumin is known to inhibit amyloid Aβ1–40 and Aβ 1–42 oligomer formation and cell toxicity at micromolar concentrations in vitro [193, 194] and binds to senile plaques, reducing amyloid levels in vivo [186]. In vitro, curcumin inhibited Aβ aggregation (IC50 = 0.8 μM) and disaggregated fibrillar Aβ40 (EC50 = 1 μM) while lower IC50 values were reported by others [193, 195, 196]. Interestingly, *curcumin, Congo Red,* and *chrysamine G* (another organic dye which binds with high affinity to Aβ) all share a similar chemical scaffold; they contain two substituted aromatic groups separated by a rigid, planar backbone (Figure 4.8). Several groups

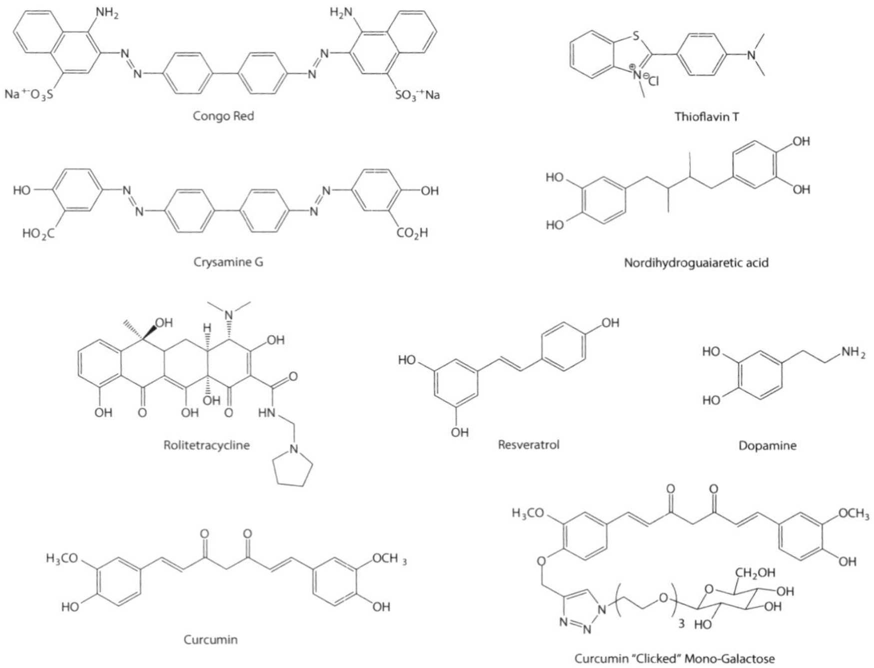

Figure 4.8 Structure of some characteristic small molecule Aβ aggregation inhibitors.

have reported that other curcumin-like ligands are also inhibitors of Aβ aggregation [156, 186, 187, and 197]. Based on curcumin, the most important factors that influence the structure–activity relationship of amyloid β–aggregation inhibitors were suggested to be: (a) the presence of the two aromatic end groups; (b) the substitution pattern of these aromatics; and (c) the length and flexibility of the linker region [198]. This opinion is under negotiation, since other compounds (like *Resveratrol*), which did not follow these "rules" were found active Aβ inhibitors, suggesting a possible alternative mechanism for inhibition or a possible different binding site on the amyloid peptide. This was also emphasized by Mason *et al.* [199], who highlighted the lack of structural similarity of selected Aβ inhibitors presented therein, suggesting that they bind to different amyloid sites, in contrast to most drugs, which bind to a single active site. Such findings result in inferring conclusions in terms of structure/activity relationships, and make rational drug design difficult. In accordance to this argument, the synthesis of a water/plasma soluble, noncytotoxic, "clicked" sugar-derivative of curcumin with amplified bioefficacy in modulating amyloid-β and tau peptide aggregation at concentrations as low as 8 nM and 0.1 nM, respectively, was recently reported. This derivative (curcumin-"clicked" mono-galactose) is a monofunctional curcumin derivative (at the one of the two phenol groups of curcumin) (Figure 4.8). In comparison to curcumin, this conveniently synthesized Alzheimer's drug candidate is a more powerful antioxidant.

It is of great importance that some derivatives were found to inhibit: (a) Aβ oligomerization but not fibrillization [o-Vanilin, Tetradecyl trimethyl ammonium bromide Class I, R(-) Norapomorphine hydrobromide, 1,2-Napthoquinone, Meclocycline sulfosalicylate, Juglone, Indomethacin, Hexadecyl trimethyl ammonium bromide Class I, Apomorphine, Rolitetracycline, NDGA, Myricetin, Congo Red, Thioflavin T, Basic blue 41]; (b) Both oligomerization and fibrillization [Tetradecyl trimethyl ammonium bromide Class II, Neocuproine, Hexadecyl trimethyl ammonium bromide Class II, Ferulic acid, Rodamine B, Rifamycin SV, Hemin, Hematin, 2,2'-Dihydroxybenzophenone, Phenol red]; and (c) Fibrillization but not oligomerization [Apigenin, Orange G, Direct Red 80, Chicago Sky Blue 6B].

Peptide Ligands: Several peptide inhibitors of Aβ aggregation have been investigated during the last decade for their applicability as new therapeutic compounds. It must be emphasized that only very few, and more specifically: iAβ5 [220–222], Aβ12–28P [223], LPYFDa [224, 225], trp-Aib[226], D-4F [227] and D3 [228–230] were proven to be effective in rodent mouse models and only one compound, **PPI-1019**, is being tested in clinical trials.

The use of amyloid ligands as contrast agents (or for contrast agent targeting to plaques) can be a promising approach for the detection and characterization of amyloid plaques in the brain. Some ligands have been already studied clinically. Small Aβ binding peptides could be coupled to radionucleides, other markers, or nanoparticles for imaging of amyloid plaques in living AD patients. In 2003, Kang *et al.* identified two 20-amino acid peptides that can be specifically bound to the amyloid form of Aβ1–40, but not to monomeric Aβ. One of the peptides (DWGKGGRWRLWPGASGKTEA) bound Aβ1–40 amyloid with a Kd of 60 nM and stained as well as the chemically synthesized version. Both versions specifically stained amyloid plaques in brain tissue slices of AD patients. It is known that D-enatiomeric peptides are highly resistant to proteases, which can dramatically increase serum and saliva half-life and can be absorbed systemically after oral administration. Also D-peptide immunogenicity is reported to be reduced in comparison to L-peptides [230, 231]. For these reasons peptides containing D-amino acids are usually selected. The most representative peptide in the selection procedure, D1, was demonstrated to bind Aβ with an affinity in the submicromolar range. Employing surface plasmon resonance, binding to Aβ oligomers and fibrils, but not to monomers could be demonstrated. D1 also stained amyloid plaques in brain tissue sections derived post-mortem from AD patients, whereas other, non-Aβ amyloidogenic deposits, were not stained [231]. Its in vivo binding to Aβ1–42 is specific as it stains all dense deposits in the brain but does not stain diffuse plaques, which consist mainly of Aβ1–40 [229], suggesting that D1 might be suitable for further development into a molecular probe for monitoring Aβ1–42 plaque-load in the living brain. Recently, Larbanoix *et al.* designed another peptide based on the Aβ1–42 amino acid sequence. Two of 26 selected clones, which had the highest binding affinities to Aβ1–42, were translated to synthetic peptides with biotin label (Pep1: LIAIMA and Pep2: IFALMG, corresponding Aβ fragment IIGLMV31–36) and presented lower Kd values (still in the micromolar range) compared to the peptides described initially. The specific interaction of both peptides with amyloid plaques in human brain tissue was proved by immune-histochemistry [232].

Tjernberget *et al.* tried to identify binding sequences within Aβ in order to synthesize Aβ-derived peptide ligands. The short Aβ fragment KLVFF (Aβ16–20) was identified to bind full length Aβ and to prevent the fibrillization. Experiments revealed that amino acids Lys16, Leu17 and Phe20 were critical for Aβ interference [233]. The full-length Aβ and the KLVFF peptide lead to the formation of a typical antiparallel β-sheet structure stabilized by Lys16, Leu17 and Phe20 [234]. The effects of KLVFF-containing or synthetic peptides was further proven by other studies [235]. Additionally, it was proposed that conjugates, containing copies of the KLVFF sequence linked to dendrimers or to branched poly(ethylene glycol) moieties, possess superior affinity and efficiency [236, 237]. In 2008, Austen et al. accomplished to add water-soluble amino acids residues to KLVFF generating the peptides OR1(RGKLVFFGR) and OR2 (RGKLVFFGR-amide). Both inhibited Aβ fibrillogenesis [238] but unlike OR2, the retro-inverso D-enantiomeric version (RI-OR2) was highly resistant to proteolysis and stable in human plasma and brain extracts [239]. Additionally OR1 and OR2 increase the inhibitory effects of the peptides [240]. In 2004, Fólop et al. developed an Aβ aggregation inhibitor, based on the Aβ31–34 sequence IIGL, which also plays a fundamental role in Aβ aggregation and cytotoxicity [241]. The strategy was again to link water-soluble amino acid residue to the original sequence RIIGL. In contrast to propionyl-IIGL, another derivative of the same sequence, PIIGL, did not self-aggregate and was non-toxic to cells in culture, while it inhibited the formation of Aβ fibrils and reduced Aβ cytotoxicity. In 2008, Fradinger et al. prepared a series of AβC-terminal fragments (Aβx-42; x = 28–39). These peptides inhibit Aβ induced toxicity by stabilizing Aβ in non-toxic oligomers [242]. Sadowski et al. investigated whether blocking the interaction between APO E4 and Aβ can have therapeutic effects. It was found that Aβ12–28P blocked Aβ-APO E4 interaction and reduced Aβ fibrillogenesis and toxicity in vitro. The peptide was BBB-permeable and inhibited Aβ deposition in AD transgenic mice [223].

Ghanta et al. (1996) designed Aβ binding peptides based on the KLVFF-binding sequence consisting of a chain of charged amino acids (KKKKKK or RRRRRR). Interestingly, some of the hybrid peptides accelerated Aβ aggregation, but reduced Aβ toxicity. [243]. In 2003, Gordon et al. replaced the amide bonds of Aβ16–20 with ester bonds in an alternating fashion. This Aβ16–20e peptide inhibited Aβ aggregation and disassembled existing fibrils. Aβ16–20e, could, however, build dimers. [244]. Other β-sheet breakers based on the KLVFF sequence are the conformationally constrained peptides AMY-1 (Figure 4.9) and AMY-2. Both contain alpha, alpha-disubstituted aminoacids and prevent Aβ fibril growth [245]. Rangachari et al. investigated Ala containing peptides P1 (KLVF-_A-I_A) and P2 (KF-_A-_A-_A-F). Both peptides inhibited Aβ aggregation [246].

Omologous peptide to the central hydrophobic region of Aβ (amino acids 17–21: LVFFA), containing the amino acid proline to prevent the formation of β-sheet structure and to inhibit Aβ amyloid formation. iA_1 (RDLPFFPVPID) did not aggregate itself, but inhibited amyloid formation and disassembled existing fibrils in vitro. Co-injection of iAβ5in 20% molar excess leads to significantly reduced plaque deposition. In

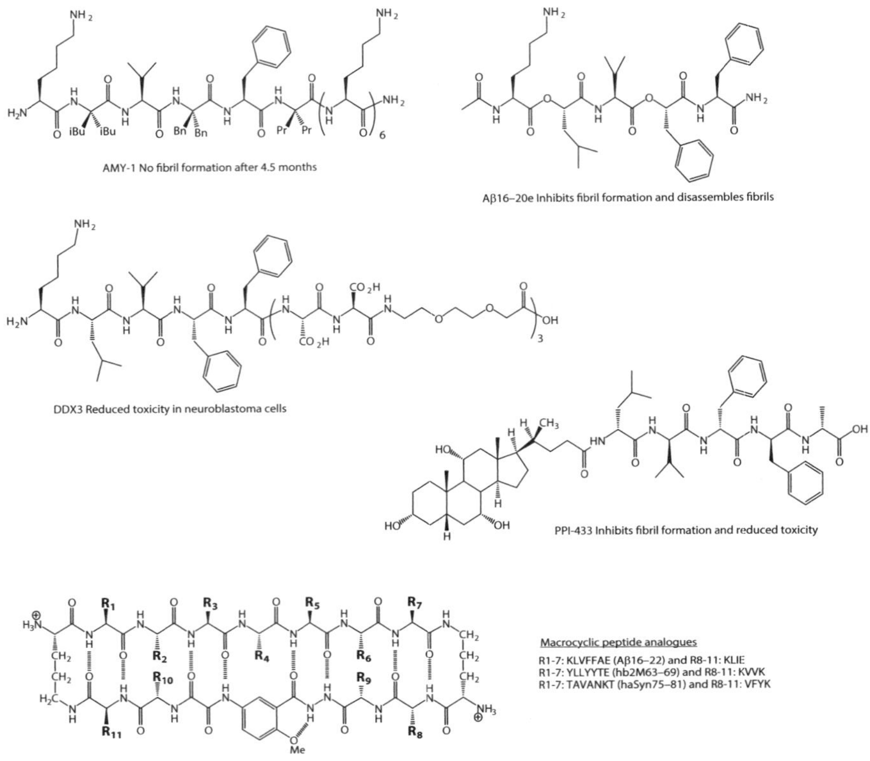

AMY-1 No fibril formation after 4.5 months

Aβ16–20e Inhibits fibril formation and disassembles fibrils

DDX3 Reduced toxicity in neuroblastoma cells

PPI-433 Inhibits fibril formation and reduced toxicity

Macrocyclic peptide analogues
R1-7: KLVFFAE (Aβ16–22) and R8-11: KLIE
R1-7: YLLYYTE (hb2M63–69) and R8-11: KVVK
R1-7: TAVANKT (haSyn75–81) and R8-11: VFYK

Figure 4.9 Structures of abeta binding peptide conjugates; peptides containing unnatural aminoacids and other complicated peptide derivatives.

addition, it was shown that iAβ5 induced disassembly of fibrils and also reduced Aβ-induced histopathological changes [220]. For chronic intraperitoneal administration of the peptide to the rat model, iAβ5 was end protected by N-terminal acetylation and C-terminal amidylation [221], as the non-protected peptide was unstable in blood and proteolytically degraded very rapidly [222]. Two different AD transgenic mouse models were used to demonstrate that intraperitoneal injected, end-protected iAβ5 reduced amyloid plaque formation, neuronal cell death and brain inflammatory processes. The mechanism of peptide action remained unclear, but administration of large doses of the peptide did not lead to antibody production in the treatment and evaluation period. Using an LPFFD based peptide (amino acid sequence LPYFD) it was found that neurite degeneration and tau aggregation were significantly decreased in vitro. The pentapeptide LPYFD-amide protected neurons against Aβ toxicity in vitro and in vivo. LPYFD-amide crossed the BBB in rats at least to a certain extent, and protected against synaptotoxic effects of Aβ up to 3.5 h after i.p. injection [224, 225]. Hughes *et al.* synthesized six N-methylated derivatives to prove that those could prevent Aβ wildtype aggregation and cytotoxicity. The localization of the N-methyl group was very critical as some of the other peptides did not prevent Aβ aggregation, but altered fibril morphology [247]. Gordon *et al.* described the synthesis and biochemical characterization of peptides based on the KLVFF sequence, containing N-methyl amino acids. One of the compounds, termed Aβ16–22m (H_2N-K(Me-L) V(Me-F)F(Me-A)E-$CONH_2$) was shown to be highly soluble in aqueous media and monomeric in buffer solution. It inhibited Aβ fibrillization and disassembled preformed Aβ fibrils, in vitro. Protease resistance of the methylated peptide was increased in comparison to the un-methylated Aβ16–22 peptides. Aβ16–20m was effective to inhibit Aβ polymerization and to disassemble preformed fibrils and it was highly water soluble despite its content in hydrophobic amino acids. The peptide passed model phospholipid bilayers and cell membranes, suggesting promising pharmacological properties [248]. The peptide "inL" was synthesized, based on KLVFF Aβ recognition element containing an additional Lys in the N-terminus, in order to increase solubility, and an N-methyl-20F in order to block Aβ aggregation. The peptide inhibited Aβ toxicity in cell culture very efficiently. In 2009, the corresponding D-peptide "inD," as well as the retro-inverso peptide "inrD," was investigated and shown to be a more effective inhibitor of Aβ aggregation than the other two peptide versions [249]. The peptidic inhibitor PPI-1019, which is derived from the D-enantiomeric Cholyl-LVFFA-NH_2, increased Aβ1–40 levels in the CSF, as a sign which is a sign of Aβ clearance. Optimized N-methylated derivatives based on the KLVFF sequence by peptide length, methylation sites, end-blocking, side chain identity and chirality were synthesized and it was found that only one methylated amino acid was essential [250]. The D-enantiomeric β-sheet breaker H_2N-D-Trp-Aib-OH, which combines an indole and an aminobutyric acid (Aib), reduced the amount of plaques in the brains of AD transgenic mice [226]. Orner *et al.* identified peptides that bind to monomeric or fibrillar Aβ, in a way that was correlated with the affinity of the peptides to the N-terminal part of Aβ [251]. Three peptides, named amyloid neutralizing agents (ANA) 1 to 3, were shown to significantly reduce Aβ SOD-like activity in cell culture. A 15-mer peptide additionally reduced Aβ toxicity in cell culture and seemed to be comparably potent as the known Aβ metal-mediated redox activity inhibitor Clioquinol [252]. Baine *et al.* selected three peptides (1A, 1B, and 2) that inhibit Aβ aggregation and one of them disaggregated preformed Aβ fibrils [253]. Recently, dominant peptide sequence RPRTRLHTHRNR was obtained, referred to as D3. D3 modulated Aβ aggregation and inhibited

Aβ toxicity in cell culture. [228, 230]. A D3 hybrid compound, JM169, demonstrated that the hybrid compound was more efficient in vitro than the sum of its components and had novel properties. In 2009, Paula-Lima *et al.* identified heptapeptides with the amino acid sequence GNLLTLD (designated GN peptide) that were found to be homologous to the N-terminal domain of mammalian apolipoprotein A-I. It was shown that purified human apoA-I and Aβ formed complexes. The interaction of apo A-I also rendered the morphology of amyloid aggregates [254]. Another group evaluated the apo A-Imimetic peptide D-4F, synthesized from D-amino acids and co-administered with pravastatin, as a treatment for AD transgenic mice [255]. D-amino acid containing peptides were efficient inhibitors of Aβ1–42 aggregation. D-KLVFF inhibited fibril formation, while a library of D-amino acid containing peptides (SEN 301–307) was reported, among which, SEN 303 and 304, significantly reduced Aβ toxicity [250]. Other interesting designed peptide inhibitors of Aβ1–42 aggregation are C-terminal modified peptides.

Recently Cheng *et al.* reported the preparation of macrocyclic peptide sequences, which efficiently inhibited amyloid aggregation (Figure 4.9) [256]. Peptide analogues containing R1–7: KLVFFAE (Aβ16–22) and R8–11: KLIE; R1–7: YLLYYTE (hb2M63–69) and R8–11: KVVK; R1–7: TAVANKT (haSyn75–81) and R8–11: VFYK have shown the best inhibitory activity.

Antibody Ligands: Antibodies against Aβ (Table 4.6) can be classified into four different types, according to the location of the epitopes that they recognize. Antibodies targeting the N-terminal region of Aβ (amino acids 1–16) can recognize all Aβ structural forms (monomers, oligomers, protofibrills and fibrils) as well as the APP, suggesting that this epitope is always exposed during Aβ fibrillogenesis [259]. The prototypical anti-Aβ N-terminus antibody is Bapineuzumab (humanized version of the clone 3D6), which is currently in phase III human trials. Other antibodies targeting N-terminal region are also shown in Table 4.6. Antibodies against the central region can only bind to Aβ monomers so act via the peripheral sink mechanism by sequestrating Aβ monomers into the blood stream and thus disrupting the Aβ equilibrium between the brain and periphery [260]. The best characterized antibody against the central region is Solenazumab, from Eli Lilly (humanized version of the clone m266). The mechanistic action of antibodies targeting the C-terminal region of Aβ (amino acids 33–42) is not yet fully understood. These antibodies may facilitate Aβ clearance by entering the CNS as well as acting in the periphery [261]. Wilcock *et al.* have a deglycosylated version of an anti-C-terminal antibody (clone 2H6). This humanized version of 2H6, Ponezumab, is now in phase II clinical trials initiated by Pfizer but efficacy data are not yet available [262]. Other antibodies targeting C-terminal region are also shown in table 4.6. The last category of antibodies against Aβ, target conformational epitopes (oligomers and protofibrils), which are the major culprits in AD pathogenesis [263, 264]. Abbott has succeeded in producing the most promising antibody yet, the A-887755 [265]. This antibody is specific for Aβ oligomers and improves cognition in an AD Tg mouse model. F(ab0)2 fragments of mAb 3D6 and mAb IIA2 were administered intraperitoneally or intracranially to PDAPP Tg mice [266].

4.4.3   Is Passing the BBB Always Needed?—Sink Theory

According to the sink theory [286], there is a possibility that molecules with high binding affinities for Aβ peptides (in the form present in circulation) may be able to extract the peptide from the blood and reverse the equilibrium that exists between blood and brain. This would potentially stop, or at least slow down, further formation of plaques in the brain (by reducing the levels of Aβ peptides in the brain). Some more optimist scientists believe that this method could also result in decrease of plaques and/or disaggregation of already formed Aβ aggregates, however at the time being no definite proof about the validity of the sink theory has been provided, and the matter is still controversial. Therapeutic strategies for Aβ clearance from the brain to blood across the BBB have been increasingly developed. The "peripheral sink" approaches are now challenged by anti-Aβ antibodies, the agents with high affinity to Aβ, and the modification of molecules that influence Aβ transport across the BBB [267, 287].

4.4.4   Functionalization of Ligands to NPs

NPs are used in diagnosis and/or therapy because they can deliver imaging agents and/or drugs in a targeted way (combining passive and active delivery mechanisms). Targeted nanoparticles are undergoing clinical development as diagnostics, molecular imaging probes, and therapeutic delivery vehicles [288]. Targeting is typically achieved through the surface display of multiple high affinity ligands, such as antibodies, peptides, or natural products. Multivalent interactions between the nanoparticles and their targets can increase the affinity of target binding [289]. The functionalization of the different types of NPs by the desired targeting ligands is achieved by simple and efficient conjugation chemistry methods, some of which are schematically presented in Figure 4.10. Other methods for functionalization of NPs are described elsewhere [290, 291].

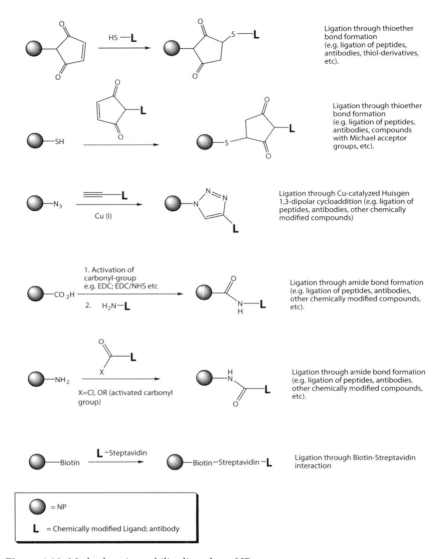

Figure 4.10 Methods to immobilize ligands on NPs.

Mourtas *et al.* [201] used two different techniques for decoration of liposomes with two different derivatives of curcumin (planar and non-planar) [(a) by incorporation of a synthesized lipid-curcumin derivative (liposomes 1); and (b) by click reaction on pre-formed lipid-azide liposomes (liposomes 2)] (Figure 4.11). Binding of liposomes 1 and 2 to Aβ1–42 was investigated by Surface Plasmon Resonance (SPR). Results revealed that planar

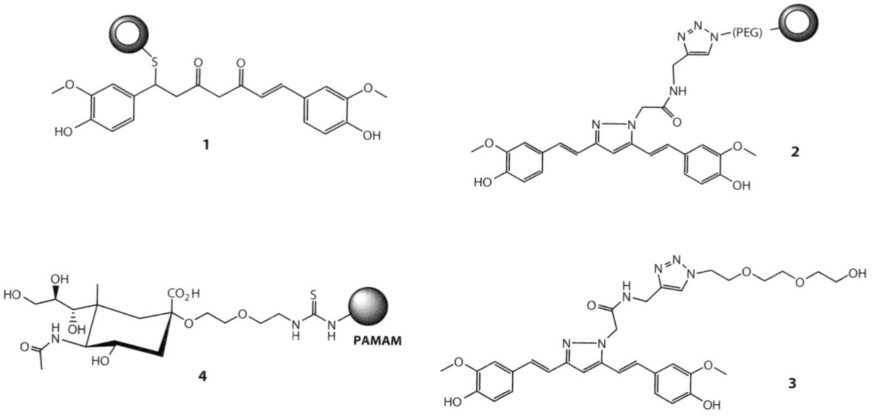

Figure 4.11 NPs with affinity for Aβ species.

curcumin derivative 3 had a much lower binding affinity (KD = 7 μM) than the corresponding curcumin functionalized liposomes 2 (KD = 1–5 nM), which has been attributed to multivalency. On the other hand, SPR analysis of the non-planar curcumin derivative showed no binding to the immobilized Aβ1–42 fibrils. The same group reported the disaggregation efficiency of nanoliposomes (decorated with the same curcumin derivatives [planar and non-planar], phosphatidic acid, cardiolipin and GM1 ganglioside) on the aggregation of the Aβ1–42. All nanoliposomes with curcumin (either planar of non-planar derivative) were able to inhibit the formation of fibrillar and/or oligomeric Aβ in vitro. The planar curcumin derivative was by far the most effective, while liposomes with lipid ligands only inhibited Aβ fibril and oligomer formation at a very high ratio of liposome to peptide. [203]. Liposomes decorated with non-planar curcumin (with curcumin exposed at the surface, Figure 4.11) down-regulated the secretion of amyloid peptide and partially prevented Aβ-induced toxicity. They also strongly labeled Aβ deposits in post-mortem brain tissue of AD patients and APP-PS1 mice. Injection in the hippocampus and in the neocortex of these mice showed that curcumin-conjugated nanoliposomes were able to specifically stain the Aβ deposits in vivo [202].

Aβ was also found to interact with membrane lipids [205] particularly ionic lipids [292] and gangliosides that have been demonstrated to act as seeds for Aβ aggregation. [293, 294]. These were also incorporated in liposomes and SLN NPs and Aβ affinity was tested by SPR, where it was found that nanoliposomes containing phosphatidic acid (PA), and cardiolipin (CL) targeted aggregated forms of Aβ1–42 with high binding affinity (22–60 nM).

On the other hand, Monosialogangliosides (GM1), exhibited much lower affinity for Aβ (0.2 μM) [206]. 2-methyl-N-(2'-aminoethyl)-3-hydroxyl-4-pyridinone (Iron chelator) [210], Sialic acid (Figure 4.11) [211, 212] and Maltose [213] were also ligated on dendrimers and the corresponding conjugated nanoparticles were found to be effective Aβ aggregation inhibitors.

4.5   Nanoparticles for Diagnosis of AD

4.5.1   Introduction

Biomarkers are measures of disease processes or of the predisposition of a living person to disease. An elevated white blood cell count signals infection, and high serum cholesterol correlates with the predisposition to atherosclerosis. The practice of clinical medicine highly depends on such essential diagnostic tools. Unfortunately, there is no simple, reliable, and reproducible blood test for any aspect of Alzheimer's disease. Many attempts to develop one have been made, but none has succeeded up to now. In order for a very early diagnosis of a complex disease like AD to be feasible we need to have an affordable, ultrasensitive, and selective molecular detection method and of course to identify a specific biomarker (or a specific combination of more than one biomarkers) which correlates with the severity of AD. The recently growing application of nanotechnology in molecular detection of biomarkers is promising for very early diagnosis of AD. From a practical point of view, one may perform a molecular detection process either inside the body (in vivo) or on the samples derived from the body (in vitro). Nanotechnology may help us to achieve early diagnosis of AD by providing us with a highly potent signal transduction approach. Signal transduction refers to the process through which a biological signal (a biomarker) transforms to a recordable signal, and is amplified enough to be recorded. This potential application of nanotechnology in molecular diagnosis is mainly based on the special physical (optical, electrical, or magnetic), chemical and biological characteristics of certain multifunctional nanoparticles. In the following section of this chapter a number of methods that have been developed for this aim, are presented and compared [295–297]. The various attempts are summarized in Table 4.7, and are presented below categorized by specific NP type.

4.5.2   Organic NPs for AD Diagnosis

Polymeric NPs or Micelles for AD Diagnosis: Several different polymeric NPs have been developed as diagnostic tools for AD, as seen in Table 4.7.

Table 4.7 NPs for AD diagnosis.

Type of NPs: Imaging agent- Ligands	Model	Result	Reference
Polymeric NPs			
Polystyrene/polybu-tylcyanoacrylate -AChE inhibitor PE154	TTG	Plaque detection by confocal laser-scanning microscopy	(295)
PBCA-NPs-Alexa-488– AβMAb (6E10)	APP/PS1	Amyloid plaque detection in brain	(298)
PBCA- polystyrene NPs- Thioflavin	In vitro	Visualization of amyloid aggregates	(299)
BCA NPs- Clioquinol radioiodinated	AD mice	AD brain section localized uptake of radiotracer	(296)
PLGA NPs-amyloid-binding aptamer Tet peptide	In vitro	binding to amyloid plaques	(200)
Liposomes			
PA-Lips-PA (Phosphatidic acid)	In vitro	high affinity (e.g., 22–60 nM) toward Aβ fibrils	(206)
PA-LIP- PA and RI7217 MAb	In vitro	targeting amyloid-β peptide	(300)
ApoE fragment 141-150	In vitro	↑uptake by cells ↑binding to Aβ peptide	(301)
Curcumin (or encap. dyes) - Curcumin	In vitro	high affinity for amyloid-β1-42 peptide.	(201)
Curcumin- nanoliposomes-Curcumin	APPxPS1	stain the Aβ deposits in vivo	(202)
methoxy-XO4	APP/ PSEN1	iv delivery of Aβ-targeted NP to plaques in AD model	(302)

(Continued)

Table 4.7 (*Cont.*)

Type of NPs: Imaging agent- Ligands	Model	Result	Reference
USPIOs- OX-26	In vitro	In vitro barrier crossing of liposomes containing USPIOs	(303)
Iron oxide [IO] NPs			
Monocrystalline IO-Abeta1-40 peptide	APP/PS1	Method to detect Abeta in AD transgenic mice	(304)
USPIO- Aβ1-42 peptide	APP/PS1	Amyloid plaques detected by T2*-weighted μMRI	(296)
SPIONs-anti-AβPP	APP/PS1	MRI of Plaques	(270)
USPIO- Aβ1-42 peptide	APP/PS1	Detection of amyloid plaques in vivo	(305)
Maghemite NPs- Congo red, rhodamine	In vitro	The hybrid system selectively marks Aβ40 fibrils	(306)
IONPs- anti-Aβ antibody (Aβ1-40 ;Aβ1-42)	In vitro	Assaying Aβ through immunomagnetic reduction	(271)
Gold NPs			
Au - Aβ1-40 antibody	In vitro	Aβ (1-42) concentrations (10 fg/mL level)	(297)
Au-sialic acid	In vitro	Amyloid-beta detection	(295)
Au- MAb specific to ADDL(amyloid derived diffused ligands)	Ex vivo	Bio-barcode assay to measure ADDL in CSF	(296)
Au-anti-tau MAb	In vitro	Two-photon scattering assay detection of tau protein	(295)

Type of NPs: Imaging agent- Ligands	Model	Result	Reference
Au- Aβ(1–40) MAb	In vitro	Detection of Aβ Antigen or aggregates	(272, 273, 307)
Quantum dots(QDs)			
PEG-QDs - Aβ1–40, 1–42 peptides	In vitro	Study of the Aβ peptide aggregation	(295)
QDs-Gold NP-BACE1 peptide	In vitro	Visualize BACE1 activity in living cells	(308)
QDs- Aβ peptides	In vitro	Visualisation of Aβ plaques and inhibition of fibrillation	(309)
Other NPs			
Manganese oxide NPs- Aβ(1–40)MAb	APP/PS1	Detect amyloid plaque deposition in AD mouse models	(310)
Microbubbles-Gadolinium	APP/PS1	MRI imaging of amyloid plaques in the brain	(311)
Co@Pt-Au core-shell nanoparticles	In vitro	MRI to monitor the structural evolution of Abeta assemblies	(312)
Silver NPs - ADDL antibody	In vitro	Antigen detection; determination of ADDL concentration	(297)
Silica nanoparticles anti-tau MAb	In vitro	Detection of tau at 10pg/mL in cerebral spinal fluid (CSF)	(313)

Some of these are described in more detail below. The in vivo staining and detection of Aβ plaques by confocal laser-scanning microscopy with the use of polystyrene/polybutylcyanoacrylate (PS/PBCA)-NPs encapsulating the fluorescent biomarker PE154, was reported [295]. Briefly PE154 a heterodimeric AChE inhibitor that allows histochemical staining of cortical Aβ plaques in triple-transgenic (TTG) mice was encapsulated in biodegradable core-shell PS/PBCA-NPs and in vivo labelling of the plaques was demonstrated. Furthermore, it was shown that PE154 targeted only the Aβ plaques but not tau tangles nor reactive astrocytes (which surrounded the plaques). PBCA-dextran NPs coated with polysorbate 80 were recently proposed as systems to deliver BBB-impermeable molecular imaging probes into the brain [298]. It was indeed demonstrated in vivo (in a mouse model of AD) by using a non-BBB-permeable dye (Alexa-488) that these PBCA-dextran NPs accomplish visualization of amyloid plaques by successful targeting after conjugation to a specific AβMAb (6E10). Additionally, the brain delivery of gadolinium-based contrast agents from similar NPs was proved by MRI. In fact the first study reporting amyloid staining with PBCA NPs was from 2006 [299]. In that study Thioflavin (a ligand for fibrillar Aβ peptides)-encapsulating PBCA NPs, showed significantly stronger fluorescent staining compared to the free fluorophore. Indeed, after intracerebral injection, NP-thioflavin selectively targeted fibrillar Abeta (following biodegradation-induced release from the NPs), in the cortices of APP/PS1 mice with age-dependent beta-amyloidosis. Another study demonstrated that encapsulation of (125)I-clioquinol (CQ, 5-chloro-7-iodo-8-hydroxyquinoline) to PBCA nanoparticles improved its transport to the brain and its retention on amyloid plaques. (125)I-CQ-PBCA NPs successfully stained plaques on post-mortem frontal cortical sections of AD patients, and enhanced its retention in AD-mouse brains [296]. This combination makes the specific NPs a promising delivery vehicle for in vivo single photon emission tomography (SPECT) ((123) I) or PET ((124)I) amyloid imaging agent. Recently the development of Tet-1 targeted **PLGA** NPs encapsulating curcumin was reported [200]. It was observed that Tet-1 increased the neuronal uptake of curcumin from curcumin-PLGA NPs, compared to the non-targeted ones. These NPs were able to attach to the amyloid aggregate surface and decrease the size of the aggregates within 12 hours of co-incubation.

Liposomes for Diagnosis of AD: As mentioned above, liposomes are colloidal, vesicular structures composed of one or more lipid bilayers. Their properties make them ideal candidates for the delivery of drugs or use for diagnosis of AD. Some liposomes with affinity for Aβ peptides have been demonstrated to stain amyloid plaques in vitro and (in some cases) also in vivo, while on some NPs also BBB targeting ligands have been added with the aim to target brain-located Aβ species or plaques. However, specific diagnostic evaluations have not been yet carried out, and most studies are only in preclinical stage. As examples: liposomes and SLNs incorporating phosphatidic acid (PA) or cardiolipines (CL) as a way to target Aβ peptides were prepared, and Surface Plasmon Resonance (SPR) investigations demonstrated that both PA/CL-containing liposomes and SLNs displayed high affinity (e.g., 22–60 nM) towards chip-immobilized Aβ fibrils, likely due to multivalent interactions [206]. The PA incorporating liposomes were subsequently further functionalized with a monoclonal antibody (mAb) [RI7217] against the transferrin receptor, for BBB targeting [300]. SPR experiments revealed high affinity of these nanoliposomes for Aβ-plaques and higher uptake and permeability of the targeted liposomes (compared to non-targeted ones) by cells which are good in vitro BBB models. In a similar approach the same PA liposomes were further functionalized with synthetic ApoE derived peptides in order to enhance their BBB targeting via the LDL receptor [301]. Results confirmed enhanced targeting of human microvascular brain capillary endothelial cells in vitro (60% higher compared to the non-functionalized liposomes), while the functionalization with ApoE-derived peptide does not affect their previously reported ability to bind amyloid-β. Additionally, in a recent study the functionalization of azido-decorated liposomes with an alkyne-derivatized curcumin was reported [201]. SPR experiments demonstrated that the later liposomes had the highest affinity constant (in the 1–5 nM range) reported up-to-date for Aβ fibrils when decorated with a planar curcumin conjugate, while non planar curcumin-decorated liposomes did not show any binding. However, the later liposomes type (with non-planar curcumin conjugate) was recently reported to label amyloid deposits in postmortem tissue of transgenic mice (APP-PS1) after IV administration (in vivo) and down-regulate the secretion of Aβ peptide in cells overexpressing hAPP (in vitro) [202]. Thereby, the question is open about the importance of planar curcumin structure (as also mentioned before in section 4.4). Others recently reported the preparation of Aβ-targeted stealth liposomal nanoparticles using the Aβ-targeted lipid conjugate DSPE-PEG-XO4 [302]. These liposomal NPs maintain similar binding profiles to Aβ (1–40) as the free XO4 ligand in vitro. They selectively bind to amyloid deposits in brain tissue sections of APP/PSEN1 transgenic mice in vitro. Ex vivo analyses of treated brain tissue show that when injected into mice, the targeted particles efficiently bind both parenchymal plaques and CAA-associated amyloids throughout the brain. In vitro immunohistochemistry verified co-localization of both the liposome encapsulated and bilayer membrane components on brain tissue sections obtained from treated animals, confirming the ability of the particles to traverse the BBB and bind amyloid-β plaque deposits. Finally, in another study the encapsulation of USPIOs in liposomes (and formation of magnetoliposomes, MLs) functionalized with OX-26 antibody, which enables the NPs to pass through the BBB, has been reported [303]. In a second step, curcumin derivatives are additionally immobilized on the ML surface for amyloid affinity (A. Skouras *et al*, unpublished results). Ongoing in vitro and in vivo studies are currently carried out in order to demonstrate the potential of such multifunctional MLs to target Aβ plaques in the brain.

4.5.3 Inorganic NPs for AD Diagnosis

Iron Oxide NPs: Some iron oxide NPs such as Feridex (SPIO; Feridex-USA; Endorem-Europe) have already been FDA-approved as magnetic resonance imaging (MRI) contrast agents for cells of the reticulo-endothelial system. The use of iron oxide magnetic nanoparticles for various bio-medical applications, e.g., hyperthermia, diagnostic, cell-labelling and sorting, DNA separation, MRI contrast agents and drug delivery has already been demonstrated. Wadghiri *et al.* was one of the first groups to report the detection of Aβ plaques by Magnetic Resonance Imaging (MRI). Monocrystalline iron oxide NPs (MIONs) were synthesized and covalently tethered to the N-terminus of Aβ1–40 peptide for targeting and imaging of senile plaques. This MRI agent was able to recognize Aβ plaques with high affinity in the brain of amyloid precursor protein (APP) and APP/PS1 transgenic mice when co-injected with mannitol (used to transiently open the BBB) [304]. Ultrasmall superparamagnetic iron oxide (USPIO) NPs were additionally conjugated with Aβ1–42 peptides as means to detect amyloid depositions (after co-injection of mannitol) [296], and Aβ plaques were detected by T2-weighted MRI. More recently, a novel anti-Aβ peptide able to penetrate the BBB and bind to Aβ plaques was covalently conjugated to superparamagnetic iron oxide nanoparticles (SPIONs) for use as an in vivo agent for MRI detection of plaques [270]. Following injection in transgenic mice, the conspicuity of the plaques increased from an average Z-score of 5.1 ± 0.5 (in control AD mice) to 8.3 ± 0.2 (in AD mice treated with SPIONs) and the number of MRI-visible plaques per brain increased from 347 ± 45 to 668 ± 86 in the SPION, respectively. Bi-functionalized USPIOs conjugated with an Aβ1–42 peptide for the selective binding of Aβ plaques and also PEG to improve BBB permeability [305] were intravenously injected in transgenic mice, and amyloid plaques were detected by T2-weighted MRI. In another study, a novel method for the selective labeling of Aβ1–40 fibrils with non-fluorescent or fluorescent (rhodamine-tagged or Congo Red-encapsulated) magnetic γ-Iron Oxide NPs, even under competitive conditions (e.g., in the presence of human serum albumin) [306] was described; Moreover, the ability of these iron oxide NPs to readily remove fibrils from solubilized Aβ samples by simple use of an external magnetic field, was reported. Recently, magnetic NPs biofunctionalized with antibodies against Aβ-40 and Aβ-42 demonstrated that immunomagnetic reduction signals of Aβ-40 and Aβ-42 in plasma from normal humans and AD patients showed significant differences. Such results may find future applications in AD diagnosis [271].

Gold NPs: Gold nanoparticles (AuNPs) have extraordinary optical, electronic, and molecular-recognition properties. Electronic microscopy is one of the areas where gold nanoparticles have been extensively used as contrast agents. They can be associated with many traditional biological probes such as antibodies, lectins, superantigens, glycans, nucleic acids and receptors. Because gold particles have various sizes they can be easily spotted in electron micrographs, while it is possible for multiple experiments to be conducted simultaneously. These advantages make AuNPs one of the main NPs used in studies for AD diagnosis, mainly for sensitive measurement of Aβ concentrations. Some examples of recently proposed techniques follow. Recently, a research group developed an ultrasensitive electrical detection method for Aβ1–42 using scanning tunneling microscopy (STM) [297]. For this, a monoclonal antibody (mAb) fragment with high affinity for Aβ1–42 was immobilized onto a gold surface and the sample was deposited onto the mAb-Functionalized surface, leading to its capture. Subsequently, mAb-Au NP complex was reacted and resulted in the formation of "sandwich-like" structures. The resulting chip was finally analyzed by STM. It was shown that the surface density of the Au NPs correlated with the number of Aβ-Antigen binding events and that a successful Aβ detection was achievable at a concentration 10 fg/mL. Another interesting procedure for Aβ detection was based on electrochemical sensing of saccharide-protein interactions [295]. The densely packed sialic acid domains were able to capture the Aβ peptides as a result of specific interactions, and the method enabled the detection of non-labeled Aβ down to sub-micromolar concentrations. Remarkable results towards the development of new approaches for biomarker detection have been proposed by Georganopoulou *et al.*, who developed an ultrasensitive NPs-based bio-barcode assay capable of detecting AD soluble biomarkers in CSF [275]. The key feature of the system relies on the isolation of antigens by means of a "sandwich process" involving oligonucleotide (DNA-barcode) modified Au NPs, and magnetic microparticles (MMPs), both functionalized with antibodies specific to the ADDLs. Practically, an excess of Au NPs and MMPs (when compared to the ADDLs concentration) are mixed in a CSF sample; the recognition of the antigen from both particles leads to the formation of sandwiches that are then purified by magnetic separation. The strands of a dehybridized double-stranded DNA are isolated and easily quantified by a scanometric method using DNA microarray. The efficient antigen sequestration in solution and the amplification process resulting from the large number of DNA strands released for each antigen recognition, allowed the system to identify ADDLs at sub-femtomolar concentrations, thus improving the ELISA test sensitivity by 6 orders of magnitude. Neely *et al.* designed Au NPs coated with mAb specific to τ protein and employed the NPs in a two-photon Rayleigh scattering assay, which enabled the detection of τ protein at concentrations greater than about 1 pg.mL-1. This concentration was about 2 orders of magnitude lower than typical τ protein concentration values (i.e., 195 pg.mL-1) in CSF. Moreover, the two-photon Rayleigh scattering assay showed a strong sensitivity for τ protein and was able to discriminate other proteins such as bovine serum albumin [276]. El-Said *et al.* developed a method to detect Aβ plaques using surface-enhanced Raman scattering. Briefly gold nanoparticles (Au NPs) were electrochemically deposited on an indium tin oxide (ITO) substrate [273]. Aβ antibodies were immobilized on the Au-NP-coated ITO substrate, after which the interactions between the antigen and the antibody were determined via SERS spectroscopy. The SERS responses had a good linear relationship that corresponded to the change in the concentration of the antigen with high sensitivity (100 fg/ml). Recently, Sakono *et al.* reported a simple method to detect Aβ aggregates. Au NPs modified with the Aβ antibody were treated with bovine serum albumin to stabilize their dispersibility in buffer [307]. After adding appropriate concentrations of Aβ a red-coated precipitate could be observed by naked eye. The precipitate is only observed when oligomers or fibrils are added, but not in the presence of monomers. Another group fabricated a dense Au nano dot array (ca. 60 nm) on indium tin oxide (ITO). Aβ1–42 antibody is allowed to immobilize on the Au dots followed by its target protein and Au NP-antibody complex is prepared and then applied to preimmobilized protein arrays to get the pulse like current peak under STM. STM derived profiles show a logarithmic increase of the current with an increase of the Aβ1–42 concentration, successfully detecting concentrations as low as 100 fg/mL (23 fM) [272].

Quantum Dots (QD): A nanoprobe for amyloid-β aggregation and oligomerization using PEG coated QDs as Aβ42 labels, was recently developed [295]. The oligomerization behavior of Aβ42 in solution and on intact cells was compared, as well as the ingestion manner of microglia for Aβ42 monomers and oligomers. In order to use this diagnostic technology to monitor Aβ42 biochemical behavior in vivo, in addition to QD safety considerations, special attention should be paid to the successful passage of QD-Aβ nanoprobes through the BBB. QDs conjugated with transferrin were recently reported to be able to successfully transmigrate through an in vitro model of the BBB. Another interesting approach [308] utilized a novel fluorogenic nanoprobe prepared from the assembly of CdSe/ZnS QDs and gold (Au) NPs in which QD was conjugated with a specifically designed β-secretase (BACE1) substrate peptide. This coordination-mediated binding of the QD with Au nanoparticles via Ni-NTA-histidine (His) interaction resulted in highly efficient quenching of QD fluorescence through a distance-dependent fluorescence resonance energy transfer (FRET) phenomenon. The prequenched QD-Au assembly recovered the fluorescence in the presence of the BACE1 enzyme after incubation in vitro. The high quenching efficiency of AuNP and robust QD fluorescence signal recovery upon BACE1 enzymatic digestion enabled the visualization of BACE1 activity in living cells. These results show the potential application of QD-AuNP nanoparticles as an efficient probe to identify active molecules in BACE1-related diseases such as AD. Zhang *et al.* reported the labeling of Aβ with QDs (QDs-Aβ) by a simple mixing-incubation strategy through which Aβ became wrapped on the surface of QDs. Such QDs inhibited the formation of β-folding, and the fibrillation of Aβ [309]. The QDs-Aβ retained dispersivity and fluorescence properties. When coincubated with astrocytes, the QDs-Aβ were endocytosed, without separation of QDs and Aβ; Aβ was degraded in 24 h, indicating that labeling with QDs did not affect the uptake and degradation of Aβ by astrocytes, providing a possible new method for visualization of the the Aβ elimination process.

4.5.4   Other NP-Types for Diagnosis of AD

Besides well-established NP-types (mentioned in Table 4.1), other NPs or hybrid systems have also been proposed for AD diagnosis. Some examples follow. The use of hollow manganese oxide NPs (HMONs) for detection of amyloid plaques by MRI has been proposed [310]. For this, HMONs are conjugated with an antibody against Aβ1–40 peptide and injected to APP/PS1 transgenic and wild-type mice. The T1-weighted MRI images obtained were useful to detect brain regions with amyloid plaque deposition. In another approach for imaging of Aβ plaques with MRI, a non-targeted contrast agent (Gd-DOTA, Dotarem®) was intravenously injected to APP/PS1 transgenic mice and the BBB was transiently opened with unfocused ultrasound (1MHz) and microbubbles. Amyloid plaques were detected with high sensitivity and good resolution [311]. Another interesting approach regarding MRI imaging was that of Choi *et al.* (2008) with the use of core-shell NPs [312]. Cobalt platinum alloy(core)-gold(shell) NPs were prepared with a high magnetization value, making them appropriate for T2-weighted spin echo magnetic resonance measurements. MRI of the NPs after neutravidin conjugation and labeling with biotinylated Aβ1–40 peptides showed contrast changes dependent on peptide concentration. Haes *et al.* designed a localized **surface plasmon resonance** nanosensor (**LSPR**) to detect Amyloid derived diffused ligands (ADDLs) as a potential biomarker for AD [297]. The signal transduction mechanism of the LSPR nanosensor was based on its sensitivity to local refractive index changes near the surfaces of the metal-(**Au, Ag**)-NPs. The resulting biosensors were incubated with samples containing ADDLs, washed and incubated with a polyclonal antibody solution specific to ADDLs to enhance the shift response. The LSPR nanosensor was finally demonstrated to be sensitive enough for the detection of ultralow concentrations of ADDLs in biological samples. In the last example, a **silica**-NP-gold capped LSPR chip with a monoclonal anti-tau antibody (tau-mAb) immobilized on the surface, was developed for tau detection. This immune-chip enabled detection of tau at 10 pg/mL, lower than the cut-off value of 195 pg/mL (for AD) for tau protein in cerebral spinal fluid (CSF), while BSA showed no interference with tau detection [313].

4.6 Nanoparticles for Therapy of AD

The therapeutic potential of nanotechnology for AD includes neuroprotective and neuroregenerative approaches. Several NP systems have been studied in recent years to increase the bioavailability and efficacy of different AD therapeutic agents. The two main sources of neurotoxicity in AD are Aβ oligomers and free radicals. Some of the nanotechnology-based approaches are capable of protecting neurons from Aβ toxicity by preventing amyloid oligomerization (disaggregation strategy) and/or accumulation of Aβ oligomeric species. Other nanotechnology neuroprotective approaches include those that protect neurons from oxidative stress of free radicals. Targeted drug delivery is an important application of nanomedicine due to the additional obstacle of the blood–brain barrier (BBB) for molecule transport to the CNS. Up-to-date information about NP-types that have been recently demonstrated to have potential to cross the BBB and demonstrate therapeutic efficacy against AD is presented in this subchapter [295, 314].

4.6.1 Polymeric NPs for Therapy of AD

Different types of polymeric NPs have been developed as systems to enhance BBB transport of drugs or as therapeutics for AD, as presented below:

Table 4.8 NPs for therapy of AD.

Type of NPs-Ligands/Molecules	Model	Result	Reference
Liposomes			
Click-LIP-Curcumin (TREG)	In vitro	↑ affinity for Aβ	[201]
PA LIP ; cardiolipin-LIP	In vitro	↑ affinity Aβ fibrils	[206]
LIP-Curcumin, PA, cardiolipin, GM1	In vitro	↓ Aβ secretion and toxicity, staining Aβ	[202, 203]
Anti-Aβ Lips- AβMAb	In vitro	↑ binding to Aβ1-42 peptides	[267, 268]
PA LIP + MAb	In vitro	↑ binding to Aβ1-42 peptides	[300]
LIP+negatively charged PLs+ApoE	In vitro	↑ uptake by brain cells, ↑binding to Aβ peptides	[301]
LIP-Curcumin, PA, cardiolipin, GM1	In vitro	↓ aggregation of Aβ	[203]
Polymeric NPs			
Polystyrene NPs-- Iron chelator-MAEHP	In vitro	↓ Aβ fibril formation- ↓ Aβ-related toxicity	[295, 314]
NP-chelator conjugate (Nano-N2PY)-MAEHP	In vitro/ in vivo	↓ Aβ- toxicity, ↓ Aβ aggregation	[295, 314]
PBCA-NPs - Quinoline derivatives	In vitro	Aβ plaques staining [APP/PS1]	[295, 314]
Polysorbate-80 coated PBCA NPs - Tacrine/Rivastigmine	Rats	↑ brain tacrine or rivastigmine (3.82 fold)	[316]
NPs - ApoE 3 PBCA-NPsCurcumin-ApoE 3	In vitro	increased uptake of curcumin; ↓Aβ toxicity	[257]
PLGA NPs - VIP peptide	Mice	↑ brain delivery of estradiol	[217]
PBCA- NPs-Curcuminoids	Mice	2.53x ↑ accumulation in brain	[295, 314]

(*Continued*)

Table 4.8 (*Cont.*)

Type of NPs-Ligands/ Molecules	Model	Result	Reference
Graphene oxide GO-polymeric NPs	In vitro	Inhibition of Aβ aggregation	[331]
PEG-PACA [poly- alkyl cyanoacrylate] NP-Selegiline	In vitro	No positive effect	[204]
CopolymericNiPAM: BAM NPs	In vitro	↓nucleation step of Abeta fibrillation	[318]
Polymeric nanostructures	In vitro	↓Aβ aggregation and ↓ cytotoxicity	[319]
PACA and PLA NPs- (PEG) corona and ApoE3	In vitro	Capture Aβ1–42 in circulation ("sink effect")	[258]
Angiopep-conjugated PEG-co-poly (ε-caprolactone) (ANG-PEG-NP)- Rhod-B	In vitro	↑ NP accumulation in cortical layer, lateral ventricle, third ventricle and hippocampus	[315]
PBCNPs - Curcumin	Mice	2.53-fold in brain	[295, 314]
PLGA/PBCA- Rivastigmine tartrate (RT)	In vivo	Regain memory loss	[323]
PLGA - curcumin NPs + Aβ-binding aptamer	In vitro	↓ Aβ aggregates, anti-oxidative effect	[200]
Chitosan NPs -Tacrine	In vitro	↑ bioavailability in brain	[320]
TrimethylatedCH-PLGA NPs- Coenzyme Q_{10} (co-Q_{10})	APP/PS1	↑memory restoring	[321]
Estradiol–loaded NPs	Rats	↑ estradiol in CNS	[322]
CH-PLGA NPs-novelAβMAb	in vitro	↑ uptake/targeting BBB; Aβ	[269]
Solid Lipid Nanoparticles (SLNs)			
Ferulic acid (FA)-SLN	In vitro	↓ ROS; ↓ cytochrome c -apoptosis	[324]
Quercetin-SLN	In vitro	↑ therapeutic efficacy of quercetin	[325]

Type of NPs-Ligands/ Molecules	Model	Result	Reference
Cardiolipin - SLN	In vitro	↑ affinity for Aβ fibrils	[206]
Dendrimers			
PAMAM dendrimers-SA; 3 PAMAM - +KLVFF peptide	In vitro	↓ Aβ-induced toxicity; ↓ peptide aggregation; ↑ affinity	[212-214, 236]
Gold NPs			
+CLPFFD peptide	In vitro/ in vivo	↓ amyloidogenic process; ↑ permeability of BBB; ↓ Aβ fibrillization	[328, 330]
(-) NPs-Congo Red	In vitro	↓ Aβ Fibrillization and Fibril Dissociation; ↓ Neurotoxicity	[329]
AuNPs-Mifepristone	Mice	↑ in drug bioavailability in brain	[295, 314]
Nanocapsules			
Eudragit S100 NC- Melatonin	Mice	↓ lipid peroxidation ↑ antioxidant reactivity (hippocampus)	[295, 314]
Lipid Core NC - Indomethacin	In vitro	↓ Aβ-induced cell death; ↓ Aβ1-42 neuroinflammation ↓ glial activation	[326]
Other NPs			
PDP-NPs-Copper chelator-penicillamine	In vitro	↓ Aβ (1–42) accumulation	[295, 314]
SWCN -Achetylcholine	AD mice	↑ Acetylcholine to brain	[295, 314]
EGCG [(-)-Epigallocatechin-3] NPs +a-secretase inhibitor	AD mice	↑ alpha-secretase; ↓ plaques	[295, 314]
Ceria NPs-Cerium oxide		Specific for Aβ	[332]
Cholesterol-pullulanNanogels	AD mice	↓ aggregation-cytotoxicity of Aβ	[333]

(Continued)

Table 4.8 (*Cont.*)

Type of NPs-Ligands/Molecules	Model	Result	Reference
Pegylated micelles	In vitro	↓ aggregation of Aβ1-42	[295, 314]
Core Shell NPs-Thioflavin T -S	APP/PS1	Aβ clear brains	[299]
Albumin NPs- ApoE 3	SV 129	↑ brain NP uptake	[81-83]
Nanocurcumin powder	Tg2576	↑memory; ↑curcumin in brain	[334]
Core shell NPs - Cu		↓ Aβ levels	[335]
Modified NPs -TGN + NAP peptide	Mice	↑ spatial learning; no Aβ plaques	[336]
PEG-GSH-NPs-Glutathione (GSH)	In vitro	Protect neuronal cells from oxid. stress	[295, 314]
Fullerene (or derivative)-Malonic acid	In vitro	↓ Aβ-related toxicity; Aβ(1-42) in fibrils brain	[337, 338]
C60 fullerene- NPs- Malonic acid	In vitro	↓ Aβ(25-35)-toxicity	[339]
Fluorinated complexes	In vitro	Induce α-helix structure; prevent fibrils	[295, 314]
Inorganic NPs- CdTe NPs	In vitro	↓ Aβ fibrillation	[295, 314]
Pullulancholesteryl NPs	In vitro	↓ Aβ toxicity	[295, 314]
Self-Assembled PeptidePolyoxometalate NPs	Mice	↓Aβ aggregation	[295, 314]

Angiopep-conjugated poly(ethylene glycol)-co-poly(ε-caprolactone) NPs (ANG-PEG-NPs), in which the angiopeptide BBB-targeting ligand significantly enhanced BBB transport (compared to control PEG-NPs) as evidenced by in vitro and in vivo studies [315]. Caveolae- and clathrin-mediated endocytosis, involving a time-dependent, concentration-dependent and energy-dependent mode, was evidenced. The brain coronal section showed a higher accumulation of ANG-PEG-NPs in the cortical layer, lateral ventricle, third ventricles and hippocampus than that of PEG-NPs. In another study poly(n-butylcyanoacrylate) (PBCA) NPs coated with 1% polysorbate 80 significantly increased tacrine and rivastigmine concentration in the brain (compared to uncoated NPs or free drug) [316].

NP-surface engineering with lectins opened a novel pathway for brain absorption of drugs. Drugs loaded in biodegradable poly(ethylene glycol)-poly(lactic acid) (PLGA) NPs are absorbed in the brain, following intranasal administration. Vasoactive intestinal peptide (VIP), a neuroprotective peptide, was efficiently incorporated in PLGA NPs modified with wheat germ agglutinin and its brain uptake and neuroprotective effect were significantly increased (compared to non surface modulated NPs) [317]. Improvements in spatial memory in ethylcholine aziridium-treated rats were also observed following intranasal administration of 25 µg/kg and 12.5 µg/kg of VIP-loaded by unmodified nanoparticles and wheat germ agglutinin-modified NPs, respectively. These results clearly indicated wheat germ agglutinin-modified NPs might serve as promising carriers especially for biotech drugs such as peptides.

Aβ fibril formation was hindered by poly (N-isopropylacrylamide)-co-poly (N-tert-butylacrylamide) (PNIPAAM-co-PtBAM) NPs with a diameter of 40 nm [318]. Interestingly, it was found that the oligomerization of the peptide could be sufficiently reversed when mature fibrils start forming. Additionally, the NPs also interfered with the aggregation process by delaying, or even blocking, the nucleation step, but no influence on the elongation step was noticed. In other words, the NPs introduced a "lag phase" between nucleation and elongation steps of fibrillation. The "lag phase" was observed to be strongly dependent on NP concentration, surface properties and physicochemical characteristics. Such results help to better elucidate the Aβ aggregation process and design NPs with optimized surface properties for AD treatment.

Biodegradable PEGylated poly (alkyl cyanoacrylate) NPs consisting of poly[methoxypoly(ethylene glycol)-co-(hexadecyl cyanoacrylate)] [P(MePEGCA-co- HDCA)] co-polymer were able to bind Aβ and inhibit its aggregation as proven by capillary electrophoresis coupled to laser-induced fluorescence (CE-LIF) [258]. The crucial role of PEG chains on NPs surface for their interaction with Aβ was noticed and studies are continuing in order to clarify the exact role of PEG chains and to develop functionalized NPs for AD therapy.

Current therapies for Alzheimer disease (AD) such as acetylcholinesterase inhibitors and the latest NMDA receptor inhibitor, Namenda®, provide moderate symptomatic delay at various stages of AD. Metal-mediated oxidative damage is a significant contributor to the disease progression, since metals such as iron, aluminum, zinc, and copper are disregulated and/or increased in AD brain tissue, creating a pro-oxidative environment. This role of metal ion-induced free radical formation in AD makes chelation therapy an attractive method for decreasing the oxidative stress burden in neurons. Chelator-conjugating-NPs have shown a unique ability to cross the BBB, chelate metals, and exit the brain (through the BBB) with the complexed metal ions. This method can provide a safe and effective method of reducing the neural tissue metal load, attenuating the harmful effects of oxidative damage. Chelating agents that selectively bind to and remove and/or "redox silence" transition metals have been long considered as attractive therapies for AD. However, the BBB and neurotoxicity of many traditional metal chelators has limited their utility in AD or other NDs, so it has been suggested that iron chelator-conjugating-NPs may have the potential to deliver chelators into the brain and overcome such issues as chelator bioavailability and toxic side-effects [295]. For this, a prototype chelator-conjugating-NP (Nano-N2PY) was synthesized, and its ability to protect human cortical neurons from Aβ-associated oxidative toxicity was demonstrated. In addition, Nano-N2PY effectively inhibited Aβ aggregation.

Curcumin-loaded PBCA NPs, which were coated with apolipoprotein E3 (ApoE3-C-PBCA) demonstrated enhanced therapeutic efficacy against Aβ-induced cytotoxicity in SH-SY5Y neuroblastoma cells compared to plain curcumin solution [257]. The activity of curcumin was enhanced with ApoE3-C-PBCA compared to plain curcumin solution suggesting enhanced cell uptake and sustained drug release. Additionally a synergistic effect between ApoE3 and curcumin (ApoE3 also possesses antioxidant and antiamyloidogenic activity) may exist.

When designing ligand decorated NPs for Aβ peptide targeting, it is important to use methodologies that ensure presence of the ligand on the NP-surface. Selegiline-functionalized polymeric NPs were synthesized with selegiline and fluorescent poly (alkyl cyanoacrylate) (PACA), but no increase in the interaction between these functionalized and non-functionalized NPs with Aβ(1–42) peptides was observed, highlighting the lack of availability of the ligand at the surface of the nanoparticles [204]. Adsorption of Aβ peptides to NPs was proposed as a strategy to inhibit Aβ aggregation (by re-conversion of Aβ conformation). NPs, synthesized by sulfonation and sulfation of polystyrene, and forming microgels and latexes, affected the conformation of Aβ inducing an unordered state with reduced cytotoxicity and significantly reduced oligomerization [319].

Tacrine-loaded chitosan NPs were proven to be an optimized delivery system of tacrine (a drug with potential significance in AD) to the brain [320].

Trimethylated chitosan (TMC) surface-modified PLGA NPs (TMC/PLGA–NP) were developed as a drug carrier for brain delivery [321] and were observed to deliver 6-coumarin at high amounts in the cortex, paracoele, third ventricle, and choroid plexus epithelium, while no brain uptake of coumarin loaded in control PLGA–NP was observed. Behavioral tests showed that coenzyme-Q10-loaded TMC/PLGA–NP greatly improved memory impairment, restoring it to a normal level; the efficacy of Q10-loaded-PLGA–NP, without TMC conjugation was substantially lower. Senile plaque and biochemical parameter tests confirmed the brain-targeted effects of TMC/PLGA–NP.

Vitamin E2-loaded chitosan NPs (prepared by ionic gelation of chitosan) [322] were administered IV and intranasally in male wister rats. The plasma levels of E2 after intranasal administration (32.7±10.1 ng/ml; tmax 28±4.5 min) were significantly lower than those obtained after IV administration (151.4±28.2 ng/ml), but the CSF concentrations of the drug achieved after intranasal administration (76.4±14.0 ng/ml; tmax 28±17.9 min) were significantly higher than those after IV administration (29.5±7.4 ng/ml; tmax 60 min), showing that when using E2-loaded chitosan NPs, E2 is directly transported from the nasal cavity into the CSF, significantly improving the levels of E2 being transported into the brain.

Chitosan-coated PLGA NPs conjugated with a novel anti-Aβ antibody showed enhanced uptake in BBB cell models and better targeting of Aβ deposits in the CAA model in vitro (compared to control NPs) [269]. Another strategy for treatment of AD is the delivery of antioxidant species to the brain, because of their ability to quench the reactivity of reactive oxygen species. The in vivo acute antioxidant effect of i.p.-delivered melatonin-loaded polysorbate 80-coated nanocapsules with that of melatonin aqueous solution in mice brains (frontal cortex and hippocampus) and liver was compared. Melatonin aqueous solution had no antioxidant activity, while melatonin-loaded NP formulations caused a marked reduction on lipid peroxidation levels in all tissues studied. [314].

NP formulations of rivastigmine tartrate (RT), consisting of PLGA and PBCA (optimized [using factorial design]) were evaluated for brain targeting and memory improvement in scopolamine-induced amnesic mice using Morris Water Maze Test. Results demonstrated faster regain of memory loss in amnesic mice with both PLGA and PBCA NPs indicating rapid and higher extent of transport of RT into the mice brains proving the suitability of both NPs as potential carriers for brain delivery of RT [323].

Curcumin-loaded PLGA NPs decorated with Tet-1 peptide were able to destroy amyloid aggregates and exhibit anti-oxidative properties, while being non-cytotoxic [200]. The amyloid-binding aptamer conjugated to these NPs enhances their bind affinity to Aβ peptides, making such NPs potential systems to target amyloids in the plasma (AD therapy in accordance to the "sink" theory).

Polymer colloids composed of a polystyrene core and a degradable **PBCA** [poly(butyl-2-cyanoacrylate)] shell, that bind fibrillar Aβ peptides were proposed as carriers to target thioflavin T and thioflavin S to amyloid-plaques in the brain [299].

4.6.2 Lipidic NPs for Therapy of AD

Different types of Lipids NPs, such as SLNs, Liposomes, Lipidic micelles etc., have been proposed as therapeutic systems for AD. Most of the latest approaches are mentioned below:

Solid Lipid NPs (SLNs): Ferulic acid (FA) treatment, in particular if loaded into SLNs, decreased ROS generation, restored mitochondrial membrane potential and reduced cytochrome c release and intrinsic pathway apoptosis activation. Further, FA modulated the expression of peroxiredoxin, an anti-oxidative protein, and attenuated phosphorylation of ERK1/2 activated by Aβ-oligomers [324].

Solid lipid nanoparticles (SLNs) of quercetin were constructed for IV administration in order to improve quercetin's permeation across the BBB, and eventually to improve its therapeutic efficacy in AD. The optimized formulation was subjected to various in vivo behavioral and biochemical studies in Wistar rats, which demonstrated markedly improved memory-retention compared to pure quercetin-treated rats. [325]. Finally, as presented above, various SLN formulations functionalized to target Abeta(1–42) demonstrated high affinity to various Aβ species [206].

Micelles: The ability of **PEGylated phospholipidic micelles** to interact with Aβ1–42 and to moderate its in vitro neurotoxicity was attributed to a double mechanism: in the extracellular medium, such micelles would first interact with the peptide so as to bury its hydrophobic domains in their hydrophobic core via a favored α-helical conformation that prevents its self-aggregation and then PEGylated micelles would shield the exposed hydrophobic domains of small Aβ1–42 aggregates with their hydrophobic acyl chains, thus avoiding further formation of aggregate-aggregate or aggregate-monomeric Aβ1–42 [295].

While the antioxidant properties of green tea polyphenol (EGCG) are well known, it has been shown to be able to promote the non-amyloidogenic processing of APP by upregulating α-secretase and thus preventing Aβ plaque formation. Nanolipidic green tea polyphenol [(-) epigallocatechin-3-gallate (EGCG)] particles (NanoEGCG) were prepared and tested in murine neuroblastoma cells that were transfected with the human APP gene (APP; SweAPP N2a cells). It was observed that among all the formulations tested, some NanoEGCG formulations promoted enhanced levels of α-secretase activity even at the lowest EGCG concentration tested. The prepared formulations were also tested in male Sprague Dawley rats and were delivered via oral gavage at a dosage of 100 mg EGCG/kg body weight. The oral bioavailability of the formulations in vivo was observed to be enhanced by more than two-fold as compared to the free EGCG in 10% ethanol solution after 5 and 10 minutes. Prevention of amyloidogenic processing of amyloid precursor protein with the use of natural phytochemicals capable of enhancing alpha-secretase activity may be a therapeutic approach for treatment of AD [295].

Bernardi *et al.* investigated the potential protective effect of indomethacin-loaded lipid-core nanocapsules (IndOH-LNCs) against cell damage and Aβ-induced neuroinflammation in AD models [326]. IndOH-LNCs attenuated Aβ-induced cell death and were able to block the neuroinflammation triggered by Aβ1–42 in organotypic hippocampal cultures. Additionally, IndOH-LNC treatment was able to increase interleukin-10 release and decrease glial activation and c-jun N-terminal kinase phosphorylation. As a model of Aβ-induced neurotoxicity in vivo, animals received a single intracerebroventricular injection of Aβ1–42, and 1 day after Aβ1–42 infusion, they were administered either free IndOH or IndOH-LNCs (intraperitoneally) for 14 days. Only the treatment with IndOH-LNCs significantly attenuated the Aβ-induced impairments. Further, treatment with IndOH-LNCs was able to block the decreased synaptophysin levels induced by Aβ1–42 and suppress glial and microglial activation. These findings might be explained by the increase of IndOH concentration in brain tissue.

Liposomes: As mentioned above (in subchapter 4.4) various types of liposomal formulations with nanodimensions and high affinity for Aβ peptides (monomer, oligomers, or fibrils) have been recently deveopled. The ligands used for Aβ targeting varied from various lipids as cardiolipin or phospatidic acid [206], curcumin or curcumin derivatives [201–203, 327] as well as a monoclonal antibody against Aβ1–42 [268]. In some cases the liposomes were further engineered to incorporate a second functionality in order to also enhance their transport across the BBB and thus target Aβ plaques in the brain. For this a peptide derivative of ApoE 3 (known to target the apolipoprotein receptor overexpressed on BBB cells) [301], or an anti-transferin receptor monoclonal antibody [267, 300], were used. All of the previous liposome types were indeed demonstrated to have increased Aβ binding affinity (compared to the free ligands and non-ligand control liposomes) explained by multivalent effects and most of them significantly retarded or blocked Aβ aggregation and could stain depositions on post-mortem brain slices of AD patients. Several of these nanoliposome types are currently under in vivo investigation for their therapeutic effect on AD.

4.6.3   Other NP Types

Dendrimers: The possibility of functionalizing the peripheral groups of dendrimers with ligands of interest is an attractive strategy to study the physical interactions of these macromolecules with Aβ peptides. Several studies on the aggregation process of Aβ identified the critical peptidic sequence involved in amyloid aggregate formation. The hydrophobic core from residues 16–20 of Aβ, the so-called KLVFF sequence, is crucial for the formation of β-sheet structures. It was also demonstrated that this peptidic region binds to its homologous sequence in Aβ and prevents its aggregation into amyloid fibrils. This sequence has been employed as a key compound for the development of agents for preventing Aβ aggregation in vivo. KLVFF-functionalized dendrimeric scaffolds demonstrated marked inhibitory effect on Aβ1–42 aggregation, and were also able to disassemble preexisting amyloid aggregates [236]. These nanodevices also exploited the multivalency feature of dendrimers to drastically enhance the affinity and specificity of KLVFF sequence toward Aβ.

Several independent studies suggested that Aβ is able to bind cells via an interaction with glycolipids or glycoproteins present at the external surface of the cellular membrane. It was also shown that the interaction affinity increased when gangliosides or sialic acid molecules were clustered on the cell surface. Sialic acid-conjugated polyamidoamine (PAMAM) dendrimers as membrane-cluster-mimetics had increased affinity to Aβ and also significantly reduced Aβ-induced toxicity compared to nontreated control cells and cells treated with free sialic acid [212]. The positioning of the covalent bond between the dendrimer and sialic acid was crucial regarding the modulation of the biological activity of the resulting conjugates. The effect of polypropyleneimine dendrimers on the formation of amyloid fibrils as a function of pH was studied, in order to gain further insight in the aggregation mechanism and its inhibition [213, 214]. The level of protonation of His, Glu, and Asp residues is important for the final effect, especially at low dendrimer concentration when their inhibiting capacity depends on the pH. At the highest concentrations, dendrimers were very effective against fibril formations from amyloid peptides.

Gold NPs: A novel approach that offers a strategy to remotely inhibit amyloidogenic process has been proposed, where peptide-gold NPs which selectively bind to Aβ aggregates are irradiated with microwave. This results in inhibition of amyloidogenesis and restoration of the amyloidogenic potential [328]. More specifically, the irradiation effect on the amyloidogenic process was studied and amyloidogenic aggregates rather than amyloid fibrils seemed to be better targets for the treatment. It was also found that bare AuNPs inhibited Aβ fibrillization to form fragmented fibrils and spherical oligomers. By adding bare AuNPs to preformed Aβ fibrils, it was evidenced that AuNPs bind preferentially to fibrils [329]. Similar results were observed with carboxyl- but not amine-conjugated AuNPs, where co-incubation of negatively charged AuNPs with Aβ, reduced Aβ toxicity to neuroblastoma cells. Another interesting approach is the destruction of toxic β-amyloid aggregates by CLPFFD–peptide-conjugated gold NPs [330]. The THRPPMWSPVWP peptide sequence was introduced into the CLPFFD-conjugated-gold NPs. This peptide sequence interacts with the transferrin receptor present in the microvascular endothelial cells of BBB causing an increase in the permeability of the NPs into the brain, as seen in vitro and in vivo.

Single-wall Carbon Nanotubes (SWCN): It was recently shown that lysosomes are the pharmacological target organelles for single-walled carbon nanotubes (SWCNTs) and that mitochondria are the target organelles for their cytotoxicity [295]. SWCNTs delivered orally were lysosomotropic but also entered mitochondria at large doses. Indeed, SWCNT administration resulted in collapse of mitochondrial membrane potentials, overproduction of reactive oxygen species, and finally damage of mitochondria, followed by lysosomal and cellular injury. Based on these findings, SWCNTs were successfully used to deliver acetylcholine into the brain for treatment of experimentally induced AD, by precisely controlling the doses in order to ensure that the NTs preferentially enter lysosomes, and not mitochondria where they induce cytotoxicity.

The effect of **graphene oxide** (GO) and their protein-coated surfaces on the kinetics of Aβ fibrillation in aqueous solution was investigated [331]. GO and their protein-covered surfaces delay the Aβ aggregation process via adsorption of amyloid monomers on the large available surface of GO sheets. The inhibitory effect of the GO sheet was increased when GO concentration was increased from 10% (in vitro) to 100% (in vivo).

Nanoceria Particles: Nanocerias (i.e., mixed-valence-state cerium) were used to drastically reduce the reactive oxygen intermediate intracellular concentration in vitro and in vivo, so as to prevent the loss of vision due to light-induced degeneration of photoreceptor cells. These results indicated that nanoceria particles were active for the inhibition of reactive oxygen intermediateintermediate- mediated cell death that is involved, among other species, in AD pathogenesis [332].

Nanogels represent a promising class of drug delivery devices because of their high loading capacity, their high stability, as well as their responsiveness to environmental factors, such as ionic strength, pH, and temperature. An original use of cholesterol-bearing pullulan (CHP) nanogels with a diameter of 20–30 nm as artificial chaperone systems for controlling the aggregation and cytotoxicity of Aβ1–42 was suggested [333]. These colloidal nanomaterials were able to efficiently incorporate the monomeric peptide and to inhibit its aggregation, thus suppressing its related toxicity against PC12 cells. Recently, the ability of these nanogels to interact with the Aβ1–42 oligomeric forms and to reduce their toxicity on primary cortical and microglial cells was evaluated. In vitro experiments indicated that CHPs prevented Aβ1–42 oligomer toxicity and did not accumulate into lysosomes within the first 30 minutes. Further experiments on transgenic animals mimicking conditions of the AD neurological disorder are continuing, even if the ability of these nanostructures to surpass the BBB is still unproved. The concept developed with these CHP nanogels is very interesting if one considers the internalization of more specific Aβ-targeted ligands within the gel network.

The Cu (I) chelator D-penicillamine was covalently conjugated to nanoparticles via a disulfide bond in order to investigate whether NP-conjugated chelators could act to reverse metal ion induced protein precipitation [314]. Release of D-penicillamine from the nanoparticles was achieved using reducing agents. Nanoparticles treated only under reducing conditions that released the conjugated D-penicillamine were able to effectively resolubilize copper-Abeta (1–42) aggregates.

In another study, a stable curcumin NP formulation was developed and tested in vitro and in Tg2576 mice [334]. Before and after treatment, memory was measured by radial arm maze and contextual fear conditioning tests. Nanocurcumin produced significantly better cue memory in the contextual fear conditioning test than placebo, and tendencies toward better working memory in the radial arm maze test than ordinary curcumin or placebo. Amyloid plaque density, pharmacokinetics, and Madin–Darby canine kidney cell monolayer penetration were measured to further understand in vivo and in vitro mechanisms. Nanocurcumin produced significantly higher curcumin concentrations in plasma and in brain than ordinary curcumin.

Studies in animals have reported that normalized or elevated Cu levels can inhibit or even remove Alzheimer's disease-related pathological plaques and exert a desirable amyloid-modifying effect. Engineered nanocarriers composed of diverse core-shell architectures to modulate Cu levels under physiological conditions through bypassing the cellular Cu uptake systems were developed [335]. Two different nanocarrier systems were able to transport Cu across the plasma membrane of yeast or higher eukaryotic cells, **CS-NPs** (core-shell nanoparticles) and **CMS-NPs** (core-multishell nanoparticles). Intracellular Cu levels could be increased up to 3-fold above normal with a sublethal dose of carriers. Both types of carriers released their bound guest molecules into the cytosolic compartment where they were accessible for the Cu-dependent enzyme SOD1. In particular, CS-NPs reduced Aβ levels and targeted intracellular organelles more efficiently than CMS-NPs. Fluorescently labeled CMS-NPs unraveled a cellular uptake mechanism, which depended on clathrin-mediated endocytosis in an energy-dependent manner. In contrast, the transport of CS-NPs was most likely driven by a concentration gradient.

NAP-loaded TGN-(phage-display peptide)-modified nanoparticles (TGN-NP/NAP) showed better improvement in spatial learning than NAP solution and non-modified NAP-loaded nanoparticles in Morris water maze experiment [336] after intravenous administration. The crossing number of the mice with memory deficits recovered after treatment with TGN-NP/ NAP in a dose dependent manner (same also observed in AChE and ChAT activity), and no morphological damage and or detectable Aβ plaques were found in mice hippocampus and cortex treated with TGN-NP/NAP.

Specific carbon-based nanostructures have shown some promising therapeutic effects in AD. For instance, radical scavenging entities, such as carboxy-fullerenes (C60) could trap multiple radicals and have been consequently exploited as "radical sponges." In this view, researchers investigated the ability of water-soluble C60 carboxylic acid derivatives, containing three malonic acid groups per molecule, to reduce the apoptotic neuronal death induced by exposure to Aβ1–42 [337].

Another hypothesis about AD stipulates that calcium channels may play an important role in mediating Aβ activity on neurons. More precisely, the neurodegeneration could be mediated by an increase of Ca2+ influx caused by Aβ aggregates that would be able to create membrane channels permeable to Ca2+. The antioxidative effect of fullerenol-1 on the in vitro reduction of Aβ-related toxicity was proven [338]. Fullerenol-1 was found to be able to attenuate the increase of intracellular Ca2+ concentration promoted by Aβ aggregates, either by interacting with the membrane lipid components and thus changing the membranes permeability, or by altering the lipid peroxidation and the membrane composition. PEG-C(60)-3, a C(60) fullerene derivative incorporating poly(ethylene glycol), and its pentoxifylline-bearing hybrid (PTX-C(60)-2) were investigated against β-amyloid (Aβ) (25–35)-induced toxicity toward Neuro-2A cells [339]. PEG-C(60)-3 and PTX-C(60)-2 significantly reduced Aβ(25–35)-induced cytotoxicity, with comparable activities in decreasing reactive oxygen species and maintaining the mitochondrial membrane potential. Cytoprotection by PEG-C(60)-3 and PTX-C(60)-2 was partially diminished by an autophagy inhibitor, indicating that the elicited autophagy and antioxidative activities protect cells from Aβ damage and PTX-C(60)-2 was more effective than PEG-C(60)-3 at enduring the induced autophagy. Sodium fullerenolate Na (4)[C(60)(OH) (~30)] (NaFL), a water soluble polyhydroxylated (60)fullerene derivative, destroys amyloid fibrils of the Aβ(1–42) peptide in the brain and prevents their formation in in vitro experiments [340]. The cytotoxicity of NaFL was found to be negligibly low with respect to nine different culture cell lines. At the same time, NaFL showed a very low acute toxicity in vivo. The maximal tolerable dose (MTD) and LD50 for NaFL correspond to 1000 mg kg(-1) and 1800 mg kg(-1), respectively, as revealed by in vivo tests in mice.

CdTe nanoparticles (NPs) can efficiently prevent fibrillation of amyloid peptides. The process is based on the multiple binding of Aβ oligomers to CdTe NPs. The molar efficiency and the inhibition mechanism displayed by the NPs are analogous to the mechanism found for proteins responsible for the prevention of amyloid fibrillation in the human body.

The latest novel strategy reported concerns self-assembled polyoxometalate-peptide (POM-P) hybrid NPs as bifunctional Aβ inhibitors [341]. The two-in-one bifunctional POM-P NPs show an enhanced inhibition effect on amyloid aggregation in mice cerebrospinal fluid. Incorporating a clinically used Aβ fibril-staining dye, Congo Red (CR), into the hybrid colloidal spheres, the NPs can act as an effective fluorescent probe to monitor the inhibition process of POM-P via CR fluorescence change. This flexible organic-inorganic hybrid system may prompt the design of new multifunctional materials for AD treatment.

4.7   Summary of Current Progress and Future Challenges

As presented above, many nanotechnological approaches have been proposed for the diagnosis and therapy of AD. A lot of progress has been made in the area of development of extremely sensitive analytical techniques to measure AD-related biomarkers in vitro, as Aβ peptides and tau proteins, and such methodologies have decreased the lowest measurable concentrations by many orders of magnitude. Such techniques will definitely contribute towards the development of reliable diagnostic tools for AD in the near future, and help in the development of successful therapies, since it will become easier and faster to reliably monitor the diagnostic and/or therapeutic effect of each proposed approach. This will also contribute in a potentially faster delivery of proposed therapies from the lab to the clinic and to the market.

In the area of in vivo diagnostics and therapeutics with nanotechnological approaches, different systems have been proposed and some of them have been found to be promising in preclinical in vitro and—in some cases—also in vivo investgations. Nevertheless, a very big challenge for such systems to be successful is the barrier between blood and brain, which has not been solved up to now, regardless of the numerous methodologies developed and proposed.

One method to overcome the problem of the BBB is to target AD-related pathologies in the blood, knowing that this might reduce brain levels, as is the case for Aβ peptides, although this is still a controversial issue. Nevertheless, recently some nanoparticulate systems with high affinity for Aβ peptides have been developed and are currently under in vivo investigation in order to detect possible reduction in amyloid loads in the brain as a result of the increased blood clearance. Other approaches include the "opening" of BBB with various chemicals; however, the toxicity related issues with such approaches might be severe. If one decides to develop NPs to target neurodegenerative diseases in the brain (in which case transport of the NPs across the BBB is required), in the view of the authors of this chapter, it should become a priority to focus on nanoparticulate systems that have known, and clinically tested (for many years) biocompatibility. This is the brain, which is the "CPU" (central proccessing unit) of human beings; thereby toxicity/compatibility issues are expected to be exaggerated (compared to other biological tissues/organs). Thereby, the priority should be given to liposomes, albumin nanoparticulates and perhaps PLGA carriers, which have been used in clinical practice and tested for years.

AD is a serious ND disease with high social and economical burden, which is expected to geometrically increase in the next decade, due to the increased life expectancy and therefore aging population. Today there is no accurate diagnosis and no therapy. Because of this, the funding provided for AD is constantly increasing; therefore, we believe that if this funding is distributed and used appropriately it will be possible to soon have safe diagnostic and perhaps also good therapeutic approaches.

Acknowledgments

Funding from the European Community's Seventh Framework Programme (FP7/2007–2013) under grant agreements n° 212043, is acknowledged.

References

1. J.E. Riggs, *Neurobiol Aging*, Vol. 1, p. 3, 2001.
2. G. Modi, V. Pillay, Y. E. Choonara, V.M.K. Ndesendo, L.C. Du Toit, D. Naidoo, *Prog Neurobiol*, Vol. 88, p. 272, 2009.
3. H. Koo, M.S. Huh, I.-C Sun,S.H. Yuk, K. Choi, K. Kim, I.C. Kwon, *Accounts of Chemical Research*, Vol. 44, p. 1018, 2011.

4. Y. Liu, H. Miyoshi, and M. Nakamura, *Int. J. Cancer*, Vol. 120, p. 2527, 2007.

5. T. M. Allen, and P. R Cullis, *Science*, Vol. 201, p. 1818, 2004.

6. A H. Faraji and P. Wipf, *Bioorganic & Medicinal Chemistry*, Vol. 17, p. 2950, 2009.

7. J. Xie, S.Lee, and X. Chen, *Adv. Drug Delivery Rev*, Vol. 62, p. 1064, 2010.

8. R.A. Petros and J.M DeSimone, *Nat. Rev. Drug Discovery* Vol. 9, p. 615, 2010.

9. F. Danhier, O.Feron and V. Preat, *J. Controlled Release*, Vol. 148, p. 135, 2010.

10. H. Maeda, *Adv. Enzyme Regul*, Vol 41, p. 189, 2001.

11. R. Cheng, F. Meng, C. Deng, H-A. Klok and Z. Zhong, *Biomaterials* Vol 34, p. 3647, 2013.

12. Y. Wang, J.D. Byrne, M.E. Napier and J.M. DeSimone, *Adv Drug Deliv Rev*, Vol. 64, p. 1021, 2012.

13. S.J. Lee, H. Koo, D-E. Lee, S. Min, S. Lee, X. Chen, Y. Choi, J.F. Leary, K. Park, S. Jeong, I.C. Kwon, K. Kim, and K. Choi, *Biomaterials*, Vol. 32, p. 4021, 2011.

14. K. Kim, J.H. Kim, H. Park, Y-S. Kim, K. Park, H. Nam, S. Lee, J.H. Park, R-W. Park, I-S. Kim, K. Choi, S.Y. Kim, K. Park, I.C. Kwon, *J. Control Rel*, Vol. 146, p. 219, 2010.

15. R.M. Pearson, S. Sunoqrot, H-J. Hsu, J.W. Bae and S. Hong, *Therapeutic Delivery* Vol. 3, p. 941, 2012.

16. a) V. P. Torchilin, *Pharm. Res*, Vol. 24 p. 1, 2007; b) S. Biswas, V.P. Torchilin, *Pharmaceuticals* Vol. 6, p. 161, 2013.

17. A. Maham, Z. Tang, H. Wu, J. Wang and Y. Lin, *Small*, Vol. 5, p. 1706, 2009.

18. H.L. Wong, X.Y. Wu, and R. Bendayan, *Adv Drug Deliv Rev*, Vol. 64, p. 686, 2012.

19. N.T. Huynh, C. Passirani, P. Saulnier and J.P. Benoit, *Int. J. Pharm.* Vol. 379, p. 201, 2009.

20. L. Fenart, A. Casanova, B. Dehouck, C. Duhem, S. Slupek, R. Cecchelli and D. Betbeder, *J. Pharmacol. Exp. Ther,* Vol. 291, p. 1017, 1999.

21. a] T.M. Goppert and R.H.Muller, *J. Drug Target*, Vol. 13, p. 179, 2005; b] B.A. Kerwin, *J. Pharm. Sci,* Vol 97, p. 2924, 2008.

22. D. Peer, J.M. Karp, S. Hong, O.C. Farokhzad, R. Margalit and R. Langer, *Nat. Nanotechnol*, Vol. 2, p. 751, 2007.

23. a) S.G. Antimisiaris, D. Fatouros and P. Kallinteri, Liposome Technology In *Pharmaceutical Manufacturing Handbook*, Chapter 5.3, Eds. John Wiley and Sons, p. 443, 2008; b) S.G. Antimisiaris. Nanosized Liposomes in Drug Delivery, In *Encyclopedia of Nanoscience and Nanotechnology*, Edited by H.S. Nalwa, Vol. 18, p. 455, 2011 (ISBN: 1-588883-167-1)

24. Y.W. Jun, J.H. Lee and J. Cheon, *Angew. Chem., Int. Ed*, Vol. 47, p. 5122, 2008.

25. Y. Liu, M. Shi, M. Xu, H, Yang, C. Wu, *Expert Opinion on Drug Delivery*, Vol. 9, p. 1197, 2012.

26. J.T. Jang, H. Nah, J.H. Lee, S.H. Moon, M.G. Kim and J. Cheon, *Angew. Chem., Int. Ed.*, Vol. 48, p. 1234, 2009.

27. K. Stojanov, I.S. Zuhorn, R.A.J.O. Dierckx and E.F.J De Vries, *Pharm Research*, Vol. 29, p. 3213, 2012.

28. S.-F. Lai, C.-C. Chien, W.-C. Chen, H.-H. Chen, Y.-Y. Chen, C.-L. Wang, Y. Hwu, (...), and G. Margaritondo, *Biotechnology Advances* Vol. 31, P. 362, 2013.

29. I.C. Sun, D.K. Eun, J.H. Na, S. Lee, I.J. Kim, I.C. Youn, C.Y. Ko, H.S. Kim, D. Lim, K. Choi, P.B. Messersmith, T.G. Park, S.Y. Kim, I. C. Kwon, K. Kim and C.H. Ahn, *Chem.–Eur. J.*, Vol. 15, p. 13341, 2009.

30. X.M. Qian and S.M. Nie, *Chem. Soc. Rev*, Vol. 37, p. 912, 2008.

31. A.J. Mieszawska, W.J.M. Mulder, Z.A. Fayad, and D.P. Cormode, Molecular Pharmaceutics Vol. 10, p. 831, 2013.

32. G. B. Braun, A. Pallaoro, G.H.Wu, D. Missirlis, J. A. Zasadzinski, M. Tirrell and N. O. Reich, *ACS Nano*, Vol. 3, p. 2007, 2009.

33. Y. Xing, Z. Xia, and J. Rao, *IEEE Trans Nanobiosci*, Vol. 8, p. 4, 2009.

34. M. Benezra, O. Penate-Medina, P. B. Zanzonico, D. Schaer, H. Ow, A. Burns, E. DeStanchina, V. Longo, E. Herz, S. Iyer, J. Wolchok, S.M. Larson, U. Wiesner and M. S. Bradbury, *J. Clin.Invest*, Vol. 121, p. 2768, 2011.

35. H. Kim, S. Kim, C. Park, H. Lee, H. J. Park and C. Kim, *Adv. Mater*, Vol. 22, p. 4280, 2010.

36. T. Wang, L. Zhang, Z. Su, C. Wang, Y. Liao and Q. Fu, *ACS Appl. Mater. Interfaces*, Vol. 3, p. 2479, 2011.

37. J. H. Park, G. von Maltzahn, L. L. Ong, A. Centrone,T. A. Hatton, E. Ruoslahti, S. N. Bhatia and M. J. Sailor, *Adv.Mater.* Vol. 22, p. 880, 2010.

38. Z. Liu, W. Cai, L. He, N. Nakayama, K. Chen, X. Sun, X. Chen and H. Dai, *Nat. Nanotechnol*, Vol. 2, p. 47, 2007.

39. M. S. Dresselhaus and P. T. Araujo, *ACS Nano*, Vol. 4, p. 6297, 2010.

40. H. Kobayashi and M. W. Brechbiel, *Adv. Drug Delivery Rev*, Vol. 57, p. 2271, 2005.

41. H.S. Choi, W. Liu, P. Misra, E. Tanaka, J. P. Zimmer, B. Itty Ipe, M. G. Bawendi and J. V. Frangioni, *Nat. Biotechnol*, Vol. 25, p. 1165, 2007.

42. J. Kolosnjaj, H. Szwarc and F. Moussa, *Adv. Exp. Med. Biol*, Vol. 620, p. 181, 2007.

43. G.P.A.K. Michanetzis, Y.F. Missirlis, and S.G. Antimisiaris, *J. Biomed. Nanotechnol*, Vol. 4, p. 218, 2008.

44. P. Mayeux, *Annu Rev Neurosci*, Vol. 26, p. 81, 2003.

45. J.L. Gilmore, X. Yi, L. Quan, and A.V. Kabanov, *J Neuroimmune Pharmacol*, Vol. 3, p. 83, 2008

46. A.G.de Boer, and P.J.Gaillard, *Annu Rev Pharmacol Toxicol*, Vol. 47, p. 323, 2007.

47. J. R. Kanwar, X. Sun, V. Punj, B. Sriramoju, R. R. Mohan, S-F. Zhou, A. Chauhan, and R. K. Kanwar, *Nanomedicine: Nanotechnology, Biology, and Medicine*, Vol. 8, p. 399, 2012.

48. R.H. Swartz, S.E. Black, and P. St George-Hyslop, *Can J Neurol Sci,* Vol. 26, p. 77, 1999.

49. Z. Cui, P.R. Lockman, C.S. Atwood, C.H. Hsu, A. Gupte, D.D. Allen, *et al. Eur J Pharm Biopharm,* Vol. 59, p. 263, 2005.

50. B.P. Imbimbo, J. Lombard, and N. Pomara. *Neuroimaging Clin N Am*, Vol. 15, p. 727, 2005.

51. G Benzi, and A Moretti. *Neurobiol Aging*, Vol. 16, p. 661, 1995.

52. L.M. Ittner, J.Götz, *Nat Rev Neurosci*, Vol. 12, p. 6572, 2011.

53. A.P. Sagare, R.D.Bell, and B.V Zlokovic, *Cold Spring Harbor perspectives in medicine*, Vol. 2, p. 10, 2012.

54. A. García-Osta, M. Cuadrado-Tejedor, C. García-Barroso, J. Oyarzábal, R. Franco, *ACS Chemical Neuroscience*, Vol. 3, p. 832, 2012.

55. RJ. Boado, *Adv Drug Deliv Rev*, Vol 15, p. 73, 1995.

56. Y. Persidsky, S.H. Ramirez, J. Haorah, and G.D. Kanmogne, *J. Neuroimmune, Pharmacol.*, Vol. 1, p. 223, 2006.

57. H. Wolburg, and A. Lippoldt, *Vascul. Pharmacol.*, Vol. 38, p. 323, 2002.

58. B.T. Hawkins, and R.D. Egleton, *Curr. Top. Dev. Biol.*, Vol. 80, p. 277, 2008.

59. N. J Abbott, L Rönnbäck and E Hansson, *Nat Rev Neuroscience*, Vol. 7, p. 41, 2006.

60. F.L. Cardoso, D. Brites, and M.A. Brito, *Brain Res. Rev.*,Vol. 64, p. 328, 2010.

61. J. Bernacki, A. Dobrowolska, K. Nierwinska, and A. Malecki, *Pharmacol. Rep.*, Vol. 60, p. 600, 2008.

62. P.A. Stewart, *Cell. Mol. Neurobiol.*, Vol. 20, p. 149, 2000.

63. N.J. Abbott, *Cell. Mol. Neurobiol.*, Vol. 25, p. 5, 2005.

64. Hawkins, B.T., Davis, T.P. *Pharmacological Reviews*, Vol. 57 (2) , p. 173, 2005.

65. M. Ramsauer, J. Kunz, D. Krause, R. Dermietzel, and J. Cereb, Blood Flow Metab, Vol. 18, p. 1270, 1998.

66. M. Ramsauer, D. Krause, and R. Dermietzel, *FASEB J.*, Vol. 16, p. 1274, 2002.

67. S. Dohgu, F. Takata, A. Yamauchi, S. Nakagawa, T. Egawa, M. Naito, T. Tsuruo, Y. Sawada, M. Niwa, and Y. Kataoka, *Brain Res.* Vol. 1038(2), p. 208, 2005.

68. H. Wekerle, J. *Infect. Dis.*, Vol. 186 (Suppl 2), p. S140, 2002.

69. R. Daneman and M. Rescigno, *Immunity*, Vol. 31, p. 722, 2002.

70. K. Williams, X. Alvarez, A.A. Lackner, and K. Glia Vol. 36, p. 156, 2001.

71. W. J. Streit, J. R. Conde, S. E. Fendrick, B. E. Flanary, and C. L. Mariani, *Neurol. Res.*, Vol. 27, p. 685, 2005.

72. Y. Chen and L Liu, *Advanced Drug Delivery Reviews*, Vol. 64, Issue 7, p. 640, 2012.

73. W.A. Vandergrift, S.J. Patel, J.S. Nicholas, and A K. Varma, *Neurosurg.* Focus 20 (4), p. E10. 2006.

74. L.S. Ashby, and T.C. Ryken, *Neurosurg. Focus*, Vol. 20 E3, 2006.

75. D.D. Allen, P.R. Lockman, K.E. Roder, L.P. Dwoskin, and P.A. Crooks, *J. Pharmacol. Exp. Ther.*, Vol. 304, p. 1268, 2003.

76. W.M. Pardridge, *Mol. Interv.*, Vol. 3, p. 90, 2003.

77. Y. Zhang and W.M. Pardridge, *J. Pharmaco. Exp. Therap.*, Vol. 313, p. 1075, 2005.

78. Y.Zhang and W.M. Pardridge, *Brain Research*, Vol. 1111 (1), p. 227, 2006.

79. M.Z. Kounnas, Moir, G.W. Rebeck, A.I. Bush, W.S. Argraves, R.E. Tanzi, B.T. Hyman, and D.K. Strickland, *Cell*, Vol. 82, p. 331, 1995.

80. J. Hertz, and D.K Strickland, *J. Clin. Invest.*, Vol. 108, p. 779. 2001.

81. J. Kreuter, P. Ramge, V. Petrov, S. Hamm, S.E. Gelperina, B. Engelhardt, R. Alyautdin, H. von Briesen, and D.J. Begley, *Pharma Res.*, Vol. 20, p. 409, 2003.

82. J. Kreuter, *Int. Congr. Ser.*, Vol. 1277, p. 85, 2005.

83. S.C.J Steiniger, J. Kreuter, A.S. Khalansky, I.N. Skidan, A.I. Bobruskin, Z. S.Smirnova, S.E. Severin, R. Uhl, M. Kock, K.D. Geigerand, and S.E.Gelperina, *Int. J. Cancer*, Vol. 109, p. 759, 2004.

84. M.J. Coloma, H.J. Lee, A. Kurihara, E.M. Landaw, R.J. Boado, Morrison, and W.M. Pardridge, *Pharm. Res.*, Vol. 17, p. 266, 2000.

85. A. R. Jones, and E.V. Shusta, *Pharm Res*, Vol. 24, p. 1759, 2007.

86. R.J. Boado, Y. Zhang, Y. Zhang, Y. Wang, and W.M. Pardridge, *Biotech. Bioeng*, Vol. 100, p. 387, 2007.

87. P.J. Gaillard, C.C. Visser, and A.G. de Boer, *Expert Opinion on Drug Delivery*, Vol. 2 (2), p. 299, 2005.

88. Raab, G., Klagsbrun, M., Bioch. *Bioph. Acta*, Vol. 1333, p. F179, 1997.

89. K. Michaelis, M.M. Hoffmann, S. Dreis, E. Herbert, R.N. Alyautdin, M. Michaelis, J. Kreuter, and K. Langer, *J. Pharmacol. Exp. Ther.*, Vol. 317, p. 1246, (2006)

90. W.M. Pardridge, *Adv. Drug Deliv. Rev.*, Vol. 15, p. 5, 1995.

91. E.N. Lerner, E.H. van Zanten, and G.R. Stewart, *J. Drug Target.*, Vol. 12, p. 273, 2004.

92. S.V. Dhuria, L.R. Hanson, and W.H. Frey III, *J. Pharm. Sci.*, Vol. 99, p. 1654, 2010.

93. L. Illum, P.Watts, A.N. Fisher, M. Hinchcliffe, H. Norbury, I. Jabbal-Gill, R. Nankervis, and S.S. Davis, *J. Pharmacol. Exp. Ther.*, Vol. 301, p. 391, 2002.

94. Y.C. Wong, and Z. Zuo, *Pharm. Res.*, Vol. 27, p. 1208, 2010.

95. A. Reichel, and D. J Begley, *Pharm. Res.*, Vol. 15, p. 1270, 1998.

96. L. Di, E. H. Kerns, K. Fan, O. J. McConnell, and G. T. Carter, *Eur. J. Med. Chem.*, Vol. 38, p. 223, 2003.

97. S. Nakagawa, M.A. Deli, S. Nakao, M. Honda, K. Hayashi, R. Nakaoke, Y. Kataoka, and M. Niwa, *Cell. Mol. Neurobiol.*, Vol. 27 , p. 687, 2007.

98. R. Cecchelli, B. Dehouck, L. Descamps, L. Fenart, V.V. Buee-Scherrer, C. Duhem, S. Lundquist, M. Rentfel, G. Torpier, and M.P. Dehouck, *Adv. Drug Deliv. Rev.*, Vol. 36, p. 165, 1999.

99. P. Garberg, M. Ball, N. Borg, R. Cecchelli, L. Fenart, R.D. Hurst, T. Lindmark, A. Mabondzo, J.E. Nilsson, T.J. Raub, D. Stanimirovic, T. Terasaki, J.O. Oberg, and T. Osterberg, *In Vitro*, Vol. 19, p. 299, 2005.

100. K. Biernacki, J.P. Antel, M. Blain, S. Narayanan, D.L. Arnold, and A. Prat, *Arch. Neurol.*, Vol. 62, p. 563, 2005.

101. C. Coisne, L. Dehouck, C. Faveeuw, Y. Delplace, F. Miller, C. Landry, C. Morissette, L. Fenart, R. Cecchelli, P. Tremblay, and B. Dehouck, *Lab. Invest.*, Vol. 85, p. 734, 2005.

102. J. Meyer, U. Mischeck, M. Veyhl, K. Henzel, and H.J. Galla, *Brain Res.*, Vol. 514, p. 305, 1990.

103. N. Perriere, S. Yousif, S. Cazaubon, N. Chaverot, F. Bourasset, S. Cisternino, X. Decleves, S. Hori, T. Terasaki, M. Deli, J.M. Scherrmann, J. Temsamani, F. Roux, and P.O. Couraud, *Brain Res.*, Vol. 1150, p. 1, 2007.

104. Y. Persidsky, M. Stins, D. Way, M.H. Witte, M. Weinand, K.S. Kim, P. Bock, H.E. Gendelman, and M. Fiala, *J. Immunol.*, Vol. 158, p. 3499, 1997.

105. K.A. Stanness, J.F. Neumaier, T.J. Sexton, G.A. Grant, A. Emmi, D.O. Maris, and D. Janigro, *Neuroreport*, Vol. 10, p. 3725, 1999.
106. A. Zozulya, C. Weidenfeller, and H.J. Galla, *Brain Res*, Vol. 1189, p. 1, 2008.
107. M. Aschner, V.A. Fitsanakis, A.P. dos Santos, L. Olivi, *Methods Mol. Biol.*, Vol. 34, p. 11, 2006.
108. F. Roux, O. Durieu-Trautmann, N. Chaverot, M. Claire, P. Mailly, J.M. Bourre, A.D. Strosberg, and P.O. Couraud, *J. Cell. Physiol.*, Vol. 159, p. 101, 1994.
109. F. Roux, and P.O. Couraud, *Cell. Mol. Neurobiol.*, Vol. 25, p. 41, 2005.
110. I.E. Blasig, H. Giese, M.L. Schroeter, A. Sporbert, D.I. Utepbergenov, I.B. Buchwalow, K. Neubert, G. Schönfelder, D. Freyer, I. Schimke, W.E. Siems, M. Paul, R.F. Haseloff, and R. Blasig, *Microvasc Res.* Vol 62(2), p. 114, 2001.
111. O. Durieu-Trautmann, C. Fédérici, C. Créminon, N. Foignant-Chaverot, F. Roux, M. Claire, A.D. Strosberg, and P.O. Couraud, *J Cell Physiol.* Vol. 155(1), p. 104, 1993.
112. M. Laschinger, and B. Engelhardt, *J. Neuroimmunol.*, Vol. 102, p. 32, 2000.
113. M. Teifel, and P. Friedl, *Exp. Cell Res.*, Vol. 228, p. 50, 1996.
114. P.V. Afonso, S. Ozden, M.C. Prevost, C. Schmitt, D. Seilhean, B. Weksler, P.O. Couraud, A. Gessain, I.A. Romero, and P.E. Ceccaldi, *J. Immunol.*, Vol. 179, p. 2576, 2007.
115. G. Schreibelt, G. Kooij, A. Reijerkerk, R. van Doorn, S.I. Gringhuis, S. van der Pol, B.B. Weksler, I.A. Romero, P.O. Couraud, J. Piontek, I.E. Blasig, C.D. Dijkstra, E. Ronken, and H.E. de Vries, *FASEB J.*, Vol. 21, p. 3666, 2007.
116. B.B. Weksler, E.A. Subileau, N. Perriere, P. Charneau, K. Holloway, M. Leveque, H. Tricoire-Leignel, A. Nicotra, S. Bourdoulous, P. Turowski, D.K. Male, F. Roux, J. Greenwood, I.A. Romero, and P.O. Couraud, *FASEB J.*, Vol. 19, p. 1872, 2005.
117. S. Santaguida, D. Janigro, M. Hossain, E. Oby, E. Rapp, and L. Cucullo, *Brain Res.* 1109, p. 1, 2006.
118. L. Cucullo, P.O. Couraud, B. Weksler, I.A. Romero, M. Hossain, E. Rapp, and D. Janigro, *J. Cereb. Blood Flow Metab.*, Vol. 28, p. 312, 2008.
119. K. A. Stanness, L. E Westrum, E. Fornaciari, P. Mascagni, J. A. Nelson, S. G. Stenglein, T. Myers, and D. Janigro,. *Brain Res.* Vol. 771, p. 329, 1997.
120. L. Cucullo, M. S. McAllister, K. Kight, L. Krizanac-Bengez, M. Marroni, M. R. Mayberg, K. A. Stanness, and D. Janigro, *Brain Res.* Vol. 951, p. 243, 2002.
121. Q. R. Smith, S. Nag, Ed. Humana, and N. J. Totowa, *Biology and Research Protocols*, p. 193, 2003.
122. X. Liu, and C. Chen, *Curr. Opin. Drug Disc. Dev.,* Vol. 8, p. 505, 2005.
123. J. A. Gratton, M. H Abraham, M. W Bradbury, and H. S. Chadha, *J. Pharm. Pharmacol.*, Vol. 49, p. 1211, 1997.
124. D. K. Hansen, *J. Pharm. Biomed. Anal.*, Vol. 27, p. 945, 2002.
125. M. A. Chishti , D. S. Yang , C. Janus , A. L. Phinney , P. Horne , J. Pearson , R. Strome , N. Zuker, J. Loukides, J. French, S. Turner, G. Lozza, M. Grilli , S. Kunicki , C. Morissette , J. Paquette, F. Gervais , C. Bergeron , P. E. Fraser , G. A. Carlson , P. S. George-Hyslop , and D. Westaway , *J. Biol. Chem.*, Vol. 276, p. 21562, 2001.

126. B. T. Lamb , K. A. Bardel , L. S. Kulnane , J. J. Anderson, G. Holtz, S. L. Wagner, S. S. Sisodia, and E. J. Hoeger , *Nat. Neurosci.*, Vol. 2, p. 695, 1999.

127. Q. Guo , W. Fu , B. L. Sopher , M. W. Miller , C. B. Ware , G. M. Martin , and M. P. Mattson , *Nat. Med.*, Vol. 5, p. 101, 1999.

128. K. Duff , H. Knight , L. M. Refolo , S. Sanders , X. Yu , M. Picciano , B. Malester , M. Hutton , J. Adamson , M. Goedert , K. Burki , and P. Davies, *Neurobiol. Dis.*, Vol. 7, p. 87, 2000.

129. P. M. Sullivan, H. Mezdour, S. H. Quarfordt, and N. J. Maeda, *Clin. Invest.*, Vol. 102, p. 130, 1998.

130. D. Schenk, R. Barbour, and W. Dunn, G. Gordon, H. Grajeda, T. Guido, *et al.*, *Nature*, Vol 400, p. 173, 1999.

131. K. Hsiao , P. Chapman , S. Nilsen , C. Eckman, Y. Harigaya , S. Younkin , F. Yang , and G. Cole , *Science*, Vol. 274, p. 99, 1996.

132. C. Sturchler-Pierrat , D. Abramowski , M. Duke , K. H. Wiederhold , C. Mistl , S. Rothacher , B. Ledermann , K. Bürki , P. Frey , P. A. Paganetti , C. Waridel , M. E. Calhoun , M. Jucker , A. Probst , M. Staufenbiel , and B. Sommer, *Proc. Natl. Acad. Sci. USA,* Vol. 94, p. 13287, 1997.

133. L. Holcomb, M. N. Gordon , E. McGowan, X. Yu, S. Benkovic, P. Jantzen, K. Wright, I. Saad, R. Mueller, D. Morgan, S. Sanders, C. Zehr, K. O'Campo, J. Hardy, C. M. Prada, C. Eckman, S. Younkin, K. Hsiao, and K. Duff, *Nat. Med.*, Vol. 4, p. 97, 1998.

134. M. E. Calhoun , K. H. Wiederhold , D. Abramowski , A. L. Phinney , A. Probst, C. Sturchler-Pierrat , M. Staufenbiel, B. Sommer, and M. Jucker, *Nature*, Vol. 395, p. 755, 1998.

135. A. Probst , J. Götz, K. H. Wiederhold, M. Tolnay, C. Mistl, A. L. Jaton, M. Hong, T. Ishihara, V. M. Lee, J. Q. Trojanowski, R. Jakes, R. A .Crowther, M. G. Spillantini, K. Bürki , and M. Goedert, *Acta Neuropathol. (Berl.),* Vol. 99, p. 469, 2000.

136. J. Lewis, E. McGowan, J. Rockwood, H. Melrose, P. Nacharaju, M. Van Slegtenhorst, K. Gwinn-Hardy, M. Paul Murphy , M. Baker , X.Yu, K. Duff, J. Hardy, A. Corral, W.L. Lin, S.H. Yen, D.W. Dickson, P. Davies, and M. Hutton , *Nat. Genet.*, Vol. 25, p. 402, 2000.

137. K. Spittaels, C. Van den Haute, J. Van Dorpe, K. Bruynseels, K. Vandezande, I. Laenen, H. Geerts, M. Mercken, R. Sciot, A. Van Lommel, R. Loos, and F. Van Leuven, *Am. J. Pathol.*, Vol. 155, p. 2153, 1999.

138. C. Andorfer , Y. Kress , M. Espinoza , R. de Silva , K. L. Tucker , Y. A. Barde , K. Duff , and P. Davies, *J. Neurochem.*, Vol. 86, p. 582, 2003.

139. S. Oddo, A Caccamo, J. D. Shepherd, M. P. Murphy, T. E. Golde, R. Kayed, R. Metherate, M. P. Mattson, Y. Akbari, and F. M. LaFerla., *Neuron*, Vol. 39, p. 409, 2003.

140. D. Games, D. Adams, R. Alessandrini, R. Barbour, P. Berthelette, Blackwell, T. Carr, J. Clemens, T. Donaldson, and F. Gillespie, *Nature*, Vol. 373, p. 523, 1995.

141. M. C. Irizarry, F. Soriano, M. McNamara , K. J. Page, D. Schenk , D. Games, and B. T. Hyman , *J. Neurosci.*, Vol. 17, p. 7053, 1997.

142. H. Yu, C. A. Saura, S. Y. Choi, L. D. Sun, X. Yang, M. Handler, T. Kawarabayashi, L. Younkin, B. Fedeles, M. A. Wilson, S. Younkin, E. R. Kandel, A. Kirkwood, and J. Shen, *Neuron*, Vol. 31, p. 713, 2001.

143. M. S. Wolfe, *J. Neurochem.*, Vol. 76, p. 1615, 2001.

144. D.M. Holtzman, K. R. Bales, T. Tenkova, A. M. Fagan, M. Parsadanian, L. J. Sartorius, B. Mackey, J. Olney, D. McKeel, D. Wozniak, and S. M. Paul, *Proc. Natl. Acad. Sci. USA*, Vol. 97, p. 2892, 2000.

145. J. Götz, A. Probst, M. G. Spillantini, T. Schäfer, R. Jakes, K. Bürki, and M. Goedert, *Science,* Vol. 293, p. 1487, 2001.

146. a) H. W. Querfurth, and F. M. LaFerla, *N Engl J Med*, Vol. 362, p. 329, 2010.
b) H. Potschka, *Handb Exp Pharmacol*, Vol. 197, p. 411, 2010.

147. G. A. Silva. *Ann N YAcad Sci*, Vol. 1199, p. 221, 2010.

148. G. Amadoro, M. T. Ciotti, M. Costanzi, V. Cestari, P. Calissano, and N. Canu, *Proc. Natl. Acad. Sci. U.S.A.*, Vol. 103, p. 2892, 2006.

149. E. D. Roberson, K. Scearce-Levie, J. J. Palop, F. Yan, I. H. Cheng, and T. Wu, *Science*, Vol. 316, p. 750, 2007.

150. S. Khatoon, I. Grundke-Iqbal, and K. Iqbal, *J. Neurochem.*, Vol. 59, p. 750, 1992.

151. L. Martin, A. Magnaudeix, F. Esclaire, C. Yardin, and F. Terro, *Brain Res.*, Vol. 1252, p. 66, 2009.

152. B. Bulic, M. Pickhardt, E.-M. Mandelkowb, and E. Mandelkowb, *Neuropharmacology* 59, p. 276, 2010

153. B. Bulic, M. Pickhardt, I. Khlistunova, J. Biernat, E. M. Mandelkow, E. Mandelkow, and H. Waldmann, *Angew. Chem. Int. Ed. Engl.*, 46, p. 9215, 2007.

154. M. Pickhardt, G. Larbig, I. Khlistunova, A. Coksezen, B. Meyer, E. M. Mandelkow, B. Schmidt, and E. Mandelkow, *Biochemistry*, 46, p. 10016, 2007.

155. M. Pickhardt, J. Biernat, I. Khlistunova, Y. P. Wang, Z. Gazova, E. M. Mandelkow, and E. Mandelkow, *Curr. Alzheimer Res*, Vol. 4, p. 397, 2007.

156. S. Taniguchi, N. Suzuki, M. Masuda, S. Hisanaga, T. Iwatsubo, M. Goedert et al., *J Biol Chem*, Vol. 280, p. 7614, 2005.

157. E. E. Congdon, Y. H. Figueroa, L. Wang, G. Toneva, E. Chang, J. Kuret, C. Conrad, and K. E. Duff, *J. Biol. Chem*, Vol. 284, p. 20830, 2009.

158. S. A. Sievers, J. Karanicolas, H. W. Chang, A. Zhao, L. Jiang, O. Zirafi, J. T. Stevens, J. Munch, D. Baker, and D. Eisenberg, *Nature*, Vol. 10154, 2011.

159. D. J. Selkoe, *Physiol. Rev.*, Vol. 81, p. 741, 2001.

160. M. R. Sawaya, *et al.*, *Nature* 447, p. 453, 2007.

161. J. Zheng, C. Liu, M. R. Sawaya, B. Vadla, S. Khan, R. J. Woods, D. Eisenberg, W. J. Goux, and J. S. Nowick, *J Am Chem Soc.*, Vol. 133 (9), p. 3144, 2011.

162. P. N. Cheng, C. Liu, M. Zhao, D. Eisenberg and J. S. Nowick, *Nature Chemistry*, Vol. 4, 2012.

163. D.S. Himmelstein, S. M. Ward, Jody K. Lancia, Kristina R. Patterson, Lester I. Binder, *Pharmacology & Therapeutics*, Vol. 136, p. 8, 2012.

164. V. Cavallucci, M. D'Amelio and F. Cecconi, *Mol Neurobiol*, Vol. 45, p. 366, 2012.

165. S. G. Younkin, *J Physiol Paris*, Vol. 92, p. 289, 1998.

166. P. H. Reddy, M. Manczak, P. Mao, M. J. Calkins, A. P. Reddy, and U. Shirendeb, *J. Alzheimers Dis,* Vol. 20, p. S499, 2010.

167. R. Deane, Z. Wu, A. Sagare, J. Davis, S. Du Yan, K. Hamm, F. Xu, M. Parisi, B. LaRue, H. W. Hu, P. Spijkers, H. Guo, X. Song, P.J. Lenting, W. E. Van Nostrand, B. V. Zlokovic, *Neuron,* Vol. 43, p. 333, 2004.

168. S. Vogelgesang, G. Jedlitschky, A. Brenn, L.C. Walker, *Curr. Pharm. Des.,* Vol. 17, p. 2778, 2011.

169. H. Xiong, D. Callaghan, A. Jones, J. Bai, I. Rasquinha, C. Smith, K. Pei, D.Walker, L. F. Lue, D. Stanimirovic, and W. Zhang, *J. Neurosci.,* Vol. 29, p. 5463, 2009.

170. J. B. Jaeger, S. Dohgu, J. L. Lynch, M. A. Fleegal-DeMotta, and W. A. Banks, *Brain Behav. Immun.,* Vol. 23, p. 507, 2009.

171. A. M. S. Hartz, D. S. Miller, and B. Bauer, *Mol. Pharmacol.,* Vol. 77, p. 715, 2010.

172. R. D. Bell, A. P. Sagare, A. E. Friedman, G. S. Bedi, D. M. Holtzman, R. Deane, B. V. Zlokovic, *J. Cereb. Blood Flow Metab,* p. 909, 2007.

173. J. B. Owen, R. Sultana, C. D. Aluise, M. A. Erickson, T. O. Price, G. Bu, and W. A. Banks, D. A., *Free Radic. Biol. Med.,* Vol. 49, p. 1798, 2010.

174. S. A. Farr, H. F. Poon, D. Dogrukol-Ak, J. Drake, W. A. Banks, E. Eyerman, D. A. Butterfield, and J. E. Morley, *J. Neurochemisrty,* Vol. 84, p. 1173, 2003.

175. S. Ito, S. Ohtsuki, Y. Nezu, Y. Koitabashi, S.Murata, T. Terasaki, *Fluids Barriers CNS.,* Vol. 8, p. 20, 2011.

176. J. A. Esteban, *Trends Neurosci,* Vol. 27, p. 1, 2004.

177. A. Kontush , *Free Radic Biol Med,* Vol. 31, p. 1120, 2001.

178. S. A. Frautschy, A. Baird, and G. M. Cole , *Proc Natl Acad Sci,* Vol. 88, p. 8362, 1991.

179. S. T. Liu, G. Howlett, and C. J. Barrow, *Biochemistry,* Vol. 38, p. 9373, 1999.

180. A. K. Ghosh, N. Kumaragurubaran, J. Tang, *Curr Topics Med Chem,* Vol. 5, p. 1609, 2005.

181. W. B. Stine Jr., K. N. Dahlgren, G. A. Krafft, and M. J. LaDu, *J Biol Chem.,* Vol. 278, p. 11612, 2003.

182. F. M. LaFerla, K. N. Green, and S. Oddo, *Nat Rev Neurosci,* Vol 8, p. 499, 2008.

183. I. A. Mastrangelo, M. Ahmed, T. Sato2, W. Liu, C. Wang, P. Hough and S. O. Smith *J. Mol. Biol,* Vol. 358, p. 106, 2006.

184. S. Barghorn, V. Nimmrich, A. Striebinger, C. Krantz, P. Keller, and B. Janson, *J Neurochem,* Vol. 95, p. 834, 2005.

185. M.A. Findeis. *Curr Top Med Chem.,* Vol. 2(4), p. 417, 2002.

186. Y. Porat, A. Abramowitz, and E. Gazit, *Chem. Biol. Drug Des.,* Vol. 67, p. 27, 2006.

187. M. Necula, R. Kayed, S. Milton, C. G. Glabe, *J Biol Chem,* Vol. 282, p. 10311, 2007.

188. K. Ono, K. Hasegawa, Y. Yoshiike, A. Takashima, M. Yamada, H. Naiki, *J. Neurochem.,* Vol. 81, p. 434, 2002.

189. K. Ono, K. Hasegawa, H. Naiki, M. Yamada, *Neurochem. Int.,* Vol. 48, p. 275, 2006.

190. K. Ono, Y. Yoshiike, A. Takashima, K. Hasegawa, H. Naiki, M. Yamada, *J. Neurochem.*, Vol. 87, p. 172, 2003.

191. G. Forloni, L. Colombo, L. Girola, F. Tagliavini, M. Salmona, *FEBS Let*, Vol. 487, p. 404, 2001.

192. Y. Shen, L.C. Yu, *Neurochem Res*, Vol. 33, p. 2112, 2008.

193. K. Ono, K. Hasegawa, H. Naiki, M. Yamada, J. *Neurosci. Res*, Vol. 75, p. 742, 2004.

194. F. Re, C. Airoldi, C. Zona, M. Masserini, B. la Ferla, N. Quattrocchi, and F. Nicotra, *Current Medicinal Chemistry*, Vol. 17 (27), p. 2990, 2010.

195. F. Yang, G. P. Lim, A. N. Begum, O. J. Ubeda, M. R. Simmons, S. S. Ambegaokar, P. P. Chen, R. Kayed, C. G. Glabe, S. A. Frautschy, and G. M. Cole, *J Biol Chem*, Vol. 280, p. 5892, 2005.

196. K. Ono, M. Hirohata, and M. Yamada, *Biochem. Biophys. Res. Commun*, Vol. 341, p. 1046, 2006.

197. M. Masuda, N. Suzuki, S. Taniguchi, T. Oikawa, T. Nonaka, T. Iwatsubo, G. Hisanaga, M. Goedert, and M. Hasegawa, *Biochemistry*, Vol. 45, p. 6085, 2006.

198. A. Reinke, and J. E. Gestwicki, *Chem. Biol. Drug Des.*, Vol. 70, p. 206, 2007.

199. J. M. Mason, N. Kokkoni, K. Stotty, and A. J. Doig, *Current Opinion in Structural Biology*, Vol. 13, p. 1, 2003.

200. A. Mathew, T. Fukuda, Y. Nagaoka, T. Hasumura, H. Morimoto, Y. Yasuhiko, M. Toru, V. Kizhikkilot, and D. S. Kumar, *PLoS ONE*, Vol. 7(3), p. 32616, 2012.

201. S. Mourtas, M. Canovi, C. Zona, D. Aurilia, A. Niarakis, B. La Ferla, M. Salmona, F. Nicotra, M. Gobbi, and S. G. Antimisiaris, *Biomaterials,* Vol. 32, p. 1635, 2011.

202. A. N. Lazar, S. Mourtas, I. Youssef, C. Parizot, A. Dauphin, B. Delatour, S. G. Antimisiaris, and C. Duyckaerts, *Nanomedicine*, Vol. 9, p. 712, 2013.

203. M. Taylor, S. Moore, S. Mourtas, A. Niarakis, F. Re, C. Zona, B. La Ferla, F. Nicotra, M. Masserini, S. G. Antimisiaris, M. Gregori, and D. Allsop, *Nanomedicine*, Vol. 7(5), p. 541, 2011.

204. B. Le Droumaguet, J. Nicolas, D. Brambilla, S. Mura, A. Maksimenko, L. De Kimpe, E. Salvati, C. Zona, C. Airoldi, M. Canovi, M. Gobbi , N. Magali, B. La Ferla, F. Nicotra, W. Scheper, O. Flores, M. Masserini, K. Andrieux, and P. Couvreur, *ACS Nano*, Vol. 6(7), p. 5866, 2012

205. K. Matsuzaki, *Biochim Biophys Acta*, Vol. 1768(8), p. 1935, 2007.

206. M. Gobbi, F. Re, M. Canovi, M. Beeg, M. Gregori, S. Sesana, S. Sonnino, D. Brogioli, C. Musicanti, P. Gasco, M. Salmona, and M. E. Masserini, *Biomaterials,* Vol. 31(25), p. 6519, 2010.

207. A. Kakio, S. Nishimoto, K. Yanagisawa, Y. Kozutsumi, and K. Matsuzaki, *Biochemistry*, Vol. 41(23), p. 7385, 2002.

208. T. Ariga, K. Kobayashi, A. Hasegawa, M. Kiso, H. Ishida, and T. Miyatake, *Arch Biochem Biophys*, Vol. 388(2), p. 225, 2001.

209. M. S. Lin, H. M. Chiu, F. J. Fan, H. T. Tsai, S. S. Wang, and Y. Chang, *Colloids Surf B Biointerfaces*, Vol. 58(2), p. 231, 2007.

210. G. Liu, P. Men, W. Kudo, G. Perry, and M.A. Smith, *Neurosci Lett*, Vol. 455(3), p. 187, 2009.

211. M. P. Williamson, Y. Suzuki, N. T. Bourne, T. Asakura, *Biochem J*, Vol. 397, p. 483, 2006.

212. D. A. Patel, J. E. Henry, and T. A. Good, *Brain Res*, Vol. 1161, p. 95, 2007.

213. O. Klementieva, N. Benseny-Cases, A. Gella, D. Appelhans, B. Voit, and J. Cladera, *Biomacromolecules*, Vol. 12(11), p. 3903, 2011.

214. T. Wasiak, M. Ionov, K. Nieznanski, H. Nieznanska, O. Klementieva, M. Granell, J. Cladera, J. P. Majoral, A. M. Caminade, and B. Klajnert, *Mol Pharm*, Vol. 9(3), p. 458, 2012.

215. J. Li, M. Zhu, A. B. Manning-Bog, D. A. Di Monte, and A. L. Fink, *FASEB J.*, Vol. 18, p. 962, 2004.

216. K. Ono, K. Hasegawa, H. Naiki, and M. Yamada, *Biochim. Biophys. Acta*, Vol. 1690, p. 193, 2004.

217. K. Ono, Y. Yoshiike, A. Takashima, K. Hasegawa, H. Naiki, and M. Yamada, *Exp. Neurol.*, Vol. 189, p. 380, 2004.

218. C. Airoldi, F.Cardona, E. Sironi, L. Colombo, M. Salmona, A. Silva, F. Nicotra, and B. La Ferla, *Chem Commun*, Vol. 47, p. 10266, 2012.

219. C. I. Stains, K. Mondal, and I. Ghosh, *ChemMedChem*, Vol. 2, p. 1674, 2007.

220. E. M. Sigurdsson, B. Permanne, C. Soto, T. Wisniewski, and B. Frangione. *J Neuropathol Exp Neurol.*, Vol. 59(1), p. 11, 2000

221. M. A. Chacon, M. I. Barria, C. Soto, and N. C. Ine strosa. *Mol Psychiatry*, Vol. 9 (10), p. 953, 2004.

222. B. Permanne, C. Adessi, G. P. Saborio, S. Fraga, M. J. Frossard, and J. Van Dorpe, *FASEB J*, Vol. 16 (8), p. 860, 2002.

223. M. Sadowski, J. Pankiewicz, H. Scholtzova, J. A. Ripellino, Y. Li, and S. D. Schmidt, *et al. Am J Pathol*, Vol. 165 (3), p. 937, 2004.

224. G. Juhasz, A. Marki, G. Vass, L. Fulop, D. Budai, and B. Penke, *J Alzheimers Dis*, Vol. 16(1), p. 189, 2009.

225. V. Szegedi, L. Fulop, T. Farkas, E. Rozsa, H. Robotka, and Z. Kis, *Neurobiol Dis*, Vol. 18(3), p. 499, 2005.

226. A. Frydman-Marom, M. Rechter, I. Shefler, Y. Bram, D.E. Shalev, and E. Gazit. Angew *Chem Int Ed Engl*, Vol. 48(11), p. 1981, 2009.

227. S. P. Handattu, D. W. Garber, C. E. Monroe, T. van Groen, I. Kadish, and G. Nayyar, *et al. Neurobiol Dis*, Vol. 34(3), p. 525, 2009.

228. T. van Groen, K. Wiesehan, S. A. Funke, I. Kadish, L. Nagel-Steger, and D. Willbold. *ChemMedChem*, Vol. 3(12), p. 1848, 2008.

229. T. van Groen, I. Kadish, K. Wiesehan, S. A. Funke, D. Willbold, *ChemMedChem*, Vol. 4(2), p. 276, 2009.

230. S. A. Funke, T. van Groen, I. Kadish, D. Bartnik, L. Nagel-Steger, and O. Brener, *Acs Chem Neurosci.*, Vol. 1(9) , p. 639, 2010.

231. K. Wiesehan and D. Willbol, *Chembiochem a European journal of chemical biology*, Vol. 4(9), p. 811, 2003.

232. L. Larbanoix, C. Burtea, E. Ansciaux, S. Laurent, I. Mahieu, and L. Vanderelst, *Peptides,* Vol. 32, p. 1232, 2011.

233. L. O. Tjernberg, J. Naslund, F. Lindqvist, J. Johansson, A. R. Karlstrom, and J. Thyberg, *J Biol Chem*, Vol. 271(15), p. 8545,1996

234. L. O. Tjernberg, C. Lilliehook, D. J. Callaway, J. Naslund, S. Hahne, and J. Thyberg, *J. Biol Chem*, Vol. 272(19), p. 12601,1997.

235. Y. Matsunaga, A. Fujii, A. Awasthi, J. Yokotani, T. Takakura, and T. Yamada, *Regul Pept*, Vol. 120(1–3), p. 227, 2004.

236. S. M. Chafekar, H. Malda, M. Merkx, E. W. Meijer, D. Viertl, and H. A. Lashuel, *Chembiochem*, Vol. 8, p. 1857, 2007.

237. G. Zhang, M. J. Leibowitz, P. J. Sinko, and S. Stein. *Bioconjug Chem*, Vol. 14(1), p. 86, 2003

238. B. M. Austen, K. E. Paleologou, S. A. Ali, M. M. Qureshi, D. Allsop, and O. M. El- Agnaf. *Biochemistry*, Vol. 47(7), p. 1984, 2008.

239. M. Taylor, S. Moore, J. Mayes, E. Parkin, M. Beeg, and M. Canovi, *et al.* *Biochemistry*, Vol. 49(15), p. 3261, 2010.

240. B. Matharu, O. El-Agnaf, A. Razvi, and B. M. Austen. *Peptides*, Vol. 31(10), p. 1866, 2010.

241. L. Folop, M. Zarandi, Z. Datki, K. Soos, and B. Penke, *Biochem Biophys Res Commun.*, Vol. 324(1), p. 64, 2004.

242. E. A. Fradinger, B. H. Monien, B. Urbanc, A. Lomakin, M. Tan, and H. Li, *Proc Natl Acad Sci U S A*, Vol. 105(37), p. 14175, 2008.

243. T. J. Gibson , and R. M. Murphy. *Biochemistry*, Vol. 44, p. 8898, 2005.

244. D. J. Gordon, K. L. Sciarretta, and S. C. Meredith. *Biochemistry*, Vol. 40, p. 8237, 2001.

245. M. A. Etienne, J. P Aucoin, Y. Fu, R. L. McCarley, and R. P. Hammer. *J Am Chem Soc.*, Vol. 128 (11), p. 3522, 2006.

246. V. Rangachari, Z. S. Davey, B. Healy, B. D. Moore, L. K. Sonoda, and B. Cusack, *Biopolymers*, Vol. 91 (6), p. 456, 2009.

247. E. Hughes, R. M. Burke, A. J. and Doig. *J Biol Chem.*, Vol. 275 (33), p. 25109, 2000.

248. D. J. Gordon, R. Tappe, and S. C. Meredith. *J Pept Res.*, Vol. 60 (1), p. 37, 2002.

249. D. Grillo-Bosch, N. Carulla, M. Cruz, L. Sánchez, R. Pujol-Pina, S. Madurga, F. Rabanal, E. Giralt *et al*, *ChemMedChem*, Vol. 4 (9), p. 1488, 2009.

250. N. Kokkoni, K. Stott, H. Amijee, J. M. Mason, and A. J. Doig, *Biochemistry*, Vol. 45, p. 9906, 2006.

251. B. P. Orner, L. Liu, R. M. Murphy, L. L. Kiessling. *J. Am Chem Soc.*, Vol. 128 (36), p. 11882, 2006.

252. K. Taddei, S. M. Laws, G. Verdile, S. Munns, K. D'Costa, and A. R. Harvey, *Neurobiol Aging*, Vol. 31 (2) , p. 203, 2008.

253. M. Baine, D. S. Georgie, E. Z. Shiferraw, T. P. Nguyen, L. A. Nogaj, and D. A. Moffet. *J Pept Sci.*, Vol. 15 (8), p. 499, 2009.

254. A. C. Paula-Lima, M. A. Tricerri, J. Brito-Moreira, T. R. Bomfim, F. F. Oliveira, and M. H. Magdesian, *Int J Biochem Cell Biol.*, Vol. 41 (6), p. 1361, 2009.

255. S. P. Handattu, D. W. Garber, C. E. Monroe, T. van Groen, I. Kadish, and G. Nayyar, *et al. Neurobiol Dis.*, Vol. 34 (3), p. 525, 2009.

256. P. N. Cheng, C. Liu, M. Zhao, D. Eisenberg and J. S. Nowick, *Nature Chemistry*, Vol. 4, 2012.

257. R. S. Mulik, J. Mönkkönen, R. O. Juvonen, K. R. Mahadik, and A. R. Paradkar. *Mol Pharm*, Vol. 7 (3), p. 815, 2010.

258. D. Brambilla, R. Verpillot, B. Le Droumaguet, J. Nicolas, M. Taverna, J. Kóňa, B. Lettiero, S. H. Hashemi, L. De Kimpe, M. Canovi, M. Gobbi, V. Nicolas V, W. Scheper, S. M. Moghimi, I. Tvaroška, P. Couvreur, and K. Andrieux. *ACS Nano*. Vol. 6 (7), p. 5897, 2012.

259. a) N. Ida, T. Hartmann, J. Pantel, J. Schroder, R. Zerfass, H. Forstl, R. Sandbrink, C. L. Masters, and K. Beyreuther, *J. Biol. Chem.*, Vol. 271, p. 22908–22914, 1996; b) R. Robert, M. P. Lefranc, A. Ghochikyan, M. G. Agadjanyan, D. H. Cribbs, W. E. Van Nostrand, K. L. Wark, and O. Dolezal, *Mol. Immunol.*, Vol. 48, p. 59, 2010.

260. R. B. DeMattos , K. R. Bales , D. J. Cummins , J. C. Dodart , S. M. Paul , and D. M. Holtzman, *Proc Natl Acad Sci U.S.A.*, Vol. 98(15), p. 8850, 2001.

261. A) A. J. Gray, G. Sakaguchi, C. Shiratori, A. G. Becker, and J. LaFrancois, P. S. Aisen, K. Duff, and Y. Matsuoka, *NeuroReport*, Vol. 18, p. 293, 2007. b) P. Lichtlen, M. H. Mohajeri, *J. Neurochem.*, Vol. 104, p. 859, 2008.

262. G. B. Freeman, J. C. Lin, J. Pons, and N. M. Raha, *J. Alzheimers Dis.*, p. 531, 2011.

263. M. P. Lambert, A. K. Barlow, B. A. Chromy, C. Edwards, R. Freed, M. Liosatos, T. E. Morgan, I. Rozovsky, B. Trommer, K. L. Viola, P. Wals, C. Zhang, C. E. Finch, G. A. Krafft, and W. L. Klein, *Proc. Natl. Acad. Sci. USA*, Vol. 95, p. 6448, 1998.

264. D. M. Walsh, I. Klyubin, J. V. Fadeeva, M. J. Rowan, and D. J. *Biochem Soc Trans*, Vol. 30, p. 552, 2002.

265. H. Hillen, S. Barghorn, A. Striebinger, B. Labkovsky, R. Muller, V. Nimmrich, M. W. Nolte, C. Perez-Cruz, I. van der Auwera, F. van Leuven, M. van Gaalen, A. Y. Bespalov, H. Schoemaker, J. P. Sullivan, and U. Ebert, *J. Neurosci.*, Vol. 30, p. 10369, 2010.

266. Y. Tamura, K. Hamajima, K. Matsui, S. Yanoma, M. Narita, N. Tajima, K. Q. Xin, D. Klinman, and K. Okuda, *Neurobiol. Dis.*, Vol. 20, p. 541, 2005.

267. E. Markoutsa, K. Papadia, C. Clemente, O. Flores, and S. G. Antimisiaris, *European Journal of Pharmaceutics and Biopharmaceutics*, Vol. 81 (1), p. 49, 2012.

268. M. Canovi, E. Markoutsa, A. N. Lazar, G. Pampalakis, C. Clemente, F. Re, S. Sesana, M. Masserini, M. Salmona, C. Duyckaerts, O. Flores, M. Gobbi, and S. G. Antimisiaris. *Biomaterials*, Vol. 32 (23), p. 5489, 2011.

269. K. M. Jaruszewski, S. Ramakrishnan, J. F. Poduslo, and K. K. Kandimalla, *Nanomedicine: Nanotechnology, Biology, and Medicine*, Vol. 8 (2), p. 250, 2012.

270. L. O. Sillerud, N. O. Solberg, R. Chamberlain, R. A. Orlando, J. E. Heidrich, D. C. Brown, C. I. Brady, T. A. Vander Jagt, M. Garwood and D. L. Vander Jagt, *Journal of Alzheimer's Disease*, Vol. 34, Number 2, 2013.

271. C. C. Yang, S. Y. Yang, J. J. Chieh, H. E. Horng, C. Y. Hong, H. C. Yang, K. H. Chen, B. Y. Shih, T. F. Chen and M. J. Chiu, *ACS Chem Neurosci.*, Vol. 2(9), p. 500, 2011.

272. J. W. Choi, A. T. Islam, J. H. Lee, J. M. Song and B. K. Oh, *J Nanosci Nanotechnol.*, Vol. 11(5), p. 4200, 2011.

273. W. A. El-Said, T-H Kim, C-H Yea, H Kim, and J-W Choi, *J Nanosci Nanotechnology,*Vol. 10, p. 1, 2010.

274. J. H. Kim, T. L. Ha, G. H. Im, J. Yang, S. W. Seo, I. S. Lee and J. H. Lee. *Neuroreport*, Vol. 24(1), p. 16, 2013.

275. D. G. Georganopoulou, L. Chang, J. M. Nam, C. S. Thaxton, E. J. Mufson, W. L. Klein and C. A. Mirkin, *Proc Natl Acad Sci U.S.A.*, Vol. 102(7), p. 2273, 2005.

276. A. Neely, C. Perry, B. Varisli, A. K. Singh, T. Arbneshi, D. Senapati, Kalluri and P. C. Ray. *ACS Nano.*, Vol. 3(9), p. 2834, 2009.

277. F. Bard, C. Cannon, R. Barbour, R. L. Burke, D. Games, H. Grajeda, T. Guido , K. Hu, J. Huang, K. Johnson-Wood, K. Khan, D. Kholodenko, M. Lee, I. Lieberburg, R. Motter, M. Nguyen, F. Soriano, N. Vasquez, K. Weiss, B. Welch, P. Seubert, D. Schenk, and T. Yednock, *Nat Med.*, Vol. 6 (8), p. 916, 2000.

278. J. C. Dodart, K. R. Bales , K. S. Gannon, S J. Greene, R. B. DeMattos, C. Mathis, C. A. DeLong, S. Wu, X. Wu , D. M. Holtzman and S. M. Paul, *Nat Neurosci.*, Vol. 5(5), p. 452, 2002.

279. B. Solomon, R. Koppel, E. Hanan, and T. Katzav, *Proc. Natl. Acad. Sci.*, Vol. , p. 452, 1996.

280. D. M. Wilcock, J. Alamed, P. E. Gottschall, J. Grimm, A. Rosenthal, J. Pons, V. Ronan, K. Symmonds, M. N. Gordon, and D. Morgan, *J. Neurosci.*, Vol. 26, p. 5340, 2006.

281. Y. Horikoshi, G. Sakaguchi, A. G. Becker, A. J. Gray, K. Duff, P. S. Aisen, H. Yamaguchi, M. Maeda, N. Kinoshita, and Y. Matsuoka, *Biochem. Biophys. Res.Commun.*, Vol. 319, p. 733, 2004.

282. M. H. Mohajeri, K. Saini, J. G. Schultz, M. A. Wollmer, C. Hock, and R. M. Nitsch, *J. Biol. Chem.*, Vol. 277, p. 33012, 2002.

283. H. Hillen, S. Barghorn, A. Striebinger, B. Labkovsky, R. Muller, V. Nimmrich, M. W. Nolte, C. Perez-Cruz, I. van der Auwera, F. van Leuven, M. van Gaalen, A. Y. Bespalov, H. Schoemaker, J. P. Sullivan, and U. Ebert, *J. Neurosci.*, Vol. 30, p. 10369, 2010.

284. A. Lord, A. Gumucio, H. Englund, D. Sehlin, V. S. Sundquist, L. Soderberg, C. Moller, P. Gellerfors, L. Lannfelt, F. E. Pettersson, and L. N. Nilsson, *Neurobiol.Dis.,* Vol. 36, p. 425, 2009.

285. B. O'Nuallain, and R. Wetzel, *Proc. Natl. Acad. Sci. USA*, Vol. 99, p. 1485, 2002.

286. Yamada, K., Yabuki, C., Seubert, P., Schenk, D., Hori, Y., Ohtsuki, S., Terasaki, T., (...), Iwatsubo, T. *Journal of Neuroscience*, Vol. 29 (36), p. 11393, 2009.

287. H. Kuwahara, Y. Nishida, and T. Yokota, *Brain and Nerve*, Vol. 65 (2), p. 145, 2013.

288. M. De, P. S. Ghosh, and W. Rotello, *Adv. Mater.*, Vol. 20, p. 4225, 2008.

289. X. Montet, M. Funovics, K. Montet-Abou, R. Weissleder, and L. Josephson, *J. Med. Chem.*, Vol. 49, p. 6087, 2006.
290. M. Shi, J. Lucd, and M. S. Shoichet, *J. Mater. Chem.*, Vol. 19, p. 5485, 2009.
291. M. K. Yu, J. Park, and S. Jon, *Theranostics*, Vol. 2(1), p. 3, 2012.
292. A. Chauhan, I. Ray, and V. P. Chauhan, *Neurochem Res*, Vol. 25, p. 423, 2000.
293. C. Aisenbrey, B. Bechinger, and G. Gröbner, *J Mol Biol*, Vol. 375, p. 376, 2008.
294. K. Yanagisawa, *Biochim Biophys Acta*, Vol. 1768, p. 1943, 2007
295. D. Brambilla, B. L. Droumaguet, J. Nicolas, S. H. Hashemi, L. P. Wu, S. M. Moghimi, P. Couvreur and K. Andrieux, *Nanomedicine: Nanotechnology, Biology, and Medicine*, Vol. 7, p. 521, 2011.
296. F. Re, M. Gregori and M. Masserini, *Maturitas*, Vol. 73, p. 45, 2012.
297. A. Nazem and G. Ali Mansouri, *Insciences J.*, Vol. 1(4), p. 169, 2011.
298. R. M. Koffie, C. T. Farrar, L. J. Saidi, C. M. William, B. T. Hyman and T. L. Spires-Jones, *Proc Natl Acad Sci USA.*, Vol. 108(46), p. 18837, 2011.
299. T. Siegemund, B. R. Paulke, H. Schmiedel, N. Bordag, A. Hoffmann, T. Harkany, H. Tanila, J. Kacza and W. Härtig, *Int J Dev Neurosci.*, Vol. 24(2–3), p. 195, 2006.
300. E. Salvati, F. Re, S. Sesana, I. Cambianica, G. Sancini, M. Masserini and M. Gregori, *Int J Nanomedicine*, 2013, in press
301. F. Re, I. Cambianica, S. Sesana, E. Salvati, A. Cagnotto, M. Salmona, P. O. Couraud, S. M. Moghimi, M. Masserini and G. Sancini, *J Biotechnol. Dic*, Vol. 156(4), p.341, 2011
302. E. A. Tanifum, I. Dasgupta, M. Srivastava, R. C. Bhavane, L. Sun, J. Berridge, H. Pourgarzham, R. Kamath, G. Espinosa, S. C. Cook, J. L. Eriksen and A. Annapragada, *PLoS ONE*, Vol. 7(10), p. e48515, 2012.
303. A. Skouras, S. Mourtas, E. Markoutsa, M. C. De Goltstein, C. Wallon, S. Catoen and S. G. Antimisiaris, *Nanomedicine: Nanotechnology, Biology, and Medicine*, Vol. 7, p. 572, 2011.
304. Y. Zaim Wadghiri, E. M. Sigurdsson, M. Sadowski, J. I. Elliott, Y. Li, H. Scholtzova, C. Y. Tang, G. Aguinaldo, M. Pappola, K. Duff, T. Wisniewski and D. Turnbull, *Magn Reson Med*, Vol. 50, p. 293, 2003.
305. Y.Z. Wadghiri, J. Li, J. Wang, D.M. Hoang, Y. Sun, H. Xu, W. Tsui, Y. Li, A. Boutajangout, A. Wang, M. de Leon and T. Wisniewski, *PLoS One.*, Vol. 8 (2),p. e57097, 2013.
306. H. Skaat and S. Margel, *Biochem Biophys Res Commun.*, Vol. 386, p. 645, 2009.
307. M. Sakono, T. Zako and M. Maeda, *Anal Sci.*, Vol. 28 (1), p. 73, 2012.
308. Y. Choi, Y. Cho, M. Kim, R. Grailhe and R. Song, *Anal Chem.* 16,84 (20), p. 8595, 2012.
309. J. Zhang, X. Jia, H. Qing, H-Y Xie and H.W. Li, *Nanotechnology*, Vol 24 (3), art. no. 035101, 2013.
310. J. H. Kim, T. L. Ha, G. H. Im, J. Yang, S. W. Seo, I. S. Lee and J. H. Lee, *Neuroreport*, Vol. 24 (1), p. 16, 2013.
311. M. D. Santin, T. Debeir, S. L. Bridal, T. Rooney and M. Dhenain, *Neuroimage*. 2013.

312. J. S. Choi, H. J. Choi, D. C. Jung, J. H. Lee, J. Cheon, *Chem Commun*, p. 2197, 2008.

313. M. Vestergaard, K. Kerman, D. K. Kim, M. H. Ha and E. Tamiya, *Talanta,* 15;74(4), p.1038, 2008.

314. J. K. Sahni , S. Doggui, J. Ali, S. Baboota, Lé Dao , C. Ramassamy, *J Control Rel* , Vol. 152, p. 208, 2011.

315. H. Xin, X. Sha, X. Jiang, L. Chen, K. Law, J. Gu, Y. Chen, X. Wang, and X. Fang. *Biomaterials.* Vol. 33(5), p. 1673, 2012.

316. B. Wilson, M. K. Samanta, K. Santhi, K P. Kumar, N. Paramakrishnan, and B. Suresh, *Eur. J. Pharm. Biopharm*, Vol. 70 (1), p. 75, 2008.

317. X. Gao, B. Wu, Q. Zhang, J. Chen, J. Zhu, W. Zhang, Z. Rong, H. Chen, and X. Jiang, *J. Control. Release* Vol. 121 (3), p. 156, 2007.

318. C. Cabaleiro-Lago, F. Quinlan-Pluck, I. Lynch, S. Lindman, A. M. Minogue, E. Thulin, D. M. Walsh, K. A. Dawson, and S. Linse. *J Am Chem Soc.,* Vol. 130(46), p. 15437, 2008.

319. A. M. Saraiva, I. Cardoso, M. J. Saraiva, K. Tauer, M. C. Pereira, M. A. Coelho, H. Möhwald, and G. Brezesinski. *Macromol Biosci.* Vol. 10(10), p. 1152, 2010.

320. B. Wilson, M. K. Samanta, K. Santhi, K. P. S. Kumar, M. Ramasamy, and B. Suresh, *Nanomed. Nanotechnol. Biol. Med.* Vol. 6 (1), p. 144, 2010.

321. Z. H. Wang, Z. Y. Wang, C. S. Sun, C. Y. Wang, T. Y. Jiang, and S. L. Wang, *Biomaterials*, Vol. 31 (5), p. 908, 2010.

322. X. Wang, N. Chi, and X. Tang, *Eur. J. Pharm. Biopharm.* Vol. 70 (3), p. 735, 2008.

323. S. A. Joshi, S. S. Chavhan, and K. K. Sawant. *Eur J Pharm Biopharm.* Vol. 76(2), p. 189, 2010.

324. P. Picone, M. L. Bondi, G. Montana, A. Bruno, G. Pitarresi, G. Giammona, and M. Di Carlo. *Free Radic Res.* Vol. 43(11), p. 1133, 2009.

325. S. Dhawan, R. Kapil, and B. Singh. *J Pharm Pharmacol.* Vol. 63(3), p. 342, 2011.

326. A. Bernardi, R. L. Frozza, A. Meneghetti, J. B. Hoppe, A. M. Battastini, A. R. Pohlmann, S.S. Guterres, and C. G. Salbego. *Int J Nanomedicine.* Vol. 7, p. 4927, 2012.

327. A. N. Lazar, S. Mourtas, I. Youssef, C. Parizot, A. Dauphin, B. Delatour, S. G. Antimisiaris, and C.Duyckaerts, *Nanomedicine: Nanotechnology, Biology, and Medicine*, 2013, in press

328. E. Araya, I. Olmedo, N. G. Bastus, S. Guerrero, V.F. Puntes, E. Giralt, and M. J. Kogan, *Nanoscale Research Letters*, Vol.3 , 2008.

329. Y. H. Liao , Y. J. Chang , Y. Yoshiike , Y. C. Chang, and Y. R. Chen, *Nano small*, Vol. 8, p. 3631, 2012.

330. R. Prades, S. Guerrero, E. Araya, C. Molina, E. Salas, E. Zurita, J. Selva, G. Egea, I. C. López, M. Teixidó, M. J. Kogan, and E. Giralt. *Biomaterials.* Vol. 33(29), p. 7194, 2012.

331. M. Mahmoudi, O. Akhavan, M. Ghavami, F. Rezaee, and S.M. Ghiasi. *Nanoscale,* Vol. 4(23), p. 7322, 2012.

332. A. Cimini, B. D'Angelo, D. Soumen, and S. Seal, Nanoparticles Of Cerium Oxide Targeted To An Amyloid-Beta Antigen Of Alzheimer's Disease And

Associated Methods US Patent Application No: 2012/0070,500, March 22, 2012.

333. K. Ikeda, T. Okada, S. Sawada, K. Akiyoshi,and K. Matsuzaki. *FEBS Lett.* Vol. 580(28–29), p. 6587, 2006

334. K. K. Cheng, C. F. Yeung, S. W. Ho, S. F. Chow, A. H. Chow, and L. Baum. *AAPS J.* Vol. 15(2), p. 324, 2013.

335. C. Treiber, M. A. Quadir, P. Voigt, M. Radowski, S. Xu, L. M. Munter, T. A. Bayer, M. Schaefer, R. Haag, and G. Multhaup. *Biochemistry,*Vol. 48(20), p. 4273, 2009.

336. J. Li, C. Zhang, J. Li, L. Fan, X. Jiang, J. Chen, Z. Pang, and Q. Zhang. *Pharm Res* 2013 in press

337. L. L. Dugan, D. M. Turetsky, C. Du, D. Lobner, M. Wheeler, and C. R. Almli, *Proc Natl Acad Sci U.S.A,* Vol. 94, p. 9434, 1997.

338. H. M. Huang, H. C. Ou, S. J. Hsieh, and L. Y. Chiang. *Life Sci,* Vol. 66, p. 1525, 2000.

339. C. M. Lee, S. T. Huang, S. H. Huang , H. W. Lin, H. P. Tsai, J. Y. Wu, C. M. Lin, and C. T. Chen, *Nanomedicine: Nanotechnology, Biology, and Medicine,* Vol. 7(1), p. 107, 2011.

340. A. G. Bobylev, A. B. Kornev, L. G. Bobyleva, M. D. Shpagina, I. S. Fadeeva, R. S. Fadeev, D. G. Deryabin, J. Balzarini, P. A. Troshin, and Z. A. Podlubnaya. *Org Biomol Chem,* Vol. 9(16), p. 5714, 2011.

341. M. Li ,C. Xu, L. Wu, J. Ren, E. Wang, and X. Qu. *Small,* 2013 *in press*

Part 2
POINT-OF-CARE DIAGNOSTICS

5

Novel Biomaterials for Human Health: Hemocompatible Polymeric Micro- and Nanoparticles and Their Application in Biosensor

Chong Sun, Xiaobo Wang, Chun Mao* and Jian Shen

Jiangsu Key Laboratory of Biofunctional Materials, Biomedical Functional Materials Collaborative Innovation Center, College of Chemistry and Materials Science, Nanjing Normal University, Nanjing, P. R. China

Abstract

In the past two decades, the development of nanomaterials for the ultra-sensitive detection of biological species has received great attention because of their unique optical, electronic, chemical, and mechanical properties. Different nanomaterials were investigated to determine their properties and possible applications in biosensors. Novel nanomaterials for use in bioassay applications represent a rapidly advancing field. The strategy for decorating electrode of glucose biosensor with the hemocompatible polymeric micro- and nanoparticles that exhibit excellent antibiofouling property was suggested by our research group. This smart strategy demonstrates a methodology for the incorporation of actively antibiofouling moieties onto a passively antibiofouling electrode, and thus expands the range of applications of electrochemical biosensors, especially in whole blood.

Keywords: Nanomaterials, hemocompatible materials, biosensors

5.1 Introduction

Blood is a bodily fluid in animals that delivers necessary substances such as oxygen and nutrients to the cells and transports metabolic waste products

---

Corresponding author: maochun127@aliyun.com

Ashutosh Tiwari (ed.) Advanced Healthcare Materials, (181–202) 2014 © Scrivener Publishing LLC

away from them. It plays a very important role in the body. Because improved detection of blood disease can save lives, blood tests have been used for approximately 50 years to detect substances that are present in the blood that indicate either disease or a future risk of the development of a disease. Blood tests detect substances that normally are not present or measure substances that, when elevated above normal levels, indicate disease [1].

At present, the clinical conditions of diabetes mellitus are well known and well understood, yet remain a growing concern as the prevalence of the disease increases worldwide at an alarming rate [2]. Accurate blood glucose values especially play an important role in the diagnosis of diabetes. The primary methods of detecting blood glucose concentration are performed by biochemical analyzer and glucose meter [3–5]. For biochemical analyzer, the quantification of the concentration of glucose is mainly involved in serum samples, which are isolated from whole blood separation by centrifugation process, but not untreated whole blood. The test results are influenced by the different model numbers of test instruments and detection reagents, treatment processes of blood samples, factitious operations, especially additional centrifuge and too long a measure of time from collecting blood specimens to examination. This method for the diagnosis of diabetes is not recommended. Further, the red blood cells have a higher concentration of protein (e.g., hemoglobin) than serum, and serum has higher water content and consequently more dissolved glucose than whole blood. To convert from whole-blood glucose, multiplication by 1.15 has been shown to generally give the serum/plasma level. In principle, blood glucose values should be given in terms of whole blood, but most hospitals and laboratories now measure and report the serum glucose levels.

As for commercial glucose meters, there are some defects that cannot be ignored during its operation. For example, the blood samples are obtained from fingertip peripheral but not vein, and doped easily with tissue fluid. So the accurate results of glucose concentration cannot be provided by commercial glucose meters.

However, it is very difficult to design and prepare an electrochemical biosensor that can be used in whole blood just because the biofouling of electrode surface can be developed by platelet, fibrin and blood cell adhesion in the complex environment of whole blood media. And the biofouling of electrode surface will bring catastrophic damage to the electron transfer between enzyme and electrode redox center. So the development of novel glucose biosensors for antifouling, rapid, highly sensitive, and selective detection is of paramount importance for blood glucose concentration monitoring in whole blood samples.

5.2   Design and Preparation of Hemocompatible Polymeric Micro- and Nanoparticles

Polymeric micro- or nanoparticles are successively employed for delivery of conventional drugs, recombinant proteins, vaccines and nucleotides [6–19]. The polymeric matrixes should be compatible with the body in the terms of adaptability (non-toxicity) and (non-antigenicity) and should be biodegradable and biocompatible [20]. Moreover, one of the important issues in developing biomedical materials with small scale that contact blood is improving their blood compatibility [21]. How can design of hemocompatible polymeric micro- and nanoparticles? Three methods are suggested as follows.

(1) Polymeric micro- and nanoparticles loaded anticoagulant drug

As we know, to improve the blood compatibility of the polymeric films, heparin has been used to modify their surface [22–25]. It is one of the most intensively studied glycosaminoglycans (GAGs) as a result of its antico-agulant properties. It is a potent anticoagulant agent that interacts strongly with antithrombin III to prevent the formation of fibrin clot. So heparin immobilized to a surface enhances various surface properties, improving blood compatibility and biocompatibility. The immobilized forms include soluble heparin and heparin immobilized to supporting matrices by physical adsorption, by covalent chemical methods, and by photochemical attachment [22]. The efficiency, stability, and activity of heparin are determined by the different immobilized methods and support materials [26]. Besides its anticoagulant property, heparin also has great significance in regulating many biological pathways including cell-cell recognition, signal transduction, growth processes, coagulation cascade and the cellular interaction of growth factors [27]. So the development of novel heparin-loaded microspheres for anticoagulant drug-release, hemocompatible coating, and selective cellular interaction is of paramount importance for the biomedical application of biomaterials or bio-devices in whole blood environment. Our research group reported a novel kind of heparin-loaded polyurethane microsphere (Hep-PU MS), which was synthesized by a single-step phase separation method. The morphology of the Hep-PU MS was depicted in Fig. 5.1a. As shown in Fig. 5.1a, the Hep-PU MS were formed. The average particle size of Hep-PU MS was about 540 nm. With the simplicity of the loaded method, the excellent hemocompatibility and the slow-release of heparin, the Hep-PU MS with desirable bioproperties can be readily tailored to cater to various biomedical applications.

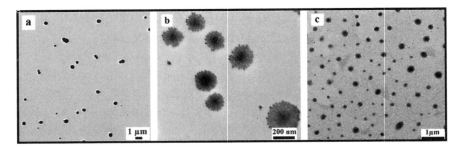

Figure 5.1 Representative TEM images of (a) PU-Hep MS, (b) PU-F127 hybrid nanospheres, (c) HBPE-SO$_3$ nanospheres.

(2) Polymeric micro- and nanoparticles hybrided with hydrophilic molecules

Surface modification with increasing hydrophilicity is believed to be a useful method for improving blood compatibility, and various polymer materials have been modified by water-soluble polymer for biomedical use such as poly(ethylene glycol) (PEG) or poly(ethylene oxide) (PEO) that can prevent plasma protein adsorption, platelet adhesion, and thrombus formation by the steric repulsion mechanism. Steric repulsion by surface-bound water-soluble polymer chains occurs as a result of overlapping polymer layers that could lead to loss in configurational entropy because of volume restriction and/or osmotic repulsion between interdigitated polymer chains. The accepted mechanism for preventing protein adsorption by the grafted PEO chains is that such a technique decreased interfacial free energy and the steric repulsion force between PEO chains and the proteins [28]. Similarly, hydrophilic polysulfone membranes, polyvinylpyrrolidone-polysulfone (PVP-PSf), were prepared from PSf membranes covalently conjugated with PVP on the surface. It was found that PVP-PSf membranes gave lower protein adsorption from a plasma solution than PSf membranes. This is also attributed to the hydrophilic surface of the PVP-PSf membranes [29].

How can prepare hemocompatible polymeric micro- and nanoparticles by surface modification with increasing hydrophilicity? It became a target for our research group. Pluronic F127 (triblock copolymer PEO106PPO70PEO106) has good blood compatibility [30, 31]. The preparation and hemocompatibility of PU-F127 nanospheres by a spontaneous emulsion solvent diffusion method were investigated by our research group. As displayed in Fig. 5.1b, typical TEM photograph of the PU-F127 hybrid nanospheres showed the average diameter was about 200 nm. It is very interesting that every PU-F127 hybrid nanosphere has a porous structure. The results of blood test indicate the PU-F127 hybrid nanospheres have good blood compatibility.

(3) Polymeric micro- and nanoparticles surface modified by special functional groups

Hyperbranched polymers have attracted significant interests because of their unique architecture and novel properties that include good solubility, special viscosity behavior, and high density of their functional groups [32, 33]. Owing to the multifunctionality in hyperbranched polymers, the physical properties can be adjusted to a large extent by the chemical modification of the end-groups [34, 45]. The use of hyperbranched polymers by the chemical modification has attracted increasing attention in recent years [36–39]. These features of hyperbranched polymers have been used extensively in diverse fields, such as coatings, additives, blends, nonlinear optics, composites, and copolymers [40–43]. Especially, hyperbranched polymers hold great potential as drug delivery agents because of their three-dimensional shapes and availability of a large number of surface functional groups amenable to various modification chemistries for drug conjugation and targeting purposes [44–46].

Herein, water-soluble nanoparticles were synthesized by the chemical modification of aliphatic hyperbranched polyester (HBPE) with sulfonic acid functional groups (HBPE-SO$_3$). TEM photographs of the nanoparticles showed an average diameter of 210 nm (Fig. 5.1c). The blood compatibility and cytotoxicity of HBPE-SO$_3$ nanospheres were investigated by a series of specialized blood experiments. The results showed these hemocompatible polymeric micro- and nanoparticles provide a promising platform of blood circulation system for diagnosis and therapy with the help of the drug-loaded capacity of hyperbranched polyester micelles. Similarly, hemocompatible and water-soluble nanoparticles were also synthesized by the chemical modification of aliphatic HBPE with carboxylic acid functional groups.

As to the actual operation for preparation of polymeric micro- and nanoparticles, many methods in the past ten years were used to prepare polymeric micro- and nanoparticles, such as micro-emulsification [47–49], phase conversion [50–54], template method [55–57], surface modification by special functional groups [58, 59], and ATRP technique [60–62].

5.3 The Biosafety and Hemocompatibility Evaluation System for Polymeric Micro- and Nanoparticles

When polymeric micro- and nanoparticles enter biosystems, they interact with various biomolecules, especially proteins, forming a protein corona on the surface. Understanding how polymeric micro- and nanoparticles interact with biomolecules is crucial for bioapplications and for the bio-safety of polymeric micro- and nanoparticles.

In this paper, Hep-PU MS we prepared were chosen as an example for investigating their hemocompatibility and cytotoxicity in the blood and cell experiments mentioned below [50].

5.3.1 In vitro Coagulation Time Tests

The activated partial thromboplastin time (APTT), prothrombin time (PT), and thromboplastin time (TT) were widely used for the clinical detection of the abnormality of blood plasma. In recent times, they were applied in the evaluation of in vitro antithrombogenicity of biomateri-als [63–65]. All the samples were dispersed in phosphate buffer solution (PBS) at the pre-determined concentrations. The effect on coagulation in the presence of the Hep-PU MS was studied after mixing anticoagulated rabbit blood plasma with the sample solution in the cuvette strips at 37 °C for 5 min before adding the coagulation reagents. PBS acted as the con-trol. All the assays were performed and measured by using a Semi auto-mated Coagulometer (RT-2204C, Rayto, USA). All the coagulation tests were performed in triplicate. The APTT, PT, and TT values of control for a healthy blood plasma were 21.3±0.2, 6.7±0.1 and 16.5±0.1 s, respectively. The effect of Hep-PU MS can be observed by comparing the APTT, PT, and TT with those of control. The results showed that the data of APTT/PT/TT were statistically longer in tests than in controls after the injection of Hep-PU MS, indicating that Hep-PU MS have excellent anticoagulant effects (Fig. 5.2a).

5.3.2 Complement and Platelet Activation Detection

The hemocompatibility of the Hep-PU MS was also evaluated by mea-suring complement and platelet activation under in vitro conditions. Complement activation is a key indicator of both adaptive and innate immunity, thus hemocompatibility when foreign material is introduced into blood [66–68]. We have used an enzyme immunoassay kit (C3a kit) for measuring the complement activation. C3a is an anaphylatoxin pro-duced during the complement cascade and its concentration in the plasma is a measure of the extent of complement activation. Hep-PU MS was incu-bated with platelet poor plasma (PPP) at 37 °C for 1 h and the amount of C3a produced was measured. Hep-PU MS were found to be neutral to the complement system in that the amount of C3a (27.1 ng/mL) produced

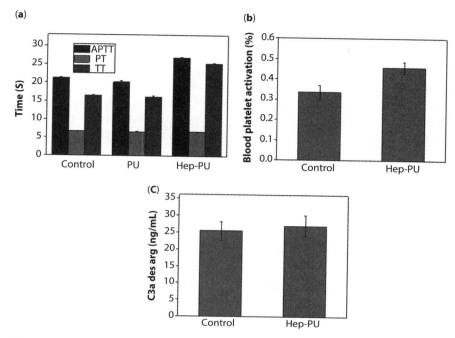

Figure 5.2 Hemocompatibility of Hep-PU MS. (a) APTT/PT/TT of PU MS and Hep-PU MS; (b) Complement activation upon interaction of Hep-PU MS with PPP for 1 h at 37 °C; and (c) Platelet activation upon interaction of Hep-PU MS with PRP for 30 min at 37 °C. Plasma incubated with saline as a negative control. Data are presented as means ± standard deviation (n = 3).

by the Hep-PU MS. From the data, it is obvious that the Hep-PU MS do not activate the complement system (Fig. 5.2b). It is known that negatively charged sulfate and sulfonate groups of heparin play a major role in the biocompatibility acticity of heparin, which reduce the inflammatory potential of an external additive [69, 70].

Platelet activation upon interaction with samples is another major barrier in blood incompatibility and can lead to thrombotic complications under in vivo conditions [67]. We have measured the platelet activation after incubating the Hep-PU MS in platelet rich plasma (PRP) for 30 min at 37 °C using flow cytometry. Platelet activation is expressed as the percentage of platelets positive for both of the bound antibodies (CD 62P and CD42). There was no significant difference between the Hep-PU MS and the control plasma sample (Fig. 5.2c). No effect in platelet activation that occurred with the Hep-PU MS was also attributed to heparin.

5.3.3   Percent Hemolysis of RBCs

Hemolysis occurs when cells swell to the critical bulk to break up the cell membranes. Meanwhile, released adenosine diphosphate from the broken red blood cells intensifies the assembly of blood platelets, which accelerates the formation of clotting and thrombus. So hemolysis of the blood is a very important problem associated with the bio-incompatibility of material [71]. Less than 5% hemolysis was regarded as a nontoxic effect level [64].

The experimental process was as follows. First, 5 mL of rabbit blood sample was added to 10 mL of PBS, and then red blood cells (RBCs) were isolated from whole rabbit blood by centrifugation at 1500 rpm for 10 min. The RBCs were further washed five times with 10 mL of PBS solution. The RBCs were diluted to 50 mL of PBS. Herein, RBCs incubation with doubly distilled deionized water (DI water) and PBS were used as the positive and negative controls, respectively. Then 1 mL of diluted 2% RBCs suspension was added to 1 mL Hep-PU MS solutions at required concentration. All the sample tubes were kept in static condition at 37 °C for 3 h. Finally, the mixtures were centrifuged at 1500 r/min for 10 min, and 200 μL of supernatants of all samples were transferred to a 96-well plate. The absorbance values of the supernatants at 570 nm were determined by using a microplate reader. The percent hemolysis of RBCs was calculated using the following formula.

Percent hemolysis % = ((sample absorbance – negative control absorbance)/(positive control absorbance – negative control absorbance)) × 100

In this case, Hep-PU MS did not cause any hemolysis (0.89±0.10%) on rabbit erythrocyte comparing with the negative control (normal saline, haemolysis rate is 0%) and the positive control (water, haemolysis rate is 100%)

5.3.4   Morphological Changes of RBCs

Morphologically aberrant forms of RBCs can give insights into the diagnosis of various medical conditions such as hemolytic anemia. The RBC pellet was resuspended in 0.9% saline solution. The Hep-PU MS were diluted to a required concentration in cell suspension. Observation of morphological changes by light microscopy was captured after 1.5 h of the Hep-PU MS exposure to analyze the morphological variation at the early stages of haemolysis. The pellet obtained after centrifugation was diluted in 0.9% saline solution and mounted on clean glass slides covered with cover slips and observed under Olympus BX41 microscope and photographed with an Olympus E-620 camera (Olympus Ltd., Japan). As shown in Fig. 5.3a, the untreated RBCs in saline appeared in a normal biconcave shape. It is known to all, exposure to materials with the bad blood compatibility will

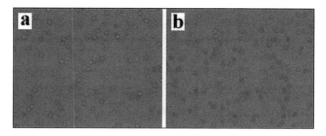

Figure 5.3 Optical images of (a) saline solution-treated RBCs, and (b) Hep-PU MS treated RBCs.

induce appearance of morphological aberrant forms for RBCs such as echinocyte-like forms with numerous surface spikes, the RBCs appear swollen, and the phenomena of ghost cells (lysed RBCs) [72]. The RBCs treated by Hep-PU MS was imaged as shown in Fig. 5.3b. The result showed that there were no obvious changes of cell morphology. Such a result was in agreement with the above hemolysis analysis.

5.3.5 Cytotoxic Assessment

Cytotoxicity of Hep-PU MS was evaluated by MTT (3-[4,5-dimethylthiazolyl-2]-2,5-diphenyltetrazolium bromide) assays. Briefly, human embryonic kidney 293 (HEK 293) cells were seeded to a 96-well culture plate and the cells would come to about 50% confluence after 24 h of culture. The media were changed by fresh ones, and the mixtures containing required concentration of Hep-PU MS samples were added to the wells. The cells of positive control were only incubated with equal Dulbecco's modified eagle's medium (DMEM) (10% fetal bovine serum) and the cell viability was set as 100%. All of the cells were allowed to grow for 24 h before 10 mL MTT (5 mg/mL) was added to each well. Then, the cells were incubated at 37 °C for an additional 4 h until the purple precipitates were visible. The medium was replaced by 100 mL dimethyl sulfoxide (DMSO) and the cell plate was vibrated for 15 min at room temperature to dissolve the crystals formed by the living cells. Finally, the absorption at 490 nm of each well was measured by an ultramicroplate reader [73]. The effect of the Hep-PU MS on cell viability ($90\pm1.5\%$) was compared to nontreated cells (control sample), which are considered as 100%. According to the relationship between cell proliferation rate and cytotoxicity grade of United States pharmacopeia (USP), we got the conclusion that Hep-PU MS have no cytotoxicity, which is very important for their potential use in vivo [74–76].

5.4   Construction of Biosensor for Direct Detection in Whole Blood

As we know, when in direct contact with blood, the foreign materials are prone to initiate the formation of clots, as platelets and other components of the blood coagulation system are activated. So it is very difficult to design and prepare an electrochemical biosensor that can be used in whole blood directly just because the biofouling of electrode surface can be developed by platelet, fibrin and blood cell adhesion in the complex environment of whole blood media. As for the glucose biosensor that used in whole blood directly, the biofouling of electrode surface will bring catastrophic damage to the electron transfer between enzyme and electrode redox center. In Reichert's review paper, he suggested that biofouling is one of several causes for the failure of in vivo biosensors due to the accumulation of proteins, cells and other biological materials on the surface of biosensors [77]. Anti-biofouling is the process of removing or preventing these accumulations include platelets and other components of the blood from forming. So improving the hemocompatibility and anti-biofouling property of biomaterials or biodevices has become a very important task for biomedical material scientists. In the past years, different approaches were taken by various research groups to combat the challenge of forming anti-biofouling and fouling release surfaces. Particular attention was given to the chemical structures produced and the various techniques utilized to demonstrate these surfaces inherent anti-biofouling character.

Nowadays, nanomaterials of polymers are found to have superior performance compared with conventional polymeric materials due to their much larger exposed surface area and better biocompatibility. In this case, Hep-PU MS were prepared and chosen as an example to modify the glassy carbon electrode (GCE) for its antibiofouling effect that due to the coherence between blood compatibility and antifouling property of Hep. Herein, using hemocompatible polymeric nanospheres to do the study of anti-biofouling of the glucose biosensor was attempted by our research group.

5.4.1   Evaluation of GOx/(Hep-PU) Hybrids

The maintenance of enzyme activity on the supporting materials is crucial in the biosensor designs, because the secondary conformational variations of the enzyme can affect its activity markedly. Herein, circular dicroism (CD) were utilized to check the secondary conformation variations of the polypeptide chain of GOx on the Hep-PU MS [78]. The CD spectra in the far-UV (with the range from 185 to 260 nm) were measured on a JASCO

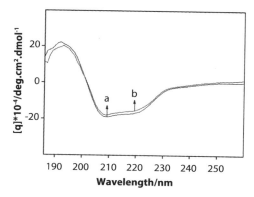

Figure 5.4 CD spectra of (a) pure GOx, and (b) GOx/(Hep-PU) in 0.1 M PBS (pH=7.4) in the wavelength region of 190–250 nm.

J-715 spectropolarimeter using a 1 cm quartz cuvette. The contents of α-helix, β-sheet, β-turn, and the random coil conformation were calculated using the JASCO710 program.

Fig. 5.4 showed the CD spectra of GOx in PBS without (curve a) and with Hep-PU MS (curve b) in the far-UV region (185–260 nm). The positive bands at 191 nm in both curves correspond to the π–π* transition of the amide groups in the GOx peptide chain. Two negative bands at 209 and 219 nm are in accordance to the π–π* and n–π* transition of the amide groups of the GOx polypeptide chain, respectively [79]. In the presence of Hep-PU MS, the intensity of the positive band at 191 nm significantly decreases compared with that of pure GOx, indicating the interactions between the amino acid residues of GOx and Hep-PU nanocomposites. Additionally, the intensity of the dual bands at 209 and 219 nm had a slight decrease for the GOx/(Hep-PU), indicating that the secondary structure of GOx was not obviously changed.

In addition, the contents of α-helix and β-turn conformation of GOx/(Hep-PU) changed 4.3% and 0.6% compared with pure GOx. All results indicate that Hep-PU MS could essentially maintain the native conformation of GOx [80]. Thus, the secondary structure of GOx was well maintained in the prepared biosensor and Hep-PU MS indeed have good biocompatibility.

5.4.2   Evaluation of Whole Blood Adhesion Tests

The whole blood adhesion test has already become a recognized technique to estimate the blood compatibility or anticoagulation of a prepared material. Thus, our work also employed this test to evaluate the blood compatibility of the surface of electrode that modified by Hep-PU MS. The blood was obtained from a healthy adult volunteer anticoagulated with sodium citrate solution. The blank substrate without Hep-PU MS and GOx/(Hep-PU) modified substrate were immersed in PBS for 24 h before they were placed in the 24-well microplates. Each well was added with 1.0 mL of whole blood. After being incubated for 60 min at 37 °C in humidified air, the samples were taken out, and rinsed by PBS for three times to remove the physically attached blood cells. After that, the adhered blood cells were fixed with 2.5% glutaraldehyde in PBS for 30 min. Finally, the sample was washed with PBS and dehydrated with a series of ethanol/water mixtures of increasing ethanol concentration (50, 60, 70, 80, 90, 95 and 100% of ethanol) for 10 min in each mixture respectively. The surfaces membranes were air-dried, coated with gold and blood cells visualized by scanning electron microscope (SEM, JEOL JSM Model 6300, Japan).

Fig. 5.5 showed SEM photographs of blood cells that adhered to the surfaces of the substrates with or without Hep-PU MS. Numerous adherent blood cells and some aggregates were observed on the blank GCE (Fig. 5.5a), while the blood cells and platelets adhering were remarkably suppressed on the surface of electrode substrate modified with GOx/(Hep-PU) (Fig. 5.5b and c). Based on the above observations, it was believed that thrombus was difficult to form onto the surface of Hep-PU MS without fused blood cells and platelets, which was caused by the anti-biofouling effect. The results strongly display that the anticoagulation of Hep-PU MS could efficiently suppress blood-cell adhesion and increase the microenvironment for GOx to undergo facile electron-transfer reactions.

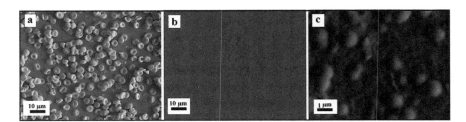

Figure 5.5 Representative SEM images of (a) blank electrode substrate, (b) electrode substrate modified with GOx/(Hep-PU), and (c) enlarged view of (b) exposed to human whole blood for 60 min, respectively.

5.4.3 Direct Electrochemistry of GOx/(Hep-PU)/GCE and Calibration Curve

All electrochemical experiments were performed on a CHI 760D electrochemical workstation (Chenhua Co. Ltd., China) in a three-electrode configuration. A saturated calomel electrode (SCE) and a platinum electrode served as reference and counter electrode, respectively. All potentials given below were relative to the SCE. The working electrode was a GCE. Hep-PU MS suspension was dropped onto the electrode surface and dried in air. After that, GOx solution was dropped onto the surface of (Hep-PU)/GCE and kept overnight at 4 °C, then the GCE modified with GOx/(Hep-PU) was obtained and the same modified method was used to get the (Hep-PU)/GCE. When not in use, the electrodes were stored at 4 °C in a refrigerator.

Cyclic voltammogram (CV) measurements were conducted in a 5 mL PBS cell at room temperature and the solution was purged with high purity nitrogen firstly and blanked with nitrogen during the electrochemical experiments.

Fig. 5.6 showed the electrochemical behaviors of (Hep-PU)/GCE, GOx/GCE and GOx/(Hep-PU)/GCE in PBS. There was no apparently redox process in CVs of the (Hep-PU)/GCE (curve a) and GOx/GCE (curve b), while GOx/(Hep-PU)/GCE displayed a pair of well-defined and quasi-reversible CV peak with a formal potential value (E^0) of -0.408 V (curve b). The good electrochemical response of GOx/(Hep-PU)/GCE indicated

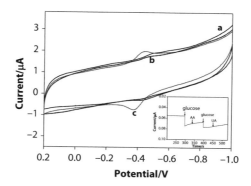

Figure 5.6 CVs of (a) (Hep-PU)/GCE, (b) GOx/GCE, and (c) GOx/(Hep-PU)/GCE in 0.1 M PBS (pH=7.4). Scan rate: 100 mV/s. The insert: Amperometric responses of the biosensor upon additions of glucose (1.0 mM), AA (0.1 mM), glucose (1.0 mM) and UA (0.1 mM), respectively, in PBS. The biosensor was biased on the potential of -0.41 V.

Hep-PU MS played an important role in facilitating the electron exchange between the electroactive center of GOx and GCE, and provided a mild environment so that the bioactivity of GOx could be retained. In addition, the anodic (Epa) and cathodic (Epc) peak potential were detected at −0.374 V and −0.442 V, respectively, and the separation of peak potentials (ΔEp) was 68 mV, at a scan rate of 100 mV/s. The ratio of anodic to cathodic peak currents was about 0.85. These results indicate that GOx underwent a quasi-reversible redox process (FeIII/FeII redox couple) at the GCE modified with Hep-PU MS.

An important analytical parameter for a biosensor is its ability to discriminate between the interfering species commonly present in similar physiological environment and the target analyte [81]. Electrochemical response of the GOx/(Hep-PU)/GCE was examined in the presence of some electroactive interfering substances like ascorbic acid (AA) and uric acid (UA) at higher levels than normal physiological [81]. The experiment was carried out by adding 1.0 mM glucose followed by 0.1 mM AA, 1.0 mM glucose and 0.1 mM UA. As shown in the insert of Fig. 5.6, the electroactive species did not cause interference significantly for the determination of glucose. The results imply that the GOx/(Hep-PU)/GCE has a good anti-interference ability.

Differential pulse voltammetry (DPV) measurements were carried out with pulse amplitude of 0.05 V and pulse width of 0.2 s. Different concentrations of glucose solution were added under intensive stirring, then CV was performed until the currents did not change any more, and DPV was immediately carried out. Then correlation between response currents and different concentrations of glucose solution was obtained. Fig. 5.7 showed typical DPVs for the detection of glucose in PBS and whole blood at the GOx/(Hep-PU)/GCE. Firstly, the GOx/(Hep-PU)/GCE was applied in PBS and the linear response range was from 0.2 mM to 20 mM. The linear regression equation was $I (\mu A) = 0.074c (mM) + 0.122$ ($R = 0.9990$), where I was current and c was the glucose concentration. The enhanced linear range expands the applications of the biosensor, especially in the glucose determination in whole blood samples. Then the proposed biosensor was used to determine blood glucose in whole blood. Blood samples were supplied by volunteer, within sodium fluoride to prevent glucose metabolism by blood cells prior to glucose determination [82, 83]. The current response was determined in 5 mL of 0.1 M, pH 7.4 PBS containing whole blood sample of 500 μL. It can be observed in Fig. 5.7B that with the increasing of glucose concentration, an anodic peak at -0.48 V emerges gradually. The calibration curve by plotting the current response with glucose concentration was presented in the insert of Fig. 5.7B. The linear relation had a

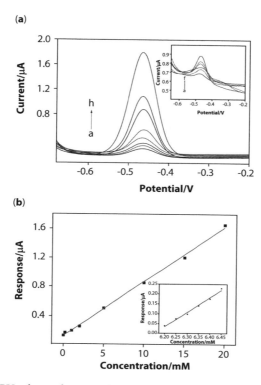

(a)

(b)

Figure 5.7 (A) DPVs obtained at GOx/(Hep-PU)/GCE in 0.1 M PBS (pH=7.4) with the concentration of glucose (from a to h) 0, 0.2, 1, 2, 5, 10, 15, 20 mM; and the insert: DPVs obtained at GOx/(Hep-PU)/GCE in whole blood samples at 25 °C with glucose of (a) 6.20 mM, (b) 6.25 mM, (c) 6.30 mM, (d) 6.35 mM, (e) 6.40 mM, (f) 6.45 mM. (B) Relationship between the peak current and the concentration of glucose in PBS; and the insert: relationship between the peak current and the concentration of glucose in whole blood samples.

regression equation of I (μA) = 0.074c (mM) + 0.122 (R = 0.9990) with a correlation coefficient (R) of 0.9939 and the calculated detection limit is 1.4×10^{-5} M (S/N = 3) in whole blood, which is lower than those obtained for GOx immobilized on poly(pyrrole propylic acid)/Au nanocomposite films (5×10^{-5} M) [84] and PtPd-MWCNTs films (3.1×10^{-5} M) [85].

5.4.4 Human Blood Samples Measurement

In an attempt to explore the GOx/(Hep-PU)/GCE for practical applications, the biosensor was applied to determine glucose in human blood

Table 5.1 Determination of glucose in whole blood samples using the GOx/(Hep-PU)/GCE.

NO.	Referenced values[a] (mM)	Determined values[b] (mM)
1	12.2±0.29	12.1±0.31
2	11.6±0.33	11.4±0.37
3	8.9±0.35	8.7±0.36
4	5.2±0.25	5.0±0.29
5	4.9±0.28	4.8±0.32

[a] Referenced values were provided by the hospital biochemistry laboratory. [b] The values were determined by the GOx/(Hep-PU)/GCE; they were average values of five measurements for each sample

samples of diabetic and healthy people. The glucose concentration in the serum of the diabetic is as high as 12.2 mM, while it is only 4.7–5.3 mM for the healthy ones [86]. Table 5.1 displayed the determination results of five samples including three diabetic and two healthy people.

The samples were first analyzed in the hospital biochemistry laboratory, then reanalyzed with the GOx/(Hep-PU)/GCE in our laboratory. The values of glucose was determined by the proposed biosensor are very close with the data provided by hospital. Furthermore, the values we measured direct in whole blood by the GOx/(Hep-PU)/GCE are more closed to the real values probably compared to the values that were measured in serum samples. Deeper investigation will be performed by our group in future.

5.5   Conclusion and Prospect

Considering the significance of blood in the area of biomedical and health for life, this paper has practical importance. This idea and technique of innovative hemocompatible polymeric micro- and nanoparticles provide a promising platform for the development of novel electrochemical biosensors that can be directly used in whole blood-contact system for illness diagnosis. In other words, the integration of nanotechnology, anticoagulant design and biosensor that will, without doubt, bring significant input to practical biomedical devices relevant to diagnostics and therapy of interest for human health.

References

1. J. Adams and F. Apple, *Circulation*, Vol. 109, pp. e12–e14, 2004.
2. J. Wang, *Chem. Rev.*, Vol. 108, pp. 814–825, 2008.
3. A. Kulkarni, M. Saxena, G. Price, M. J. O'Leary, T. Jacques and J. A. Myburgh, *Intens. Care Med.*, Vol. 31, pp. 142–145, 2005.
4. C. Voulgari and N. Tentolouris, *Diabetes Technol. Ther.*, Vol. 12, pp. 529–535, 2010.
5. M.H. Lin, M.C. Wu and J. Lin, *J. Clin. Biochem. Nutr.*, Vol. 47, pp. 45–52, 2010.
6. A. Schatz, M. Hager and O. Reiser, *Adv. Funct. Mater.*, Vol. 19, pp. 2109–2115, 2009.
7. C. Mao, L.C. Jiang, W.P. Luo, H.K. Liu, J.C. Bao, X.H. Huang and J. Shen, *Macromolecules*, Vol. 42, pp. 9366–9368, 2009.
8. X.Y. Zhang, E.M. Hicks, J. Zhao, G.C. Schatz and R.P. VanDuyn, *Nano Lett.*, Vol. 5, pp. 1503–1507, 2005.
9. H. Hu, X.B. Wang, S.L. Xu, W.T. Yang, F.J. Xu, J. Shen and C. Mao, *J. Mater. Chem.*, 2012, 22, 15362–15369.
10. L. Marcon, O. Melnyk and D. Stievenard, *Biosens. Bioelectron.*, Vol. 23, pp. 1185–1188, 2008.
11. C. Mao, C.X. Liang, W.P. Luo, J. C. Bao, J. Shen, X.M. Hou and W.B. Zhao, *J. Mater. Chem.*, Vol. 19, pp. 9025–9029, 2009.
12. R. Krishnan and M J. Heller, *J. Biophoton.*, Vol. 2, pp. 253–261, 2009.
13. B.S. Harrison and A.A. Atala, *Biomaterials*, Vol. 28, pp. 344–353. 2007.
14. A.S. Barnard, N.P. Young, A.I. Kirkland, M. A. V. Huis and H. F. Xu, *ACS Nano*, Vol. 3, 1431–1436, 2009.
15. Q. Xu, C. Mao, N. N. Liu, J. J. Zhu and J. Sheng, *Biosens. Bioelectron.*, Vol. 22, pp. 768–773, 2006.
16. C. H. Hou, S.M. Hou, Y.S. Hsueh, J. Lin, H. C. Wu and F.H. Lin, *Biomaterials*, Vol. 30, pp. 3956–3960, 2009.
17. J.M. Karp and R. Langer, *Curr. Opin. Biotechnol.*, Vol. 18, pp. 454–459, 2007.
18. Y. Fang, Y.L. Ni, G.H. Zhang, C. Mao, X.H. Huang and J. Shen, *Bioelechemistry*, Vol. 88, pp. 1–7, 2012.
19. F. Crisante, I. Francolini, M. Bellusci, A. Martinelli, L. D'Ilario and A. Piozzi, *Eur. J. Pharm. Sci.*, Vol. 36, pp. 555–564, 2009.
20. B.V.N. Nagavarma, H.K.S. Yadav, A. Ayaz, L.S. Vasudha and H. G Shivakumar, *Asian J. Pharm. Clin. Res.*, Vol. 5, pp. 16–23, 2012.
21. J. Yuan, C. Mao, J. Zhou, J. Shen, S.C. Lin, W. Zhu and J.L. Fang, *Polym. Int.*, Vol. 52, pp. 1869–1875, 2003.
22. S. Murugesan, J. Xie and R.J. Linhardt, *Curr. Top. Med. Chem.*, Vol. 8, pp. 80–100, 2008.
23. Q. Lv, C.B. Cao and H.S. Zhu, *Biomaterials*, Vol. 24, pp. 3915–3919, 2003,
24. C. Minelli, A. Kikuta and A. Yamamoto, *Open Biotechnol. J.*, Vol. 2, pp. 43–50, 2008.

25. I.K. Kang, O. H. Kwon, Y. M. Lee and Y. K. Sung, *Biomaterials*, Vol. 17, pp. 841–847, 1996.

26. S. Thorslund, J. Sanchez, R. Larsson, F. Nikolajeff and J. Bergquist, *Colloids Surf. B.*, Vol. 45, pp. 76–81, 2005.

27. A. B. Schreiber, J. Kenney, W. J. Kowalski, R. Friesel, T. Mehlman and T. Maciag, *Proc. Nat. Acad. Sci.*, Vol. 82, pp. 6138–6142, 1985.

28. W. Norde and D. Gage, *Langmuir*, Vol. 20, pp. 4162–4167, 2004.

29. A. Higuchi, K. Shirano, M. Harashima, B. Yoon, M. Hara, M. Hattori and K. Imamura, *Biomaterials*, Vol. 23, pp. 2659–2666, 2002.

30. C. Mao, C.X. Liang, Y.Q. Mao, L. Li, X.M. Hou and J. Shen, *Colloids Surf. B*, Vol. 74, pp. 362–365, 2009.

31. J. H. Lee, Y.M. Ju and D.M. Kim, *Biomaterials*, Vol. 21, 683–691, 2000.

32. Y.B. Kim, H.K. Kim, H. Nishida and T. Endo, *Macromol. Mater. Eng.*, Vol. 289, pp. 923–926, 2004.

33. M. Ahmed and R. Narain, *Biomaterials*, Vol. 33, pp. 3990–4001, 2012.

34. Y.L. Xiao, H. Hong, A. Javadi, J.W. Engle, W. J. Xu, Y. A. Yang, Y. Zhang, T. E. Barnhart, W. B. Cai and S. Q. Gong, *Biomaterials*, Vol. 33, pp. 3071–3082, 2012.

35. D. Foix, A. Serra, L. Amparore and M. Sangermano, *Polymer*, Vol. 53, pp. 3084–3088, 2012.

36. X.Y. Zhu, L. Chen, D.Y. Yan, Q. Chen, Y.F. Yao, Y. Xiao, J. Hou and J.Y. Li, *Langmuir*, Vol. 20, pp. 484–490, 2004.

37. D.L. Wang, H.Y. Chen, Y. Su, F. Qiu, L.J. Zhu, X.Y. Huan, B.S. Zhu, D.Y. Yan, F.L. Guo and X.Y. Zhu, *Polym. Chem.*, 2013, 4, 85–94.

38. X.H. He, X.M. Wu, X. Cai, S.L. Lin, M.R. Xie, X.Y. Zhu and D.Y. Yan, *Langmuir*, Vol. 28, pp. 11938–11947, 2012.

39. J.Y. Liu, Y. Pang, J. Chen, P. Huang, W. Huang, X.Y. Zhu and D.Y. Yan, *Biomaterials*, Vol. 33, pp. 7765–7774. 2012.

40. A.V. Ambade and A. Kumar, *J. Polym. Sci. Pol. Chem.*, Vol. 42, pp. 5134–5145, 2004.

41. C. Gao and D. Yan, *Prog. Polym. Sci.*, Vol. 29, pp. 183–275, 2004.

42. C.R.Yates and W. Hayes, *Eur. Polym. J.*, Vol. 40, pp. 1257–1281 , 2004.

43. M. Seiler, C. Jork, A. Kavarnou, W. Arlt and R. Hirsch, *AICHE J.*, Vol. 50, pp. 2439–2454, 2004.

44. H. Y. Hong, Y.Y. Mai, Y. F. Zhou, D.Y. Yan and Y. Chen, *J. Polym. Sci. Pol. Chem.*, Vol. 46, pp. 668–681, 2008.

45. W. Ajun and Y.X. Kou, *J. Nanopart. Res.*, Vol. 10, pp. 437–448, 2008.

46. C. Kontoyianni, Z. Sideratou, T. Theodossiou, L.A. Tziveleka, D. Tsiourvas and C.M. Paleos, *Macromol. Biosci.*, Vol. 8, pp. 871–881, 2008.

47. W.B. Zhao, Y.L. Ni, Q. S. Zhu, M. Liu, K. Wang, X.H. Huang and J. Shen, *Biosens. Bioelectron.*, Vol. 44, pp. 1–5, 2013.

48. S. Bhushan, V. Kakkar, H.C. Pal, S.K. Guru, Ajay Kumar, D.M. Mondhe, P.R. Sharma, S. C. Taneja, I.P. Kaur, J. Singh and A.K. Saxena, *Mol. Pharmaceutics*, Vol. 10, 225–235, 2013.

49. S. Tsunekawaa, R. Saharaa, Y. Kawazoea, K. Ishikawa, *Appl. Surf. Sci.*, Vol. 152, pp. 53–56, 1999.

50. F.Y. Tong, X.Q. Chen, L.B. Chen, P.Y. Zhu, J.F. Luan, C. Mao, J.C. Bao and J. Shen, *J. Mater. Chem. B*, Vol. 1, pp. 447–453, 2013.

51. C. Sun, L.B. Chen, F.J. Xu, P.Y. Zhu, J.F. Luan, C. Mao and J. Shen, *J. Mater. Chem. B*, Vol. 1, pp. 801–809, 2013.

52. F. Wang, Y. Han, C.S. Lim, Y. Lu, J. Wang, J. Xu, H. Chen, C. Zhang, M. Hong and X. Liu, *Nature*, Vol. 463, pp. 1061–1065, 2010.

53. H. Qian, W. T. Eckenhoff, Y. Zhu, T. Pintauer and R. Jin, *J. Am. Chem. Soc.*, Vol. 132, pp. 8280–8281, 2010.

54. J. Ming, Y. Wu, L.Y. Wang, Y. Yu and F. Zhao, *J. Mater. Chem.*, Vol. 21, pp. 17776–17782, 2011.

55. C. Sun, X.B. Wang, M. Zhou, Y.L. Ni, C. Mao, X. H. Huang and J. Shen, *J. Biomed. Nanotechnol.*, Vol. 9, pp. 1–8, 2012.

56. J. Han, Y. Liu and R. Guo, *Adv. Funct. Mater.*, Vol. 19, pp. 1112–1117, 2009.

57. H. Su, L. Jing, K. Shi, C. Yao and H. Fu, *J. Nanopart. Res.*, Vol.12, pp. 967–974, 2010.

58. C. Sun, X. H. Chen, Q. R. Han, M. Zhou, C. Mao, Q.S. Zhu and J. Shen, *Anal. Chim. Acta*, Vol. 776, pp. 17–23, 2013.

59. R.A. Sperling, and W. J. Parak, *Phil. Trans. R. Soc. A*, Vol. 368, pp. 1333–1383, 2010.

60. J.C. Boyer, M.P. Manseau, J.I. Murray and F.C.J.M. van Veggel *Langmuir*, Vol. 26, pp. 1157–1164, 2010.

61. X.B. Wang, M. Zhou, Y.Y. Zhu, J.J. Miao, C. Mao and J. Shen, *J. Mater. Chem. B*, Vol. 1, pp. 2132–2138, 2013.

62. P.L. Golas, S. Louie, G.V. Lowry, K. Matyjaszewski and R. D. Tilton *Langmuir*, Vol. 26, pp. 16890–16900, 2010.

63. X.Y. Sun, S.S. Yu, J.Q. Wan and K.Z. Chen, *J. Biom. Mat. Res.*, Vol. 101A, pp. 607–612 2013.

64. S.E. Skrabalak, J. Chen, L. Au, X. Lu, X. Li and Y.N. Xia, *Adv. Mater.*, Vol. 19, pp. 3177–3184, 2007.

65. D.K. Kweon, S.B. Song and Y.Y. Park, *Biomaterials*, Vol. 24, pp. 1595–1601, 2003.

66. M.B. Gorbet and M. V. Sefton, *Biomaterials*, Vol. 25, pp. 5681–5703, 2004.

67. R.K. Kainthan, M. Gnanamani, M. Ganguli, T. Ghosh, D.E. Brooks, S. Maiti and J.N. Kizhakkedathu, *Biomaterials*, Vol. 27, pp. 5377–5390, 2006.

68. R.K. Kainthan, S.R. Hester, E. Levin, D.V. Devine and D. E. Brooks, *Biomaterials*, Vol. 28, pp. 4581–4590, 2007.

69. F.R. Gong, X.Y. Cheng, S.F. Wang, Y.C. Zhao, Y. Gao and H.B. Cai, *Acta Biomater.*, Vol. 6, pp. 534–546, 2010.

70. L.C. Wang, J.R. Brown, A. Varki and J.D. Esko, *J. Clin. Invest.*, Vol. 110, pp. 127–136, 2002.

71. D. Shim, D.S. Wechsler, T.R. Lloyd and R.H. Beekman, *Catheter. Cardiovasc Diagn*, Vol. 39, pp. 287–290, 1996.

72. P.V. Asharani, S. Sethu, S. Vadukumpully, S.P. Zhong, C.T. Lim, M.P. Hande and S. Valiyaveettil, *Adv. Funct. Mater.*, Vol. 20, pp. 1233–1242, 2010.

73. D. Li, G.P. Li, P.C. Li, L.X. Zhang, Z.J. Liu, J. Wang and E.K. Wang, *Biomaterials*, Vol. 31, pp. 1850–1857, 2010.

74. J.D. Li, X. B. Zhang, L.B. Hao, Q.C. Xing and J.F. Wang, *J. Clin. Rehabil. Tissue Eng. Res.*, Vol. 14, pp. 559–562, 2010.

75. F.X. Soriano, J.L. Galbete and G. Forloni, *Neurochem. Int.*, Vol. 43, pp. 251–261, 2003.

76. USP XXII 24. Bilolgical Tests/Biological Reactivity Tests, *In vivo*.

77. N. Wisniewski and M. Reichert, *Colloids Surf. B*, Vol. 18, pp. 197–219, 2000.

78. X. Wu, P. Du, P. Wu and C. Cai, *Electrochim. Acta*, Vol. 54, pp. 738–743, 2008.

79. G. Zoldák, A. Zubrik, A. Musatov, M. Stupák and E. Sedlák, *J. Biol. Chem.*, Vol. 279, pp. 47601–47609, 2004.

80. X. Ren, D. Chen, X. Meng, F. Tang, X. Hou, D. Han and L. Zhang, *J. Colloid Interface Sci.*, Vol. 334, pp. 183–187, 2009.

81. H. Zheng, H. Xue, Y. Zhang and Z. Shen, *Biosens. Bioelectron.*, Vol. 17, 541–545, 2002.

82. D.B. Sacks, D. E. Bruns, D. E. Goldstein, N. K. Maclaren, J. M. McDonald and M. Parrott, *Clin. Chem.*, Vol. 48, pp. 436–472, 2002.

83. A.Y. Chan, R. Swaminathan and C.S. Cockram, *Clin. Chem.*, Vol. 35, 315–317, 1989.

84. M. Şenel and C. Nergiz, *Curr. Appl. Phys.*, Vol. 12, pp. 1118–1124, 2012.

85. K.J. Chen, C.F. Lee, J. Rick, S.H. Wang, C.C. Liu and B.J. Hwang, *Biosens. Bioelectron.*, Vol. 33, pp. 75–81, 2011.

86. L. C. Jiang and W. D. Zhang, *Biosens. Bioelectron.*, Vol. 25, pp. 1402–1407, 2010.

6

The Contribution of Smart Materials and Advanced Clinical Diagnostic Micro-Devices on the Progress and Improvement of Human Health Care

F.R.R. Teles[1] and L.P. Fonseca[2,*]

[1]*Laboratory of Mycology / Microbiology Unit and Centre for Malaria and Tropical Diseases (CMDT), Institute of Hygiene and Tropical Medicine (IHMT), Universidade Nova de Lisboa (UNL), Lisboa, Portugal.*
[2]*Department of Bioengineering, Instituto Superior Técnico (IST), Institute for Biotechnology and Bioengineering (IBB), Centre for Biological and Chemical Engineering, Lisboa, Portugal.*

Abstract

Nowadays, effective and rapid diagnostic tests constitute an essential approach to manage and combat human diseases. Clinical diagnostic micro-devices constitute a new paradigm for efficient human disease diagnosis, in line with well-known goals of personalized medicine and point-of-care diagnosis. The combination of the traditional expertise of clinicians and biotechnologists with the advances in bioelectronics and very often in combination with new materials brought to light an enormous portfolio of new proof-of-concept detection schemes, with similar or even improved performances compared to conventional bioassays. Some of these new clinical diagnostic micro-devices, with enhanced sensing properties, have already come to the commercial scenario, but they still face obstacles from the technical, marketing, and regulatory point of view. This chapter examines the current situation and progresses on the developments in the field of clinical diagnostic micro-devices and of smart materials for biosensing, in particular based on electrochemical detection, for ultimate diagnosis of human diseases.

Keywords: Diagnostic tests, biosensors, microfluidics, affinity sensors, nanomaterials, lab-on-a-chip

*Corresponding author: luis.fonseca@ist.utl.pt

Ashutosh Tiwari (ed.) Advanced Healthcare Materials, (203–236) 2014 © Scrivener Publishing LLC

6.1   Introduction

Diagnostic testing outputs influence approximately 70% of health care decisions [1]. This means that diagnostics are essential tools for managing numerous health care conditions. These range from infectious (e.g., human immunodeficiency virus infection/acquired immunodeficiency syndrome (HIV/AIDS), tuberculosis, malaria, diarrheal diseases, and lower respiratory infections) to chronic diseases (e.g., diabetes, cardiovascular diseases, respiratory diseases). In fact, non-communicable diseases are, by far, the leading causes of death in the world, representing 63% (36 million) of annual deaths [2].

Accurate medical diagnosis by conventional means usually requires trained health care workers to recognize symptoms or interpret clinical and biological analyses' results which, due to growing health demands and population aging in the developed world, and to financial limitations in developing countries, will continue to pose enormous difficulties to individual and national health systems. Recently, the great progresses observed in molecular analysis, molecular biology, and genome sequencing have opened a window of opportunity to improve the detection of pathogens, toxic compounds and physiologically relevant molecules (biomarkers) in body fluid samples, especially towards early disease diagnosis and disease follow-up [3]. In particular, biomarkers are biomolecules or other bioelements that can be objectively measured and assessed as indicators of normal or abnormal (disease) physiological processes, as well as of body responses to pharmacological therapeutic intervention. They can be present in different clinical specimens such as whole blood, serum/plasma, saliva, and urine.

Nowadays, clinical analyses are no longer carried out exclusively in bulky, centralized clinical chemistry laboratories; the use of point-of-care (POC) micro-devices has allowed caregivers to provide medical diagnoses outside hospital settings, and even, ultimately, by patients themselves, at home. These new clinical diagnostic tests, usually based on biosensing methods, are being increasingly exploited as promising alternatives to classical, "heavy" lab instrumentation when complying with the challenges of the evolving regulatory issues, as well as with the new business and economic realities. Indeed, in recent years, research and development of new medical diagnostic micro-devices has focused in accomplishing the basic theoretical requirements for a successful POC testing, namely rapidity, accuracy, portability, ease of operation, reusability and cost-effectiveness [4]. They should also be suitable for mass production and for in-the-field applications [5]. In the particular case of detection of infectious agents, disposability rather than reusability is also an advantage; it not only eliminates the problem of the biohazards nature of infected body fluid wastes, but also avoids the need for washing steps between sample preparations, a serious burden in many tropical regions poorly supplied with clean water. The particular contexts of remote places and tropical environments make the use of these new easy-to-use diagnostic tools the preferred (or even unique) option, as recently reviewed by Teles and Fonseca [6].

Many current commercial diagnostic clinical tests rely on dipstick or paper immune-chromatographic formats; as a particular case, common electrochemical devices for blood glucose monitoring in diabetic patients are very suitable for single-case diagnosis in field applications. In parallel, the enormous amount of genetic information brought by extensive genome sequencing has led to a need for multiplexed analyte detection using high-throughput miniaturized analytical devices. In this regard, microchips (also called "microarrays" or simply "biochips") constitute a major achievement towards the research of novel diagnostic systems, by allowing the monitoring of thousands of biomolecular interactions using very small reagent volumes dotted on a single glass slide. Current microarray technology is too expensive for valuable POC diagnosis of infectious diseases, despite being ideal for pathologies of more complex etiology, such as cancer. However, microarray efficiency is usually hampered by the large size of biological samples and by the complex treatments required (which also makes it difficult to obtain real-time outputs), difficulty of scaling down the array density, limited resolution, and strong sample concentration dependence [7].

The unprecedented opportunities for innovations in new clinical diagnostic micro-devices have benefited from the increased potential for the discovery of novel diagnostic biomarkers through genomics and proteomics, better understanding of pathogen virulence factors and host gene expression, and developments in cutting-edge materials science and nanotechnology. The ultimate goal of decentralized clinical testing may be the fabrication of lab-on-a-chip (LOC) devices, essentially an adaptation of microchips with channels and chambers for biomolecular recognition events, integrating, in a single platform, modules for specimen loading, sample processing, biomarker(s) amplification, biochemical reactions and product(s) detection. The success of clinical diagnostic micro-devices has run in parallel with the recent enormous developments in materials science and nanotechnology.

New biomaterials with enhanced and unusual physicochemical properties have been used for designing sensing surfaces with improved biorecognition performances and as matrices for the construction of the sensing solid platforms as well. In this chapter, the main concepts underlying diagnostic micro-devices for medical care will be reviewed, regarding both more consolidated schemes and novel devices and applications. It will be given particular attention to electrochemical sensing platforms and electroconductive nanomaterials. Finally, some of the main existing methods and devices for medical diagnosis of some of the main worldwide pathologies will be described in dedicated sections.

6.2   Physiological Biomarkers as Targets in Clinical Diagnostic Bioassays

Bioassays usually involve a biochemical reaction between a target (bioelement under measurement) and a probe (which binds specifically to the target through an affinity reaction). The chemical or biological method used for the detection may be carried out in solution (homogeneous assay) or onto a solid surface (heterogeneous assay), where a specific physicochemical change occurs during the biorecognition event.

Conventional clinical chemistry relies on the detection and measurement of a wide range of body substances, especially in the blood; these include gases, electrolytes, hemoglobin, hydrogen ion, enzymes, metabolites, lipids, hormones, vitamins, inflammatory markers, cytokines, coagulation proteins, therapeutic drugs and drugs of abuse [2]. Many of these clinical biomarkers have relevant roles as predictors of specific diseases, being further useful for monitoring disease follow-up and the effect of therapeutic drugs, including the assessment of eventual drug resistance effects. The major types of clinically relevant biomarkers as targets in clinical diagnostic tests are described below [2, 3].

6.2.1   Small Analytes

Analyzing common electrolytes in body fluids (e.g., H^+, Na^+, K^+, Cl^-, Ca^{2+}), blood gases (e.g., CO_2, O_2), and a number of small biomolecules (e.g., urea, glucose, cholesterol, ketones) is useful to routinely monitor, within clinical settings, important health parameters and, as such, prevent many disease conditions. Currently, the detection methods more widely employed for their detection are of electrochemical nature, especially potentiometry, amperometry and conductance [7].

6.2.2   Antigens and Antibodies

One of the most common types of bioassays in clinical diagnosis is the immunoassay, which is based in the high specific affinity between an antigen and its complementary antibody. Such affinity ensures that the binding event can occur in spite of the presence of nonspecific interactions or interferences from other molecules that may be present in the reaction medium [8]. The analytes can be detected and quantified by a single tagged antibody or by using, in addition to a primary label-free antibody, a secondary, labeled antibody ("sandwich" assay). The reference standard format is the enzyme-linked immunosorbent assay (ELISA), composed by multiwell plates that can be automated for high-throughput processing at well-equipped central laboratories. ELISA methods rely on colorimetric or chemiluminescent detection, characterized by very low detection limits as a result of enzyme-mediated signal amplification and serial washing steps. Unfortunately, the traditional ELISA method format requires expensive and bulky instrumentation (for liquid handling and signal detection) and trained workers, being therefore unsuited for portable POC diagnostics devices. For diagnostic purposes, immunoassays can be used to detect protein markers (antigens or antibodies) from either pathogens or the human host. The range of pathogens includes viruses (e.g., anti-HIV antibodies and p24 antigen for HIV), bacteria (e.g., anti-treponemal antibodies for syphilis and the early secretory antigenic target 6 for tuberculosis) and parasites (e.g., histidine-rich protein 2 for malaria) [3].

Wide applications have also occurred for the diagnosis of biomarkers related with cancer, cardiovascular and other types of genetic, chronic, and non-infectious diseases. Concerning cancer diagnosis, both monoclonal and polyclonal antibodies have been used to specifically target cancer cells. Utilization of monoclonal antibodies usually results in more specific and reproducible tests. However, monoclonal antibody-based immunoassays are also more expensive and take more time to be produced; this is the reason why polyclonal antibodies are more suitable for utilization in high-throughput tests. Micro-devices for immunodiagnostics rely on heterogeneous assays formats where the biomarkers are captured at the device surfaces previously modified with specific protein molecule probes.

6.2.3 Nucleic Acids

The hybridization ability between complementary, single-stranded polynucleotide chains of deoxyribonucleic acid (DNA) or ribonucleic acid (RNA) is the basis for highly specific affinity assays for nucleic acid detection. Nucleic acid-based detection methods are able to detect trace amounts of microorganism genomes, ultimately at the level of a single nucleotide change within the polynucleotide structure (i.e., discrimination between single base-pair mismatches). Common molecules used to label single-stranded chain probes include fluorophores, magnetic nanobeads, and redox enzymes [9]. The annealing between the two complementary chains can be detected and eventually quantified by microscopy, magneto-resistivity, or electrochemistry. Alternatively, as in immunoassays, a "sandwich" strategy can be tailored by employing an unlabeled capture nucleic-acid chain probe complementary to a terminal segment of the target chain and a second, labeled nucleic-acid chain probe complementary to the opposite segment of the target [9].

Nucleic acid-based prenatal diagnosis of inherited disorders has been extensively undertaken, as well as forensic analysis. Unfortunately, the high sensitivity and specificity standards achieved in these types of nucleic acid-based analysis have only been possible after previous amplification of nucleic-acid biomarkers by polymerase-chain reaction (PCR)-based methods and after complex sample processing steps. The window period between the occurrence of an infection and its detectability by a diagnostic test can be reduced significantly by using available clinical tests based on nucleic-acid detection, being remarkable examples those for detection and viral load monitoring of HIV, influenza A - subtype H1N1, tuberculosis, and group B streptococcal infection [10].

These diagnostic techniques are mainly used in hospitals and centralized laboratories, requiring complex operation steps for sample treatment, through the use of complex and expensive instrumentation. In order to achieve the goals of decentralized and POC diagnosis, nucleic-acid detection with portable clinical micro-devices designed as fully integrated systems has been attempted. Some of these advanced devices operate under a microfluidic format, with compartmentalized modules for sample collection, cell isolation and lysis, and nucleic-acid extraction, purification, pre-concentration, amplification, and detection of the amplified product.

6.2.4   Whole Cells

Intra- and extracellular biomarkers characteristic of cell metabolism, viability and respiration, including surface proteins (e.g., enzymes), can be used as bioreceptors either for signaling the presence of the cell or viral particle, or to detect, and possibly quantify, physiological molecular markers or toxic compounds.

Conventional cell-based assays are often used for hematological analysis and infectious disease diagnosis (including the identification of multiresistant strains). The most widely tested hematology parameters are hemoglobin and hematocrit, frequently for monitoring of anemia, red blood cell transfusion therapy, acute hemodilution during surgery, coagulation/clotting time (e.g., for cardiac surgical and catheterization procedures) and CD_4^+ lymphocyte counting (for monitoring the progression of HIV/AIDS infections) [11].

These traditional assays often require trained personnel, suitable facilities for cell culturing and ground electric power, which is not fully compatible with the requisites for operating portable diagnostic micro-devices. Some of the most recent diagnostic micro-devices rely in on-chip cell sorting and counting techniques from heterogeneous cell suspensions (such as whole blood) in straightforward formats for miniaturization. Other schemes make use of dielectrophoresis, which enables differential response of cells to electric fields due to their specific densities and physiological/metabolic states [12].

6.3 Biosensors

6.3.1 Principles and Transduction Mechanisms

Biosensors can be considered as small, portable and easy-to-use devices for selective molecular detection, identification and eventual quantification of biochemical or biological agents. The concept of "biosensor" can be considered equivalent, in the clinics, to that of "rapid test." In general, despite the technical and conceptual controversies around these concepts, we can also define a biosensor, or rapid test, as any diagnostic scheme or device especially tailored for POC and in-the-field applications, whilst complying with the general requirements of the World Health Organization defined by the acronym ASSURED, meaning: affordable, sensitive, specific, user-friendly, rapid and robust, equipment free and deliverable to end-users [2].

The first biosensor was reported by Clark and Lyons [13] for blood glucose monitoring. They coupled the glucose oxidase enzyme to an amperometric electrode for pO_2 measurement in solution. The enzyme-catalyzed oxidation of glucose consumed O_2, proportionally to the glucose concentration in the sample. By acting as biocatalytic elements, the enzymes can exhibit satisfactory or high detection limits when reaction is accompanied by the consumption or production of species such as CO_2, NH_3, H_2O_2, H^+, O_2 or toxic compounds (frequently correlated with clinically relevant parameters).

The major processes involved in biosensor operation are analyte recognition, signal transduction, signal amplification, signal processing and readout displaying. The recognition event is based on the interaction of the analyte with the bioreceptor (e.g., enzyme, antibody, protein, oligonucleotide, cell, tissue), designed to produce a physicochemical change measured by a transducer and converted it into a measurable signal. The transducer is the key element in biosensing devices; it converts the input signal into an electrical signal that is further amplified, processed, and displayed in a visual readout.

Related with the nature of the input signal, the main signal transduction mechanism principles can be classified as follows [3]:

a. Optical—it is the most diversified way of detection because many different types of optical principles can be employed, such as absorption, fluorescence, phosphorescence, Raman, surface-enhanced Raman scattering (SERS), refraction, and dispersion spectrometry. Indeed, these spectroscopic methods measure different optical properties, such as energy, polarization, amplitude, decay time, etc. Amplitude is the most commonly measured parameter, as it can be easily be correlated with the concentration of the analyte of interest [14].

b. Electrochemical—it measures changes in electrochemical properties of the system under study, resulting from redox reactions involving electroactive chemical species present onto, or near, an electrode surface selective for the species under measurement. Amperometric and potentiometric transducers are the most common electrochemical methods. Amperometric transducers, which operate by applying a constant potential between the working and the reference electrodes, monitor the resulting electric current (proportional to the rate of production/consumption of the electroactive species); they are more attractive for bioanalytical purposes for having higher sensitivity and wider linear range than potentiometric systems. Instead of bulk solution properties, interfacial electrochemical parameters can also be monitored, through capacitance or impedimetry (Figure 6.1).

c. Mass-sensitive (microgravimetric)—according to this principle, bioanalytical measurements can detect small mass changes caused by the binding of bioelements to small piezoelectric crystals. The bare crystal has a specific frequency of oscillation that changes upon immobilization of a bioprobe onto its surface and even further after selective binding of the analyte of interest to the probe.

   d. Thermal—temperature variations resultant from a chemical or enzymatic reaction between a probe and a suitable analyte can be measured. Such changes can be correlated with the amount of consumed reactants or formed products.

   e. Magnetic—these systems (e.g., spin-valve magnetic field sensors) are capable of detecting and quantifying magnetic fields created by paramagnetic nano- or micro-bead particles used as labels in biochemical processes [9].

6.3.2   Immunosensors *vs.* Genosensors

Concerning the nature of the biomolecules involved in the biorecognition process, the two major types of biosensors are nucleic-acid biosensors (DNA and RNA biosensors, also called genosensors) and antigen or antibody biosensors (also called immunosensors). Both exhibit the advantages of high sensitivity and selectivity inherent to the use of complementary affinity ligand interactions. DNA is the only biomolecule that can be easily copied and amplified by PCR and other related techniques, a great advantage for genosensors compared to immunosensors. Additionally, monitoring nucleic-acids allows earlier diagnosis than immunological assays. Moreover, genosensing does not exhibit the typical problems of cross-reactivity frequently observed in immunosensing. Unlike antibodies, DNA forms biological recognition layers easily synthesizable and reusable after simple thermal melting of the DNA duplex. Nucleic-acid arrays are also more suitable than their protein counterparts for direct synthesis onto chip surfaces, without the need, for instance, to produce and purify antibodies [15]. A major drawback of genosensors when compared to immunosensors is the fact that, in vivo, microorganism genomes are tightly packaged inside the cell biomembrane, thus, their concentration may be too low for successful diagnosis without prior amplification by PCR. Despite the ultimate goal of directly determining DNA traces in clinical samples, only very low levels of nucleic acids exist in biological fluids, otherwise undetectable, if previous PCR amplification is not performed. Ongoing efforts with new DNA-based diagnostics aim the production of PCR-free DNA detection systems, but this has not been fully achieved yet in commercially available devices [15]. Salt-dependent electrostatic effects greatly influence the stability, structure, reactivity and binding of nucleic acids in solution or at an interface. Biosensor schemes for single copy detection have already been developed, but amplification has the benefit of increasing

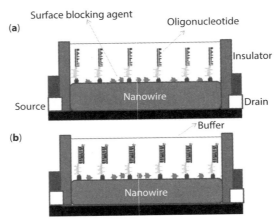

Figure 6.1 Lateral view scheme of an impedimetric genosensor based on nanowire (a) covalently modified with oligonucleotides by single-point attachment using a cross-linker, and (b) upon hybridization with the complementary single stranded target after a washing step. The hybridization event increases the electrical resistance of the surface between the two electrodes.

the signal-to-noise (S/N) ratio of an assay, thus lowering the probability of interferences by sample contaminants.

6.3.3   Optical vs. Electrochemical Detection

Electrochemical and optical biosensors are, by far, the most promising and readily available detection methods employed in the research and development of clinical diagnostic micro-devices. Most of the described catalytic biosensors (e.g., enzyme-based) rely on electrochemical transduction, whereas affinity biosensors (e.g., protein and nucleic acid-based) are usually more suitable for coupling to optical methods. In general, electrochemical biosensors are relatively simpler, faster, less costly, less power-demanding and more amenable for miniaturization and mass-production than optical biosensors. However, electrochemical transduction is prone to interferences from electrochemically active substances and troublesome electron-transfer pathways between the biosensing layer and the surface of the bare electrode. Additional challenges for the development of fully portable clinical diagnostic micro-devices for POC applications include [16]:

- Decreasing the complexity and unitary cost of schemes comprising electrodes integrated within disposable cartridges.

This has already been partially achieved by the use of mass-produced screen-printed electrodes;

- Efficient and cost-effective multiplexing abilities for simultaneous and accurate analysis of thousands of analytes within a single array electrode;
- Incompatibilities between current microelectronic fabrication processes and specific features of electrochemical sensors, with regard to dimensions, materials and passivation layer;
- Environmental dependence of many electrochemical processes, namely in relation to temperature and ionic concentrations (e.g., H^+), due to redox by-product accumulation near the electrodes and to washings.

The success of glucose biosensors and of other commercial rapid tests relies on their suitability for single-case diagnosis in field applications. However, diabetes causes significant visual disabilities, which may lead to misinterpretation of results by the patients. As a consequence, micro-devices with digital output-based displays have leveraged the development of other forms of signal generation.

Concerning optical transduction, fiber-optic sensors seem particularly suitable for decentralized testing format, for the absence of electrical interferences, inherent small size, high transmission efficiency and chemical stability and flexibility, thus opening the way to miniaturization and remote sensing. Furthermore, the harmless nature of optical signals, the biocompatibility, and the easy sterilization make them attractive for in vivo measurements. Nonetheless, their propensity to interference from biological samples and the intrinsic complexity and expensiveness of most optical transducers and components have hindered wider applications in the field [17].

Optical detection is generally seen as having superior performance. It is the most widely employed detection method in centralized laboratories of clinical chemistry. Currently, optical detection, in general, has more favorable per-test costs than electrochemical detection. Optical detection is especially amenable for multiplexed applications; in particular, commercial charged-coupled device (CCD) and complementary metal-oxide semiconductors (CMOSs) image sensors can detect hundreds or thousands of different compounds. While optical detection is quite straightforward in a laboratory environment where microscopes, lasers, spectrophotometers, lenses, and filters can be precisely set-up and calibrated, this is not the case of micro-devices for in-the-field applications and decentralized diagnosis. This explains, at least in part, the only modest success of optical detection for such applications. The high frequency of optical signals—compared, for instance, with electrical ones—is an undeniable advantage considering the enormous amount of information that can be carried by optical devices. This has been decisive for the superior commercial success, in the last few years, of many optical sensing platforms (used in centralized clinical laboratories) compared to those based on other transduction mechanisms [18].

Some of the most commercially successful platforms include: GeneChipR high-density (high spatial resolution of individual probes) microarray from Affymetrix (Santa Clara, CA), with fluorescence-based detection coupled to a confocal readout; fluorescence-detecting microbead-based BeadXpressR array system, from Illumina (San Diego, CA); BIAcore (real-time biospecific interaction analysis) surface plasmon resonance (SPR)-based platforms, of Pharmacia Biosensor (Uppsala, Sweden); NanoLantern, from Lighthouse Biosciences (West Henrietta, NY), that reaches exquisite sequence specificity by using DNA hairpin probes.

A particularly recent and attractive application of optical detection systems in remote areas and low-resource conditions has arisen under the context of telemedicine, which links in-the-field sample collection and analysis with data processing and interpretation within a central well-equipped facility, usually far away from the testing place. However, one challenge of telemedicine based on wireless communication is to transmit images with sufficient resolution for accurate interpretation at the data processing setting, as the intensity of the colors in an image varies according to the lighting conditions, the resolution of the camera and the focus of the picture. This is still a challenge in optical detectors, but also a field of continuous improvements [19].

The high sensitivity, specificity, simplicity, and inherent proneness for miniaturization allow modern electrochemical detectors to compete with more established and diversified optical detection-based devices, even in terms of multiplexed detection. The ongoing advances in both areas will certainly bring, in the near future, a new portfolio of devices adapted for the POC, with improved flexibility and performance over the existing ones.

6.3.4   Merging Electrochemistry with Enzyme Biosensors

Over the past three decades, a tremendous research activity in the area of biosensors has been carried out. However, as in the past of biosensor development, the most common commercial biosensors are those for glucose measurement, usually based on immobilized redox enzymes.

Unlike other types of biomolecules (e.g., nucleic-acids and antigens/antibodies), they are more amenable to transduce electrical signals arising from charged substrates and products involved in catalytic reactions. A single enzyme or cascades of two or more enzymes for detection of a specific target analyte are the more frequently employed schemes. Glucose oxidase or glucose dehydrogenase enzymes have been the most commonly employed. Glucose biosensors generally make use of electrochemical signal transduction, because it provides good specificity and reproducibility. Moreover, electrochemical biosensing is simple, reliable, and provides low detection limits and wide dynamic ranges, owing to the fact that the reactions involving electroactive species occur at electrode – solution interfaces. Due to their specificity, speed, portability, and low cost, electrochemical biosensors offer exciting opportunities for numerous decentralized clinical applications, including "alternative-site" testing (e.g., physician's office), emergency-room screening, bedside monitoring and home self-testing [20].

Substantial amounts of published works about enzyme-based biosensors may be related to the cost-effectiveness of electrochemical transducers and to the relevant accumulated knowledge already existent concerning enzyme-based sensing. However, research on enzyme biosensors still needs to overcome some persistent technical and operational difficulties, namely the modest enzyme stability when immobilized onto solid supports, especially under long storage periods. A particularly popular layout refers to disposable enzyme electrodes, such as those obtained by screen-printing technology, due to simple large-scale and low-cost manufacturing. Screen-printed electrodes (SPEs) are produced by printing different conductive inks and/or graphite materials onto the surfaces of insulating plastic or ceramic substrates. SPEs work successfully when coupled with hand-held, battery-operated clinical analyzers, in single-use (disposable) cartridges, being extremely useful for rapid POC measurements of multiple electrolytes and metabolites [21]. A remarkable example of a commercial device is the iSTAT, from the Abbott Laboratories (Abbott Park, IL, USA).

6.3.5   Strip-Tests and Dipstick Tests

In these formats, which are similar to those of cheap, rapid and common immunoassay pregnancy tests, the clinical liquid sample migrates across the surface of a paper or nitrocellulose membrane by capillary action. The results of these single-case diagnostic tools are essentially qualitative or semi-quantitative by comparison with a printing calibration chart. In this type of clinical diagnostic tools, the reagent solutions, in the order of 0.1–1.0 μL, are spotted by hand or inkjet printing in the test zones, and dried afterwards. Most of the work to date on strip-tests has relied in colorimetric assays, enabling, in just a few minutes, easy reading of the results by medical personnel or even by the patients themselves. Most configurations operate by naked-eye detection, although more complex and sophisticated readouts are also available [18]. Very often, these tests are also called lateral-flow immunoassays (LFIs) or immunochromatographic assays, due to the sample capillary flow and to the visual readout system. These strip tests have been intensively used for detection of small molecules (e.g., indicators) and in enzymatic sensors, some of the most common being the urinalysis dipsticks. There is nowadays a wide portfolio of commercially available LFIs for inexpensive and simple POC testing, including for pregnancy, cardiac diseases and infectious diseases. Their sensitivities are still modest when compared to those of conventional immunological laboratory assays (such as ELISA). Nevertheless, LFIs are ideal for situations where the analyte is relatively abundant, when complex sample preparation is not needed and in cases where a simple "yes/no" diagnostic response is enough [22].

6.3.6   Biosensor Arrays and Multiplexing

The determination of the levels of multiple biomarkers in the human body can be crucial to properly help in the diagnosis of early disease and to assist clinicians in decision-making. A remarkable example is in cancer diagnosis, due to the complex etiology of this disease. With this purpose, DNA and protein sensor arrays (biochips), with fabrication inspired by planar silicon-based electronics, can be highly benefic. Biochips are essentially very dense microband sensor arrays coated with different probes for simultaneous detection of multiple targets (with or without a label) printed on the chip by conventional photolithography. They allow monitoring of thousands of biomolecular interactions using very small reagent volumes dotted on a single glass slide. They may be built to detect either protein signature patterns or multiple DNA mutations [23].

Unfortunately, conventional multiplex diagnostic tests, such as ELISA, too often do not reach the ultrasensitivity standards required to detect cancer biomarkers in early stages of disease development, when prognostic odds are much more favorable. This bottleneck has been the driving force for the development of new diagnostic technologies, aimed at detecting multi-biomarkers present in body fluid samples at very low concentrations [21]. This has been possible by the great progresses observed in device miniaturization, which allows packing huge amounts of very tiny and compacted sensors in high-density arrays. Commercially available biochips for clinical diagnosis encompass viral (e.g., anti-HIV antibodies, antibodies against influenza A/B virus and rotavirus antigens), bacterial (e.g., antibodies against *Streptococcus* A and B, *Chlamydia trachomatis and Treponema pallidum*), and parasitic (e.g., histidine-rich protein for *P. falciparum*, trichomonas antigens) infections, as well as non-infectious diseases (e.g., PSA for prostate cancer, C-reactive protein for inflammation and HbA1c for plasma glucose concentration) [24]. Unfortunately, such arrays require proper spatial separation of the individual transducers to eliminate cross-talking as a result of diffusional processes from enzymatically generated electroactive species [21].

6.3.7   Microfluidic-Based Biosensors

Microfluidic biosensors are electrochemical devices for fluid flow along a sequence of very small compartments imprinted within a solid platform. These components are usually microscopic (from 1 to 500 μm) and include filters, channels, pumps and valves, controlled by integrated electronics (electronic chip). Advantages over tests in centralized laboratories include small reagent consumption, rapid and parallel analysis, small size, low power consumption and functional integration of multiple components. As such, microfluidic devices are amenable for being low cost and portable—two critical assets for implementation of clinical diagnostic microdevices [8].

Microfabrication techniques initially made use of glass as the essential component of chip devices, an unsuitable material for biological applications owing to its high cost and low biocompatibility. In second-generation biochips, glass was replaced by the polymer polydimethylsiloxane (PDMS), which exhibits excellent optical properties (low autofluorescence and high transmissibility) and is biocompatible. However, it is still a relatively expensive material for mass-production of such devices and hence for their commercial viability [25]. For these reasons, efforts have been put in the development and use of polymeric materials amenable for mass-production, such as elastomers and hard plastics.

In microfluidic systems, it is crucial to have a rigorous fluid control, achieved by precise fluid actuation and delivery, for accurate analytical detection. Pneumatic-based actuation of fluids and hand pumps can be used for moving the fluid throughout the microchannels flowing in microfluidic systems. Another group of microfluidic devices are those based on injection-molded centrifugal-based platforms (CDs), where automated spinning of the CD-like device allows the fluid to move from a central spot to the periphery, along the imprinted microchannels.

Today, special attention has been given to passive approaches for fluid movement; they are based on variations of micro-channel geometries in capillary systems, causing fluid propelling without the need for external power and moving off-chip moieties. Microfluidic devices have already demonstrated efficiency in the detection of C-reactive protein, botulin neurotoxin, anti-HIV antibodies, anti-treponemal antibodies, and hepatitis B virus (antigen and antibody) [4].

Paper has been extensively used as a solid support in analytical and clinical chemistry. Paper chromatography, in particular, is commonly employed to separate and identify mixtures of small molecules, amino acids, proteins, and antibodies, making use of the ubiquity and inexpensiveness of paper. Moreover, paper material allows passive transport of fluids without active pumping. Hence, it is not surprising the interest that this material has risen for the development of simple diagnostics, particularly for application in resource-limited contexts. This began with the well-known LFIs and, more recently, has gained the interest from developers of microfluidic-based devices, in the form of the so-called microfluidic paper-based analytical devices (μPADs). These systems combine the capabilities of conventional microfluidic devices with the simplicity of diagnostic strip-tests. In a pioneer configuration, the μPAD is designed with a central channel for sample introduction; this channel then branches into different test zones on the same device, without cross-contamination of the reagents, allowing easy multiplexed analyte detection. In each test zone, specific reagents for each particular and pre-defined assay are spotted and dried during fabrication of the device [26]. Proof-of-concept μPAD devices have been applied in the detection of biotoxins, e.g., ricin, Shiga toxin I and Staphylococcal enterotoxin B; other applications include the detection of endogenous matrix metalloproteinase-8 in saliva and anti-HIV antibodies [26].

Based on this concept, it was recently developed and proposed the use of paper as a platform in microarray format for multiplexed assays. This was carried out by immobilizing capture antibodies onto the surface of an optical waveguide, in stripes resembling "bar codes." Each strip in the "bar codes" was directed against a different analyte of interest. The testing sample was then loaded perpendicularly to the orientation of the bar codes using flow chamber modules. This enabled each analyte to find and bind its specific "bar code" [27].

More recent developments have focused on new fabrication methods for μPADs and integration of components such as electrodes, valves, filters, mixers, and coatings, in order to considerably expand their capabilities. Another research line has led to the development of a new generation of smart paper-based strip-tests containing carbon or metal fibers for enhancement of conductive or magnetic responsive properties [28]. Examples have been reported for glucose, lactate, and uric acid electrochemical detection.

In this field, it was recently proposed the use of paper and other fibers coated with a thin-film of Ion-jelly®, with the aim of adding electroconductive properties to such supports (Figure 6.2). Ion Jelly® technology confers tailor-made properties resultant from the modification of gelatin with different ionic liquids to optimize its properties for biosensing applications [29]. By manipulating the hidrophilicity and polarity of selected ionic liquids, it is possible to design new biomaterials with new or improved properties, such as flexiblility in different forms (e.g., films, fibers, microparticles, monoblocks), transparency, electroconductance when submitted to an electric potential, and more compatible microenvironments for enzymes/proteins. Ion-jelly®-impregnated paper filter functions as strip test, showing excellent compatibility with glucose oxidase and horseradish peroxidase for colorimetric detection of glucose [30]. For this reason, this new thin-film material holds great promise for the development of a new generation of smart paper-based strip tests.

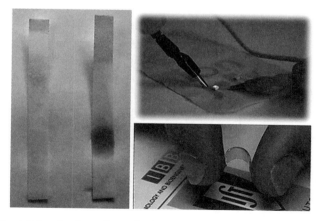

Figure 6.2 Ion Jelly® technology allows the design of new composite biomaterials for coating paper and other fiber supports, conferring them electroconductive properties when submitted to an electric potential and enhanced biocompatibility, e.g., for enzyme-based detection of glucose (red staining).

6.3.8   Lab-on-a-chip (LOC)

In tight relation with microfluidic devices, especial attention has been devoted, over the past years, on the development of fully integrated tests for bioanalysis, able to perform, in a single and compact miniaturized device, all the sequential unitary operations normally undertaken in a laboratory setting, which has led to the concept of "lab-on-a-chip" (LOC) devices. These novel biomedical diagnostic devices are characterized by high sensitivity, accuracy, specificity, reproducibility, speed of response, automation, portability, and simplicity of operation. The recent advances in microfluidics-related physical sciences and in their applications towards the construction of microfluidic prototypes has naturally created an opportunity for the emergence of LOC devices, carrying individual compartments interconnected by microchannels for fluid flow, and where different steps of a complete bioanalytical procedure take place. They usually incorporate sample preparation from raw biological and clinical samples (e.g., whole blood, urine or saliva) and detection sensitivity comparable to those of conventional methods [18].

An interesting and recent application of LOC in bioanalysis consists in platforms for cell sorting and counting. However, for measurements of intracellular components requiring previous sample processing (preparation) steps, the development of LOC devices for POC testing has been more troublesome; indeed, sample pretreatment has been a relatively neglected step in this regard, compared to the downstream bioanalytical steps, because of its intrinsic complexity. For instance, cell lysis and subsequent extraction (with eventual labeling) of target nucleic-acid sequences significantly burdens the automated assay, which is the reason why many efforts have been made to develop rapid and reliable methods to break apart the hard coating that packages genetic material, namely within many pathogenic microorganism cells. Sometimes these systems require off-line sample preparation and reagent handling, being therefore unavailable for routine home testing or in remote and resource-limited places [31]. In such cases, sample preparation must be performed off-chip, using bulky laboratory equipment, whilst amplification and detection are undertaken within the microfluidic system. Nonetheless, ongoing efforts to integrate pretreatment steps in the microfluidic devices may lead to improvements in sensitivity, since less amounts of sample are lost between steps, and also leading to reduced probability of contamination.

Detection of DNA hybridization events in DNA chips is one of most interesting and important applications for clinical diagnostics. One of the first microfluidic-based DNA purification procedures envisaged single-cell mRNA extraction and analysis via cDNA synthesis [32]. Miniaturized automated bioanalytical devices present many advantages for PCR-based processes, such as decreased cost of fabrication and operation, decreased reaction time for DNA amplification, reduced cross-talk of the PCR reaction and ability to perform large numbers of parallel amplification analyses. Isothermal techniques for nucleic-acid amplification without thermal cycling have also been used in LOC devices [33]. Their use, in alternative to PCR-based methods, allows the production of simpler, cheaper, and more user-friendly instrumentation for POC usage [3]. Some commercially available POC platforms for DNA-based diagnostics are based in helicase-dependent (HDA; BioHelix), transcription-mediated (TMA; Gen-Probe), nucleic-acid sequence-based (NASBA; BioMerieux), strand displacement (SDA; Becton Dickinson), loop-mediated (LAMP; Eiken) and recombinase polymerase (RPA; TwistDx) amplification. On the other hand, fluorescence-based techniques remain the most commonly employed due to the intrinsic high level of sensitivity and low background noise of fluorescent methods. Cepheid's GeneXpert platform (Cepheid Inc.), for example, uses real-time PCR (with fluorescently labeled probes) and has shown promising results for detecting tuberculosis, particularly drug-resistant strains.

6.4   Advanced Materials and Nanostructures for Health Care Applications

The emergence of nanoscale technologies and structures has driven a next generation of bioanalytical methods and devices with improved sensitivity and reduced costs. A brand new range of electronic devices and biosensor nanoplatforms has emerged as a consequence of the inherent small size and unusual physicochemical properties of these nanostructures, unlike those of bulk materials. As the system decreases below 10 nm, many physicochemical properties such as plasticity, thermal and optical parameters, reactivity, catalytic ability, electron/ion transport and quantum mechanical properties become more pronounced. By conjugating specific biomolecules with these nanostructures, it is possible to engineer innovative biological functionalities. These new nanomaterials have revealed useful properties for numerous electrochemical detection applications [34]. Moreover, with an appropriate transducing method, the selectivity of nanobiosensors and nanobiochips may be tuned as a result of signal dependence on nanostructures morphologies, with remarkable sensitivity and multiplexing capability [34]. Additionally, when the electrode size decreases down to the nanometer level, radial diffusion becomes dominant as the contribution of convective transport is negligible and steady state or quasi-steady state currents are rapidly achieved. In this case, redox reactions are limited by the rate of electron transfer [35].

Based on their dimensions, three main types of nanomaterials have been usually employed in biosensor design: dimensionless (e.g., nanoparticles and nanospheres), one-dimension (e.g., nanotubes and nanowires) and two-dimension (polymeric thin-films) nanostructures (Figure 6.3). Many nanomaterials have been used as intermediate layers for merging electrodes with biomolecules (enzymes, proteins, antibodies, etc.), aiming the development of electrochemical nanobiosensors. Traditionally, nanostructures involved in capillary driven microflow within plastic supports have been used to scale-up disposable microfluidic devices, envisaging mass production through known molding techniques. Inversely, it is possible to foresee mass fabrication of future nanochips and nanofluidic systems as an extension of current mechanical methods for production of microsensing devices, most often based on organic polymers and gels, especially PDMS frames. However, the need for high-throughput analysis is creating a demand for a broader range of less costly and easier fabrication methods. The sudden rise in the expected cost/benefit of miniaturizing photolithographically-produced microsystems for

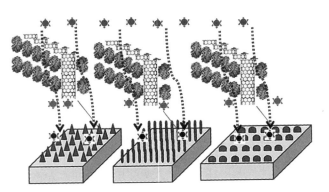

Figure 6.3 Schemes of biosensing nanoplatforms combining electroactive nanospheres (0D), carbon nanotubes (1D) and 2D nanostructures.

high-throughput production of disposable chip devices has led to the assembly of micromechanical systems and functional biomimetic structures in the 5–100 nm range [36].

Biosensing probe immobilization is the key step in biosensor and biochip construction but, regardless their peculiar advantages, the conventional methods for biomolecule immobilization (physical adsorption, covalent binding, crosslinking and entrapment in gels or membranes) have, in general, low reproducibility and poor spatially-controlled deposition, a crucial problem for the development of commercial miniaturized sensors. The rapid growth in biomaterials, especially the availability, application and combination of a wide range of polymers with new sensing techniques, has generated remarkable innovations in the design and construction of biochips and in the sensor response as well. In fact, polymers are becoming inseparable from biomolecule immobilization strategies and design of biochip platforms. Their original role as electrical insulators has been progressively substituted by their electrical conductive abilities, which opens a new and broad scope of applications, especially for electrochemical transduction schemes. Polymers are usually advantageous in terms of lightweight, flexibility, corrosion-resistivity, high chemical inertness and ease of processing, as well as for providing a better balance between chemical, physical, and mechanical properties than metal or ceramic materials. Additionally, clinical diagnostics require disposable devices to avoid contamination and, therefore, the use of cheap microfabricated polymer devices may show a significant commercial viability in the near future in relation to glass and ceramics materials [8].

Concerning the process for immobilization of biomolecules in polymeric films, one-step electropolymerization is a simple and rapid method to deposit biomolecules into confined spaces, but usually requires large amounts of both monomer and biomolecule. A related approach is electrogeneration of functionalized polymers followed by biomolecule attachment to the polymer surface. Conducting polymers (CPs) have been widely studied, especially polypyrrole (PPy) and polyaniline (PANI). They are especially amenable for electrochemical biochip development for providing biomolecule immobilization and for rapid electron transfer. Films of PPy and derivatives have good conductivity, selectivity, stability, and efficient polymerization at neutral pH. Electrochemical synthesis of polymers enables judicious control of the polymer layer thickness and of the material structure, based on the measurement of the electrical potential or current that crosses the interface during the electropolymerization step. Most of the electrochemically-deposited polymers used for biomolecule immobilization are CPs, since their formation—unlike non-conducting polymers—is not restricted to the synthesis of very thin films. The state-of-the-art and main novelties about the use of polymers for immobilization of biomolecules in electrochemical biosensor platforms were recently reviewed by Teles and Fonseca [37].

Dimensionless nanostructures, usually of metallic nature, can be conductive, semiconductive and insulating, depending on their electrical properties. Basically, their roles in biosensors include surface grafting of probe biomolecules, reaction catalysis, electron transfer enhancement, reactant in biochemical reactions and probe labeling [38]. Specifically referring to electrochemical detection, the enhancement of electron transfer is an extremely important challenge in the case of enzymatic biosensors, because very often a protein shell insulates the redox-active site of most enzymes. Thus, these nanomaterials may be able to promote direct electron transfer between enzymes and electrode surfaces, thus obviating the need for mediators or co-substrates [39].

Gold nanoparticles (AuNPs), one of the most common types of nanoparticles, can be employed for biomolecule immobilization or labeling, causing enhancement of signal responses and lowering the detection limits. There are many reported biosensors involving antigen/antibody and DNA detection with AuNPs [40]. AuNPs can also be strongly attached to electrode surfaces through self-assembled monolayers (SAMs) of alkanethiol molecules, through strong S–Au bonding, which creates large surface areas for immobilization of biomolecules [41, 42]. Indeed, highly-ordered SAMs allow obtaining very homogeneous surface coverage and hence more sensitive and reproducible sensors. The use of AuNP tags (e.g., redox tracers) in stripping voltammetry for electrochemical detection of DNA hybridization and antibody–antigen interactions was already demonstrated with remarkable sensitivity [43]. In such protocols, after the event of biomolecular recognition, electrochemical stripping measurement of the metal tracer occurs. Another electrochemical detection methodology based on the use of nanoparticles relies on nanoparticle-induced changes in the conductivity across a microelectrode gap for highly sensitive electronic detection of DNA hybridization [44]. The capture of the nanoparticle-tagged DNA targets by probes immobilized in the gap between two closely spaced microelectrodes, and subsequent silver precipitation, resulted in a conductive metal layer across the gap, and led to a measurable conductivity signal.

Inorganic nanocrystals offer an enormous diversity of electrical tags for biosensing, as needed for multiplexed clinical testing, and already shown for multi-target electronic detection of proteins [45] and DNA [46]. Four sulfide nanoparticles (of cadmium, zinc, copper, and lead) were used to differentiate the signals of four proteins and DNA targets, in connection with sandwich immunoassay and hybridization assay, respectively, along with stripping voltammetry of the corresponding metals. Each binding event thus yielded a distinct voltammetric peak, whose size and position permitted the identification and quantification of the corresponding targets. The concept can be readily scaled-up and multiplexed by using a parallel high-throughput automated microwell operation, with each microcavity capable of carrying out multiple measurements [47].

1D-nanostructures can be considered as elongated versions of dimensionless nanostructures, being nanotubes and nanowires the most commonly used in biosensing. Because of their high surface-to-volume ratio and improved electron transport properties, their electronic conductance is strongly influenced by minor surface perturbations (e.g., binding of biomolecules), hence indicating great promise for label-free and real-time detection. Due to the extreme smallness of these nanomaterials, it is possible to pack a large number of antibody-functionalized nanotubes and nanowires onto a remarkably small footprint of an array device. Such 1D-material thus offers usefulness for assays of multiple disease biomarkers in ultrasmall sample volumes. Several studies already indicated the potential of functionalized nanotubes and nanowires for highly sensitive real-time biodetection [21].

Carbon nanotubes (CNTs) constitute a new allotrope of carbon, originated from the fullerene family. They behave like microcrystals with molecular dimensions and have unique electrical properties, for instance, with respect to electron transportability, apart serving as excellent matrices for high-density biomolecule immobilization. CNTs can be produced in the form of cylindrical holed multi- (MWCNTs) or single-walled carbon nanotubes (SWCNTs). Plus, they can display extremely high sensitivity in response to very small changes in environmental conditions. In tightly packed nanostructure networks, each nanotube may act as an individual nanoelectrode, with sufficient free space between neighboring nanotubes, thus preventing the overlap of their diffusion layers, therefore yielding high signal-to-noise ratios and improved detection limits. The nanodimensions of CNTs guarantee a very large active surface area, well suited for the conception of miniaturized sensors [48]. In order to integrate biomolecules with CNTs, chemical and electrochemical treatments have to be performed for introduction of oxygenated functionalities, such as hydroxyl groups, thus providing sites for covalent linking of biomolecules [49]. In short, their impressive physicochemical properties, especially in terms of electroconductivity, make them very suitable for label-free heterogeneous assays [50]. The addition of CNTs to CP matrices results in higher decrease of the

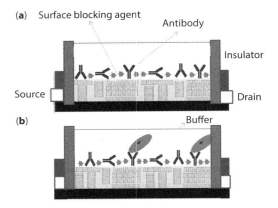

Figure 6.4 Lateral view scheme of a CNT-based immunonanosensor for detection of pathogenis microorganisms. a) The CNT network is directly adsorbed to the CP layer. Specific monoclonal antibodies are then immobilized onto the CNT-CP layer and a blocking agent (e.g., albumin) used to cover the unprotected gaps of the layer, thus preventing non-specific binding of other cells or proteins. b) Exposure to a biological sample containing the target pathogen causes a significant decrease of the CNT conductance, as a result of antibody-pathogen binding.

charge transfer resistance and of the mass transfer impedance compared to other nanomaterials (Figure 6.4).

It was found that SWNT-DNA adducts selectively hybridize with complementary DNA strands, with minimal non-specific interactions from non-complementary sequences. This suggests that the majority of SWNT-immobilized single-stranded DNA is covalently attached rather than physisorbed to nanotubes. Not only SWNTs, but also MWNTs, can be functionalized with DNA; CNTs modified with PPy/single-stranded DNA seem very promising for electrochemical DNA sensor design, particularly when coupled to magnetic nanoparticles [51]. CNT-CP composites are one of the most common approaches for the preparation of electrochemical sensors [52, 53]. The combination of the well-known characteristics of CPs (good stability, reproducibility, strong adherence and homogeneity in electrochemical deposition) with those of CNTs leads to the improvement of the sensing performance. Composites of CPs and CNTs have been synthesized either by chemical or electrochemical polymerization [54–55]. Some recent applications are the incorporation of CNTs in PPy-modified electrodes [56], electrochemical sensors based on CNT-PANI [57] and solubilization of CNTs in poly(vinyl alcohol) (PVA) [58].

It has also been shown DNA hybridization and single virus detection in connection to p-type silicon nanowires functionalized with peptide nucleic acid (PNA) probes or anti-influenza antibodies, respectively [59]. Discrete conductance changes, characteristic of the binding event, were observed at extremely low target concentrations. The use of an antibody-functionalized silicon nanowire sensor array for multiplexed label-free and real-time monitoring of cancer markers in undiluted serum samples has been demonstrated [21].

The ultra-high sensitivity of such nanoparticle-based sensing protocols opens the possibility for detecting biological markers, especially towards cancer, that cannot be measured by conventional methods, thus making possible early disease diagnosis. The successful clinical realization of these ultrasensitive bioelectronic detection schemes requires proper attention to non-specific adsorption that frequently limits the performance of bioaffinity assays. Modern clinical diagnostic devices based on nanoscale processes and structures have the potential to greatly improve methods for detecting biomarkers of human diseases in a cheaper, faster and easier-to-use way than conventional diagnostic methods. Furthermore, they also hold great promise for integration in POC and real-time detection regimens.

6.5 Applications of Micro-Devices to Some Important Clinical Pathologies

6.5.1 Diabetes

About 3% of the world population suffers from diabetes, a leading cause of death, with an incidence that is growing fast. Diabetes is a disordered metabolism syndrome that results in abnormally high blood sugar levels. Under normal physiological conditions, the concentration of fasting plasma glucose is in the range of 6.1–6.9 mM, and hence the variation of the blood glucose level can be a sign of diabetes mellitus, among other pathological conditions. Consequently, quantitation of glycaemia is of extreme importance for diagnosing diabetes. Diagnostic testing of glucose in the blood is still, by far, the most important application of current commercialized rapid tests for in vitro clinical diagnosis of human diseases. Such commercial driving force may be due, in part, to the conjugation of two factors not easily seen in other diseases: a potentially very broad target public and the need for multiple daily screenings.

Cheap and easy-to-use-self testing pocket-size meters, coupled with strip-tests based on screen-printed enzyme electrodes, have dominated,

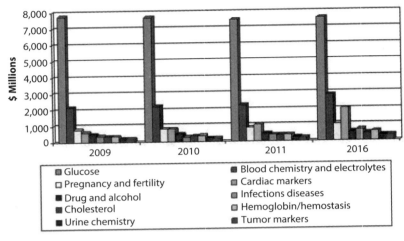

Figure 6.5 Forecast of the global POC market (in $ millions) by test type in the 2009–2016 period. Source: BCC Research (with authorization), report title "Point of Care Diagnostics" (report id HLC043C) (January 2012).

and will continue to dominate, the nearly $7.5 billion/year diabetes monitoring market throughout the years [60] (Figure 6.5). These electrochemical biosensors have played a major role towards truly simplified testing, with generation of home-use disposable devices rendering results within 5–10 and requiring as few as 0.5–10 μL fingerstick blood drops. Through education and simple training, regular at-home blood glucose monitoring revolutionized the lifestyles of those suffering from diabetes and reduced significantly many diabetes complications (e.g., cardiovascular diseases, blindness, kidney failure, and risk for neurological disorders). Thus, patients are now able to self-monitor their glycaemia and self-administer insulin injections as required, with tolerable invasive standards. Even so, future glucose sensors will most likely become increasingly non-invasive and sensitive to rapid changes in glucose concentrations. GlucoWatch (Cygnus, Inc.) developed an in vivo glucose detector that operates by reverse iontophoresis; it makes use of glucose-containing interstitial fluid that is lured to the skin surface by a small current passing between two electrodes. Hydrogel pads contain glucose oxidase molecules for enzyme-mediated measurement of glucose concentration in the interstitial fluid. Although interstitial fluid and blood concentrations are similar at steady state, there is a significant delay between them when the blood glucose concentration rapidly changes after a meal. Therefore, in this configuration, improvements aimed at minimizing the lag time between glucose level variations in the interstitial fluid and the corresponding changes in the blood must still be addressed [8].

In parallel, there is also the goal of increasing the miniaturization of amperometric-based biosensors to enable subcutaneous implantation for glucose measurements in blood or tissue. In these continuous glucose monitoring systems (CGMSs), a needle-type amperometric enzyme electrode is implanted in the subcutaneous tissue, to measure interstitial fluid glucose concentrations, and coupled to a portable data log. Today's clinical biosensors for blood glucose diagnostics can already be integrated with a personal digital assistant (PDA) for wireless communication, thus meeting the desired goals of telemedicine.

6.5.2 Cholesterol and Cardiovascular Disease

Cardiovascular diseases are highly preventable, yet they remain major causes of human mortality. One of the most important reasons for the increasing incidence of cardiovascular diseases is hypercholesterolemia, i.e., abnormal high level of blood cholesterol. Hence, estimation of cholesterol blood level is of utmost importance in the clinics [60]. As a consequence, it is not surprising that biosensors for cholesterol measurement constitute most of the published scientific articles in the field of cardiovascular diseases' biosensing. In the design of biosensors for free and total cholesterol, cholesterol oxidase and cholesterol esterase have been the most commonly used sensing elements, coupled to electrochemical transducing [61].

There is nowadays a big interest and clinical relevance about early assessment of several biomarkers for acute coronary syndrome. At present, cardiac troponin, which is specific for myocardial tissue, is the only biomarker routinely used for that purpose, being a valuable complement to electrocardiographic and patient history for therapeutic decision-making. Silva *et al.* [62] incorporated streptavidin polystyrene microspheres into a SPE surface in order to increase the analytical response of the cardiac troponin T. While Park *et al.* [63] used a highly sensitive and selective assay for assessment of acute myocardial infarction, based on virus nanoparticles targeted for troponin I.

Determination of non-myocardial tissue-specific biomarkers can also be helpful, as in the case of myeloperoxidase, copeptin, growth differentiation factor 15 and C-reactive protein (CRP) [64]. CRP, which reflects different aspects of the development of atherosclerosis and acute ischemia, is a known "acute-phase" plasma protein, whose level rises dramatically during body inflammatory processes. Such increment is due to a rise in the plasma concentration of interleukin 6 (IL-6), which is produced predominantly by macrophages, as well as adipocytes. CRP levels can rise as high as 1000-fold during inflammation. It was found that CRP is the only marker of inflammation that independently predicts the risk of a heart attack. In biosensor schemes, CRP measurements rely mainly on immune reactions with optical, through electrochemical and acoustic transducing [2]. As an example, a saliva-based nano-biochip immunoassay has been developed to detect a panel of C-reactive protein, myoglobin, and myeloperoxidase in acute myocardial infarction patients [65].

6.5.3   Cancer

Cancer can be caused by a huge range of factors, both genetic and environmental. The existing methods for cancer screening are heavily based on cell morphology, using staining and microscopy, which are invasive techniques. Furthermore, tissue removal can miss cancer cells at the early onset of the disease. Recently, electrochemical detection of rare circulating tumor cells has shown potential to provide clinicians with a stand-alone system to detect and monitor changes in cell numbers throughout therapy, aimed at efficient cancer treatment. However, this technique can only be used at an advanced stage of cancer proliferation, the reason why an early diagnostic method is essential [66].

Genome sequencing has allowed the detection of inherited disease-causing point mutations and human pathogens through their specific nucleic acid sequences. Today, it is consensual that no single gene is universally associated with cancer disease, but rather a complex set of them, creating difficulties for the selection of a single, well-defined biomarker for accurate diagnosis and prognosis. Therefore, a range of distinct proteins with concentration profiles correlated with specific types of cancer can play the desired role in approaching the disease diagnosis. These molecular signatures can be produced either by the tumor itself or by the body, in response to the presence of the tumour. Such biomarkers include DNA (e.g., oncogenes), RNA, proteins (e.g., enzymes, glycoproteins, hormones), molecules of the immune system, etc. A vast range of biomarkers has been identified for different types of cancers, which was recently reviewed [67].

Affinity electrochemical biosensors have been usually applied to detect gene mutations and protein biomarkers of cancer. In particular, electrochemical affinity sensors based on antibodies, developed originally in the late 1970s for tumor markers, offer great selectivity and sensitivity for early cancer diagnosis [68]. Immunoassays have also been frequently reported for detection of prostate cancer, via measurement of prostate specific antigen blood levels, as is the case of the system developed by OPKO Diagnostics (formerly Claros Diagnostics). Some success has also been observed among methods based on DNA hybridization for detection of cancer gene mutations. However, for cancer diagnosis, detection methods based on protein biomarkers (mainly in the form of immunoassays) seem more attractive than those based in genetic markers, due to the high protein abundance in common biological samples, to easy protein recovery and to favorable cost-effectiveness, essential requisites for successful clinical diagnostic micro-devices.

6.6   Conclusions and Future Prospects

Human healthcare and technological advances in the area of in vitro diagnostics have a history of undeniable parallel knowledge interchange that has strongly benefited individuals and health systems. The merging between the advent of human genome sequencing and the promises of personalized medicine have created endless opportunities and future novelties and advances in both areas. The concepts of POC and decentralized diagnosis, especially in times of severe economic constraints and of fast evolutions in the traditional paradigms of health care provision, have opened roads for sustainable research, development and marketing of new diagnostic micro-devices, essentially cheaper, simpler, and more user-friendly than conventional laboratory instrumentation for clinical assays. It is essential to take advantage of the ongoing advances in materials science, surface engineering, bioelectronics and other related fields, but also to engage commercial manufacturers and suppliers in such developments, not only of scientific nature, in a way that, ultimately, may be useful for patients and for society as a whole.

Acknowledgment

Luís P. Fonseca thanks the financial support to the project PTDC/EBB-BIO/114288/2009 from Foundation for Science and Technology—Portugal.

References

1. S. Murg, B. Moore, M. Lovell, R. Wolf, A.B. Bejjany., A.H.B. Wu and S. Kelles, "Business Strategy for Molecular Diagnostics" in the *Lab. Report Summary*, 2nd ed., Washington G-2 Reports, Kennedy Information, Peterborough, NH, USA, 2009.

2. C. Gouvea, "Biosensors for health applications", Pier Andrea Serra, ed., in the *Biosensors for Health, Environment and Biosecurity*, InTech, pp. 71–86, 2011.

3. C.D. Chin, S.Y. Chin, T. Laksanasopin and Samuel K. Sia, "Low-Cost Microdevices for Point-of-Care Testing", D. Issadore and R.M. Westervelt, eds., in the *Point-of-Care Diagnostics on a Chip*, Springer-Verlag, pp. 3–21, 2013.

4. S. Carrara, S. Ghoreishizadeh, J. Olivo, I. Taurino, C. Baj-Rossi, A. Cavallini, M. de Beeck, C. Dehollain, W. Burleson, F.G. Moussy, A. Guiseppi-Elie, and G. De Micheli, "Fully Integrated Biochip Platforms for Advanced Healthcare," *Sensors*, Vol. 12, pp. 11013–11060, 2012.

5. R.W. Peeling, P.G. Smith and P.M.M. Bossuyt, "Evaluation Diagnostics: A guide for diagnostic evaluations," *Nat. Rev. Microbiol.*, Vol. 4, pp. S2–S6, 2006.

6. F.R.R. Teles, L.P.T. Tavira and L.P. Fonseca, "Biosensors as rapid diagnostic tests for tropical diseases," *Crit. Rev. Clin. Lab. Sci.*, Vol. 47, pp. 139–169, 2010.

7. C.D. Chin, V. Linder and S.K. Sia, "Lab-on-a-chip devices for global health: Past studies and future opportunities," *Lab Chip*, Vol. 7, pp. 41–57, 2007.

8. C.D. Chin, L. Linder and S.K. Sia, "Commercialization of microfluidic point-of-care diagnostic devices", *Lab Chip*, Vol. 12, pp. 2118–2134, 2012.

9. V.C. Martins, F.A. Cardoso, J. Germano, S. Cardoso, L. Sousa, M. Piedade, P.P. Freitas and L.P. Fonseca, "Femtomolar Limit of Detection with a Magneto-resistive Biochip," *Biosens. Bioelectron.*, Vol. 24, pp. 2690–2695, 2009.

10. A. Niemz, T.M. Ferguson and D.S. Boyle, "Point-of-care nucleic acid testing for infectious diseases," *Trends Biotechnol.*, Vol. 29, pp. 240–250, 2011.

11. G.J. Kost (ed.), in the *Principles and Practice of Point-of-Care Testing*, Lippincott Williams & Wilkins (LWW), Philadelphia, USA, 2002.

12. P.P. Freitas, F.A. Cardoso, V.C. Martins, S.A.M. Martins, J. Loureiro, J. Amaral, R.C. Chaves, S. Cardoso, L.P. Fonseca, A.M. Sebastião, M. Pannetier-Lecoeur and C. Fermon C., "Spintronic platforms for biomedical applications," *Lab Chip*, Vol. 12, pp. 546–557, 2012.

13. L.C. Clark Jr. and C. Lyons, "Electrode systems for continuous monitoring in cardiovascular surgery," *Ann. NY Acad. Sci.*, Vol. 102, pp. 29–45, 1962.

14. T. Kubik, K. Bogunia-Kubik and M. Sugisaka, "Nanotechnology on duty in medical applications," *Curr. Pharm. Biotechnol.*, Vol. 6, 17–33, 2005.

15. F.R.R. Teles and L.P. Fonseca, "Trends in DNA biosensors," *Talanta*, Vol. 77, pp. 606–623, 2008.

16. S.K. Sia and L.J. Kricka, "Microfluidics and point-of-care testing," *Lab Chip*, Vol. 8, pp. 1982–1983, 2008.

17. T.M. Mohan, N. Nath and S. Anand, "Detection of filarial antibody using a fiber optics immunosensor (FOI)," *Indian J. Clin. Biochem.*, Vol. 12, pp. 17–21, 1997.

18. F.B. Myers and L.P. Lee, "Innovations in optical microfluidic technologies for point-of-care diagnostics," *Lab Chip*, Vol. 8, pp. 2015–2031, 2008.

19. A.W. Martinez, S.T. Phillips, E. Carrilho, S.W. Thomas III, H. Sindi and G.M. Whitesides, "Simple Telemedicine for Developing Regions: Camera Phones and Paper-Based Microfluidic Devices for Real-Time, Off-Site Diagnosis," *Anal. Chem.*, Vol. 80, pp. 3699–3707, 2008.

20. J. Newman and A.P.F. Turner, "Home blood glucose biosensors: a commercial perspective," *Biosens. Bioelectron.* Vol. 20, pp. 2435–2453, 2005.

21. J. Wang., "Electrochemical biosensors: Towards point-of-care cancer diagnostics," *Biosens. Bioelectron.*, Vol. 21, pp. 1887–1892, 2006.

22. S. Su, M.M. Ali, C.D. Filipe, Y. Li and R. Pelton, "Microgel-based inks for paper-supported biosensing applications," *Biomacromolecules*, Vol. 9, pp. 935–941, 2008.

23. M.I. Pividori, A. Merkoçi and S. Alegret, "Electrochemical genosensor design: immobilisation of oligonucleotides onto transducer surfaces and detection methods," *Biosens. Bioelectron.*, Vol. 15, pp. 291– 303, 2000.

24. P. Madhivanan, K. Krupp, J. Hardin, C. Karat, J.D. Klausner and A.L. Reingold, "Simple and inexpensive point-of-care tests improve diagnosis of vaginal infections in resource constrained settings," *Trop. Med. Int. Health*, Vol. 14, pp. 703–708, 2009.

25. N.K. Guimard, N. Gomez and C.E. Schmidt, "Conducting polymers in biomedical engineering," *Prog. Polym. Sci.*, Vol. 32, pp. 876–921, 2007.

26. A.W. Martinez, S.T. Phillips, G.M. Whitesides and E. Carrilho, "Diagnostics for the Developing World: Microfluidic Paper-Based Analytical Devices," *Anal. Chem.*, Vol. 82, pp. 3–10, 2010.

27. F.S. Ligler, C.R. Taitt, L.C. Shriver-Lake, K.E. Sapsford, Y. Shubin and J.P. Golden, "Array biosensor for detection of toxins," *Anal. Bioanal. Chem.*, Vol. 377, pp. 469–477, 2003.

28. W. Dungchai, O. Chailapakul and C. S. Henry, "Electrochemical Detection for Paper-Based Microfluidics", *Anal. Chem.* 81, 5821–5826, 2009.

29. P. Vidinha, N.M.T. Lourenço, C. Pinheiro, A.R. Brás, T. Carvalho, T.S. Silva, A. Mukhopadhyay, M.J. Romão, J. Parola, M. Dionisio, J.M.S. Cabral, C.A.M. Afonso and S. Barreiros, "Ion Jelly: a tailor-made conducting material for smart electrochemical devices, *Chem. Commun.*, Vol. 5842, 2008.

30. N.M.T. Lourenço, J. Oesterreicher, P. Vidinha, S. Barreiros, C.A.M. Afonso, J.M.S. Cabral and L.P. Fonseca, "Effect of gelatin-ionic liquid functional polymers on glucose oxidase and horseradish peroxidase kinetics," *React. Funct. Polym.*, Vol. 71, 489–495, 2011.

31. M.A. Dineva, L. Mahilum-Tapay and H. Lee, "Sample preparation: a challenge in the development of point-of-care nucleic acid-based assays for resource-limited settings," *Analyst*, Vol. 132, pp. 1193–1199, 2007.

32. P. Chomczynski, "A reagent for the single-step simultaneous isolation of RNA, DNA and proteins from cell and tissue samples," *BioTechniques*, Vol. 15, pp. 532–537, 1993.

33. P. Gill and A. Ghaemi, "Nucleic acid isothermal amplification technologies: a review," *Nucleos. Nucleot. Nucl.*, Vol. 27, pp. 224–243, 2008.

34. B. Bohunicky and S.A. Mousa, "Biosensors: the new wave in cancer diagnosis," *Nanotechnol. Sci. Appl.*, Vol. 4, pp. 1–10, 2011.

35. D. Wei, M.J.A. Bailey, P. Andrew and T. Ryhanen, "Electrochemical biosensors at the nanoscale," *Lab Chip*, Vol. 9, pp. 2123–2131, 2009.

36. G.S. Fiorini and D.T. Chiu, "Disposable microfluidic devices: fabrication, function, and application," *BioTechniques*, Vol. 38, pp. 429–446, 2005.

37. F.R.R. Teles and L.P. Fonseca, "Applications of polymers for biomolecule immobilization in electrochemical biosensors," *Mat. Sci. Eng. C*, Vol. 28, pp. 1530–1543, 2008.

38. X.L. Luo, A. Morrin A., A.J. Killard and M.R. Smyth, "Application of nanoparticles in electrochemical sensors and biosensors," *Electroanalysis*, Vol. 18, pp. 319–326, 2006.

39. K. Habermüller, M. Mosbach and W. Schuhmann, "Electron-transfer mechanisms in amperometric biosensors," *Fresenius' J. Anal. Chem.*, Vol. 366, pp. 560–568, 2000.

40. M.T. Castañeda, S. Alegret and A. Merkoçi, "Electrochemical Sensing of DNA Using Gold Nanoparticles," *Electroanalysis*, Vol. 19, pp. 743–753, 2007.

41. P. D'Orazio, "Biosensors in clinical chemistry-2011 update," *Clin. Chim. Acta*, Vol. 412, pp. 1749–1761, 2011.

42. A. Merkoci, "Electrochemical biosensing with nanoparticles," *FEBS Journal*, Vol. 274, pp. 310–316, 2007.

43. A. Kawde and J. Wang, "Amplified electrical transduction of DNA hybridization based on polymeric beads loaded with multiple gold nanoparticle tags," *Electroanalysis*, Vol. 16, pp. 101–105, 2004.

44. S. Park, T.A. Taton and C.A. Mirkin, "Array-based electrical detection of DNA with nanoparticle probes," *Science*, Vol. 295, pp. 1503–1506, 2002.

45. G. Liu, J. Wang, J. Kim and M.R. Jan, "Electrochemical coding for multiplexed immune-assays of proteins," *Anal. Chem.*, Vol. 76, pp. 7126–7130, 2004.

46. J. Wang , G. Liu and A. Merkoci, "Electrochemical coding technology for simultaneous detection of multiple DNA targets," *J. Am. Chem. Soc.*, Vol. 125, pp. 3214–3215, 2003.

47. Wang J., "Nanomaterial-based electrochemical biosensors", *Analyst,* 130, 421–426, 2005.

48. J.S. Ahammad; J.-J. Lee and M.A. Rahman, "Electrochemical sensors based on carbon nanotubes," *Sensors*, Vol. 9, pp. 2289–2319, 2009.

49. L. Agüí, P. Yáñez-Sedeño and J.M. Pingarrón, "Role of carbon nanotubes in electroanalytical chemistry: A review," *Anal. Chim. Acta*, Vol. 622, pp. 11–47, 2008.

50. T. Kurkina and K. Balasubramanian, "Towards in vitro molecular diagnostics using nanostructures," *Cell. Mol. Life Sci.*, Vol. 69, pp. 373–388, 2012.

51. A. Ramanavicius, A. Ramanaviciene and A. Malinauskas, "Electrochemical sensors based on conducting polymer—polypyrrole," *Electrochim. Acta*, Vol. 51, pp. 6025–6037, 2006.

52. M. Baibarac and P. Gómez-Romero, "Nanocomposites based on conducting polymers and carbon nanotubes: From fancy materials to functional applications," *J. Nanosci. Nanotechnol.*, Vol. 6, pp. 289–302, 2006.

53. R.T. Ahuja and D. Kumar, "Recent progress in the development of nanostructured conducting polymers/nanocomposites for sensor applications," *Sens. Actuators B Chem.*, Vol. 136, pp. 275–286, 2009.

54. S. Cosnier, "Biosensors based on electropolymerized films: New trends," *Anal. Bioanal. Chem.*, Vol. 377, pp. 507–520, 2003.

55. A. Guiseppi-Elie, "Electroconductive hydrogels: Synthesis, characterization and biomedical applications," *Biomaterials*, Vol. 31, pp. 2701–2716, 2010.

56. Y. Li, P. Wang, L. Wang and X. Lin, "Overoxidized polypyrrole film directed single-walled carbon nanotubes immobilization on glassy carbon electrode and its sensing applications," *Biosens. Bioelectron.*, Vol. 22, pp. 3120–3125, 2007.

57. W.S. Yuan, G.L. Li, K.G. Neoh and E.T. Kang, "Glucose biosensor from covalent immobilization of chitosan-coupled carbon nanotubes on polyaniline-modified gold electrode," *ACS Appl. Mater. Interface*, Vol. 2, pp. 3083–3091, 2010.

58. Y.-C. Tsai, J.-D. Huang and C.-C. Chiu, "Amperometric ethanol biosensor based on poly(vinyl alcohol)–multiwalled carbon nanotube-alcohol dehydrogenase biocomposite," *Biosens. Bioelectron.*, Vol. 22, pp. 3051–3056, 2007.

59. J. Hahm and C. Lieber, "Direct ultrasensitive electrical detection of DNA and DNA sequence variations using nanowire nanosensors," *Nano Lett.*, Vol. 4, pp. 51–55, 2004.

60. BCC Research, report title "Point of Care Diagnostics," report id (HLC043C) and date of publication (January 2012).

61. SK Arya, M. Datta and B.D. Malhotra, "Recent advances in cholesterol biosensor," *Biosens. Bioelectron.*, Vol. 23, pp. 1083–1100, 2008.

62. B.V. Silva, I.T. Cavalcanti, A.B. Mattos, P. Moura, M.P. Sotomayor and R.F. Dutra, "Disposable immunosensor for human cardiac troponin T based on streptavidin microsphere modified screen-printed electrode," *Biosens. Bioelectron.*, Vol. 26, pp. 1062–1067, 2010.

63. J.S. Park, M.K. Cho, E.J. Lee, K.Y. Ahn, K.E. Lee, J.H. Jung, Y. Cho, S.S. Han, Y.K. Kim and J. Lee, "A highly sensitive and selective diagnostic assay based on virus nanoparticles," *Nat. Nanotechnol.*, Vol. 4, pp. 259–264, 2009.

64. A. Qureshi, Y. Gurbuz, S. Kallempudi and J.H. Niazi, "Label-free RNA aptamer-based capacitive biosensor for the detection of C-reactive protein," *Phys. Chem. Chem. Phys.*, Vol. 12, pp. 9176–982, 2010.

65. E. Suprun, T. Bulko, A. Lisitsa, O. Gnedenko, A. Ivanov, V. Shumyantseva and A. Archakov, "Electrochemical nanobiosensor for express diagnosis of acute myocardial infarction in undiluted plasma," *Biosens. Bioelectron.,* Vol. 25, pp. 1694–1698, 2010.

66. Y.K. Chung, J. Reboud, K.C. Lee, H.M. Lim, P.Y. Lim, K.Y. Wang, K.C. Tang, H. Ji and Y. Chen, "An electrical biosensor for the detection of circulating tumor cells," *Biosens. Bioelectron.,* Vol. 26, pp. 2520–2526, 2011.

67. M. Polanski and N.L. Andersson, "A list of candidate cancer biomarkers for targeted proteomics," *Biomarker Insights,* Vol. 2, pp. 1–48, 2006.

68. Z. Li, Y. Wang, J. Wang, Z. Tang, J.G. Pounds and Y. Lin, "Rapid and sensitive detection of protein biomarker using a portable fluorescence biosensor based on quantum dots and a lateral flow test strip," *Anal. Chem.,* Vol. 82, pp. 7008–7014, 2010.

Part 3
TRANSLATIONAL MATERIALS

7

Hierarchical Modeling of Elastic Behavior of Human Dental Tissue Based on Synchrotron Diffraction Characterization

Tan Sui[1] and Alexander M. Korsunsky[1],*

[1]Department of Engineering Science, University of Oxford, Oxford, United Kingdom

Abstract

The human tooth that mainly consists of dentine, enamel and cementum is a hierarchical mineralized tissue with a two-level composite structure. Understanding the mechanical properties related to this hierarchical structure is essential for predicting the effects of structural alterations on the performance of dental tissues and their artificial replacements. Few studies have focused on the nano-scale structure of teeth. In this chapter, we describe how two synchrotron X-ray diffraction techniques, wide and small angle X-ray scattering (WAXS/SAXS), are used to obtain information about the deformation response of human enamel and dentine subjected to *in situ* uniaxial compressive loading. An improved multi-scale Eshelby inclusion model is proposed taking into account the two-level hierarchical structure, and is validated against the experimental strain evaluation data. The achieved agreement indicates that the multi-scale model accurately reflects the structural arrangement of human dental tissue and its response to applied forces. These results provide the basis for improved understanding of the mechanical properties of hierarchical biomaterials.

Keywords: Dentine, enamel, WAXS/SAXS, Eshelby model, mechanical properties

7.1 Introduction

Human tooth is composed of hierarchical mineralized tissues that achieve versatile mechanical properties by combining two principal calcified

Corresponding author: alexander.korsunsky@eng.ox.ac.uk

Ashutosh Tiwari (ed.) Advanced Healthcare Materials, (237–268) 2014 © Scrivener Publishing LLC

tissues: dentine and enamel [1]. Dentine is a tough tissue with a hierarchical structure, which has a typical well-oriented microstructure with an arrangement of dentinal tubules with the area density within $(19-45) \times 1000/mm^2$, with the mean diameter around 0.8–2.5 μm that extend throughout the entire dentine thickness, from the amelo-dentinal junction (ADJ) to the pulp [1–3]. At the finer nano-scale level, dentine is a composite of platelet-like hydroxyapatite crystals (HAp) that have the shape of elongated pancakes (approximately 2–4 nm in thickness, 30nm in width and up to 100 nm in length) randomly embedded in a fibrous collagen matrix [4, 5]. In comparison to dentine, enamel is a harder and more brittle outer layer that covers the crown portion of the tooth, serving as an important stiff, hard and wear-resistant outer shell of the tooth exposed to mastication contact loading and grinding action. At the microstructural level the notable features present in the enamel are, in particular, the long aligned prisms (or rods) with a cross-section that looks like a keyhole with ~5 μm diameter arranged to fill the space from the interface with dentine to the outer enamel surface, with the top oriented toward the crown of the tooth [6]. At the nano-scale, each rod is thought to be a composite with ribbon-like HAp particles (approximately 25–30nm in thickness, and length thought to be more than 1000 nm [7] that may even span through the entire enamel layer [1]) held together by a protein matrix [8–10]. Understanding the mechanical properties of complex, hierarchically structured tissue helps in understanding how the internal architecture can determine the remarkable properties of dental materials, both natural and artificial.

Over half a century, the majority of research on human dental tissues has been carried out using a variety of experimental measurement methods that focused on the mechanical properties at the macro-scale, e.g., overall Young's modulus, Poisson's ratio, hardness and fracture properties [11]. More recently, an increasing number of publications considering the microstructural effects have become available [10, 12]. However, only very few studies aimed to carry out multi-scale analysis that would allow tracing the influence of the nano-scale structure on the macroscopic mechanical response, e.g. the way the crystal shape and orientation of the mineral phase nano-crystals causes the anisotropy of overall stiffness and strength [13–15]. The investigation of the nano-scale structure as well as the hierarchical structure-property relationship requires the formulation and refinement of systematic analytical models and advanced experimental techniques.

Small- and Wide-Angle X-ray Scattering (SAXS and WAXS, respectively) are advanced non-destructive X-ray diffraction techniques that enable the characterization of the nano- and sub-nano-scale structure of materials. Both techniques have been widely used to study the load transfer between two different phases in composites [16–18]. For biomaterials, SAXS/WAXS have been applied only recently to the investigation of the mechanical behavior in mineralized biological composites such as bone [19–22] and teeth [13, 14, 23]. *In situ* compression experiments, in combination with the WAXS technique, have been used to quantify the internal phase strains in mineralized biomaterials [16–18]. Deymier-Black *et al.* [13] determined the longitudinal apparent modulus (the ratio of the applied stress to the internal local strain) of hydroxyapatite (HAp) crystals in bovine dentine using synchrotron based WAXS, while strain distribution across the ADJ in bovine teeth was investigated by Almer and Stock [23]. In addition, SAXS is able to reveal quantitative nano-scale structural information, such as the orientation and degree of alignment (percentage of aligned particles) of crystals in the mineralized tissues [24]. These parameters have been identified as critical for the mechanical properties and stability of materials [25–27]. However, since earlier studies did not take into account the nano-particle distribution effects [13, 15, 23], profound understanding of the relationship between the nano-scale structure and the macroscopic mechanical behavior is still lacking. In addition, all of the studies mentioned above were carried on non-human samples. It is noted that particularities of the mineralized tissue morphology can be associated with growth history, species and race, and result in differences in the mechanical properties [28, 29].

In parallel, many models for composite materials have been proposed to describe the interaction between different phases (e.g., Voigt, Reuss [30, 31], Gottesman & Hashin [32], Jones [33], and the so-called BW models (staggered microstructural model) [34]). One widely accepted and used model is the Eshelby inclusion model [35, 36], which has recently been applied in dental research [37–40] to explain and predict the elastic response of dentine at the microscopic level. However, in previous simulations, the models were limited in terms of the range of scales considered, with no consideration given to the nano-scale structure. This is likely to lead to some discrepancies between the predictions and experimental results [13]. An improved multi-scale Eshelby model is required in order to capture and explain the relationship between the nano-scale structure and tissue response to macroscopic loading.

In order to improve the understanding of the nano-scale structure variation of the hierarchical two-level structure of human dental tissues and of its influence on the material mechanical response, in this chapter, *in situ* synchrotron X-ray techniques (simultaneous SAXS/WAXS) were used to measure the elastic crystal lattice strain using WAXS, and the statistics of crystal orientation and degree of alignment of HAp particles using SAXS. Synchrotron beams possess numerous advantages for X-ray scattering studies. High flux allows fast measurements to be carried out, and enables area mapping and time evolution studies to be undertaken. High parallelism and brightness of synchrotron beams ensures high angular resolution of scattered beams, resulting in excellent accuracy of strain determination. Finally, large amount of space available around the sample makes it possible to apply thermal and mechanical loads to the samples.

In order to make sense of the scattering data collected during *in situ* loading, an improved multi-scale model based on the original Eshelby inclusion method for a two-level composite was proposed [41, 42]. The capability of the modeling approach to capture the relationship between the nano-scale structure and macroscopic loading was demonstrated.

7.2 Experimental Techniques

7.2.1 Micro-CT Protocol

For the purpose of planning the measuring positions and determining the precise loading cross-sectional area of dentine and enamel prismatic samples, a commercial micro-CT system (SkyScan 1172 scanner, SkyScan, Kontich, Belgium) was used, with 1.9μm isotropic resolution and 40kV voltage, 120μA current and a 0.5mm Aluminium filter. The resulting slices were reconstructed using SkyScan NRECON package, and subsequent 3-D planning models were created with Fiji imaging software [43]. A reconstruction is illustrated in Figure 7.1b.

7.2.2 *In situ* X-Ray Scattering Measurements

7.2.2.1 *Mechanical Loading Setup*

Uniaxial compressive loading was carried out separately on dentine and enamel samples in the form of small $2\times2\times2mm^3$ cubes. The loading was applied to the dentine sample in the longitudinal direction with respect to the tubules and to the enamel sample in the longitudinal direction with respect to the rods, respectively. Loading was carried out using a remotely operated and monitored compression rig (Deben, Suffolk, UK), with a 5kN calibrated load cell. The rig was equipped with custom-made jaws,

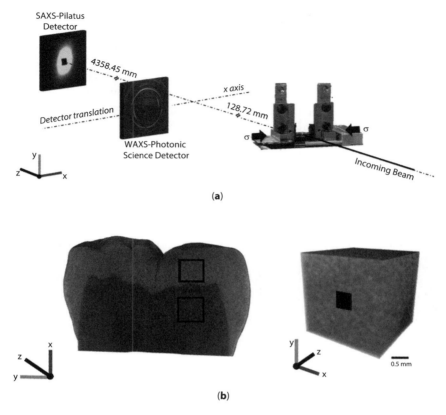

Figure 7.1 (a) Schematic diagram of experimental set-up. The sample was subjected to uniaxial compressive loading along the x direction. The monochromatic X-ray beam was travelling along the z direction perpendicular to the sample surface and to the loading direction. WAXS and SAXS diffraction patterns were recorded at each loading step at three locations on the sample. The WAXS detector was translated laterally out of the beam to expose the SAXS detector after each collection of WAXS pattern. (b) Cross-section of the reconstructed human tooth with regions for extracting the dentine and enamel samples (marked by hollow squares in Figure 7.1a), and the final $2\times2\times2mm^3$ cubic samples [41, 42]. The solid red square (online editions only) in the middle indicates the central X-ray beam position.

allowing a high-energy transmission X-ray setup to be used, as illustrated in Figure 7.1a. The samples were deformed at a displacement rate of 0.2 mm/min up to 400N (dentine) and 220N (enamel) (corresponding to about 100MPa for the dentine and 55MPa for enamel) along the x-direction. After each loading ramp, the load was maintained and the WAXS and SAXS patterns were collected.

7.2.2.2   Beamline Diffraction Setup

The experiment was carried out on B16 test beamline at Diamond Light Source, Oxford Harwell Campus, Didcot, UK. A monochromatic X-ray beam was used to illuminate the sample as illustrated schematically in Figure 7.1a. The incident beam was monochromated to the photon energy of 17.99keV, and collimated to the spot size of 0.5×0.5mm² on the sample. For statistical averaging purposes, WAXS and SAXS patterns were collected at three pre-defined locations across the sample for each load. A NIST standard silicon powder was used for the WAXS data calibration, and dry chicken collagen was used for the SAXS data calibration [44]. WAXS diffraction patterns were recorded using a Photonic Science Image Star 9000 detector (Photonic Science Ltd., UK) placed at a sample-to-camera distance of 128.72mm (Figure 7.1a). Further downstream, a Pilatus 300K detector (Dectris, Baden, Switzerland) was positioned at a distance of 4358.47mm to collect the SAXS patterns (Figure 7.1a). In order to record both the WAXS and SAXS patterns at each scanning location, the WAXS detector was translated laterally to expose the SAXS detector after each WAXS collection.

7.3   Model Formulation

7.3.1   Geometrical Assumptions

Both dentine and enamel have a hierarchical two-level composite structure. At both levels these are non-dilute systems consisting of a number of inhomogeneous inclusions. The multi-scale Eshelby model for a non-dilute system is established and used here to model the two-level composite deformation behaviour of dentine and enamel [45].

7.3.1.1   Dentine Hierarchical Structure

In dentine the first-level structure is represented by the multiple dentinal tubules within a fibrous collagen matrix, while the second level is represented by the platelet-like HAp crystals within the collagen matrix. Figure 7.2(a1) was obtained by low vacuum environmental scanning electron microscopy (ESEM) image of the dentinal tubules, while Figures 7.2(a2)-(a3) are schematic diagrams of the first-level dentine structure. Figures 7.2(a1)-(a2) show the random distribution of collagen fibrils viewed respectively along the longitudinal and transverse direction with respect to the tubules, as proposed by Bozec [46]. Figures 7.2(a4)-(a5) are schematic diagrams of the second-level dentine structure. Figure 7.2(a5) aims to illustrate how the model needs

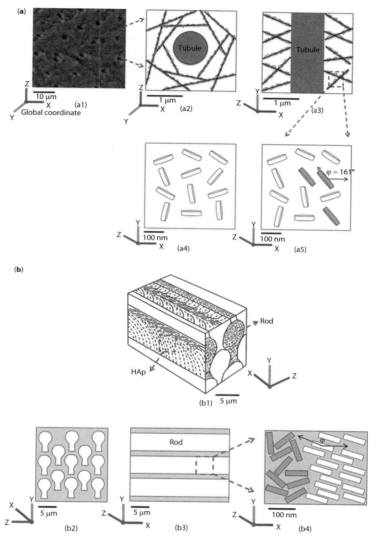

Figure 7.2 Schematic illustrations of 3D hierarchical structures of human dental tissue: (a) dentine and (b) enamel. (a1) SEM micrograph of the first level of dentine (tubules); (a2) A schematic diagram of the random distribution of collagen fibrils (black lines) around the tubules, viewed along the longitudinal direction of the tubules; (a3) The view transverse to the tubules; (a4) Randomly distributed HAp crystals viewed in the cross section in x-y plane; (a5) A schematic representation of the real structure of partially aligned platelet-like HAp crystals, where the platelets with the alignment angle of 161° with respect to the global x-axis are shown in grey. (b1) Keyhole-like rods aligned parallel to each other; (b2) A schematic of the rod arrangement viewed along their longitudinal direction; (b3) A schematic of rod arrangement viewed transversely; (b4) An illustration of the partially aligned arrangement of ribbon-like HAp crystals inside the rods [41, 42].

to represent the partial alignment of HAp crystals by the combination of a randomly distributed structure shown in Figure 7.2(a4), and a fully aligned structure. The detailed information about the degree of alignment can be obtained from SAXS data interpretation. In the multi-scale Eshelby model consistent with the two-level structure, the first-level model regards the whole dentine as composed of aligned tubules within a matrix phase, while the second-level model considers the matrix of the first level as a composite in detail, consisting of partially aligned HAp crystals and a collagen matrix.

The shape of the HAp crystals in dentine is "flagstone," i.e., that of shallow parallelepiped elongated along one of the in-plane directions. A good approximation available in classical Eshelby modeling is to use a penny-shaped inclusion to simulate each individual crystal. This ignores the non-equiaxed nature of the crystallites, but provides simple and tractable formulas for the calculation. The penny shape has two independent dimensions, the radius a_1 and the half-thickness a_3 [47]. In terms of the real dimensions of HAp crystals, the thickness of the penny-shape inclusion was taken to be equal to that of the HAp crystal (2–4nm), while the diameter of the inclusion was taken to be ~60nm, obtained by assuming the penny-shape inclusion to have an identical area with the HAp crystal (100nm in length and 30nm in width). The crystal size and the corresponding parameters of the penny-shape inclusion are listed in Table 7.1. The HAp crystal c-axis (corresponding to the [002] peak) is thought to be parallel to the long dimension of the platelet-like single crystal [48, 49]. Accordingly, in the model, the c-axis was taken to lie along the diametric direction of the penny-shape in the x-y plane, as shown in Figure 7.2(a5).

7.3.1.2   Enamel Hierarchical Structure

In enamel the first-level structure is represented by the keyhole-like rod within the protein matrix, while the second level is represented by the bundle of HAp crystals within each rod [1]. Modified from Habelitz [50], the keyhole-like microstructure of enamel is shown in Figure 7.2(b1), which demonstrates how the HAp crystals distribute within the rod in 3D. Figures 7.2(b2)-(b4) provide schematic illustrations of the geometric model derived from the enamel two-level structure. Note that although the HAp crystals are still partially (incompletely) aligned within each rod (particularly through orientation change within the sampling volume), the degree of alignment is rather high compared with that of HAp crystals in dentine.

In our approach to multi-scale Eshelby modeling of human enamel, for simplicity, both the rods and HAp crystals were assumed to be of ribbon shape (Figures 7.2(b2)-(b4)). Therefore, unlike for dentine, a cylinder-shaped inclusion is used to simulate each individual ribbon-shaped HAp crystal. In the Eshelby approach, the cylinder was approximated by a prolate spheroid described by three dimensions, a_1, a_2, and a_3. Normally $a_1 = a_2 << a_3$, i.e., the cross section is a circle [47]. In the ribbon-shaped HAp crystals, the c-axis is parallel to the a_3 axis of the cylinder [51].

In the next section, the multi-scale Eshelby inclusion model is briefly introduced. The detailed derivations are given in the Appendix.

7.3.2 Multi-Scale Eshelby Model

The two-level structures of both dentine and enamel have been explained in section 3.1. Despite the differences at the first structural level, dentine and enamel have a similar structure in the second level that consists of partially aligned HAp crystals embedded in the isotropic organic matrix. The second-level structures of the two tissues differ only in terms of over-all stiffness, volume fraction and the geometry of HAp crystals. Thus, the second-level model is expected to be similar, as is indicated below. We start with the introduction of the first-level models for the two tissues.

7.3.2.1 First-Level Eshelby Model

a) Dentine

The purpose of the first-level model for human dentine is to establish the relationship between the external stress σ^A and the dentine matrix stress σ_{M1}, which will serve as the external stress in the second-level model. The Eshelby model for a non-dilute system (Appendix, Eq. A1–A5) indicates that the mechanical response of the tubules is identical to an equivalent inclusion with the same property as the matrix and with an appropriate transformation strain.

$$\langle C \rangle_{tubule} \left(\langle \varepsilon \rangle^i + \langle \varepsilon \rangle_{M1} + \varepsilon^A \right) = C_{M1} \left(\langle \varepsilon \rangle^i + \langle \varepsilon \rangle_{M1} + \varepsilon^A - \langle \varepsilon \rangle^t \right) \qquad \text{(Eq. 7.1)}$$

where "$M1$" refers to the first-level matrix, $\langle \varepsilon \rangle^i$ is the average total strain in tubules, $\langle C \rangle_{tubule}$ is the average stiffness of tubules, $\varepsilon^A = C_{M1} \sigma^A$ is the external strain, $C_{M1} \langle \varepsilon \rangle_{M1} = \langle \sigma \rangle_{M1}$ is the "image" stress defined to satisfy the boundary conditions at the external surface of a finite composite [45, 52] and $\langle \varepsilon \rangle^t$ is the average misfit strain in the equivalent inclusion to be determined. Since the average stiffness of tubules is null, $\langle C \rangle_{tubule} = 0$, the average stress in the first-level matrix can be expressed merely in terms of the volume fraction of tubules f_{1t} (for further details see the Appendix, Eq. A7-A10)

$$\sigma_{M1} = \sigma^A + \langle\sigma\rangle_{M1} = \frac{1}{1-f_{1t}}\,\sigma^A \qquad \text{(Eq. 7.2)}$$

b) Enamel

The purpose of the first-level model for human enamel is to establish the relationship between the externally measured stress and the stress in the rod inclusions $\sigma^{inclusion} = \sigma^{rod}$. According to the Eshelby model derivation [45], the stress in the rod can be expressed as

$$\sigma^{rod} = \{T-(1-f_{1r})C_{M1}(S_r-I)\{(C_{M1}-C_{rod})[S_r-f_{1r}(S_r-I)]-C_{M1}\}^{-1}$$
$$(C_{M1}-C_{rod})T^{-T}C_{M1}-1\}\,\sigma^A \text{ or,}$$

expressed more simply

$$\sigma^{rod} = H\sigma^A \qquad \text{(Eq. 7.3)}$$

where f_{1r} is the volume fraction of rods in the whole enamel, S_r is the Eshelby tensor for the cylinder-shaped inclusion dependent only on the Poisson's ratio of the matrix, C_{M1}, C_{rod} are respectively the stiffness tensors of the protein matrix and rods, and T is the orientation matrix of rods with respect to the fixed laboratory coordinate system. The laboratory coordinate system was fixed in the present model, and it was assumed that the rods were all perfectly aligned, so that T was also fixed. The initially unknown rod stiffness C_{rod} remains to be determined from the second-level model of enamel.

7.3.2.2   Second-Level Eshelby Model

The dispersion of multiple partially aligned HAp crystals within the organic matrix forms the second-level model for both dentine and enamel. The purpose of the second-level model is to establish a relationship between the external stress from the first level and the average strain in the HAp crystals $\langle\varepsilon\rangle^{HAp}$, thus to determine the apparent modulus [53] that establishes the relationship between the global external stress and the local HAp crystal strain. The measured crystal strain corresponds to the mean strain value for all the crystals within the considered gauge volume [54]. Due to the partial alignment, the real apparent modulus is to be given by an intermediate value between the two extreme cases. For simplicity, we represent the partially aligned structure as a superposition of the two separate cases, namely, that of fully random distribution, and that of perfect alignment.

If all HAp crystals are perfectly aligned with the direction described by an orientation matrix T, the relationship between the average HAp crystal strain and the external load (the first-level stress) can be expressed as [45] (Appendix, Eq. A11):

$$\langle \varepsilon \rangle_{aligned}^{HAp} = \left\{ \{(I - C_{M2}^{-1} \langle C \rangle_{HAp})^{-1} [\langle S \rangle - f_2 (\langle S \rangle - I)]^{-1} - I\}^{-1} \right.$$
$$\left. T^{-T} + T^{-T} \right\} C_{M2}^{-1} \sigma_{M1}$$

or, expressed more simply

$$\sigma_{M1} = K_{aligned} \langle \varepsilon \rangle_{aligned}^{HAp} \qquad \text{(Eq. 7.4)}$$

where $\langle C \rangle_{HAp}$ and $\langle S \rangle$ are the average stiffness and the Eshelby tensor of HAp crystals in the gauge volume [54], $C_{M2} = C_{M1}$ is the organic phase stiffness, and f_2 is the volume fraction of HAp crystals with respect to the whole second-level structure. For perfectly aligned crystals, $\langle C \rangle_{HAp}$ and $\langle S \rangle$ can be replaced by the values for a single crystal $\langle C \rangle_{HAp} = \langle C \rangle_{HAp}, \langle S \rangle = S$. Due to the different characteristics of HAp crystals in dentine and enamel, $\langle C \rangle_{HAp}$ and $\langle S \rangle$ will be different in the two materials. Note that different orientations (different matrices T) will lead to different results of $\langle \varepsilon \rangle_{aligned}^{HAp}$. The variation of $K_{aligned}$ with different alignment angles of crystals can be calculated by changing the orientation matrix in Eq. 7.4. The anisotropy of the elastic properties of HAp crystals is an important property used by nature in dental tissue design.

If all HAp crystals are randomly distributed, the relationship between the average local HAp crystal strain and external stress is independent of the orientation matrix in Eq. 7.4.

$$\langle \varepsilon \rangle_{random}^{HAp} = \left\{ \{(I - C_{M2}^{-1} \langle C \rangle_{HAp})^{-1} [\langle S \rangle - f_2 (\langle S \rangle - I)]^{-1} - I\}^{-1} + I \right\}$$
$$C_{M2}^{-1} \sigma_{M1}$$

or, expressed more simply

$$\sigma_{M1} = K_{random} \langle \varepsilon \rangle_{random}^{HAp} \qquad \text{(Eq. 7.5)}$$

In contrast with the case of perfectly aligned crystals, $\langle S \rangle$ and $\langle C \rangle_{HAp}$ values here are not those of the single HAp crystal, but should rather be obtained from the volume average of all the randomly distributed crystals. In fact, this arrangement of crystals gives rise to isotropic stiffness and Eshelby tensors (spherical tensors) regardless of the anisotropy in the stiffness or shape of a single crystal. However, as an alternative to simplify the derivation, the averaging effect can be captured by using the single crystal relationship such as expressed by Eq. 7.4, and averaging over all the values with different orientation matrices (see Appendix, the average results obtained from each single crystal relationship as Eq. 7.4 over all possible orientations).

For the actual partially aligned HAp crystals, we express $\langle K \rangle^{HAp}_{partial\_aligned}$ as a mixture between K_{random} and $K_{aligned}$, taking into account the contributions from the two cases.

$$\langle K \rangle^{HAp}_{partial\_aligned} = (1 - f_{aligned})K_{random} + f_{aligned}K_{aligned} \qquad \text{(Eq. 7.6)}$$

where $f_{aligned}$ is the volume fraction of aligned HAp crystals with respect to all crystals, i.e., the degree of alignment of crystals revealed by SAXS measurements.

As for the determination of the enamel rod stiffness C_{rod}, the HAp crystals were assumed to be perfectly aligned in the direction almost along the rod's longitudinal direction. The slight misorientation was determined by WAXS data interpretation. The assumption of perfect alignment appears to be reasonable, because it is understood that the degree of alignment of HAp crystals in enamel is very large [55]. The expression for C_{rod} is given here without detailed derivation [45].

$$
\begin{aligned}
C_{rod} = \Big\{ C_{M2}{}^{-1} - f_2 \{ (\tilde{C}_{HAp} - C_{M2})[\tilde{S}_2 - f_2(\tilde{S}_2 - I] + C_{M2} \}^{-1} \\
(\tilde{C}_{HAp} - C_{M2})C_{M2}{}^{-1} \Big\}^{-1}
\end{aligned}
\qquad \text{(Eq. 7.7)}
$$

Here $\tilde{S}_2 = T^T_{HAp} S_2 T^{-T}_{HAp}$ is the transformed Eshelby tensor and $\tilde{C}_{HAp} = T^{-1}_{HAp} C_{HAp} T^{-T}_{HAp}$ is the transformed stiffness of the aligned HAp crystals, expressed in terms of T_{HAp}, the orientation matrix of the perfectly aligned HAp crystals.

Finally, the overall relationship between the average HAp crystal strain and the externally applied stress can be established from Eq. 7.2, Eq. 7.3 and Eq. 7.6.

$$\sigma^A = \frac{1}{1 - f_{1t}}\sigma_{M1} = \frac{1}{1 - f_{1t}}K^{HAp}_{partial\_aligned}$$

$$\langle \varepsilon \rangle^{HAp}_{partial\_aligned} = K_{dentine}\langle \varepsilon \rangle^{HAp}_{partial\_aligned} \qquad \text{(Eq. 7.8)}$$

$$\sigma^A = H^{-1}\sigma^{rod} = H^{-1}K^{HAp}_{partial\_aligned}$$

$$\langle \varepsilon \rangle^{HAp}_{partial\_aligned} = K_{enamel}\langle \varepsilon \rangle^{HAp}_{partial\_aligned} \qquad \text{(Eq. 7.9)}$$

where Eq. 7.8 is for dentine and Eq. 7.9 is for enamel.

7.4   Experimental Results and Model Validation

7.4.1   Nano-Scale HAp Distribution and Mechanical Response

Figures 7.3a&c show the WAXS patterns of dentine and enamel, respectively, consisting of a series of Debye-Scherrer rings (peaks). Note that since the enamel is usually textured, only a limited range of diffraction

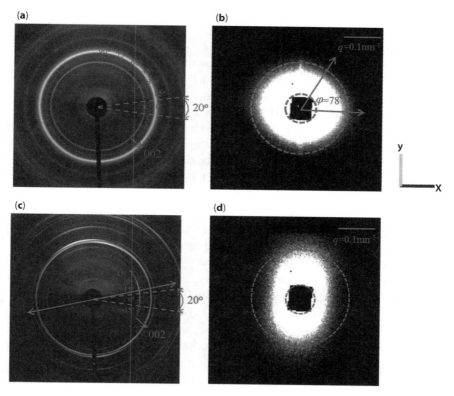

Figure 7.3 Representative SAXS/WAXS patterns of human dental tissue. (a) and (c) are WAXS patterns (Debye-Scherrer rings) of dentine and enamel with different intensities. The dark region in the centre is the beamstop. The (002) peak is marked with a small arrow. Peak shifts at different positions on the (002) ring represent the average strains of (002) planes along different directions in the laboratory coordinate system. Multiple angles with respect to the x-axis were "caked" (i.e. binned azimuthally within a 20° range) in order to examine the strain variation. The direction with the double-arrow line in (c) indicates the preferred orientation of HAp crystals in the enamel. (b) and (d) are SAXS patterns of dentine and enamel. The direction of the short axis of the ellipse pattern indicates the predominant orientation of HAp crystals. The integration over q was performed from the beamstop radius to the outer radius of the pattern as marked with the dashed circles [41, 42].

rings can be captured. The apparent radial shifts of the (002) peak center position in the pattern were measured under uniaxial compressive loading applied in the longitudinal direction with respect to the preferential tubule direction in dentine or rod direction in enamel, respectively. The preferential orientation of the c-axis and ribbon axis in HAp crystals is shown to be roughly along the mid-angle of the arc of (002) peak and the detailed value is determined from the stress-free plot of $I_{WAXS}(\varphi) \sim \varphi$. Different values of the shifts of (002) peak center position determine the elastic lattice strain variation with the angle to the loading direction. The loading areas of the samples were accurately determined from the micro-CT scans, giving 4.413 mm^2 for the dentine sample and 3.551 mm^2 for the enamel sample, respectively.

Figures 7.3b&d illustrate the SAXS patterns from the dentine and enamel samples, respectively. It should be noted that due to the dense packing of crystals in enamel, it is the electron density change occurring in the gaps between the crystalline particles that makes the principal contribution to the SAXS scattering signal [56]. It is also understood that the orientation of the gaps roughly coincides with that of the HAp crystallites within the rod [56]. Thus the SAXS pattern of enamel can be used to obtain the degree of alignment of HAp crystals in enamel. The detailed values of the degree of alignment are listed in Table 7.1. Compared with the HAp crystals in the highly textured enamel sample, the degree of alignment for HAp crystals in dentine is relatively small, indicating as expected that the distribution of HAp crystals in human dentine is close to random, but nevertheless possesses certain preferred orientation.

For the elastic lattice strain evolution, the maximum applied stress of the two samples was constrained within the elastic limit and the experimental results of applied stress vs. average HAp lattice strain of dentine and enamel are shown in Figures 7.4a&b, where the linear increasing tendency is found, as expected. The ratio of the uniaxial stress and the average HAp lattice strain gives the apparent modulus [53], the value of which is listed in Table 7.1. In addition it is worth noting that the residual (initial) strains of the two samples were found to be quite small: the two lines in Figures 7.4a&b originate from close to zero strain, and thus the initial strains can be neglected.

Meanwhile, in dentine, the (002) peak shifts along other directions were also measured by "caking" (a jargon term used to refer to the radial-azimuthal binning operation performed on a selected sector within the Debye-Scherrer pattern) with a step of 20° (in the range of 0°–360°, Figure 7.3a) to determine the azimuthal variation of the normal strain component.

Table 7.1 Experimental results from SAXS/WAXS and refined parameters in the Eshelby model of the two dentine samples [41, 42].

Parameters	dentine	Reference values	enamel	Reference values
Orientation (degree)	78		14	
Degree of alignment	0.168		0.6	
K_exp. (GPa)	24.156		124.3	
f_1	10%	$3.6 \sim 10.2\%$ [57]	97%	95% [15]
f_2	38%	30.5%, 44.4% [58]	97%	95% [15]
$C_{M1} = C_{M2}$	$E_m=0.8\text{GPa}$, $v_m=0.27$	$E_{collagen}=1\text{GPa}$, $v_{collagen}=0.30$ [37]	$E_m=0.8\text{GPa}$, $v_m=0.27$	$E_{protein}=1\text{GPa}$, $v_{protein}=0.30$ [37]
C_{HAp}	$E_{HAp}=90\text{GPa}$, $v_{HAp}=0.32$	$E_{HAp}=40\text{–}117\text{GPa}$, $v_{HAp}=0.27$ [59]	$E_{HAp}=140\text{GPa}$, $v_{HAp}=0.31$	$E_{HAp}=148.42\text{GPa}$, $v_{HAp}=0.30$ [60]
S	$a_1/a_1 = 31$	$(2\text{–}4)\times30\times100 \text{ nm}^3$ [4, 5]	$a_1/a_1 = 1$	[10]
K_model (GPa)	24.189			120.4

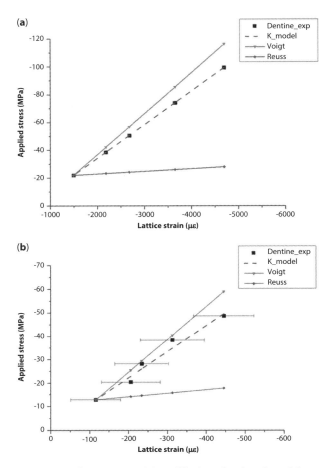

Figure 7.4 Comparison of experimental data (filled markers) and modeling results (solid and dashed lines) of the applied compressive stress vs. elastic lattice strain of HAp crystals in human dental tissue. (a) Dentine loaded to the maximum stress around 100MPa; (b) Enamel loaded to the maximum stress around 53.5MPa. The prediction results from models of Voigt (upper solid line) and Reuss (lower solid line) are also shown [41, 42].

The result is shown in polar coordinates as an azimuthal plot in Figure 7.5a, where 0° or 180° represents the loading direction and 90° or 360° represents the direction perpendicular to the loading direction. Due to the symmetry, it suffices to consider the results in the typical range of 0°–90°. It is found that positive (tensile) normal strain exists in the range 60°–90°, and negative (compressive) normal strain exists in the range 0°–60°. As for enamel, since only limited range of azimuthal angles within the rings

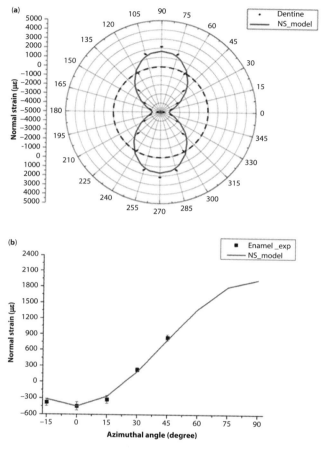

Figure 7.5 Comparison of the experimental data (filled markers) and modeling results (curves) of the variation of normal strain (NS) component with orientation distribution of HAp crystals in human dental tissues. (a) Polar coordinate plot of normal strains in dentine, where the full azimuthal angle range (0°–360°) is observable and the dashed line indicates zero strain; (b) Enamel results, where only the range (0°–45°, or 180°–225°) is observable [41, 42].

could be captured, the available directions for strain interpretation were also limited. The shifts of the (002) peak along available directions were measured by "caking" each pattern with a step of 20° along the direction given by −15°, 0°, 15°, 30° and 45° (see Figure 7.3b). The azimuthal plot of the result is shown in Figure 7.5b, where 0° represents the loading direction. Note that the normal strain component in the enamel undergoes a transition from negative to positive at around 25°.

7.4.2   Evaluation and Testing of the Multi-Scale Eshelby Model

a) Dentine

In the model for dentine, the material properties and other parameters were taken from the literature [4, 5, 13, 37–38, 57–59] and refined by fitting with the experimental data. The precise volume fraction of tubules in the first level cannot be assessed using conventional micro-CT systems, due to the polychromatic nature of the X-ray source and the limited resolution. Hence, the reported values between 3.6 and 10.2% were considered [57]. The average volume fraction of HAp crystals has been reported to lie between 30.5% and 44.4% in the human third molars, with a decreasing gradient towards the pulp [58]. As for the elastic properties of collagen, a Young's modulus of 1 GPa and Poisson's ratio of 0.30 are the values given in the literature, without taking into account the viscoelasticity and viscoplasticity [37, 38]. For polycrystalline organic HAp, high values of Young's modulus (40–117GPa) are reported, with Poisson's ratio of 0.27 [59]. However, for the purposes of the present model it was appropriate to use the stiffness matrix for a perfect HAp single crystal. Since imperfectly shaped and flawed crystals are likely to exist in complex biological mineralized composites like human dentine, this choice of elastic parameters may induce overestimation [13]. In accordance with Qin & Swain [36], 40GPa was chosen here as a combination of the intertubular modulus (35.8GPa) and peritubular modulus (66.8GPa) [37]. For the penny-shaped Eshelby tensor, only the ratio of radius and thickness (a_1/a_3) of the penny is needed for formulating the model [4, 5]. All the parameters refined to obtain a best fit are listed in Table 7.1, also with the reported values from the literature.

At the first level, the dentine matrix stress is only dependent on the volume fraction of tubules (see Eq. 7.2). Further, based on the SAXS measurement of degree of alignment (Table 7.1) in the second level, the apparent modulus can be obtained from Eq. 7.7 with the values listed in Table 7.1. A comparison of the stress/strain curve along the loading direction between the experimental data and the model prediction is shown in Figure 7.4a, where the Voigt and Reuss bound predictions are also given. The normal strain variation with the azimuthal angle is presented in Figure 7.5a, and good agreement can be noted with satisfaction.

b) Enamel

In the model for enamel, the average HAp volume fraction has been reported to be ~95% at each level. The same values of protein elastic parameters were chosen as that for the dentine model [37, 38]. The polycrystalline

HAp crystal was considered to have a transversely isotropic stiffness with five independent elastic constants [60]. To describe the shape of the rod and of the HAp crystallites for each level, the Eshelby tensor for the cylinder was used. As mentioned in section 7.3.1.2., the elliptical semi-axes a_1 and a_2 within the transverse cross-section were assumed to be the same, ($a_1/a_2=1$), but were much smaller than length a_3. The refined parameters and the reported values are also listed in Table 7.1.

The calculated apparent modulus for HAp crystals in enamel is also listed in Table 7.1. The similar comparison of the stress/strain curve as dentine is plotted in Figure 7.4b with the two bound predictions, and the improved multi-scale Eshelby model gives a satisfactory result. The comparison of normal strain variation (based on different preferred orientation angles obtained by WAXS data interpretation) is shown in the azimuthal plot in Figure 7.5b. It is found that the model prediction for the transverse tensile strain that arises under compression (the Poisson effect) falls short of the observed strain. A satisfactory agreement can be achieved by adding a pre-existing or loading-induced tensile transverse strain component (along the y axis in Figure 7.1) perpendicular to the rod or the loading direction (x axis in Figure 7.1). The result is illustrated in Figure 7.5b with the continuous curve. Obtaining a clear understanding of this remarkable effect deserves further detailed investigation.

7.5   Discussion

In this study, experiments were conducted using the penetrating power of synchrotron X-rays to provide a bulk probe for structure and strain analysis. Unlike the vast majority of studies that rely on surface characterization techniques (SEM, AFM, nanoindentation, Raman, etc.), this ensures that the effects of sample preparation (e.g., cutting and storage) are minimal, since they typically affect superficial layers of depths not exceeding ~0.05mm, i.e., a small proportion of the total sample thickness (2mm in our study).

7.5.1   Refined Parameters of the Two-Level Eshelby Model

The procedure employed to identify and refine the parameters used in the multi-scale Eshelby models of dentine and enamel was similar in both cases. The resulting best fit values are listed in Table 7.1. The values reported in the literature were found to provide a satisfactory starting point. The refinement procedure was particularly helpful in the identification of nano-scale parameters that are difficult to determine directly from other experiments. The key parameters varied in the refinement were the volume fractions of inhomogeneities at each level and the elastic constants of the HAp crystals. At the first level, the volume fractions sought were the tubules in dentine and the rods in enamel. At the second level the volume fractions refined were the HAp crystals in the first-level matrix for dentine or HAp crystals in the first-level rods for enamel.

The dentine samples considered in the present study were taken from the teeth extracted from young patients. The volume fraction of tubules was expected to be high, and assumed to be 10%. Accordingly, the volume fraction of HAp crystals was assumed to be low (~40%). Other reasons for the small volume fraction of HAp crystals may include that the cubic samples were cut from a position near the pulp chamber where the volume fraction is known to be relatively small [58]. Possible superficial demineralization effect of water storage may also be relevant [61]. For enamel, the volume fractions of rods in the whole enamel and of HAp crystallites within rods were assumed to be the same as the reported values (95%).

Besides the volume fractions, the elastic constants also exert significant influence on the result. In the optimization process, the volume fractions were firstly fixed at approximate values. Then the elastic constants, especially Young's moduli, were refined within the range reported in the literature. Subsequently, some adjustment of other parameters was also attempted, but it was found that they had a minor effect.

7.5.2   Residual Strain

During the natural growth of the teeth, and also during the preparation process using a low speed diamond saw and polishing papers, some initial residual strain may be induced at the sample surface. However, as shown in Figures 7.4a&b, the pre-existing residual strain is very small for both dentine and enamel. In addition, since only the elastic response was considered in the experiment (reflected in the linearity of the experimental stress-strain curve), the presence of initial strain only amounts to an offset that does not affect the apparent modulus. Therefore, the low value residual strain was ignored in the present analysis.

7.5.3   Normal Strain Components Variation

The normal strain components variation of the HAp crystals with respect to different azimuthal angles ($0°{\sim}90°$) at the maximum external stress, predicted by the model, is shown in Figures 7.5a&b, together with the experimental data. As mentioned earlier, the strong texture of HAp crystals in enamel leads to the scattering contribution being only accessible in the angular range 0°~45° (Figure 7.5b).

In dentine, the ratio of the normal strain components at 90° to that at 0° (absolute value) is almost the Poisson's ratio of the HAp crystals. It is interesting to note that the HAp crystals oriented at around 60° to the loading direction show no normal strain component (i.e., without any peak shift), which indicates that the crystals are subjected to pure shear stress.

In enamel, satisfactory agreement could only be obtained by incorporating an additional transverse strain. This indicates that the deformation state that arose in the sample under uniaxial loading was in fact multiaxial, with an additional tensile strain component arising in the transverse direction. A disproportionately large transverse tensile strain appears to arise within the sample, and is likely to be associated with the interaction between the material structure and the loading arrangement. A possible explanation is the barreling effect (mid-section expansion) in the enamel sample: friction against the external uniaxial compression platens causes local locking of the enamel to their surface that may be accompanied by the expansion of the sample in the section perpendicular to the loading direction in the mid-section (where the measurements were performed). In addition, micro-cracks smaller than 2μm (below the resolution of the micro-CT scan) pre-existing in the enamel may result in the debonding between rods, and thus modify the transverse strain [62]. Such mechanisms are not captured by the present model and need to be introduced externally. The validation of the precise mechanism needs direct observation using advanced ultra-high resolution (sub-micron) imaging and analysis techniques, e.g., digital volume correlation.

7.5.4 HAp Crystals Distribution Effects

The effect of the nano-scale structure (the HAp crystallite distribution) on the macroscopic mechanical response was further investigated by changing the preferential crystal orientations (changing the transformation matrix in Eq. 7.4). A schematic diagram of a 3D model of perfectly aligned crystals at the second level has been established (Figure 7.6a for dentine and Figure 7.6c for enamel), with the angle φ describing the rotation of the alignment direction around the global z axis. When all HAp crystals are aligned along the global x-direction, φ equals to 0°. By changing the alignment direction, the variation of $K_{aligned}$ in Eq. 7.4 with respect to the loading direction can be calculated. The results are visualized in Figures 7.6b&d, all of which show the transformed value of $K_{aligned}$ along the loading direction.

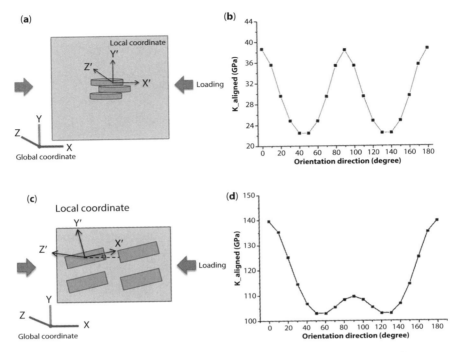

Figure 7.6 Preferential alignment effect on the apparent modulus of HAp crystals in human dental tissue. (a) and (c) show the simplified geometry of the alignment of HAp crystals in dentine and enamel. The beam direction is along the global z-axis and the alignment here represents the angle between the local x'-axis and the global x-axis (initially the alignment angle is assumed to be $\varphi = 0°$, i.e. the local x'-axis of the crystal orientation coincides with the global x-axis); (b) and (d) illustrate the variation of $K_{aligned}$ of HAp crystals in dentine and enamel under loading along the global x-axis with different preferential alignment angle (changed between $0°$–$180°$) [41, 42].

In dentine, it is found that the value of $K_{aligned}$ along the loading direction is strongly dependent on the crystal orientation direction, and the most compliant orientation observed is at $45°$ and $135°$ with respect to the loading direction. Meanwhile, since HAp crystals in dentine have a relatively low degree of alignment overall, the value of $K_{partial\_aligned}^{HAp}$ in Eq. 7.8 lies closer to K_{random}. In enamel, due to the high degree of alignment of HAp crystals, the value of $K_{partial\_aligned}^{HAp}$ lies closer to $K_{aligned}$ rather than K_{random}. The enamel displays strong microscopic elastic anisotropy. It is interesting to note that the orientation with the largest stiffness is found, as expected, around $0°$ with respect to the loading direction. However, the most compliant orientation observed is not at $90°$ (perpendicular to the loading direction), but rather lies around 50° and 130°. This effect arises due to the nature of the anisotropic (transversely isotropic) stiffness matrix of HAp crystals in enamel. It should also be noted that in practice the alignment directions of HAp crystals both in dentine and enamel are not perfect, but have some misorientation.

7.6   Conclusions

In this study, the lattice strain variation and apparent modulus of HAp crystals in human dentine and enamel were measured during *in situ* elastic compression by the combination of synchrotron WAXS/SAXS techniques. This study of the nano-scale phenomena and their influence on the macroscopic and microscopic mechanical behavior provided access to the information on both the structural and mechanical aspects of the sample and allowed progress to be made in understanding the structure-property relationships in these hierarchical biomaterials compared with the previous studies that only used WAXS [21]. Moreover, as an improvement of the earlier proposed composite model, an extended multi-scale Eshelby inclusion model was established to estimate and evaluate the elastic properties of dentine and enamel as a two-level composite in terms of their constituents. The second-level effect, i.e., the degree of alignment and orientation of HAp crystals cannot be ignored since the mechanical behavior reflects this nano-scale structure, and shows strong dependence on the crystallite orientation distribution. The models were validated by the good agreement observed between the measured and calculated lattice strains along the loading direction, and the normal strain component variation in human dentine with the general azimuthal direction. As for the normal strain component of human enamel, the model was found to underestimate the transverse tensile strain, so that the introduction of additional transverse internal strain was required for achieving agreement. The modified model result demonstrates that the deformation state in the enamel sample may not have been fully uniaxial and may have been caused by the barreling effect or de-bonding of the organic binding phase.

The systematic experimental and modeling approach reported here is able to capture the complete picture of the multi-scale structure of human dental tissue and its evolution under loading. The parameter refinement and validation approach adopted in the present study offers the possibility of the identification of nano-scale parameters, including those that may be difficult to determine otherwise. By combining the results for enamel and dentine, a comprehensive understanding of the multi-scale structural-mechanical properties within human dental tissues can be achieved, which may be useful for the development of better prosthetic materials and dental fillings and could also shed light on the mechanical property evolution associated with the multi-scale structural changes within human teeth due to disease and treatment.

Acknowledgments

AMK acknowledges the support of EPSRC through grants EP/I020691 "Multi-disciplinary Centre for In-situ Processing Studies (CIPS)," EP/G004676 "Micromechanical Modelling and Experimentation, and EP/H003215 "New Dimensions of Engineering Science at Large Facilities." Diamond Light Source is acknowledged for providing the beam time.

Appendix

A short overview of the Eshelby inclusion theory is given, leading to the derivation of the constitutive law for a non-dilute population of inhomogeneities (HAp crystals) embedded in a finite organic matrix.

1   Eshelby General Theory

1.1   Dilute System

The general geometric structure for a dilute system is an ellipsoidal inclusion embedded in an infinite matrix. "Dilute" means that the volume fraction of the inclusion is so low that it can be neglected. If there is a strain mismatch between the inclusion and the matrix, i.e. a uniform transformation strain ε^{t*} in the inclusion, then Eshelby model shows that the total strain in the inclusion ε^i is uniform and related to ε^{t*} by the Eshelby tensor S that depends only on the inclusion shape and the Poisson's ratio of the matrix.

$$\varepsilon^i = S\varepsilon^{t*} \qquad \text{(Eq. A1)}$$

Consequently, Hooke's law can be used to calculate the inclusion stress σ_I in terms of the elastic strain and the stiffness tensor C_M (the same as the surrounding matrix) of the inclusion.

$$\sigma_I = C_M(\varepsilon^i - \varepsilon^{t*}) = C_M(S - I)\varepsilon^{t*} \qquad \text{(Eq. A2)}$$

If an inhomogeneity with a different stiffness C_I is present, an equivalent inclusion method can be used by considering the inhomogeneity as equivalent to an inclusion (with identical stress and strain) with an appropriate transformation strain ε^t in the equivalent inclusion, which is to be determined from the equivalence relation:

$$\sigma_I = C_I(\varepsilon^i - \varepsilon^{t*}) = C_M(\varepsilon^i - \varepsilon^t) \qquad \text{(Eq. A3)}$$

If the material is subjected to an external load σ^A that results in the overall composite strain ε^A, then using Eq. A3 the inclusion stress can be written as

$$\sigma_I = \sigma^A = C_I(\varepsilon^i - \varepsilon^{t*} + \varepsilon^A) = C_M(\varepsilon^i - \varepsilon^t + \varepsilon^A) \qquad \text{(Eq. A4)}$$

where σ_I here is the stress caused by the elastic property mismatch between the inhomogeneity and the matrix.

1.2   Non-Dilute System

If multiple inhomogeneities are embedded in a finite matrix in which their volume fraction is not small, the composite is considered to be a non-dilute system. To satisfy the boundary conditions at the external boundaries of the finite composite, Eshelby introduced the concept of a mean "image" stress, $\langle \sigma \rangle_M = C_M \langle \varepsilon \rangle_M$ as an average of the stresses within each individual phase. Then the equivalence relation Eq. A4 becomes:

$$\langle C \rangle_I (\langle \varepsilon \rangle^i + \langle \varepsilon \rangle_M + \varepsilon^A - \langle \varepsilon \rangle^{t*}) = \langle C \rangle_M (\langle \varepsilon \rangle^i + \langle \varepsilon \rangle_M + \varepsilon^A - \langle \varepsilon \rangle^t) \qquad \text{(Eq. A5)}$$

where $\langle \cdot \rangle$ signifies the mean value, $\langle \varepsilon \rangle_M$ is the mean image strain in all the phases and $\langle C \rangle_I$ is the average stiffness of the multiple inhomogeneities.

2   First-Level Model of Human Teeth

a) Dentine

The two-level hierarchical structure of human dentine is considered to be a composite consisting of aligned tubules (inhomogeneities) within a finite collagen matrix in the first-level model. According to Eq. A5, the equivalence relation for first-level human dentine is

$$\langle C \rangle_{tubule} (\langle \varepsilon \rangle^i + \langle \varepsilon \rangle_{M1} + \varepsilon^A - \langle \varepsilon \rangle^{t*}) = \langle C \rangle_{M1} (\langle \varepsilon \rangle^i + \langle \varepsilon \rangle_{M1} + \varepsilon^A - \langle \varepsilon \rangle^t) \qquad \text{(Eq. A6)}$$

where "*M1*" means the first-level matrix, $\langle \varepsilon \rangle_{M1}$ is the mean image strain in all the phases in the first level, $\langle \varepsilon \rangle^i$ is the averaged total strain in the multiple tubules, $\langle \varepsilon \rangle^A$ is the external strain caused by the applied stress, $\langle \varepsilon \rangle^t$ is the average transformation strain for the equivalent inclusions to be determined, $\langle C \rangle_{tubule}$ is the average stiffness of the tubules, and C_{M1} is the isotropic stiffness of the collagen matrix and also the equivalent inclusion. Since $\langle C \rangle_{tubule} = 0$ and $\langle \varepsilon \rangle^i = \langle S \rangle_{tubule} \langle \varepsilon \rangle^t (\langle S \rangle_{tubule}$ is the average Eshelby tensor for the multiple tubules, Eq. A6 gives a simple expression for $\langle \varepsilon \rangle^t$

$$\langle \varepsilon \rangle^t = (1 - \langle S \rangle_{tubule})^{-1} + (\varepsilon^A + \langle \varepsilon \rangle_{M1}) \qquad \text{(Eq. A7)}$$

In a non-dilute system, the mean image stress is related to the transformation strain $\langle \varepsilon \rangle^t$ [45] in the equivalent inclusion by

$$\langle \sigma \rangle_{M1} = -f_{1d} C_{M1} (\langle S \rangle_{tubule} - I) \langle \varepsilon \rangle^t \qquad \text{(Eq. A8)}$$

where f_{1d} is the volume fraction of tubules with respect to the whole dentine. From Eq. A7, considering that $(\langle \sigma \rangle_{M1} = C_{M1} \langle \varepsilon \rangle_{M1})$,

$$\langle \varepsilon \rangle_{M1} = \frac{f_{1d}}{1 - f_{1d}} \varepsilon^A \qquad \text{(Eq. A9)}$$

Therefore, the stress in the matrix is the sum of the applied stress and the image stress

$$\sigma_{M1} = \sigma^A + \langle \sigma \rangle_{M1} = \frac{1}{1 - f_{1d}} \sigma^A \qquad \text{(Eq. A10)}$$

Eq. A10 indicates that the stress in the first-level collagen matrix is independent on the direction and detailed shape (Eshelby tensor) of the tubules.

b) Enamel

Enamel is also a two-level hierarchical structure where the first level as a non-dilute system contains protein matrix and multiple rods as the inhomogeneities. Thus the equivalence relation is the same as Eq. A5, but the average stiffness or the rods $(\langle C \rangle_{rod})$ is not zero, in contrast with that of the tubules in dentine. Based on the theory described above, in the absence of external stress, the mean stress in the rods (or equivalent inclusion) $\langle \sigma \rangle_I$ purely due to the mismatch can be expressed by [45]

$$\langle \sigma \rangle_1 = (1 - f_{1e}) C_{M1} (\langle S \rangle_{rod} - I) \langle \varepsilon \rangle^t \qquad \text{(Eq. A11)}$$

where $\langle S \rangle_{rod}$ is the average Eshelby tensor for cylindrical rods and f_{1e} is the volume fraction of the rods. When external stress is imposed, based on Eq. A5 and Eq. A8, $\langle \varepsilon \rangle^t$ can be obtained as a function of the externally applied stress

$$\langle \varepsilon \rangle^t = -\{(C_{M1} - \langle C \rangle_{rod})[\langle S \rangle_{rod} - f_{1e}(\langle S \rangle_{rod} - I)] \\ -C_{M1}\}^{-1}(C_{M1} - \langle C \rangle_{rod})C_{M1}^{-1}\sigma^A \qquad \text{(Eq. A12)}$$

Therefore, in the first-level model of enamel, with Eq. A11 and Eq. A12, the total stress in the rods can be obtained as the sum of $\langle \sigma \rangle_I$ and σ^A.

$$\langle \sigma \rangle_{rod} = \langle \sigma \rangle_I + \sigma^A \\ = \{(1 - (1 - f_{1e})C_{M1}(\langle S \rangle_{rod} - I)\{(C_{M1} - \langle C \rangle_{rod}) \qquad \text{(Eq. A13)} \\ [\langle S \rangle_{rod} - f_{1e}(\langle S \rangle_{rod} - I] - C_{M1}\}^{-1}(C_{M1} - \langle C \rangle_{rod})C_{M1}^{-1}\} \sigma^A$$

It should be noted that the average stiffness of the rods $\langle C \rangle_{rod}$ is initially unknown and needs to be determined in the second-level model.

3   Second-Level Model of Human Teeth

In contrast with the first-level model, the second-level models in dentine and enamel are similar, both considering a composite consisting of partially aligned HAp crystals and a organic matrix. Besides the different values of parameters like volume fraction, the average Eshelby tensor and the average stiffness of HAp crystals, the other difference is that the external stress in the second level of dentine is the stress of matrix in the first level, while that in the second level of enamel it is the stress in the rods. To determine the relationship between the local averaged total strain in multiple HAp crystals and the overall externally applied stress, a volume average method is introduced.

The relationship between the strain in a single HAp crystal and the external stress (matrix stress in dentine or rod stress in enamel) can be established initially based on Eq. A1 to A5

$$\varepsilon_{HAp}^{single} = T^T \left\{ \{(I - C_{M2}^{-1}C_{HAp})^{-1}[S_{HAp} - f_2(S_{HAp} - I)]^{-1} - I\}^{-1} \\ T^{-T} + T^{-T} \right\} C_{M2}^{-1}\sigma^A - K\sigma^A \qquad \text{(Eq. A14)}$$

where "*M2*" means the second-level organic matrix. C_{HAp}, S_{HAp} are the stiffness matrix and Eshelby tensor for a single HAp crystal, respectively, f_2 is the volume fraction of HAp crystals with respect to the second-level composite, I is the unity matrix and T is the orientation matrix that depends on the three Euler angles (θ, ϕ, ψ).

The partial alignment is represented as a combination of perfect alignment and random distribution. For a group of perfectly aligned HAp crystals with a certain orientation, the relationship between the strain of the group and the external stress is the same as Eq. A14, where the averaged stiffness and the Eshelby tensor are the same as the values for a single crystal $\langle C \rangle_{HAp} = C_{HAp}$, $\langle S \rangle_{HAp} = S_{HAp}$. As for a group of randomly distributed HAp crystals, the strain of the group can be determined by a volume average method, which is introduced here.

The purpose of using volume average method is to avoid the complex calculation of average Eshelby tensor and average stiffness of randomly distributed HAp crystals. In a random distribution, each crystal follows the relationship of Eq. A14 with the individual orientation matrix T. Crystals can have any possible orientation in the space, thus the volume average method is to calculate the strain value for each single crystal and average the results of all the crystals in the second level over all possible orientations.

$$\langle \varepsilon_{HAp}^{single} \rangle = \frac{1}{V} \int_v \varepsilon_{HAp}^{single} dV = \frac{\int_0^{2\pi} \int_0^{\pi} \int_0^{2\pi} K \sin\theta d\phi d\theta d\psi}{\int_0^{2\pi} \int_0^{\pi} \int_0^{2\pi} \sin\theta d\phi d\theta d\psi} \sigma^A$$

$$= \frac{\sigma^A}{2\pi^2} \int_0^{2\pi} \int_0^{\pi} \int_0^{2\pi} K \sin\theta d\phi d\theta d\psi = \langle K \rangle \sigma^A \qquad \text{(Eq. A15)}$$

References

1. A.R. Ten Cate, A.C. Dale, *Oral Histology: Development, Structure, and Function*, St. Louis: Mosby, 1980.
2. L.M. Petrovic, D.T. Spasic, T.M. Atanackovic, *Dent Mater*, Vol. 21, pp. 125–128, 2005.
3. D.H. Pashley, B. Ciucchi, H. Sano, R.M. Carvalho, C.M. Russell, *Arch Oral Biol*, Vol. 40, pp. 1190–1118, 1995.
4. E. Johansen, H.F. Parksf, *J Biophys Biochem Cy*, Vol. 7, p. 743, 1960.
5. J.C. Voegel, R.M. Frank, *J Biol Buccale*, Vol. 5, pp. 181–194, 1977.
6. G.A. Macho, Y. Jiang, I.R. Spears, *J Hum Evol*, Vol. 45, p. 81–90, 2003.

7. H. Gao, B. Ji, I.L. Jager, *et al*, *PNAS.*, Vol. 100, p. 5597–5600, 2003.
8. F.A. Siang, S. Mahnaz, V.S. Michael, *et al*, *J Mater Res.*, Vol. 27, p. 448–456, 2012.
9. B. Kerebel, G. Daculsi, L.M. Kerebel, *J Dent Res.*, Vol. 57, pp. 306–312, 1979.
10. L.H. He, *PhD thesis*, 2008.
11. J.H. Kinney, S.J. Marshall, G.W. Marshall, *Crit Rev Oral Biol Med*, Vol. 14, pp. 13–29, 2003.
12. T. Nakamura, C. Lu, C.S. Korach, *Conference Proceedings of the Society for Experimental Mechanics Series*, Vol. 1999, pp. 171–179, 2011.
13. A.C. Deymier-Black, J.D. Almer, S.R. Stock, *et al*, *Acta Biomater*, Vol. 6, pp. 2172–2170, 2010.
14. A.C. Deymier-Black, J.D. Almer, S.R. Stock, D.C. Dunand, *J Mech Behav Biomed Mater*, Vol. 5, pp. 71–81, 2012.
15. S. Bechtle, H. Özcoban, E.T. Lilleodden, *et al*, *J.R. Soc. Interface*, Vol. 9, pp. 1265–1274, 2012.
16. M.L. Young, J.D. Almer, M.R. Daymond, *et al*, *Acta Mater*, Vol. 55, pp. 1999–2011, 2007.
17. R. Mueller, A. Rossoll, L. Weber, *et al*, *Acta Mater*, Vol. 56, p. 4402–4416, 2008.
18. M.L. Young, J. DeFouw, J.D. Almer, D.C. Dunand, *Acta Mater*, Vol. 55, pp. 3467–3478, 2007.
19. A.C. Deymier-Black, F. Yuan, A. Singhal, *et al*, *Acta Biomater*, Vol. 8, pp. 253–261, 2012.
20. A. Singhal, J.D. Almer, D.C. Dunand, *Acta Biomater*, Vol. 8, pp. 2747–2758, 2012.
21. J.D. Almer, S.R. Stock, *J Struct Biol*, Vol. 152, pp. 14–27, 2005.
22. J.D. Almer, S.R. Stock, *J Struct Biol*, Vol. 157, pp. 365–370, 2007.
23. J.D. Almer, S.R. Stock, *J Biomech*, Vol. 43, pp. 2294–2300, 2010.
24. P. Fratzl, S. Schreiber, K. Klaushofer, *Connect Tissue Res*, Vol. 35, pp. 9–16, 1996.
25. H.D. Wagner, S. Weiner, *J Biomech*, Vol. 25, pp. 1311–1320, 1992.
26. J.D. Currey. *J Biomech*, Vol. 2, p. 477, 1969.
27. W. Bonfield, M.D. Grynpas, *Nature*, Vol. 270, pp. 453–454, 1977.
28. M.B. Lopes, M.A. Sinhoreti, A. Gonini Junior, *et al*, *Braz Dent J*, Vol. 20, pp. 279–283, 2009.
29. B.C.L. Nogueira, P.M. Fernandes, L.N.S. Santana, R.R. Lima, *XXIII Congress of the Brazilian Society Of Microscopy and Microanalysis*, 2011
30. J.L. Katz, *J Biomech*, Vol. 93, pp. 455–473, 1971.
31. Z. Hashin, *J Appl Mech-T Asme*, Vol. 50, pp. 481–505, 1983.
32. T. Gottesman, Z. Hashin, *J Biomech*, Vol. 13, pp. 89–96, 1980.
33. R. M. Jones, *Mechanics of composite materials. 2nd ed*, Philadelphia, Pa., Taylor & Francis, 1999.
34. S. Bechtle, H. Ozcoban, E.T. Lilleodden, *et al*, *J R Soc Interface*, Vol. 9, pp. 1265–1274, 2012.

35. P.J. Withers, W.M. Stobbs, O.B. Pedersen, *Acta Metall Mater*, Vol. 37, pp. 3061–3084, 1989.

36. Y. Takao, M. Taya, *J Compos Mater*, Vol. 21, pp. 140–156, 1987.

37. Q.H. Qin, M.V. Swain, *Biomaterials*, Vol. 25, pp. 5081–5090, 2004.

38. B. Huo, *J Biomech*, Vol. 38, pp. 587–594, 2005.

39. Y.N. Wang, Q.H. Qin, *Compos Sci Technol*, Vol. 67, pp. 1553–1560, 2007.

40. B. Huo, Q.S. Zheng, *Acta Mech Sinica*, Vol. 15, pp. 355–365, 1999.

41. T. Sui, M. A. Sandholzer, N. Baimpas, *et al*, *Acta Biomater*, Vol. 9, pp 7937–7947, 2013

42. T. Sui, M. A. Sandholzer, N. Baimpas, *et al*, *J Struct Biol*, Vol. 182, pp. 136–146, 2013

43. K.W. Eliceiri, M.R. Berthold, I.G. Goldberg, *et al*, *Nat Methods*, Vol. 9, pp. 697710, 2012.

44. Diamond Light Source Calibration, http://www.diamond.ac.uk/Home/Beamlines/small-angle/during/calibration.html, 2013.

45. T.W. Clyne, P.J. Withers, *An Introduction to Metal Matrix Composites*, Cambridge, Cambridge University Press, 1993.

46. L. Bozec, J. de Groot, M. Odlyha, B. Nicholls, *et al*, *Ultramicroscopy*, Vol. 105, pp. 79–89, 2005.

47. T. Mura, *Micromechanics of defects in solids, 2nd rev. ed.* Dordrecht, Kluwer Academic, 1987.

48. S.R. Stock, A. Veis, A. Telser, Z. Cai, *J Struct Biol*, Vol. 176, pp. 203–211, 2011.

49. H.R. Wenk, F. Heidelbach, *Bone*, Vol. 24, pp. 361–369, 1999.

50. S. Habelitz, S.J. Marshall, G.W. Marshall Jr, *et al*, *Arch Oral Biol*, Vol. 46, pp. 173–183, 2001.

51. H. Zhan, J.N. Christina, B. Jr. Pablo., I. S. Samuel, L.S. Malcolm, *Biomaterials*, Vol. 31, p. 35, 2010.

52. P.J. Withers, W.M. Stobbs, O.B. Pedersen, *Acta Metall Mater*, Vol. 37, pp. 3061–3084, 1989.

53. J.M. Powers, J.W. Farah, *J Dent Res*, Vol. 902, p. 54, 1975.

54. T.W. Chou, C.T. Sun, *Nanocomposites*, DES tech, 2012.

55. H.D. Jiang, X.Y. Liu, C.T. Lim, C.Y. Hsu, *Appl Phys Lett*, Vol. 86, 2005.

56. T. Tanaka, N. Yagi, T. Ohta, *et al*, *Caries Res*, Vol. 44, pp. 253–259, 2010.

57. A.O. Dourda, A.J. Moule, W.G. Young, *Int Endod J*, Vol. 27, pp. 184–189, 1994.

58. J.H. Kinney, S. Habelitz, S.J. Marshall, G.W. Marshall, *J Dent Res*, Vol. 82, pp. 957–961, 2003.

59. A. Marten, P. Fratzl, O. Paris, P. Zaslansky, *Biomaterials*, Vol. 31, p. 5479–5490, 2010.

60. A. Öchsner, W. Ahmed, *Biomechanics of Hard Tissues*, Weinheim, John Wiley & Sons, 2011

61. S. Habelitz, G.W. Jr Marshall, M. Balooch, S.J. Marshall, *J Biomech*, Vol. 35, pp. 995–998, 2002.

62. B. Devendra, A. Dwayne, *Acta Biomater*, Vol. 5, pp. 3045–3056, 2009.

8

Biodegradable Porous Hydrogels

Martin Pradny*, Miroslav Vetrik, Martin Hruby and Jiri Michalek

Institute of Macromolecular Chemistry AS CR, Prague, Czech Republic

Abstract

Biodegradable porous hydrogels are promising materials for various biomedical applications. Their porous structure containing interconnected pores allows cells to grow through the hydrogel material and provides sufficient permeability for nutrients and metabolites. During cell growth the biodegradable hydrogel decomposes and enables growing tissue to create integrated structure. Internal porous architecture of such hydrogels as well as degradation kinetics can be tailored according to the actual requirements. Various preparation techniques generally based on phase separation allows the control over hydrogel porosity. Adjustment of overall hydrophilic/hydrophobic balance by the choice of monomers, degradability of the main polymer chain, and of the crosslinker are the main manners to control the hydrogel degradation. This work summarizes results achieved in this field. Methods of preparation of porous hydrogels, application, and types of used materials are described.

Keywords: Hydrogels, porous structures, preparation of hydrogels, biomedical applications

8.1 Introduction

Recently, porous hydrogels have been frequently mentioned in connection with tissue engineering, cell therapy, chromatographic columns, adsorbents, and membranes. For many purposes, it is appropriate to modify the materials so that they degrade, after some time in contact with living tissue, to water-soluble substances, which can then be eliminated by the organism.

---

*Corresponding author: pradny@imc.cas.cz

Ashutosh Tiwari (ed.) Advanced Healthcare Materials, (269–294) 2014 © Scrivener Publishing LLC

Apparently, most of the articles are focused on biodegradable porous hydrogels as scaffolds for tissue engineering and cell therapy. The cell cultivation on hydrogels is usually successful due to their good biocompatibility and a high specific surface area. Moreover, biodegradable hydrogel scaffolds [1] are able to decompose after cell cultivation and, consequently, the newly created tissue is continuous.

Generally, porous hydrogels are non-homogenous organic substances of natural or synthetic origin. Non-homogeneity of these materials consists in walls separating the pores. This review chapter is focused on biodegradable porous materials based on hydrophilic crosslinked polymers prepared with communicating (interconnected) pores, which plays a key role in most of the applications in many fields, with special focus on hydrogels for tissue engineering applications.

Permanent porosity (or macroporosity) can be achieved by following methods:

1. Crosslinking polymerization in the presence of solvent that dissolves the monomers, but causes precipitation of the formed polymer.
2. Crosslinking polymerization in the presence of water-soluble substances (e.g., carbohydrates, salts), which are washed out from the hydrogel after polymerization.
3. Crosslinking polymerization in the presence of substances releasing porogen, gas bubbles which remain in hydrogel.
4. Freeze-drying (lyophilization) of the hydrogel swollen in water.
5. The use of nanofibrous materials. (Nano)fibers are also considered as macroporous materials.
6. Cryogelation, where ice crystals and/or miscibility changes serve for pore generation.

Especially for tissue engineering applications, it is often highly advantageous to have biodegradable materials, which degrade, dissolve, and can be eliminated from the body after fulfilling their task. Generally, biodegradability may be achieved by biodegradable crosslinkers crosslinking non-biodegradable linear chains or with the use of polymers degradable in the main chain.

Biodegradable crosslinkers are usually based on hydrolytically degradable diacylhydroxylamines [2], oligo(caprolactone) [3], hydrazones [4], or reductively degradable disulfides [5]. This approach usually deals with enzyme-independent degradation mechanisms.

The hydrogels formed by polymers degradable in the main chain may be of synthetic (e.g., polyesters, peptide-based) or natural (e.g., polysaccharides, proteins) origin. Enzymatic degradation is the most usual mode of their biodegradability.

Each particular system has its own benefits and drawbacks, so an appropriate choice of the chemistry and morphology of the system and tailoring the material to the particular application of interest is always a challenge. This review critically reviews structure, properties, methods of preparation, benefits, and drawbacks of such systems in order to help the reader to make proper choices for his/her purpose with special focus on porous hydrogels for tissue engineering applications.

8.2   Methods of Preparation of Porous Hydrogels

8.2.1   Crosslinking Polymerization in the Presence of Substances that are Solvents for Monomers, but Precipitants for the Formed Polymer

Porous materials may be prepared by crosslinking polymerization of monomers in solvent that is a precipitant for the resulting polymer. The polymer usually precipitates in the form of spherical particles of nano- or micrometer dimensions [6, 7], which subsequently form the walls by covalent crosslinking of the pre-formed particles in later stages of polymerization. The required distance between the discrete particles can be partly set by appropriate concentration of monomer(s). The pores are then formed by the spaces between the particles or their aggregates. The main advantage of this technique is the formation of very thin walls, which is required for, for example, super-absorbing materials [8]. The pores usually communicate and the fraction of non-communicating (disconnected) pores is very low. This method of preparation is the simplest compared with the others and generally does not require any highly sophisticated apparatus. The disadvantage of this technique is the limited possibility to adjust pore sizes, because any changes in the polymerization mixture (necessary for controlling the resulting properties) affect pore sizes only a little. For example, such hydrogel was prepared by copolymerization of N-(2-hydroxypropyl) methacrylamide (HPMA) with a crosslinker (ethylene dimethacrylate - EDMA or N,N'-methylenebisacrylamide) in acetone/dimethyl sulfoxide as a porogen mixture [6] of from 2-hydroxyethyl methacrylate (HEMA) [9, 10]. The typical morphology of such porous hydrogel based on poly(2-hydroxyethyl methacrylate) (PHEMA) [9, 10] is shown in Figure 8.1a.

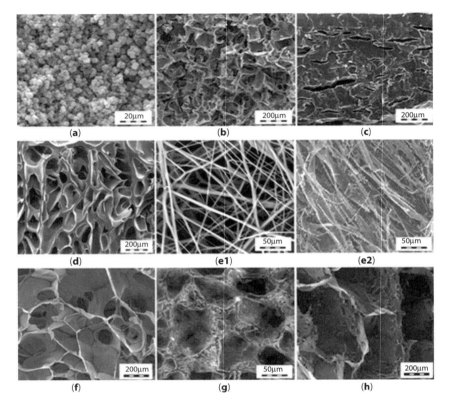

Figure 8.1 Typical morphologies of porous hydrogels prepared by methods 1.1. (a) , 1.2.(b), 1.3.(c), 1.4.(d), 1.5. (e1), 1.4.(e2), 1.6.(f), 1.7. (g), and 1.8.(h).

8.2.2   Crosslinking Polymerization in the Presence of Solid Porogen

The principle of this method is that the water-soluble fractionated particles (which are not soluble in a monomeric mixture) are added to a mixture of monomer, crosslinker, and initiator. After polymerization, the particles are washed out [1, 11–15]. The controlled size of these fractionated particles corresponds to the size of pores. An advantage of this technique is a simple adjustment of pore size by fractionating and relatively easy characterization [9]. A limitation of this method is a higher thickness of pore walls. Using this method, porous hydrogels with communicating (interconnected) or non-communicating pores can be prepared. The percentage of communicating pores increases with increasing amount of soluble particles (porogen) in the polymerization mixture [10]. However, this effect is followed by deterioration of mechanical properties. The morphology of materials prepared using this technique is obviously different from the morphology of materials prepared according to the process described in 8.2.1. The copolymerization of HEMA-co-MOETACl with EDMA as crosslinker is an example of hydrogel preparation by this method. Sodium chloride particles of average sizes 16, 40, 70, 108, and 163 μm were used as a porogen.

It was necessary to use an amount fourteen times higher of the porogen (fractionated sodium chloride particles) relative to the sum of the polymerization components to achieve communicating pores. For exclusion of bubbles and homogenization of monomers with sodium chloride oligo(ethylene glycol) (weight-average molecular weight 300 Da) was added. This component is a solvent for monomers but not for sodium chloride. After thorough mixing of all components, the viscous paste was polymerized in the apparatus depicted in our previous paper [9]. The preparation of porous hydrogels from the pre-prepared polymers is in principle possible but complicated and poorly reproducible [16]. The problem arises from sedimentation of salt particles in the polymerization mixture and, consequently, different fraction of pores may be different at the bottom and in the upper part of the mixture. This problem may be circumvented by careful adjustment of polymerization mixture-salt ratio. The typical morphology of such porous hydrogel based on PHEMA [10] is shown in Figure 8.1b.

8.2.3 Crosslinking Polymerization in the Presence of Substances Releasing a Gas

The principle of this technique is a crosslinking polymerization in the presence of components that release gases. Subsequently, the bubbles of gas evolve in the liquid polymerization mixture and are fixed in the polymerizing mass giving rise to the pores. Usually, it is necessary to add foam stabilizer into the polymerization mixture. Its function is to stabilize the foam so that the polymerization can be finished at the foam stage.

For example, porous poly(N-isopropylacrylamide) or polyacrylamide hydrogels [17, 18] with the pore size 100 μm and larger was prepared by polymerization of the monomers, crosslinker (N,N′- methylenebisacrylamide), initiator (ammonium peroxosulfate), butane-1,4-diamine hydrochloride, stabilizing agent, and sodium hydrogencarbonate. Sodium hydrogencarbonate had two functions: first, it reacted with an acid to generate CO_2 bubbles, which were essential in making the foam. Second, it increased pH of the solution and accelerated polymerization.

In this manner it is possible to prepare porous hydrogels with communicating or non-communicating pores, whereas the fraction of noncommunicating pores is similar or higher compared with the preparation techniques mentioned above. A disadvantage of this method is complicated preparation procedure. On the other hand, using this technique, porous hydrogels can be prepared even in cases where polymerization from monomers is not feasible, e.g. with polysaccharides such as chitosan and glycolchitosan [19, 20]. The porous structure was obtained by the addition of glyoxal as a crosslinker and sodium hydrogencarbonate to a solution of chitosan in acetic acid.

The block crosslinking polymerization, in the presence of a small amount of low-temperature boiling solvent, can also be considered as this technique of preparation. The gas bubbles are produced by the solvent, boiling due to the heat released at the gelation point [10]. However, this method is not commonly used because of a very low reproducibility of the final porous material and predominantly non-communicating pores. The typical morphology of such porous hydrogel based on PHEMA [10] is shown in Figure 8.1c.

8.2.4   Freeze-Drying (Lyophilization) of the Hydrogel Swollen in Water

The water-swollen hydrogel is frozen and water is sublimed under vacuum at low temperature. This technique affords only limited possibilities to adjust pore size. Because of the wide distribution of pore sizes [10, 21, 22], this technique is one of the least frequently used. Nevertheless, similarly like in the method described in 2.3., this technique enables the preparation of porous structures from the polymers whose monomers are not available. A typical structure of the material prepared in this manner is shown in Figure 8.1d. Apparently, the pore walls are very thin; even thinner than those of hydrogels prepared by precipitation polymerization (2.1.). A biodegradable porous hydrogel, poly(sodium alginate), was prepared by a three-step technique based on this principle [21]: a dilute alginate solution (2%) was transformed to a gel in homogenizer, then frozen and finally dried by lyophilization. The typical morphology of such porous hydrogel-based on poly(HEMA– co- MANa) (MANa - sodium methacrylate) [10] is shown in Figure 8.1d.

8.2.5   Fibrous Materials

Although the fibers are not usually referred to as porous materials, they possess similar properties. Walls are formed by fibers and pores by the space between them. The diameter of the nanofibers is in nano- to micrometer range. A common way to fabricate especially nanofibers is the electrospinning technique [23, 24]. With this technique, continuous thin polymeric fibers can be prepared. Schematically, the electrospinning technique can be described as follows: when a high voltage is applied to a polymeric solution against a grounded electrode, the polymeric solution becomes charged and, due to electrostatic attraction, a stream of liquid erupts from the surface (Taylor's cones), and is launched towards the grounded electrode where the substrate to be coated is placed. If the molecular cohesion of the solution is sufficiently high, a continuous thin fiber is formed and elongated following the electrical field direction. Simultaneously, the solvent is evaporated during the flight of the fiber and this nanofiber is finally deposited on the collector. Such porous material has significantly larger surface area than those prepared by previous methods and all pores are completely interconnected. The typical morphology of such porous hydrogel based on PHEMA [10] is shown in Figure 8.1e1.

Nanofibrous materials can be also obtained under certain circumstances by freeze-drying of polymer solutions. In this way, biodegradable glycogen nanofibers were obtained and were subsequently modified to glycogen-*graft*-poly(ethyl cyanoacrylate) by anionic polymerization from vapor state [25]. Morphology of the nanofibers highly depends on the starting polymer solutions—low concentrations lead to nanofibers while higher concentrations lead to sponge-like structures with interconnected pores. Morphology of such material based on glycogen is shown in Figure 8.1e2.

8.2.6 Cryogelation

This method is closely related to lyophilization (2.4). In this ice-templating process, the soluble substances (monomers, initiators, polymers), originally dissolved in the aqueous solution, are expelled from the ice crystals upon freezing of the solution. The dissolved substances are concentrated within the non-frozen liquid channels among adjacent ice crystals. As crosslinking polymerization subsequently takes place in these non-frozen channels, the ice crystals act as porogen during gelation. Removal of ice by thawing after polymerization leads to formation of structure with large and interconnected pores. Hydrogels with highly interconnected porous networks designed by this method are called cryogels. In addition to the interconnected porous structure, cryogels possess a tissue-like elasticity, are able to withstand high levels of deformations, such as elongation and torsion, and are also characterized by superfast responsiveness at water absorption [26–29].

A number of authors have demonstrated the capacity of ice-templating processes to control the morphology of the resulting macroporous structures using unidirectional freezing technique at a controlled immersion rate [30–34]. In this approach, different systems such as aqueous polymer solutions, inorganic colloidal dispersions or their hybrid composites are undirectionally frozen in cooling media as liquid nitrogen (-196 °C), so the ice crystals (or solvent crystals) form and grow undirectionally. This process leads to microchanneled structures which are well-aligned in the freezing direction with a well-patterned between channel morphology (e.g., micro-honeycomb or lamellar). Unidirectional freezing technique has been used to prepare aligned silica fibers [31], aligned porous structure from water-soluble polymers such as poly(vinyl alcohol) [28], poly(L-lactic acid) [32], poly(lactide-*co*-glycolide) [33], and recently poly(ethylene oxide) aligned porous cryogels [34].The typical morphology of such porous hydrogel based on chitosan [35] is shown in Figure 8.1f.

8.2.7   Combined Techniques

It is well known in tissue engineering that the growing cells need a specific pore volume, a specific comfort provided by the softness/rigidity of pore walls and, in any case, a good supply of nutrients through the pore walls [36]. It was observed [37, 38], that the cell growth in the gel scaffold was reduced, or even stopped after some time just because of increasingly limited nutrient supply when the large pores got blocked with cells. This means that the pore walls' permeability for nutrient molecules is one of the key issues. The problem can be resolved by creating hydrogels containing pores large enough for nesting the cells surrounded by walls being built-up of a macroporous PHEMA gel—but of much smaller pore size. To make such gel construct, the method using solid porogen removable by washing (e.g., particles of sodium chloride) was combined with dilution of the monomer mixture with a precipitant for PHEMA (1-dodecanol). The principle of double porosity in gels has a broader applicability than in tissue engineering and can be utilized wherever pore walls permeability and mechanical responsibility matters. It is worth mentioning that the formation of similar hierarchical structures were observed by Kulygin and Silverstein [39] when HEMA was copolymerized with about 20% crosslinker in an oil-in-water high internal phase emulsion. Dissolved water in the oil phase (monomer) phase was apparently the reason. The typical morphology of such porous hydrogel based on PHEMA [40] is shown in Figure 8.1g.

Although cryogelation can achieve significantly higher pore fraction than by the method described in 2.2, the problem of the impermeability for nutrients through walls remains. To solve this problem, aqueous solution of chitosan was added to fractionated particles of poly(methyl methacrylate) (PMMA) [35] and this solution was frozen. Particles were partially dispersed into the walls of arising cryogel and after the process of cryogelation accompanied by chitosan crosslinking with glutaraldehyde, PMMA particles were washed out with acetone. The resulting structure (Fig. 8.2f) indeed shows the porous walls.

8.3   Hydrogels Crosslinked With Degradable Crosslinkers

Degradability of hydrogel may be achieved by biodegradable crosslinks. Within this method, non-degradable, but water-soluble linear or branched polymers are connected with degradable links. These links may be introduced by the use of crosslinker containing degradable bond between polymerizable moieties during synthesis of the hydrogel by polymerization or by crosslinking of the linear polymer with a suitable reagent containing biodegradable bond. The 3D crosslinked structure changes during the degradation process to soluble linear (branched) polymer, which is dissolved in water and initially insoluble hydrogel disintegrates and dissolves (see Scheme 8.1). The drawback of this method is that degradability of the hydrogel is extremely sensitive to the eventual presence of non-degradable crosslinkers, which is especially a problem in the case of HEMA-based hydrogels.

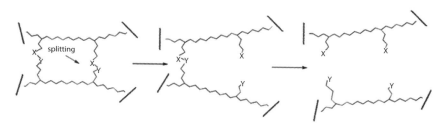

Scheme 8.1 The principle of degradation of hydrogel containing degradable crosslinker.

8.3.1   Hydrogels Degradable by Hydrolysis of the N-O Bonds

In this case, *N,O*-dimethacryloylhydroxylamine (X = O and Y = NH in Scheme 8.1) (DHMA) was used as the crosslinker for *N*-(2-hydroxypropyl) methacrylamide (HPMA) [41–44] , PHEMA [1, 45] or poly(N-isopropyl-methacrylamide) [46]. As the bond N-O is stable in acidic milieu, also the hydrogel is stable at pH < 6. However, at pH > 6 the N-O bond is hydrolytically unstable and degrades (see Scheme 8.2). The hydrogel may thus be stored in a mildly acidic milieu, but in tissue at pH 7.4, degradation starts to take place.

Porous hydrogels degradable on this principle were used for tissue engineering [1, 45]. The hydrogels containing only HPMA and DHMA degrade quickly, within 2–5 days. However, copolymerization of more hydrophobic monomer (e.g., 2-ethoxyethyl methacrylate, EOEMA) significantly prolongs degradation time. The EOEMA cannot be used as homopolymer, because it is not water-soluble, so the hydrogel would be non-degradable. In analogy, it is possible to use hydrogels with adjustable hydrophobicity and subsequently tailorable degradation rate from a combination of monomers HEMA + HPMA or EOEMA + HPMA, alone or with a positively charged comonomer (MOETACl) or a negatively charged (MA) comonomer.

For biological tests in vivo, EOEMA/HPMA hydrogels with 21% EOEMA, crosslinked with 2.1% DMHA, were used [1]. Blocks of EOEMA/HPMA hydrogel were implanted into hemisections formed in the spinal cords of laboratory brown rats. In spinal cord defects filled with hydrogel, tissue was regenerated (forming predominantly connective tissue elements) within 8 days.

The typical degradation of porous hydrogel followed by light microscopy (LM) and low vacuum scanning electron microscopy (LVSEM) is shown in Figures 8.2 and 8.3.

Degradation of the hydrogels was accompanied by their swelling, as documented by LM (Figure 8.3). At the first stage, when the crosslinks between chains break, the network density decreases and hydrogel swelling

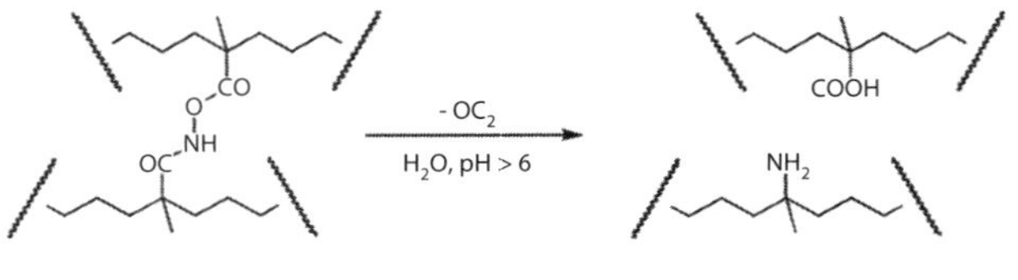

Scheme 8.2 Hydrolysis of polymer crosslinked with *N,O*-dimethacryloylhydroxylamine.

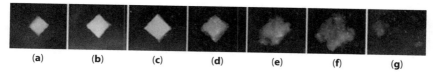

Figure 8.2 Degradation of DHMA-crosslinked EOEMA/HPMA hydrogels, followed by light microscopy (LM). The first micrograph (a) shows a small cube of hydrogel immediately after immersion in buffer at pH 7.4. The following micrographs show the sample after (b) one, (c) two, (d) three, (e) four (f) six and (g) seven days. Bar = 5 mm.

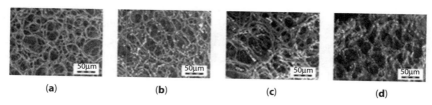

Figure 8.3 Degradation of a hydrogel followed by low vacuum scanning electron microscopy (LVSEM). The micrographs show the sample (a) immediately after immersion in pH 7.4 buffer and after incubation in the buffer for (b) one, (c) two, and (d) three days.

increases. At the moment when the network density is so low that the copolymer EOEMA/HPMA becomes soluble in water, the hydrogel disintegrates and after eight days it is completely split to water-soluble EOEMA/HPMA non-croslinked copolymer. Consequently, degradation time was possible to adjust by the ratio of EOEMA and HPMA; degradation time increased with increasing amount of hydrophobic EOEMA. LVSEM provided an opportunity to observe changes in the supramolecular structure (Figure 8.3). In the beginning, the samples exhibited a distinct structure, i.e., thin and sharp polymer walls between the pores (Figure 8.3a). In the course of time, the walls swell and later dissolve. As a result, the hydrogel structure becomes more and more diffused and fuzzy (Figure 8.3b–d). This microscopic swelling was in accordance with the macroscopic swelling observed with LM at lower magnification (Figure 8.2). After four days of degradation the samples were so soft that they could not be transferred into liquid nitrogen without being completely destroyed.

8.3.2 Hydrolytic Splitting of Crossing Chain Based on Poly(Caprolactone)

Very promising thermosresponsive degradable porous hydrogel based on poly(N-isopropyl acrylamide) was developed by Galperin *et al.* [47] as a material for tissue engineering. Porosity was achieved by method 2.2.; poly(methyl methacrylate) was used as a porogen. The degradable site was in crosslinker chain and was formed by block polycaprolactone-*block*-poly(ethylene oxide)-*block*-polycaprolactone. Also, the main chain of polymer (backbone) contained additional degradable sites arising by copolymerization of *N*-isopropylacrylamide with 2-methylene-1,3-dioxepane. This comonomer uniquely allows introduction of hydrolyzable ester bonds into the hydrocarbon main chain of the polymer formed by radical polymerization, forming an additional caprolactone fragment inside the polymer chain. A disadvantage of degradable caprolactone-based materials is a long degradation time [48] in the order of ten weeks at physiological pH (7.4). Some acceleration of the degradation can be achieved by short immersing of the sample into the solution with high pH with subsequent neutralization. Also the introduction of degradable caprolactone units in combination with 2-methylene-1,3-dioxepane into the backbone leads to a reduction of the degradation time [47].

Similar crosslinker based on caprolactone was used for the preparation of double porous poly(2-hydroxyethyl methacrylate) [49] by photopolymerization of monomer (HEMA) and crosslinker. Porosity with two pore types was achieved by combination of method 2.2. and photo-patterning process. In contrast to double porous hydrogels (Figures 8.1g, h) obtained by methods 2.7. and 2.8., here very regular structure was observed. Photopatterning process is based on manipulation of polymerization kinetics to achieve patterns in thick hydrogel. Patterned hydrogels were fabricated by creating an inverse photomask in which initiating light is allowed to pass through all areas of the photomask, but at different intensities. High light intensities caused significant deviations in the polymerization kinetics, resulting in longer polymerization times compared to lower light intensities. As a result, patterns in thick gels could be achieved.

8.3.3   Reductive Splitting of S-S Bond which is Part of Crossing Chain

Although hydrolytically biodegradable materials can be prepared from a variety of materials (sections 8.3.1., 8.3.2.), they still have many important disadvantages, such as problems with their storage in an aqueous environment and the fact that their hydrolysis is accompanied by a swelling phase, which constricts the growing cells, modifies scaffold microarchitectural properties, etc. The reductive degradability of disulfide bonds in a crosslinker by a disulfide-thiol exchange reaction or reduction with thiols omnipresent in the organism (glutathione, serum albumin, cysteine,

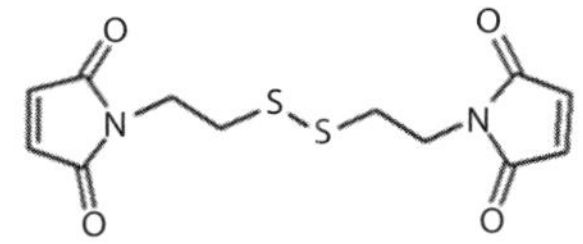

3a. DTME 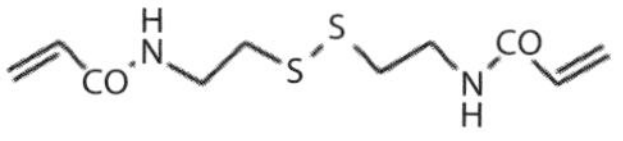   **3b. BACy**

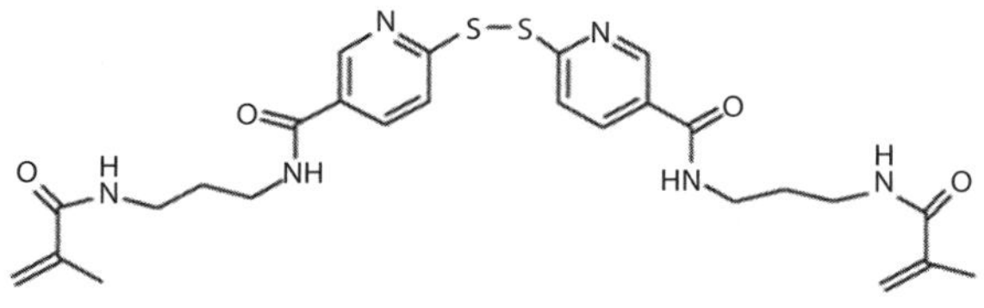

3c. MASS

Scheme 8.3 Reductively degradable crosslinkers.

homocysteine, glycylcysteine,) can be used to construct porous hydrogels. Several crosslinkers containing the reductive splitting S-S bond were described in literature in connecting with porous hydrogels:

1. Dithiobis(maleimido)ethane (*50*), DTME, Scheme 3a,
2. N,N′-Bis(acryloyl)cystamine (*51*), BACy, Scheme 3b,
3. N-[3-(methacryloylamino)propyl]   -6-{[5-({[3-(methacry-loylamino)propyl]amino}   carbonyl)-2-pyridyl]disulfanyl} nicotinamide (*5*) , MASS, Scheme 3c.

Crosslinker DTME was used for preparation of porous hydrogel based on poly(ethylene glycol) using the method of cryogelation and was intended as a scaffold for tissue engineering [50]. While the cryogels were stable under physiological conditions, complete dissolution of the cryogels into water-soluble non-toxic products was obtained in the presence of a reducing agent (glutathione) in the medium. Mechanical properties of porous hydrogel were similar as properties of soft living tissues. Cell seeding experiments and toxicologic analysis demonstrate their potential as scaffolds in tissue engineering. Degradation started in 12 h and in the first step looked similar to degradation of EOEMA/HPMA hydrogel cross-linked with DHMA (section 8.3.1.). Unfortunately the kinetics of degradation of DTME based hydrogel was not studied, so that we have no image of how the global splitting process looked.

Crosslinker BACy was used for preparation of porous hydrogel based on branched polyethylenimine using the method of lyophilization and was intended as a scaffold for drug delivery system [51]. Antibacterial ceftriaxone was chosen as model drug. Dithiothreitol (DTT) was chosen as a model reducing agent. Degradation (de-crosslinking) of studied hydrogel by means of DTT leads to polyethylenimine with thiol groups and low-molecular thiols [52]. Time for complete degradation of hydrogel was 14 days and time for ceftriaxone complete releasing was possible to adjust from several hours to 10 days [51].

Most thiols exchange with disulfides reversibly. However, some disulfides (e.g., 2,2′-dipyridyl disulfide type) are cleaved irreversibly into two separate parts due to the strong tautomeric stabilization of the thiol cleavage product into preferred non – thiol tautomer (2-thiopyridone in this case). If such disulfide bonds irreversibly cleavable by thiols are employed, it should enable to avoid the side reactions after degradation (regelation after exposure of the degraded polymer to air, reconjugation with thiol-containing proteins such as opsonins, unwanted additional swelling during degradation, etc.). HPMA-based hydrogel crosslinked with MASS is of this type. It is stable during storage in aqueous milieu and is degraded by thiols with a rate adjustable by hydrogel composition to up to 60 days [5].

8.4   Hydrogels Degradable in the Main Chain

This type of degradability requires the presence of cleavable groups (Scheme 8.4) in the main chain of the polymer. In connection with porous hydrogels, both natural (polysaccharides, peptides) and synthetic polymers (polyesters, polyethers, polyalcohols) are mentioned in the literature.

8.4.1   Polycaprolactone-Based Hydrogels

As the polycaprolactone (PCL) is a water insoluble polymer, no crosslinking for porous stable structure is necessary.

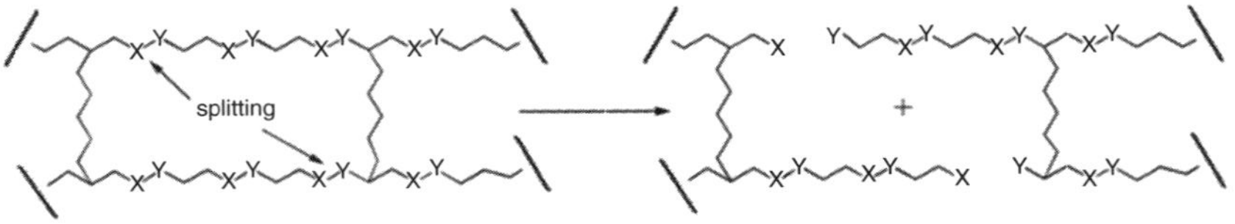

Scheme 8.4 The principle of degradation of hydrogel containing the cleavable bonds in polymer main chain.

An interesting method for preparation of porous microspheres for tissue engineering was developed by Zhang *et al.* [53]. PCL porous particles were prepared by the combination of phase separation (method described in section 2.1.) and lyophilization (2.4.). PCL and paraffin were dissolved in chloroform and the mixture was stirred at 25 °C to obtain a homogenous solution. The solution was then added drop-wisely into an 1% aqueous poly(vinyl alcohol) (PVA). The emulsion was stirred at 40°C so 300–1000 µm microspheres were formed. After lyophilization, the dried solid microspheres were obtained. The pore structure was controlled by adjusting the processing parameters. The surface pore size could be altered from 20 µm to 80 µm and the internal porosities varied from 30% to 70%. This porosity (pore fraction) is smaller compared to other methods of preparation (especially methods described in a paragraphs 2.7. and 2.8., where porosity was 90–98 %), but the obtained microspheres showed the very good cell adhesion and growth. Another advantage of this material is that it is usable in injectable form for tissue regeneration. Unfortunately, no information concerning the process of microspheres degradation is mentioned, but it is generally known that degradation of PCL takes place very slowly, ca. 3–4 years [54]. Although this material is porous, it cannot be considered as hydrogel, however, due to context it was included here.

It is possible to reach faster degradation by combination of PCL with polyethers (see also section 8.4.4.) For example, Zhang *et al.* [55] prepared biodegradable triblock copolymer poly(ε-caprolactone-*co*-lactide)-*block*-poly(ethylene oxide)-*block*-poly(ε-caprolactone-*co*-lactide). This physical hydrogel retained its integrity in vivo for a bit more than 6 weeks and then it was degraded due to hydrolysis.

8.4.2   Polysacharide-Based Hydrogels

Polysacharides [56–74] and peptides are probably the most used materials for preparation of porous biodegradable hydrogels. As most of polysaccharides are strongly hydrophilic or even soluble in water, it is commonly necessary to use them in crosslinked state. Due to the presence of hydroxyl groups or amino groups in polyaminosacharides, it is possible to use these groups for crosslinking reaction, for example, with aldehydes. Thus, Lou *et al.* [56] used for crosslinking reaction of carboxymethyl chitosan oxidized dextran (by sodium perchlorate) which is macromolecular crosslinker containing aldehyde groups. Because of the coexistence of abundant amino groups, hydroxyl groups and carboxylic groups associated with carboxymethyl chitosan and plentiful aldehyde groups as well as hydroxyl groups along the oxidized dextran, the Schiff's base and hydrogen bond formation arose as mixing these two aqueous polymer solutions. A highly porous structure was reached by mixing of dilute aqueous solutions of carboxymethyl chitosan and oxidized dextran and subsequent lyophilization. Degradation of hydrogel was observed by SEM and had the similar course as a dedradation of EOEMA/HPMA hydrogel crosslinking by N,O-dimethacryloylhydroxylamine [1] (Figure 8.3, section 8.3.1). Hydrogel degraded in pure PBS solution [56], but accelerating of degradation was reached by addition of lysozyme, where the hydrogel was completely cleaved in four weeks.

Oxidized dextrin crosslinked with adipic acid dihydrazide was used [58] for a preparation of degradable hydrogel nanoparticles with communicated pores for an injectable carrier of bioactive molecules. The principle of crosslinking was the reaction between oxidized OH groups (aldehydes groups) with aminogroups of hydrazide to give hydrolysable hydrazone bond similar to [56]. Despite their many favorable properties [58], these hydrogels also have some limitations. Their low tensile strength limits their use in load-bearing applications and, as a consequence, the premature dissolution or flow away of the hydrogel from the targeted local site can occur. Concerning drug delivery, the most important drawback of hydrogels relates to the quantity and homogeneity of drug loading, which may be limited, especially in the case of hydrophobic drugs; on the other hand, the high water content and large pores frequently result in relatively rapid drug release. The degradation time was several weeks.

A special type of polysacharide is hyaluronic acid, a very attractive material, which found wide uses in biomedical areas. Due to its high hydrophility and solubility in water, crosslinking is necessary to apply hydrogel preparation. Crosslinking is usually achieved by the reaction of an oxidized form of hyaluronic acid with diamines [57, 75, 76] to give a Schiff's base similar to other polysaccharides, and the common way to get a porosity is lyophilization. Another possibility is hyaluronic acid photo-crosslinking with a methacryloyl derivative of β-cyclodextrin [77], which enabled an encapsulation of hydrophobic drugs (hydrocortisone) due to its hydrophobic part.

8.4.3   Polylactide-Based Hydrogels

Polylactide is a thermoplastic linear aliphatic polyester, which is known for its fabrication from annually renewable resources (corn starch, tapioca root, sugar cane) and degradability to lactic acid, physiologically present in human organisms. This material is widely used in biomedical fields as scaffolds for tissue engineering and drug delivery systems. In a number of studies, polylactide was used for porous materials [78–81]. As polylactide is not soluble in water, no crosslinking is necessary to use it. The degradation time of porous polylactide is about 4–5 weeks and may be fine-tuned within certain range by copolymerization of glycolic acid or by enantiomeric composition of the mixture of monomers. Several techniques to reach porous structure were used. Scaffaro *et al.* [78] developed porous material with two communicating pore types, which was achieved by method 2.2.; sodium chloride particles and poly(ethylene oxide) were used as porogens. This material found application as a scaffold for tissue engineering. Polylactide with one type of pore prepared using method 2.2., described by Sanders *et al.* [79], solid porogen (NaCl particles) was used and the final material was used as a drug carrier. For substantial enhancement of specific pore volume, nanofibers prepared by electrospinning was developed [80, 81]. Although this material is porous, it cannot be considered as hydrogel, however, due to context it was included here.

8.4.4 Polyvinylalcohol-Based Hydrogels

Polyvinylalcohol is a water-soluble polymer, therefore it is necessary to crosslink it for a preparation of porous gel stable in water milieu. One way of crosslinking is a chemical reaction of hydroxyl groups, for example by boric acid [82]. However, a high concentration of hydroxyl pendant groups on polymer backbone makes polyvinylalcohol uniquely capable of being crosslinked physically, without the incorporation of any chemical additives [83–85]. The physical crosslinking process of PVA (known as the freeze–thaw process) may be applied by freezing the samples at a temperature around –30°C and subsequently thawing the samples at ambient temperature. The more freeze–thaw cycles applied to the sample, the more hydrogen bonds will be established among hydrogen and oxygen atoms in two parallel PVA polymeric chains. The formation of such bonding among polymer chains initiates crystal clusters known as "crystallites," which are randomly dispersed among an amorphous background. Degradation is based on the continuous formation of equilibrium between hydrogen bonds polymer-polymer and polymer-water with simultaneous washing out of the dissolved polymer. Featuring such a relative simple crosslinking mechanism, which avoids the necessity to use chemical crosslinkers, PVA is an attractive candidate for biomedical applications.

Chemical crosslinking is also possible. Soler *et al.* [86] described crosslinking of porous poly(vinyl alcohol), agar and poly(vinyl pyrrolidone) by γ-radiation for wound dressing.

8.4.5   Poly(ethylene oxide)-Based Hydrogels

As poly(ethylene oxide) is a water soluble polymer, it is necessary to cross-link it like polysaccharides. Several techniques are described in literature. Thus, poly(ethylene glycol) modified by amino groups on the ends of backbone was crosslinked by reaction with pyromellitic dianhydride [87]. The copolymer with imide rings on the main chain was then reacted with butanediamine to crosslinked hydrogel. Porous structure was achieved by lyophilization. Other way to reach crosslinking poly(ethylene oxide) is a modification of ends groups by acrylate and use this macromonomer as a crosslinker [88].

8.4.6   Peptide-Based Hydrogels

Several types of porous peptides were described [89–95]. A not so common strategy was used for the preparation of porous degradable hydrogels composed from gelatin and poly(acrylic acid) using γ radiation [89]. Aqueous solution of gelatin and acrylic acid (20 wt%) was exposed by γ-radiation. A crosslinked structure between gelatin and polyacrylic acid backbones was created. After washing with hot water to remove the soluble parts, hydrogel was dried, re-swollen to equilibrium state, frozen, and lyophilized. The hydrogel morphology strongly depended mainly on pH solution in which hydrogel was swollen. At pH = 1 non-porous structure was obtained, at pH = 3 non-communication pores was obtained and at pH = 7 fully communicated pores arose. Hydrogel was used as a drug carrier; ketoprofen was chosen as model drug. Unfortunately, no information about degradation was provided [89], but it is possible to assume that degradation time can be in the order of several weeks [90].

Gelatin was used also for preparation of electroactive porous degradable hydrogel based on anilin pentamer and gelatin [91]. Hydrogel was prepared from double carboxyl-capped aniline pentamer (AP) grafted to amino side groups of gelatin, and porous structure was achieved by lyophilization. Due to the presence of hydrophobic AP, an unusual structure was obtained. With an increase in the content of AP, the hydrogel gradually forms a porous structure, from common "honeycomb" to "bamboo raft," where pores had a rod shape and looked like bamboo rafts [91], see Figure 8.4.

Based on UV-visible spectroscopy and cyclic voltammetry results in aqueous solution, these hydrogels possessed electroactivity and a reversible

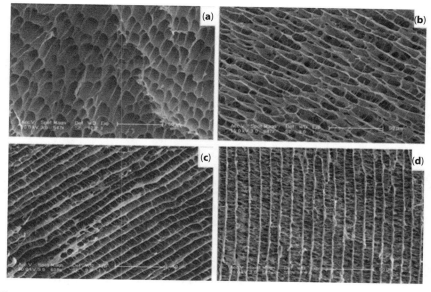

Figure 8.4 According to [91], published with permission. The scanning electron micrographs of the AP-gaft-gelatin hydrogels; (a) 0% AP, (b) 1.8% AP, (c) 4.8% AP, (d) 10.5% AP. Scale bar 50 μm.

redox property. There were three reversible redox peaks in the cyclic voltametry of the AP-graft-gelatin in aqueous solution [91].

8.5   Conclusions

Biodegradable porous hydrogels represent excellent materials for numerous applications, especially in tissue engineering applications as temporary artifical extracellular matrix. Properties and morphology of these materials may be precisely adjusted in wide range to the actual demands by changing method of preparation and choice of chemistry.

Acknowledgments

Financial support of the Grant Agency of the Czech Republic (grant # P108/10/1560) and of the Academy of Sciences of the Czech Republic (grant # M200501201) and (grant # P108/12/1538) is gratefully aknowledged.

Abbreviations

BACy	*N,N′*-Bis(acryloyl)cystamine
DHMA	N,O-dimethacryloyl hydroxylamine
DTME	dithiobis(maleinimido)ethane
EDMA	ethylene dimethacrylate
EOEMA	2-ethoxyethyl methacrylate
HEMA	2-hydroxyethyl methacrylate
HPMA	*N*-(2-hydroxypropyl)methacrylamide
MA	methacrylic acid
MaNa	sodium salt of methacrylic acid
MOETACl	[2-(methacryloyloxy)ethyl]trimethylammonium chloride
EDMA	ethylene di(methacrylate)
LM	light microscopy
LVSEM	low vacuum scaning electron microscopy
MA	methacrylic acid
MANa	sodium methacrylate
MASS	*N*-[3-(methacryloylamino)propyl] -6-{[5-({[3- (methacryloylamino) propyl]amino} carbonyl)-2-pyridyl]disulfanyl}nicotinamide
MOETACl	[2-(methacryloyloxy)ethyl]trimethylammonium chloride
PCL	polycaprolactone
PHEMA	poly(2-hydroxyethyl methacrylate)
PMMA	poly(methyl methacrylate)
PVA	polyvinylalcohol

References

1. M. Pradny, J. Michalek, P. Lesny, A. Hejcl, J. Vacik, M. Slouf and E. Sykova, *Journal of Material Science Material in Medicine*, Vol. 17, p. 1357, 2006.

2. E. O. Akala, P. Kopečková and J. Kopeček, *Biomaterials,* Vol. 19, p. 1037, 1998.

3. S. Atzet, P. Curtin, S. Trinh, B. Bryant and B. Ratner, *Biomacromolecules*, Vol 9, p. 3370, 2008.

4. M. Vetrik, M. Pradny, M. Hruby and J. Michalek, *Polymer Degradation and Stability, Vol. 96, p. 765, 2011.*

5. M. Vetrik, M. Hruby, M. Pradny and J. Michalek, *Polymer Degradation and Stability*, Vol. 96, p. 892, 2011.

6. S. Woerly, E. Pinet, L. De Robertis, M. Bousmina, G. Laroche, T. Roitback, L. Vargova and E. Sykova, *Journal of Biomaterials of Science, Polymer Edition*, Vol. 9, p. 681,1998.

7. J. Sirc, Z. Bosakova, P. Coufal, J. Michalek, M. Pradny, R. Hobzova, J. Hradil, *e-polymers*, Vol. 117, 2007

8. J. G. Omidian, K. Rocca and K. Park, . *Journal of Controlled Release.*, Vol. 102, p. 3, 2005.

9. M. Pradny, J. Fiala, J. Vacik, M. Slouf, J. Michalek and E. Sykova, *Collection of Czechoslovak Chemical Communications,* Vol. 68, p. 812, 2003.

10. M. Pradny, M. Slouf, L. Martinova and J. Michalek, *e-Polymers*, No. 043, 2010.

11. P. Lesny, M. Pradny, P. Jendelova, J. Michalek, J. Vacik and E. Sykova, *Journal of Material Science Material in Medicine*, Vol. 17, p. 829, 2006.

12. Q.H. Shi , X. Zhou and Y. Sun, *Biotechnology and Bioengineering*, Vol. 92, p. 643, 2005.

13. M. Pradny, P. Lesny, K.Jr. Smetana, J. Vacik, M. Slouf, J. Michalek and E. Sykova, *Journal of Material Science Material in Medicine,*.Vol. 16, p.767, 2005.

14. J. Michalek, M. Pradny, A. Artyukhov, M. Slouf, J. Vacik and K.Jr. Smetana, *Journal of Material Science Material in Medicine*, Vol. 16, p. 783, 2005.

15. P. Lesny, P. Jendelova, J. Michalek , J. Vacık and E. Sykova, *Collection of Czechoslovak Chemical Communications*, Vol. 10, 2003.

16. M. Pradny, *Unpublished data*, 2013.

17. F.A. Dorkoosh, M.P.M. Stokkel, D. Blok, G. Borchard, M. Rafiee-Tehrani, J.C. Verhoef and H.E. Junginger, *Journal of controlled release*, Vol. 99, p. 199, 2004.

18. J. Chen and K. Park, *Journal of Macromolecular Science:Pure and Applied Chemistry*, Vol. 36, p. 917,1999.

19. H. Park, K. Park and D. Kim, *Journal of Biomedical Materials Research Part A* , Vol. 144, p.144, 2006.

20. H. Park and D. Kim, *Journal of Biomedical Materials Research Part A* , Vol. 78A, p.662, 2006.

21. L. Shapiro and S. Cohen, *Biomaterials,* Vol. 18, p. 583, 1997.

22. H.W. Kang, Y. Tabata and Y. Ikada, *Biomaterials*, Vol. 20, p. 1339,1999.

23. F. Han, H. Zhang, J. Zhao, Y. Zhao and X. Yuan, *Polymer Engineering & Science, Vol.* 52, p. 2695, 2012.

24. M. Pradny, L. Martinova, J. Michalek, T. Fenclova and E. Krumbholcova, *Central European Journal of Chemistry*, Vol. 5, p. 779, 2007.

25. M. Vetrik, L. Kobera, M. Slouf, M. Rabyk, A. Pospisilova and Hruby M., RSC Advances, Vol. 3, p. 15282, 2013.

26. H. Kirsebom, D. Topgaard, I.Y. Galaev and B. Mattiasson, *Langmuir* Vol. 26, p. 16129, 2010.

27. M.V. Dinu, M.M. Ozmen, E.S. Dragan and O. Okay, *Polymer,* Vol. 48, p. 195, 2007.

28. N. Orakdogen, P. Karacan and O. Okay, *Reactive Functional Polymers,* Vol. 71, p. 782, 2011.

29. M.V. Dinu, M.M. Perju, E.S. Dragan, *Reactive and Functional Polymers,* Vol. 71, p. 881, 2011.

30. M.C. Gutierrez, M.L. Ferrer, F. del Monte, *Chemistry of Materials*, Vol. 20, p. 634, 2008

31. S.R. Mukai, H. Nishihara and H. Tamon, *Microporous and Mesoporous Materials,* Vol. 116, p. 166, 2008.

32. J.W. Kim, K. Taki, S. Nagamine and M. Ohshima, *Langmuir,* Vol. 25, p. 5304, 2009.

33. X.X. Hu, H. Shen, F. Yang, J. Bei and S. Wang, *Biomaterials*, Vol. 29, p. 3128, 2008.

34. J. Wu, Q. Zhao, J. Sun and Q. Zhou, *Soft Matter,* Vol. 8, p. 3620, 2012.

35. M.V. Dinu, M. Pradny, E.S. Dragan and J. Michalek, *Carbohydrate Polymers,* Vol. 94, p. 170, 2013.

36. M.S. Shoichet, *Macromolecules,* Vol. 43, p. 581, 2010.

37. P. Lesny, M. Pradny, P. Jendelova, J. Michalek, J. Vacik and E. Sykova, . *Journal of Material Science: Materials in Medicine,* Vol. 17, p. 829, 2006.

38. A. Hejcl, L. Urdzikova, J. Sedy, P. Lesny, M. Pradny, J. Michalek, M. Burian, M. Hajek, J. Zamecnik, P. Jendelova and E. Sykova,. *Journal of Neurosurgery Spine,* Vol. 8, p. 67, 2008.

39. O. Kulygin and M.S. Silverstein, *Soft Matter,* Vol. 3, p. 1525, 2007.

40. M. Pradny, M. Duskova, M. Slouf, K. Dusek and J. Michalek, *Polymer*, submitted, 2013.

41. K. Ulbrich, V. Subr, P. Podperova and M. Beresova, *Journal of Controlled Relese*, Vol. 34, p. 155, 1995.

42. P. Chivukula, K. Dusek, D. Wang, M. Duskova –Smrckova, P. Kopeckova and J. Kopecek, *Biomaterials*, Vol. 27, p. 1140, 2006.

43. K. Ulbrich, V. Šubr, L.W. Seymour and R. Duncan, *Journal of Controlled Release*, Vol. 24, p. 181, 1993.

44. E.O. Akala, P. Kopeckova and J. Kopecek, *Biomaterials*, Vol. 19, p. 1037, 1998.

45. D. Horák and O. Chaykivskyy, *Polymer Chemistry,* Vol. 40, p. 1625,2002.

46. M.H. Smith, A.B. South, J.C. Gaulding and L.L. Andrew, *Analytical Chemistry,* Vol. 82, p. 523, 2010.

47. A. Galperin, T.J. Long and B.D. Ratner, *Biomacromolecules*, Vol. 11, p. 2583, 2010.
48. M.G. Carstens, C.F. Nostrum, R. Verrijk and L.G.J. De Leede, *Journal of Pharmaceutical Sciences*, Vol. 97, p. 506, 2008.
49. S.J. Bryant, J.L. Cuy, K.D Hauch, KD and B.D. Ratner, *Biomaterials*, Vol. 28, p. 2978, 2007.
50. T. Dispinar, W. Van Camp, L.J. De Cock, B.G. De Geest and F.E. Du Prez, *Macromolecular Bioscience*, Vol. 12, p. 383, 2012.
51. H. Shou-Chen, H. Wei-Dong and L. Jian, *Journal of Polymer Science Part A: Polymer Chemistry*, Vol. 47, p. 4074, 2009.
52. L.B. Barron, K.C. Waterman, T.J. Offerdahl, E. Munson and C. Schoneich, *Journal of Physical Chemistry A*, Vol. 109, p. 9241, 2005.
53. Q.C. Zhang, K. Tan, Y.Z. Ye, Y. Zhang, W.S. Tan and M.D. Lang, *Material Science and Engineering: C. Materials for Biological Applications*, Vol. 32, p. 2598, 2012.
54. M.A. Woodruff and D.W. Hutmacher, *Progress in Polymer Science*, Vol. 35, p. 1217, 2010.
55. Z. Zheng, N. Jian, C. Liang, Y. Lin, X. Jianwei and D. Jiandong, *Biomaterials*, Vol. 32, p. 4725, 2011.
56. W.W. Lou, H.L. Zhang, J.F. Ma, D.F. Zhang, C.T. Liu, S.Q. Wang, Z.N. Deng, H.H. Xu, J.S. Liu, *Carbohydrate Polymers*, Vol. 90, p. 1024, 2012.
57. S. Nair, N.S. Remya, S. Remya and P.D. Nair, *Carbohydrate Polymers*, Vol. 85, p. 838, 2011.
58. M. Molinos, V. Carvalho and D.M. Silva, Gama, *Biomacromolecules*, Vol. 13, p. 517, 2012.
59. L. Berzina-Cimdina, D. Loca and A. Dubnika, *Proceedings of the Estonian Academy of Sciences*, Vol. 61, p. 193, 2012.
60. K. Chawla, T.B. Yu, S.W. Liao and Z. Guan, *Biomacromolecules*, Vol. 12, p. 560, 2011.
61. R. Jayakumar, K.P. Chennazhi, S. Srinivasan, S.V. Nair, T. Furuike and H. Tamura, *International Journal of Molecular Science*, Vol. 12, p. 1876, 2011.
62. K. Rinki, S. Tripathi, P.K. Dutta, J. Dutta, A.J. Hunt, D.J. Macquarrieand J.H. Clark *Journal of Materials Chemistry*, Vol. 19, p. 8651, 2009.
63. W.M. Parks and Y.B. Guo, *Materials Science and Engineering: C*, Vol. 28, p. 1435, 2008.
64. C. Chang, A. Lue and L. Zhang, *Macromolecular Chemistry and Physics*, Vol. 209, p. 1266, 2008.
65. F.M. Chen, Y.M. Zhao, H.H. Sun, T. Jin, Q.T. Wang, W. Zhou, Z.F. Wu and Y. Jin, *Journal of Controlled Release*, Vol. 118, p. 65, 2007.
66. M.V. Cabanas, J. Pena, J. Roman and M. Vallet-Regí, *Journal of Biomedical Materials Research Part A*, Vol. 78A, p. 508, 2006.
67. S. Namkung and C.C. Chu, *J. Biomater. Sci. Polym.*, Vol 17, p. 519, 2006.
68. F.M. Chen, Z.F. Wu, Q.T. Wang, H. Wu, Y.J. Zhang, YJ, X. Nie and Y. Jin, *Acta Pharmacologica Sinica*, Vol. 26, p. 1093, 2005.
69. S.E. Kim, J.H. Park, Y.W. Cho, H. Chung, S.Y. Jeong and E.B. Lee, *Journal of Controlled Release*, Vol 91, p. 365, 2003.

70. X.Z. Zhang, D.Q. Wu, G.M. Sun and C.C. Chu, *Macromolecular Bioscience,* Vol. 3, p. 87, 2003.

71. Y. Zhan and C.C. Chu, *Journal of Materials Science: Materials in Medicine,* Vol. 13, p. 667, 2002.

72. S.H. Kim and C.C. Chu, *Journal of Biomedical Materials Research,* Vol. 53, p. 258, 2000.

73. H.R. Lin, C.J. Ku, C.Y. Yang, *Materials Science Forum,* Vol. 426–4, p. 3043, 2003.

74. L. Noble, L. Sadiq and I.F. Uchegbu, *International Journal of Pharmaceutics,* Vol. 192, p. 173, 1999.

75. R. Zhang, M.Y. Xue, J. Yang and T.W. Tan, *Journal of Applied Polymer Science,* Vol. 125, p. 1116, 2012.

76. S.A. Bencherif, N.R. Washburn and K. Matyjaszewski, *Biomacromoles,* Vol. 10, p. 2499, 2009.

77. S.A. Zawko, Q. Truong and C.E. Schmidt, *Journal of Biomedical Materials Research Part A,* Vol. 87A, p. 1044, 2008.

78. S. Roberto, L. Giada, R. Salvatrice and G. Giulio, *Science and Technology of Advanced Materials,* Vol. 13, p. 045003, 2012.

79. W.G. *Sanders,* P.C. Hogrebe, D.W. Grainger, A.K. Cheung and C.M. Terry, *Journal of controlled release,* Vol. 161, p. 81, 2012.

80. T. Yang, D. Wu, L. Lu, W. Zhou and M. Zhang, *Polymer Composites,* Vol. 32, p. 1280, 2011.

81. X.G. Zhou, Q. Cai, N. Yan, X.L. Deng and X.P. Yang, *Journal of. Biomedical Materials Research,* Vol. 95A, p. 755, 2010.

82. X. Li, A. Hu and L. Ye, *Journal of Polymers and the Environment,* Vol. 19, p. 398, 2011.

83. G. Paradossi, F. Cavalieri, E. Chiessi, C. Spagnoli and M.K. Cowman, *Journal of Materials Science : Materials in Medicine,* Vol. 14, p. 687, 2003.

84. S.A. Poursamar, M. Azami and M. Mozafari,. *Colloids and Surfaces B: Biointerfaces,* Vol. 84, p. 310, 2011.

85. P. Chiarelli, A. Lanata and M. Carbone, *Materials Science and Engineering: C. Biomimetic and Supramolecular Systems,* Vol. 29, p. 899, 2009.

86. D.M. Soler, Y. Rodriguez, H. Correa, A. Moreno and L. Carrizales, *Radiation Physics and Chemistry,* Vol. 81, p. 1249, 2012.

87. J. Sun, Y. Wang, S. Dou, C. Ruan and C. Hu, *Materials Letters,* Vol. 67, p. 215, 2012.

88. A.S. Sawhney, C.P. Pathak and J.A. Hubbell, *Macromolecules,* Vol. 26, p. 581, 1993.

89. A. I. Raafat, *Journal of Applied Polymer Science, Vol.* 118, p. 2642, 2010.

90. C. Sharma, A.K. Dinda and N.C. Mishra, *Journal of Applied Polymer Science,* Vol. 127, p. 3228, 2013.

91. Y.D. Liu, J. Hu, X.L. Zhuang, P.B.A. Zhang, Y. Wei, X.H. Wang and X.S. Chen, *Macromolecular Bioscience,* Vol. 12, p. 241, 2012.

92. V. Castelletto, I.W. Hamley, C. Stain and C. Connon, *Langmuir,* Vol. 28, p. 12575, 2012.

93. N. Amosi, S. Zarzhitsky, E. Gilsohn, O. Salnikov, E. Monsonego-Ornan, R. Shahar and H. Rapaport, *Acta Biomaterialia*, Vol. 8, p. 2466, 2012.
94. T. Garg, O. Singh, S. Arora and R.S.R. Murthy, *Critical Revies in Therapeutic Drug Carrier Systems*, Vol. 29, p. 1, 2012.
95. M. Ozeki, S. Kuroda, K. Kon and S. Kasugai, *Journal of Biomaterials Applications*, Vol. 25, p. 663, 2011.

9

Hydrogels: Properties, Preparation, Characterization and Biomedical Applications in Tissue Engineering, Drug Delivery and Wound Care

Mohammad Sirousazar[1,*], Mehrdad Forough[2], Khalil Farhadi[2], Yasaman Shaabani[1] and Rahim Molaei[2]

[1]Faculty of Chemical Engineering, Urmia University of Technology, Urmia, Iran
[2]Department of Chemistry, Faculty of Science, Urmia University, Urmia, Iran

Abstract

Hydrogels are hydrophilic polymer networks that absorb large quantities of water while remaining insoluble in aqueous solutions due to chemical or physical crosslinking of individual polymer chains. Over the past few decades, advances in hydrogel technologies have spurred development in many biomedical applications, especially in tissue engineering, controlled drug delivery and wound dressing aspects due to their unique biocompatibility, flexible methods of synthesis, range of constituents, and desirable physical characteristics. The objective of this chapter is to review the properties, preparation methods, and characterization techniques of hydrogels and their biomedical applications especially in tissue engineering, drug delivery, and wound management.

Keywords: Hydrogel, biomedical application, drug delivery, tissue engineering, wound dressing

9.1 Introduction

A polymer is a natural or synthetic large macromolecule comprised of repeating units of smaller units, so-called monomers, which usually are joined by covalent or chemical bonds [1]. Natural polymers are derived

---

*Corresponding author: m.sirousazar@uut.ac.ir

Ashutosh Tiwari (ed.) Advanced Healthcare Materials, (295–358) 2014 © Scrivener Publishing LLC

from renewable resources widely distributed in nature [2]. Most commonly synthetic polymers like as polyethylene or polyfluoroethylene are non-polar and hydrophobic. However, if the monomer units contain polar or charged groups such as - OH, - CONH-, - $CONH_2$, - COOH and -SO_3H, the polymer can exhibit the ability to swell in water and retain a significant fraction of water within its structure [3]. Because of the attractive properties of these hydrophilic polymers, including high biocompatibility, biodegradability, easy availability, softness, and in most cases non-toxicity, these are suitable for biological applications [4]. Several natural polymers such as alginate, collagen, and chitosan are the biopolymers which have been extensively studied in the recent past [5–7].

Hydrogel is a blanket term for materials that are three-dimensional (3D) network structure materials derived from hydrophilic monomers, are produced by the polymerization of one or more monomers, and involve interactions such as hydrogen bonding and strong van der Waals interactions between polymeric chains [8], which have the capacity to absorb large amounts of water and swell greatly in aqueous conditions without dissolution [9]. This characteristic depends on network structure and the external environment. In the other word, the ability of hydrogels to absorb water arises from on the number of the hydrophilic functional groups which attached to the polymer backbone, while their resistance to dissolution arises from crosslinks density between the network chains [10]. According to Hoffmann [4], the amount of water present in a hydrogel may vary from 10% to thousands of times its original volume and in another report [11], hydrogels also demonstrate an extraordinary capacity (>20%) to imbibe water into their network structure. In fact, because of their swelling/deswelling properties caused by the movement of water or other biofluids, transfer of molecular and nano-scale species with water throughout the network structure of hydrogel is possible while maintaining solid-like mechanical properties. However, capillary effect and osmotic pressure are the other variables that also influence the equilibrium water uptake of hydrogels [12]. As a result of these characteristics, hydrogels can be widely used for a variety of applications such as drug delivery, biomedical and pharmaceutical fields, in the food and cosmetics industry, wound dressing, biosensors, and implantable devices in tissue engineering [13–16].

9.2 Types of Hydrogels

Natural polymers are macromolecular structures in their original state, while synthetic polymers are made from small molecules called monomers.

In contrast to natural hydrogels, synthetic hydrogel is made from carcinogenic or teratogenic monomers. Polymers are made from wide range of monomers such as vinyl acetate, acrylamide, ethylene glycol, acrylic acid (AA), and lactic acid (see Table 9.1). Hydrogels can be precisely tailored to give specific properties. Some of the monomers used in the manufacture of

Table 9.1 Monomers most often used in the synthesis of synthetic hydrogels.

Monomer	Structure
Acrylic acid	
Vinyl acetate	
Hydroxyethyl methacrylate	
Methacrylic acid	
Poly(ethylene glycol) (PEG)	
PEG methacrylate	
Glyceryl methacrylate	
Methoxyethoxyethyl methacrylate	
Hydroxyethoxyethyl methacrylate	
N-(2-hydroxypropyl) methacrylamide	

(Continued)

Table 9.1 (*Cont.*)

N-vinyl-2-pyrrolidone	
Ethylene glycol	
Methyl Methacrylate	
Dimethylacrylamide	
PEG diacrylate	
PEG dimethacrylate	
Methoxyethyl methacrylate	
Ethyleneglyc ol diacrylate	
Ethylene glycol dimethacrylate	

synthetic polymers, derived from petroleum and other monomers such as lactic acid, are extracted from natural origin such as corn and sugarcane. Also, they have a low risk of biological pathogens and they can support cellular activities with biocompatibility and biodegradability properties [17].

The researcher reports a number of classifications ways of hydrogels type, and several different points of view are presented, such as on the basis of their preparation methods, biodegradable properties, physical structure of the networks, sensitivity to surrounding environment, and also related applications [18, 19].

For example, depending on their method of preparation, hydrogels may be classified as: homopolymer (made from one type of monomer), copolymer (made from more than one type of monomer), multipolymer (more than one type of polymer), or interpenetrating polymer networks (IPNs) [20]. Homopolymer hydrogels are crosslinked networks of a single hydrophilic monomer type, whereas copolymer hydrogels are produced by crosslinking of two (or more) monomer units that at least one of which must be hydrophilic to render them swellable. Multipolymer hydrogels are produced from three or more co-monomers reacting together. Finally, IPN hydrogels are formed by preparing a first network that, after polymerization, forms a second network that is polymerized around and/or within an initial polymer network without covalent linkages between the two networks. This is typically done by immersing a pre-polymerized hydrogel into a solution of monomers and a polymerization initiator [21].

These advanced materials can also be classified into two groups based on the nature of the crosslinking reaction. If the crosslinking reaction involves formation of covalent bonds, then the hydrogels are termed as permanent hydrogel or "chemical" gels. The examples of permanent hydrogels include pMMA and pHEMA [10]. If the hydrogels are formed by the physical interactions, viz. molecular entanglement, ionic interaction and hydrogen bonding, among the polymeric chains then the hydrogels are called as physical hydrogels [4, 22]. The examples of physical hydrogels include polyvinyl alcohol-glycine hydrogels, gelatin gels and agar-agar gels. Hydrogels can also be categorized as ampholytic or ionic hydrogels based on the type of charges of their pendent groups incorporated into the gel backbone. Neutral hydrogels (no charge) such as dextran; anionic hydrogels (negative charge) such as carrageenan; cationic hydrogels (positive charge) such as chitosan; and ampholytic hydrogels (capable of behaving either positively or negatively) such as collagen, are examples from ampholytic and ionic hydrogels. The presence of ionic groups, such as carboxylic acid, along the polymer chain has a distinct effect on the solution and solid-state properties of the hydrogels. Coulombic attraction between oppositely charged sites affords formation of inter- molecular and intra-molecular ionic interactions, which can affect many of the basic properties of this class of polymeric networks, especially hydrophilicity [23]. Hydrogels are also classified based on the physical morphology of the network as amorphous, semi-crystalline, hydrogen bonded structures, super-molecular structures and hydrocolloidal aggregates [24, 25]. Polymers are mainly amorphous in nature but some of them are semi-crystalline in nature, i.e., they have both amorphous and crystalline parts. The crystalline parts in the polymers are the regions with a highly ordered structure, which act as crosslinked sites [10]. The highlighted properties of a hydrogels may vary with the change in "degree of crystallinity" and is often represented as % crystallinity or crystalline/amorphous ratio. Additionally, considering their network structures, hydrogels can be categorized in three groups as macroporous (large pores between 0.1 and 1 μm), microporous (100 and 1000 Å), or nonporous (10 and 100 Å) [26]. Another way to classify hydrogels is based on their sensitivity to surrounding environment. Hydrogels can be sensitive to some of more common external stimuli including pH, temperature, electric field, light, and so on [27]. By incorporating some stimuli-responsive co-monomers either into the backbone of the network structure or as pendant groups it is possible to prepare hydrogels with responsive properties. These hydrogels possess the ability to swell, shrink, bend or even degrade in response to a signal. These stimuli-responsive hydrogels are also called "intelligent hydrogels" and have varied applications, such as artificial muscles, controlled drug release, and sensors [28–30]. If the hydrogel is made from monomer with ionic groups, or if the hydrogel is prepared with crosslinked polyelectrolytes, which have the pendant acidic or basic groups, it displays big differences in its properties depending on the pH of the environment. The hydrogels that have the ability to respond to a change in temperature are called thermogel [31]. These types of polymers contain hydrophobic groups such as methyl, ethyl, and proyle groups [32]. Thermo-sensitive hydrogels are classified into three subgroups: positive thermo-sensitive, negative thermo-sensitive, and thermally reversible hydrogels. If the temperature increases, negative thermo-sensitive hydrogels shrink due to inter-polymer chain association through hydrophobic interactions, while positive thermo-sensitive hydrogels swell, but in the case of thermally reversible hydrogel containing non-covalently crosslinked polymer chain, a sol-gel phase transition may occur [33–35]. Electro sensitive hydrogels, as the name indicates, are another class of hydrogels. These types of gels are sensitive to electric current, which are usually made of polyelectrolytes. When the hydrogel is placed under the influence of an electric field, it undergoes shrinking or swelling [36]. Generally, the behavior of the electro responsive hydrogels depends on various parameters, including the chemical structure of hydrogels, electric field strength, the shape of the gel, position relative to the electrodes, and hydrogen ion concentration [37–39]. Light, a powerful stimulus, can be imposed instantaneously and can be delivered in specific amounts with high accuracy. As a result, light responsive hydrogel is one of the most sensitive and switchable in biological applications such as drug delivery, photo-responsive artificial muscle, and cartilage tissue engineering. Light-sensitive hydrogels can be separated into UV-sensitive and visible light-sensitive hydrogels. Unlike UV light, visible light is readily available, inexpensive, clean, safe, and easily manipulated [18].

9.3   Properties of Hydrogels

Although the mechanical stability of the hydrogel is one of the essential properties for most non-biodegradable applications, in some proposed applications for hydrogels such as in the food industry, cell culturing, super-capacitors, and sensors, considerable mechanical strength is not required [40]. In most of these applications the hydrogel acts as a scaffold, is supported by another element and does not even require sufficient strength to resist its own weight. However, in other uses, including drug delivery systems, wound dressing, actuators, and tissue engineering applications, physical and mechanical integrity of hydrogels is an important factor. To this end, the gel matrix should be able to keep its physical and mechanical strength in order to prove an effective biomaterial during the lifetime of the application, unless it has been designed to degrade. Yet, in many of these cases, the three-dimensional structure of the hydrogels must be disintegrated into harmless non-toxic products to ensure biocompatibility of the gel. Also, Food and Drug Administration (FDA) provides strict guidelines for the same depending upon the type of application. A common method for developing the strength of these materials is incorporating crosslinking agents, using co-monomers, and increasing the degree of crosslinking [41, 42].

In recent years several reports, such as topological gels, double network gels, nanocomposite gels, macro-molecular microsphere composite gels with use of layered silicates and nanoclays in the gel structure have been developed in order to design hydrogels with better mechanical properties [43–45]. Gong and co-workers have explored a class of IPNs gel with a dramatic improvement in mechanical strength and toughness, which are known as double network hydrogels [43, 46]. To be able to determine the mechanical properties of the hydrogel, several tests such as tensile testing or strip extensiometry, compression tests, bulge tests and indentation tests are recommended [47–49].

Elastic materials are assumed to deform instantaneously when stress is applied. When the stress is removed, the material completely recovers. For an ideal viscous material, the deformation is not instantaneous and is irreversible once the stress is released. The behavior of a viscoelastic material, as its name suggests, is intermediate between viscous and elastic materials.

Hydrogels are viscoelastic materials with natural (e.g., the extracellular matrix (ECM)) and manmade (e.g., contact lenses) examples. For some of common application of hydrogels like tissue engineering and food industrial applications, viscoelasticity is a main property [23]. The crosslinks between the different polymer chains results in viscoelastic and sometimes pure elastic behavior and lead to the fabrication of hardness gel in its structure, elasticity, and contribute to stickiness. This behavior is typically a result of hydrogen bonding interactions between the polymer chains facilitated by the compatible geometries of the interacting polymers. However, the concentration of polymers or their functionalized derivatives is often limited by the aqueous solubility of the gel precursors or the resulting of high viscosity of the solutions, as well as the concentration can be increased when lower molecular weight gel precursors are used [50]. Such modifications can improve the mechanical properties and viscoelasticity of these compounds for biomedical applications, for instance, increasing the total number of electrostatic interactions in this system and creating a material with novel viscoelastic properties [51].

As mentioned earlier, hydrogels are 3D polymeric networks that can absorb water and hold a large amount of water while maintaining their structure. The good biocompatibility of hydrogels emanates from their higher water content (over 70%), which causes them to become soft and to take on elastic properties (i.e., the water acts as a plasticizer). But it is also a limitation to the mechanical properties and therefore the applications [52]. Their ability to swell in a suitable solvent (commonly water) is mainly depending on many factors, such as crosslinking density, hydrophilicity of polymer chains, and the solvent nature. Network swelling and de-swelling depends on the hydrophilicity of the polymer chains that are chemically crosslinked or physically entangled, and also depends on the crosslinking density. The swelling and shrinking properties of hydrogels are currently being exploited in a number of applications including control of microfluidic flow, muscle-like actuators, filtration/separation, and drug delivery [53]. In fact, on a molecular level, water in a hydrogel either bonds to polar hydrophilic groups as "bond water" or fills the space between the network chains, pores, or voids as "free water" but does not dissolve when water or a solvent enters it under physiological conditions [11]. Anionic and cationic hydrogel networks having ionized pendant groups as a hydrophilic chain can be responsive to the pH and ionic strength of the swelling medium. By increasing of the number of the hydrophilic groups, the water-holding capacity is increased, while with an increase in the crosslinking density a decrease in the equilibrium swelling is observed due to the decline in the hydrophilic groups. As the crosslinking density increases, a subsequent increase in the hydrophobicity and a corresponding decrease in the stretchability of the polymer network occurs [10]. According to Donnan equilibrium swelling theory, the osmotic pressure gradient or ion concentration gradient between the interior of the polymer backbones and surrounding solution of the hydrogel is the driving force for swelling or shrinking response [54]. The swelling process distends the network, and is counteracted by the elastic contractibility of the stretched polymer network. Hence, the swelling pressure of non-ionic hydrogels is the result of the imbibition of solvent driven by an osmotic pressure, counteracted by the contractibility of the network, which tends to expel the solvent resulting in the equation:

$$P = \pi - p \qquad (9.1)$$

Where P is the swelling pressure of non-ionic hydrogels, π is the osmotic pressure and p is the elasticity of the stretched polymer network. At an equilibrium state, $\pi = p$, and the swelling pressure is zero ($P=0$).

Other factors that can affect water absorption are mesh size, elasticity, temperature, charge density, pKa, and pKb values [55]. The mesh size is defined as the space between macromolecular chains in a crosslinked network and swelling depends on hydrogel mesh sizes. With increasing mesh size, water uptake is increased [56]. In acidic hydrogels with pendent groups such as carboxylic or sulfonic acid, de-protonation occurs when the environmental pH is above the pKa value, which leads to the ionization of the pendent groups and increases swelling of the hydrogel. On the other hand, in basic hydrogels containing pendent groups such as amine group, ionization takes place below the pKb value and this phenomenon increases the swelling due to an increase in electrostatic repulsions [57, 58]. The influences of the fixed charge density on hydrogel swelling are also studied by several researchers [53, 59]. The published results show that with increasing the fixed charge density, the degree of swelling increases at higher pH(s) [53, 60].

As already mentioned, when the hydrogel is exposed to the solvent, it swells until it reaches its equilibrium swelling state. Coincides the absorption of the solvent, solutes entered into the hydrogel pores which this characteristic is used in various applications, especially in medical usage and drug delivery systems. As the gel swells, the crosslinked chains widen, and thus the mesh size of network increases, allowing transformation of the solute to the gel structure. Two main factors that can affect the diffusion characteristics of solutes through hydrogels are the crosslinking density of the polymer (the nature of the crosslinks) and the arrangement of water within the hydrogel (crudely represented by the degree of swelling) [61].

It has been proven that it is reasonable to assume that the free volume of water present in the hydrogel is available for diffusion of water-soluble solutes [62, 63].

The water content of hydrogels is calculated after the equilibrium swelling by the following equation:

$$\text{Water content} = (\text{Wet weight} / \text{Dry weight}) \times 100\% \qquad (9.2)$$

The degree of swelling (or swelling ratio) is directly related to the amount of solute which transfers to the hydrogel network structure. This parameter can be calculated as the following:

$$\text{Degree of swelling} = [(\text{Wet weight} - \text{Dry weight}) / \text{Dry weight}] \times 100\% \qquad (9.3)$$

Crank and Park [64] reported that the Fick's Laws for diffusion explain the release of water-soluble solutes uniformly distributed in the matrix of a fully hydrated hydrogel into aqueous sink.

If solutes cannot enter the gel significantly, their presence in the surrounding medium causes a disturbance in the swelling equilibrium, probably due to the increased osmotic value of the outside medium. This leads to a decrease in water content, so to de-swelling of the gel.

Additionally, the mesh size of the gel can be determined from the swelling experiments. The relative weight of the hydrogel, W_R, is evaluated according to the following equation:

$$W_R = W_{24h}/W_0 \qquad (9.4)$$

Where W_0 represents the weight of the hydrogel as soon as it was prepared.

Although in some cases, slow swelling is beneficial for many applications, there are many situations where a fast swelling polymer is more desirable. Therefore, a new generation of hydrogels has been developed which swell and absorb water very rapidly [65].

Many factors such as degree of crosslinking, solvent composition, hydrogen bonding, etc., can affect the swelling kinetics of the hydrogels. Several transient models to evaluate the swelling kinetics, and for diffusion of solvents into hydrogels with varying degrees of complexity, have been developed [66–69].

9.4   Preparation Methods of Hydrogels

It is well known that crosslinked networks of hydrogels are fabricated with various techniques including chemical crosslinking, physical crosslinking, grafting polymerization, and radiation crosslinking. In most cases, it seems that crosslink agents must be present in order to avoid dissolution of the hydrophilic polymer chain in aqueous solution. On the nature of the links electrostatic interactions, hydrogen bonding, donor–acceptor, van der Waals forces, or even metal–ion coordination can be described for the association of polymer-polymer, polymer-drug, or polymer-bioactive components [70]. Moreover, the physicochemical properties of hydrogels such as mesh size, shape, the swelling and permeability characteristics of the gel for various applications depend on the extension of these bonds. The general procedures for the production of chemical and physical gels are briefly described below.

9.4.1   Physical Methods

One of the major disadvantages of synthetic polymers is the usage of crosslinking agents in their structure, which must be removed before application due to how these agents affect the integrity of substances to be entrapped (e.g., cell, proteins, etc.) as well as their toxicity properties [71]. Physically crosslinking keeps a network gel by the formation of non-covalent crosslinks, hence, there is no need to use the toxic crosslinker. Especially, physically crosslinked gels have attracted particularly increased interest because of the relative ease of production, and also this kind of gel exhibits a reversible sol-gel transition depending on temperature, contributing to a large extent to the development of these materials for biomedical and pharmaceutical applications. The mechanical strength of physically crosslinked gels is generally low and this kind of gels deforms easily under stress without regaining their former shape [72]. Some of the most widely used physical methods to prepare hydrogel are summarized below.

9.4.1.1   *Crosslinking by Ionic Interactions*

Ionic interactions as a crosslinking reaction can be used to produce polymeric gels by the addition of di- or tri-valent counter-ions. Crosslinking can be carried out at normal temperature and pH(s). For example, ionic interactions cause crosslinking of Alginate and Chitosan by potassium ions [73, 74] (see Figure 9.1). In addition, anionic polymers crosslinked with metallic ions can also be obtained by complexation of polyanions and polycations [75, 76].

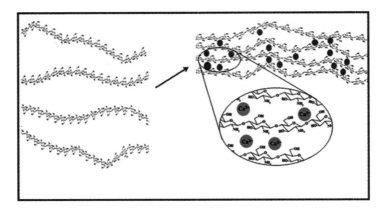

Figure 9.1 Ionotropic gelation by interaction between alginate and divalent metal ions (Ca^{2+}).

9.4.1.2   Crosslinking by Hydrogen Bonds

H-bonded hydrogel can be obtained by hydrogen bonding between the oxygen atom of monomer and the carboxylic acid group of other components. This also implies that the hydrogen bonds are only formed when the carboxylic acid groups are protonated and the swelling of these gels is pH dependent (See Figure 9.2). Examples of such hydrogels that formed by hydrogen-bonded have been reported [70, 77, 78]. Recently, researchers have shown that the mixed system molecular interaction (two or more molecules) causes a change in matrix structure due to intermolecular hydrogen bonding between them, which leads to the formation of insoluble hydrogel networks [79].

9.4.1.3   Crosslinking by Heating/Cooling

Physical crosslinked hydrogels can be prepared by subjecting semi-dilute aqueous solutions of polymer chains to successive heating-cooling cycles between below and above the melting temperature of the polymer solution. In the first step, heating-cooling cycles cause the coil conformation of chains, and in the second step, by continuing this heating and cooling cycle in the presence of salt (K^+, Na^+, etc.), double helices are aggregated further to form stable gels due to screening of repulsion of side group on the polymer chain [80] (See Figure 9.3). In some cases, hydrogel can also be obtained by simply warming the polymer solutions that cause the block copolymerization [81]. Some of the examples are Polyethylene oxide (PEO) and polyethylene glycolpolylactic acid (PLA) hydrogels [4, 71].

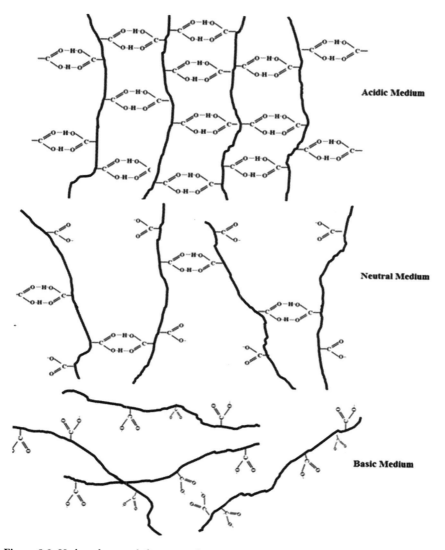

Figure 9.2 Hydrogel network formation due to intermolecular H-bonding at various pH(s).

9.4.1.4   Crosslinking by Crystallization

The choice of polymer is crucial in this regard as it should possess a high degree of tailorability for the specific application of interest. A number of techniques to produce stable hydrogels that are made through freezing-thawing cycles by the presence of crystalline regions are creating suitable properties for many applications in various medical areas. As seen

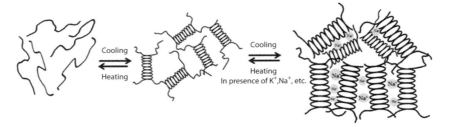

Figure 9.3 Gel formation due to aggregation of helix upon cooling a hot solution of monomer chains.

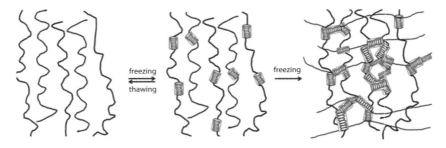

Figure 9.4 Schematic of the formation of hydrogels by crystallization process.

in Figure 9.4, upon freezing of the solution, ice forms in the amorphous region and polymer crystallites grow until they reach the facets of the other crystallites which assist in the formation of a porous network upon thawing. This is a compelling reason for this fact that these hydrogels possess better strength and stiffness than their chemically crosslinked counterparts, as the mechanical load is distributed along the crystallites in the network [82]. Examples of this type of gelation are freeze-thawed gels of polyvinyl alcohol (PVA) and xanthan [83, 84].

9.4.1.5   Crosslinking by Maturation

Arabinogalactan protein (AGP), arabinogalactan (AG), glycoprotein (GP), etc., are available in gums. If heat treatment is applied to these compounds, aggregation occurs and leads to an increase in the molecular weight of these proteinaceous components, which subsequently produces a hydrogel form with enhanced mechanical properties and water binding capability. The molecular changes that accompany the maturation process demonstrate that a hydrogel can be obtained with precisely structured molecular dimensions. The controlling feature is the agglomeration of the proteinaceous components within the molecularly disperse system that is present in of the naturally occurring gum [85].

9.4.2   Chemical Methods

Whereas physical hydrogels are formed by secondary interactions (such as hydrogen bonds, ionic bonds, hydrophobic interactions, and crystallites, etc.), chemical hydrogels consist of irreversible covalently crosslinked network. Chemical crosslinking is a direct reaction between linear polymer or branches and at least a bifunctional component with small molecular weight, called a crosslinking agent or crosslinker. This component links the polymer chains with its functional groups such as -OH, -COOH, -NH . A number of methods reported in literature to obtain chemically crosslinked permanent hydrogels are summarized below.

9.4.2.1   Crosslinking of Polymer Chains

This technique mainly involves the introduction of new molecules between the low molecular weight monomers, branched homopolymers, or polymeric chains to produce crosslinked chains. Most hydrophilic polymers have pendant hydroxyl groups, therefore such agents like aldehydes, maleic acid, oxalic acid, dimethylurea, diisocyanates, etc., that condense when organic hydroxyl groups are used as crosslinking agents. Glutaraldehyde and epichlorohydrin are two popular crosslinking agents that are used to prepare hydrogels used in biocompatible applications [86] (See Figure 9.5). This reaction is mostly carried out in solution and the solvent proposed for this reaction is usually water, but methanol, ethanol, and benzyl alcohol have also been used. These organic solvents can be

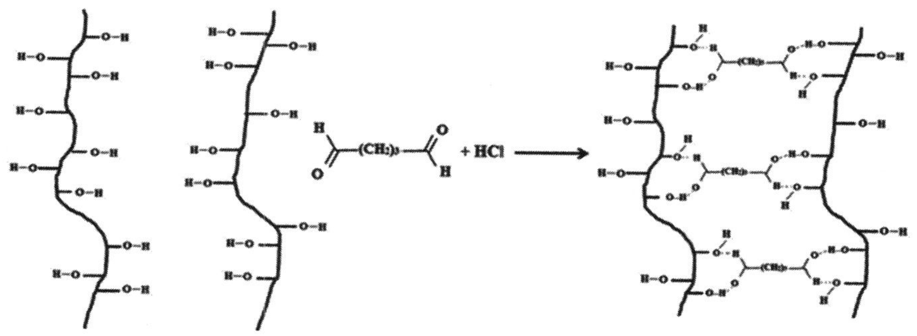

Figure 9.5 Schematic illustration of using chemical crosslinker to obtain crosslinked hydrogel network.

used only for the formation of network structure, and then the solvent can be exchanged with water.

9.4.2.2   Grafting

This method involves the polymerization of a monomer on the backbone of a preformed polymer. For this purpose, the polymer chains are activated by the action of chemical reagents, or high-energy radiation treatment. In general, initiators are used for activation as chemical reagents. The most used initiators in these reactions are radical and anionic initiators. Various initiators are used, such as Azobisisobutyronitrile (AIBN), benzoyl peroxide, etc. The growth of functional monomers on activated macro-radicals leads to branching, and further leads to crosslinking. Solvents can be added during the reaction to decrease the viscosity of the solution [87].

Unsaturated compounds polymerization can also be initiated by the use of high-energy radiation such as gamma and electron beam radiation. Water-soluble polymers derivitized with vinyl groups can be converted into hydrogels using high-energy radiation. The proposed mechanism for polymerization of gels by using high-energy radiation is in following manner: when the aqueous polymer solution is exposed to gamma or electron beam radiation, radicals are formed on the polymer. However, the presence of water promotes the diffusion of macro-radicals to combine and form a crosslinked hydrogel network. This implies that the radiolysis of water molecules has occurred and generates the hydroxyl radicals. The formed radical can attack to another polymer chain and lead to the formation of macro-radicals. Recombination of these macro-radicals on different sites, leading to the formation of covalent bonds, and finally crosslinked structure is obtained [34, 88]. During radiation, atmospheric oxygen can cause problems in the polymerization process, and as a result, radiation is performed in a nitrogen or argon atmosphere. The swelling and permeability characteristics of the gel depend on the extent of polymerization, a function of polymer and radiation dose (in general crosslinking density increases with increasing radiation dose) [89]. Polymers without additional vinyl groups can also be crosslinked via radiation. Several hydrogel preparation reports have been published that use this method to fabricate hydrogel [90, 91]. Irradiation of hydrocolloids in solid state has also been studied. During the solid-state radiolysis of hydrocolloids, scission of glycosidic bond is the dominant reaction, which eventually leads to a decrease in the molecular weight of macromolecules. In this technique, the degradation and formation of radical rates depend on the concentrations of reactants, presence of substituted group, molecular weight, and purity of hydrocolloid [92].

Performing in water under mild conditions without the use of a crosslinking agent, formation of relatively pure and residue-free hydrogels are the main advantages of using this process for gel formation. However, there are some drawbacks to using this method, namely the bioactive material must be loaded after gel formation, as well as irradiation, which might damage the agent.

9.4.2.3   Crosslinking Using Enzymes

Recent studies have shown that enzymes can be used for the synthesis of hydrogels [93]. Sperinde *et al.* [94] have reported the PEG hydrogels that they crosslinked by enzyme. First, a tetrahydroxy PEG was functionalized with the addition of glutaminyl groups and then by the addition of trans-glutaminase into the solution of PEG and poly(lysine-cophenylalanine), networks were formed. The proposed mechanism is based on the enzymatic catalyzed reaction between γ-carboxamide group of PEG and the ε- amine group of lysine to form an amide linkage between polymers. They also showed that the physical property of this kind of hydrogels depends on ratios of PEG and lysine.

9.5   Characterization of Hydrogels

In order to design hydrogels with the desired performance and structure, determination and characterization of hydrogel network parameters are of great significance. The common techniques discussed below were chosen from a larger body of methods that provide an insight into the structure of the hydrogels. Infrared spectroscopy, x-ray diffraction analysis, atomic force microscopy, electron microscopy, and many other techniques have been described to characterize the hydrogel structure.

9.5.1   Infrared Spectroscopy

Infrared spectroscopy is one of the most widely used techniques for identifying the chemical structure of polymers that is used to measure vibrational energy transitions, yielding information about the types of chemical bonds, the atoms involved, and the local chemical environment present within a material and the surface of it [95]. This technique involves many advantages, especially flexibility in terms of sampling methods [96]. The measurement mode is selected depending on what is suitable for the sample in question. The most common sampling methods include transmission (semi-thin films, and general analysis of chemical compounds), external reflection (thin films on reflecting substrates), reflection-absorption (extremely thin films on metal substrates), internal reflection (thick, soft materials and thin films prepared on such substrates, and "wet" measurements), and diffuse reflection (powders and poorly reflecting materials) [96].

9.5.2   X-Ray Diffraction Analysis (XRD)

Polymers are mainly amorphous in nature but some of them that were prepared by using freeze/thaw cycles have crystalline parts in nature that are known as semi-crystalline. The semi-crystalline polymers can be identified and characterized by XRD [97]. When the sample is being irradiated with a beam of monochromatic x-rays, a part of light is diffracted. In the XRD technique, these diffracted rays are analyzed based on angles of diffraction and the intensity of the diffracted rays to give information about structural make-up, % crystallinity and crystallite dimension, interplanar atomic spacing (d-spacing), orientation, and strains present in the polymer/ polymer blend matrix. Gels with different "degree of crystallinity" display vary in mechanical and chemical properties, which are often represented as crystallinity percentage or crystalline/amorphous ratio. The degree of crystallinity also can be determined easily with the aid of XRD [82].

9.5.3   Nuclear Magnetic Resonance (NMR)

Another common technique for the investigation of polymers and hydrogels is NMR [98]. Various modes of NMR (H-NMR, C-NMR and pulsed field gradient NMR) were applied to a survey of hydrogels in literature. H-NMR measurements were used to identify functional groups of monomer and copolymer composition and to determine the final double bond conversion at the end of the polymerization [99]. Also, completion of the polymerization process and its mechanism can be verified by H-NMR spectroscopy [100]. The proton NMR gives information about the interchange of water molecules between the so-called free and bound states [99]. Keeping this fact in mind, the pulsed field gradient NMR spectroscopy is a valuable tool for the characterization of hydrogel-based drug delivery systems [101].

9.5.4   Atomic Force Microscopy (AFM)

The monitoring of the surface map of materials can generally be done in two ways: contact profilometers or non-contact profilometers. AFM is a contact profilometer, which has opened whole new horizons for explaining biomaterial as it is able to capture images of non-conducting materials [102]. AFM is a useful and routine technique that can provide information about surface properties of the hydrogel with creates a topographic image of a hydrogel surface [103, 104]. Non-destructivity is the main advantage of this method.

9.5.5   Differential Scanning Calorimetry (DSC)

Another technique that can be widely used for hydrogel characterization is DSC. This technique is a sensitive method to measure transitions of polymers as a function of temperature through the changes in heat capacity and study of glass transition temperature, crystal structure and crystal transition temperature. Generally this method can be used to investigate the crystalline nature of the hydrogels, especially the hydrogels that were prepared by freezing-thawing processes. Also, the DSC method can be applied to determine the degrees of crystallinity and crystal size distributions of samples in the initial state (before swelling) and at various times during swelling. For more information about how to apply this technique to characterization of hydrogels, see references [83, 105].

9.5.6   Electron Microscopy

Optical methods have been applied to determine the microstructure of hydrogels, yielding a three-dimensional image of the structure. Electron microscopies, particularly scanning electron microscopy (SEM) and transmission electron microscopy (TEM), are two methods often used in the analysis of gel dispersions [106]. From the electron microscopy micrographs, a visual evaluation can be made of the size, shape and size distribution of the particles. The main feature of electron microscopy is its ability to directly observe inter-particle bridge formation [107], and in some cases [108], anomalous particle formation and presence of a fraction of smaller particles resulting from secondary nucleation. However, most other techniques do not have this advantage. Nevertheless, this technique cannot be used to view in the aqueous phase. Thus the subjected material should be dry, and therefore can be observed in the collapsed state only. Beside, these methods are not suitable for accurate measurement of the swollen diameters. Also, swollen gel may undergo aggregation under conditions of reduced pressure and electron beam irradiation [109]. It has been proved that SEM can be applied to the study of network morphology [110].

9.5.7   Chromatography

Chromatography is a physicochemical process that enables the separation of the components of a mixture. Size Exclusion Chromatography (SEC), also called gel filtration chromatography (when the mobile phase is aqueous), and gel permeation chromatography (GPC) (when the mobile phase is organic), is widely used for the determination of polymer molecular weight and molecular weight distribution (MWD) or polydispersity index (PDI) [111]. On the other hand, the process of crosslinking was monitored using GPC in concert with multi-angle laser light scattering (GPC-MALLS) [112]. Also, GPC-MALLS technique is widely used in quantifying the hydrogels of several hydrocolloids such as gum arabic, gelatine, and pullulan [113]. Al-Assaf *et al.* [113] have demonstrated that the data obtained from this technique can be used to assess the exact amount of the hydrogel.

9.5.8   Other Techniques

As discussed above, a wide variety of methods have been suggested for the characterization of the morphology and the thermodynamic properties of hydrogels. These techniques could be applied to investigation of the size, shape, and the molecular weight of the hydrogels. These methods with some of the corresponding references are tabulated in Table 9.2. A combination of different methods can provide a total "image" of the characteristic features of the hydrogels.

9.6   Biomedical Applications of Hydrogels

Potential applications of hydrogels include the food industry, cosmetics, agriculture, technical and electronic instrumentation, photography, medicine, and pharmacy, especially in the fields of drug delivery, wound dressing, artificial skin and tissue engineering, as well as other applications such as destabilizing agents, biosensors, synthetic ECM, implantable devices, separation systems, phospholipid bilayers, nano-reactors, smart microfluidics, molecular filtration, and energy-conversion systems [4, 34, 127–134]. To date, the growth of hydrogels applications has mostly occurred in the biomedical area and pharmaceuticals with more attention to different technologies. As discussed, having a porous network structure in the hydrogel is ideal for several in vivo applications. Often we can control the porosity of hydrogels by controlling the density of crosslinks or by

Table 9.2 Experimental techniques for hydrogel characterization.

Experimental technique	Measurement	Ref.
Light Scattering	Particle molecular weight	[114]
Photon Correlation Spectroscopy	Hydrodynamic size	[115]
Ultracentrifugation	Average molecular weight	[116]
Conductometric and Potentiometric titration	Surface charge	[117]
Small Angle X-ray Scattering (SAXS)	Internal structure	[118]
Small Angle Neutron Scattering (SANS)	Internal structure	[119]
High Sensitivity DSC	Thermodynamic properties	[120]
Turbidimetric methods	Stability	[121]
X-Ray Photoelectron Spectroscopy	Surface composition	[122]
Bohlin Rheometer	Rheological measurements	[123]
Confocal Laser Scanning Microscopy	Structure	[124]
Environmental SEM	Water capacity	[125]
Energy Dispersive X-ray Spectrometry	Elemental compositions	[126]

changing the swell affinity of hydrogels. The porosity property of hydrogels helps the release of drugs from hydrogels, and the release of drugs from hydrogels can be controlled by controlling the diffusion coefficient of drugs. Hence, pharmaceutical companies have developed different dosage forms of drugs with higher rate of drug release into systemic circulation from dosage forms. Hydrogels, tablets, capsules, injections, microspheres, suspensions, emulsions, nanoparticles, and transdermal patches are some examples of dosage forms with their subcategories for different routes of administration. Utilization of the existing resource of marketed and patented drug substances with known therapeutic effects, and modification of their pharmaco-therapeutic characteristics by incorporation of suitable drug delivery systems has been the target of recent pharmaceutical developments. Hydrogels have been used widely in the development of the smart drug delivery systems. Also, these compounds have been given attention as excellent candidates for controlled release devices, targetable devices, or bio-adhesive devices of therapeutic agents that can be used for oral, rectal, ocular, epidermal, and subcutaneous application. The most important hydrogels from a biomedical point of view are those sensitive to the temperature and/or pH of the surroundings. These materials, as mentioned before, are known as "stimuli-responsive" or "smart" gels and can undergo abrupt volume changes in response to small changes in environmental parameters. Their ability to swell or de-swell according to external conditions leads to a drug release profile that varies with the same specific parameters. Especially, pH-sensitive hydrogels have been most frequently used to develop controlled release formulations for oral administration and development of biodegradable drug delivery systems [135]. On the other hand, hydrogels have played a significant role in wound care systems and have successfully treated patients with skin damage related to incontinence. Furthermore, their biocompatibility, ease of fabrication, and visco-elastic properties make them highly suitable for use as constructs to engineer tissues as well as other biomedical applications.

Hydrogels, due to their flexible methods of synthesis, unique biocompatibility, range of constituents, and desirable physical characteristics, have been the material of choice for many applications in regenerative medicine. Over the past 50 years, hydrogels have been extremely useful in biomedical and pharmaceutical applications mainly due to their high capacity for water absorption, their biodegradable nature, consequent biocompatibility, and rubbery nature, which is similar to natural tissue. There is an efficient range of applications for hydrogels in the biomedical fields and an almost endless number of combinations of cells, scaffolds, and growth factors that can be adjusted for each specific need. Successful examples from hydrogel biomedical applications include soft contact lenses, wound dressings, super absorbents, and drug-delivery systems. The most recent and exciting biomedical applications of hydrogels are cell-based therapeutics [136] and soft tissue engineering. Hydrogels formed through self-assembly have attractive potential in wound treatment, tissue engineering, and drug release. In many of these applications, being able to understand and control the rheological properties of hydrogels is paramount to their effectiveness. Control of physical and biological properties of hydrogels is essential for their biomedical applications. For example, it has been shown that the rigidity of scaffold materials acts as an extracellular signal and plays a critical role in regulating cell adhesion, migration, spreading, and even survival. Anyway, in cell culture applications the hydrogel needs to be rigid enough to be self-supporting and have the correct stiffness to facilitate cell-adhesion. Out of all these applications, at the forefront of biomedical applications are hydrogel-based drug delivery and wound dressing devices. Synthetic hydrogels provide an effective and controlled way in which to administer protein and peptide-based drugs for treatment of a number of diseases. A successful drug delivery device relies not only on competent network design, but also on accurate mathematical modeling of drug release profiles. Hydrogels with well-defined chemistries yield well-defined physicochemical properties and easily reproducible drug release profiles. The network structure of hydrogels plays a key role in mesh size, stability, and diffusion behavior of incorporated drug. These polymers have thus become a premier material used for drug delivery formulations and biomedical implants, due to their biocompatibility, network structure, and molecular stability of the incorporated bioactive agent. The pioneering work on crosslinked hydrogels was done by Wichterle and Lim in 1954 [137]. From their research, and discovery of the hydrophilic and biocompatible properties of hydrogels, there emerged a new class of hydrogel technologies based on biomaterial application. Later, natural polymers such as collagen and shark cartilage were incorporated into hydrogels as wound dressings. Moreover, hydrogels now play an impressive role in tissue engineering scaffolds, biosensors and Bio-MEMS devices. Among these applications, hydrogel-based drug delivery devices have become a major area of study, and several commercially available products are already in the market [138]. Proteins, peptides, and DNA-based drugs can all be delivered via hydrogel carrier devices. The most important properties of hydrogels such as flexibility and hydro-philicity make them ideal for use as drug delivery matrix. Since hydrogels show good compatibility with blood and other bodily fluids, they are used as materials for contact lenses, burn wound dressings, membranes, and as coating applied to living surfaces. Both natural and synthetic hydrogels have applications as burn wound dressings [139], encapsulation of cells [140], and recently are being used in the new field of tissue engineering as matrices for repairing and regenerating a wide variety of tissues and organs as well as being considered as ideal matrices wound treatment [141].

9.6.1 Tissue Engineering

Tissue engineering is an interdisciplinary field that applies the principles of engineering and life sciences towards the development of biological substitutes that restore, maintain, or improve tissue function or entire organs

[142]. Inasmuch as hydrogels are made mostly of water and natural or synthetic biocompatible polymers, they are often biologically compatible. Accordingly, most hydrogels exhibit good compatibility when seeded with cells or when implanted in vivo. It is widely expected that a majority of future tissue engineering techniques will be based on hydrogel technology. No other class of materials has the flexibility or biological compatibility to enable significant advances in tissue engineering. For example, micro-scale hydrogels let engineers precisely control the cellular microenvironment, which may lead to significantly more effective stem cell differentiation techniques. Additionally, the ability to create micro-patterned, vascularized, cell-laden hydrogel scaffolds will enable more effective tissue engineering therapies. Finally, the versatile chemical and mechanical properties of micro-scale hydrogels provide a unique platform for the future development of more accurate and effective biological sensors and micro-devices. Langer and Vacanti [142] were among the first to elucidate the basic techniques used in tissue engineering to repair damaged tissues, as well as the ways polymer gels are utilized in these techniques. To date, hydrogels in regenerative medicine have been used as scaffolds to provide structural integrity and bulk for cellular organization and morphogenic guidance, to serve as tissue barriers and bio-adhesives, to act as drug depots, to deliver bioactive agents that encourage the natural reparative process, and to encapsulate and deliver cells. Tissue engineering aims to replace, repair, or regenerate tissue or organ function and to create artificial tissues and organs for transplantation [142]. Scaffolds used in tissue engineering mimic the natural ECM and provide support for cell adhesion, migration, and proliferation. They also allow for differentiated function, new tissue generation, and its 3D organization. Of course, scaffolds need to be completely biodegradable so that after tissue is grown, the resulting structures are made entirely from biological components. Hydrogels' high water content, their biocompatibility, and mechanical properties that resemble natural tissues make them particularly attractive for tissue-engineering applications. By adding cells to a hydrogel before the gelling process, cells can be distributed homogeneously throughout the resulting scaffold. Cells have been encapsulated in both natural hydrogels, such as collagen and fibrin materials, as well as in synthetic hydrogels made from PEG. Also, combinations of natural and artificial polymers can be used to provide proper scaffold degradation behavior after implantation. Fibroblasts, vascular smooth muscle cells, and chondrocytes successfully immobilize and attach to these hydrogel scaffolds. With a combination of micro-fluidic channel technology and photo-patterning of hydrogels, these scaffolds can facilitate increased growth-factor delivery and shape sculpting that is only

limited by its molded housing [143]. Desired characteristics of hydrogels include physical parameters such as mechanical strength and degradability, while biological properties include biocompatibility and the ability to provide a biologically relevant microenvironment.

9.6.2   Drug Delivery

Macromolecular drugs, such as proteins or oligonucleotides that are hydrophilic, are inherently compatible with hydrogels. By controlling the degree of swelling, crosslinking density, and degradation rate, delivery kinetics can be engineered according to the desired drug release schedule. Furthermore, photo-polymerized hydrogels are especially attractive for localized drug delivery because they can adhere and conform to targeted tissue when formed *in situ*. Drug delivery aspects in hydrogels may be used to function simultaneously with the barrier role of hydrogels to deliver therapeutic agents locally while preventing post-operative adhesion formation. These 3D network structures possess several properties that make them an ideal material for drug delivery. First, hydrogels can be tailored to respond to a number of stimuli [34]. In this case, researchers have dedicated much attention to the stimuli-responsive and environment-sensitive hydrogels. In these systems, a polymeric matrix can protect drugs from hostile environments (such as low pH and enzymes), while controlling drug release by changing the gel structure in response to environmental stimuli. The stimuli that hydrogel respond to may be physical or chemical. Drug release is triggered by various mechanisms as described in Table 9.3 [36]. Hydrogels also can exhibit dramatic changes in their swelling behavior, network structure, permeability, or mechanical strength in response to different internal or external stimuli. This enables sustained drug delivery corresponding to external stimuli such as pH or temperature. External stimuli are produced with the help of different stimuli-generating devices, whereas internal stimuli are produced within the body to control the structural changes in the polymer network and to exhibit the desired drug release [144].

These sensitive gels are useful in oral drug delivery as they can protect proteins in the digestive track.

Second, hydrogels can also be synthesized to exhibit bio-adhesiveness to facilitate drug targeting, especially through mucus membranes, for non-invasive drug administration [146]. Finally, hydrogels also have a stealth characteristic in vivo circulation time of delivery device by evading the host immune response and decreasing phagocytic activity [147]. The mechanism of hydrogel swelling is one of the most important factors

Table. 9.3 Various stimuli used for triggering drug release and related mechanisms of release from hydrogels and their potential applications [36, 145].

Environmental stimuli	Hydrogel	Mechanism	Applications
Electrical signal	Polyelectrolyte hydrogel	Change in charge distribution causes swelling and drug release	Actuator, artificial muscle, on/off drug release.
Thermal	Thermo-responsive hydrogel, e.g., poly (Nisopropylacrylamide)	Change in polymer–polymer and polymer–water interactions cause swelling and drug release	On/off drug release, squeezing device
pH	Acidic or basic hydrogel e.g. PAA, PDEAEM	Ionization of polymer chain upon pH change; pH change causes swelling and release of drug	pH-dependent oral drug delivery
Magnetic fields	Magnetic particles dispersed in the hydrogel matrix	Applied magnetic field causes pores in gel and swelling followed by drug release	Controlled drug delivery while the magnetic particles used form medical therapy.
Ionic strength	Ionic hydrogel	Change in concentration of ions inside the gel cause swelling and release of drug	Biosensor for glucose, used for medical therapy
Chemical	Hydrogel containing electron-accepting groups e.g. Chitosan -PEO	Formation of charge-transfer complex cause swelling and release of drug	Controlled drug delivery

Environmental stimuli	Hydrogel	Mechanism	Applications
Enzyme–substrate	Hydrogel containing immobilized enzymes	Product of enzymatic conversion causes swelling and release of drug	Modulated drug release in the presence of a specific antigen; sensor for immunoassay and antigen.
Light	Copolymer of PNIPAAm	Temperature change via the incorporated photosensitive molecules; dissociation into ion pairs by UV irradiation	Optical switches, ophthalmic drug delivery

in drug release phenomena. This mechanism of drug release occurs when diffusion of an active agent is faster than hydrogel swelling [34, 14, 149]. Researchers have engineered their physical and chemical properties at the molecular level to optimize their properties, such as permeability, environ-responsive nature, surface functionality, biodegradability, and surface bio-recognition sites (e.g., targeted release and bio-adhesion applications), for controlled drug-delivery applications (Figure 9.6).

In swelling-controlled system, hydrogels may undergo a swelling-driven phase transition from a glassy state to rubbery state. This transition occurs

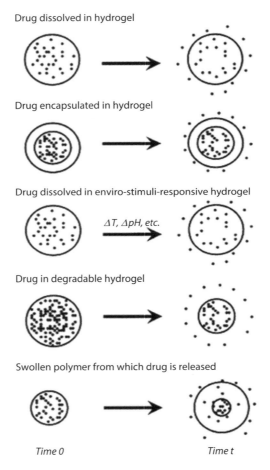

Drug dissolved in hydrogel

Drug encapsulated in hydrogel

Drug dissolved in enviro-stimuli-responsive hydrogel

$\Delta T, \Delta pH, etc.$

Drug in degradable hydrogel

Swollen polymer from which drug is released

Time 0   Time t

Figure 9.6 Various delivery and release mechanisms of hydrogels [127]. Reprinted from *(Adv Mater)*, Vol. 18, Copyright (2006) with permission from (WILEY-VCH Verlag GmbH & Co. KGaA, Weinheim).

when the characteristic glass-rubber polymer transition temperature is lower than the temperature of fluid that surrounds the drug delivery matrix. In the glassy state, entrapped molecules remain immobile, but in the rubbery state, dissolved drug molecules rapidly diffuse to the fluid through the swollen layer of polymer. Released fluid molecules contact the external layer of hydrogel. This forms a moving front that divides the hydrogel matrix into a glassy and swollen region. In these systems the rate of molecule release depends on the rate of gel swelling. In the swelling-controlled delivery system, the following phenomena take place [150]:

a. The length of drug diffusion way increases, which leads to a decrease of drug concentration gradient and a decrease of drug release rates.
b. The mobility of drug molecules increases, which causes an increase of drug release rates.

When the drug bearing hydrogel comes in contact with the aqueous medium, water penetrates into the system and dissolves the drug. Diffusion is the main phenomena by which the dissolved drug diffuses out of the delivery systems to the surrounding aqueous medium. This phenomenon is defined as the movement of the individual molecules from the region of high solute concentration to a region of low concentration when the systems are separated by a polymeric membrane. The delivery systems employing hydrogels for controlled release can be categorized into reservoir and matrix devices. As mentioned earlier, hydrogels are 3-dimensionally crosslinked polymer networks and hence act as a permeable matrix and/or membrane for the drug, thereby governing the release rate of the drug. The diffusion of the drug through the hydrogels may be affected by the property of the hydrogel depending on the chemistry of the hydrogels. Delivery devices based on hydrogel can be used for epidermal, oral, ocular, and subcutaneous application. Figure 9.7 displays various sites that are available for the application of hydrogels for drug delivery. Drug delivery through the oral route has been the most popular method in pharmaceutical applications. Drug carrier systems may be designed to release drugs in a controlled manner at the desired site. Gastrointestinal, GI, tract with its large surface area for systemic absorption is an attractive site of targeting drugs whereas colon-specific delivery has significance for the delivery of peptide and proteins [151]. Despite the huge attraction centered towards the novel drug delivery systems based on the environment-sensitive hydrogels in the past and current times, these systems have a number of disadvantages. The most considerable drawback of stimuli-sensitive hydrogels is their

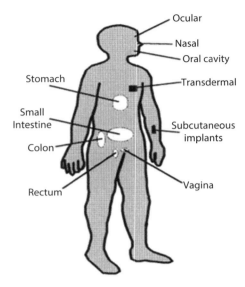

Figure 9.7 Tissue locations applicable for hydrogel based on drug delivery systems [151]. Reprinted from *(Eur. J. Pharm. Biopharm)*, Vol. 50 Copyright (2000) with permission from (Elsevier).

significantly slow response time, with the easiest way to achieve fast-acting responsiveness being to develop thinner and smaller hydrogels, which, in turn, bring about fragility and the loss of mechanical strength in the polymer network [36].

On the other hand, in drug delivery aspect, the problem of loading hydrogels with hydrophobic drugs may also be considered. This problem can be approached by copolymerization of hydrophilic hydrogels with hydrophobic segments, or by the use of amphiphilic IPNs in which a hydrophobic component is interpenetrated by a hydrophilic one. Hydroxypropyl methylcellulose (HPMC) is the most widely used hydrogel for drug delivery. It gradually swells in the aqueous medium and controls drug release by both diffusion and erosion. These types of hydrogels are non-crosslinked and ultimately dissolve over time in the presence of sufficient water or the swelling medium. The diversity of the types of hydrogels available means that a gel can be tailored to match the requirements of almost any device, requiring the control and modulation of solute transport or release, volumetric displacement, or generation of a triggering signal in response to changes in an aqueous environment. This makes hydrogels central to the development of many advanced drug delivery systems. Controlled release systems can generally be divided into three sections depending on their mode of release: diffusion-controlled, chemical erosion, and solvent activated [152]. In a diffusion-controlled device, the drug is surrounded by an inert barrier and diffuses from a reservoir, or the drug is dispersed throughout a polymer and diffuses from the polymer matrix. In a chemical erosion device, the drug is dispersed in a bio-erodible polymer system or is covalently linked to a polymer backbone via a hydrolyzable linkage. As the polymer or hydrolyzable link degrades, the drug is released. In a solvent-activated device, the drug is dispersed within polymeric matrix and the device is swelled with a suitable solvent. As the device swells, the drug is released.

Human skin is an easily accessible surface for drug delivery. Also, it receives about one-third of the blood circulating through the body. For these reasons, transdermal drug delivery represents an attractive alternative to oral delivery of drugs as well as to hypodermic injection since is a non-invasive technique and can be self-administered.

9.7 Hydrogels for Wound Management

One of the most important applications of hydrogels is in wound and burn management such as novel wound dressing devices called hydrogel wound dressings. In this section, we will focus on this application of hydrogels, after brief introduction to wound, wound care, and types of wound dressings.

9.7.1 Wound Care and Wound Dressings

A wound is defined as a defect or break in the skin, resulting from physical or thermal damage or as a result of the presence of an underlying medical or physical condition. Based on the nature and repair process of wounds, they can be classified as chronic wounds and acute wounds [153]. The primary causes of acute wounds are mechanical injuries (friction contact between skin and hard surfaces), burns, and chemical injuries. In the case of burns, the temperature of the source and time of exposure is important to decide the degree of wound. Burn wounds normally need special care because of associated trauma [153]. Wound healing is a dynamic process and the performance requirements of a dressing can change as healing progresses. In fact, the term "dressing" refers more correctly to the primary layer in contact with the wound. Prior to 1960s, wound dressings were considered to be only passive products that had a minimal role in the healing process. Winter [154] in 1962 initiated the concept of an optimal environment for wound repair and the active involvement of a wound dressing in establishing and maintaining such an optimal environment. This awareness revolutionized approaches to wound care and paved the way for development of wound dressings from the passive materials to the more functionally active ones [155]. In an ideal condition, a desirable wound dressing should: (i) provide and maintain a moist environment; (ii) protect the wound from secondary infections; (iii) absorb the wound fluids and exudates; (iv) reduce the wound surface necrosis; (v) prevent the wound desiccation; (vi) stimulate the growth factors; and also be (vii) elastic; (viii) cost-effective; (ix) non-antigenic; and (x) biocompatible [155–157]. Also, wound healing is affected by many variables, which must be assessed before treatment commences. These include patient factors (e.g., mental status, age, pain, mobility, nutritional status, and comorbidities), wound factors (e.g., peri-wound skin status, vasculitis, cellulitis, wound color, and incontinence), as well as the macroscopic and microscopic environments [158]. Evaluation of necrosis, infection, nutrition, pressure, perfusion, and tissue moisture balance are also paramount. Fluid balance in burn injury is very important since heavy loss of water from the body by exudation and evaporation may lead to a fall in body temperature and an increase in the metabolic rate. Besides this, dressings should have certain other properties like ease of application and removal, and proper adherence so that there will not be any area of non-adherence left to create fluid-filled pockets for the proliferation of bacteria [159]. While plain gauze is still the most commonly used dressing in hospitals today, new wound technologies have produced advanced products that help the body achieve the ideal moist, warm, protected wound healing environment (Table 9.4). Plain gauze

Table 9.4 Advanced wound dressings [160]. Reprinted from (Clinics in Plastic Surgery), vol. 34 Copyright (2007) with permission from (Elsevier).

Protective dressings	Descriptions
Gauze	Inexpensive; readily available
Impregnated gauze	Non-adherent; preserves moisture
Antimicrobial dressings	–
Antibacterial ointments	Reapply often to maintain moisture
Iodine based	Absorbent; not for use with thyroid disorders
Silver based	Many forms; broad spectrum; low resistance
Autolytic debridement	
Films	Occlusive; allows exchange of gasses

Protective dressings	Descriptions
Hydrocolloids	Not for exudative or infected wounds
Hydrogels	Rehydrates to soften dry wounds
Chemical debridement	–
Papain/urea	Availability issues in US
Collagenase	Selective debridement
Absorbent dressings	-
Foam	Absorbs moderate exudates
Hydrogels	Absorbs minimal exudates
Hydrofibers	Absorbs heavy exudates
Alginates	Absorbs heavy exudates

certainly has its place as it is inexpensive, readily available, and appropriate for a large number of wounds. Impregnated gauze improves upon this by adding zinc, iodine, or petrolatum to help prevent desiccation and provide non-adherent coverage.

9.7.2   Types of Wound Dressings

Dressings can be classified in a number of ways. They can be classified based on their function in the wound (antibacterial, absorbent), type of material employed to produce the dressing (collagen, hydrocolloid), physical form of the dressing (ointment, film and gel), traditional and modern dressings. Some dressings can be placed in several classifications because they fit criteria in several groups. The simplest classification is as traditional and modern dressings, and particular focus will be given to hydrogels, one of the most common modern dressings. Dressings may be either adherent or non-adherent, and are subdivided into occlusive or nonocclusive, depending on whether or not they allow the exchange of gases or water. Adherent dressings are generally used early on in wound management, when debridement of the wound is needed. These may be gauze pads (swabs), or specialized adherent dressing. Sterile gauze pads may be used dry on wounds, or may be soaked with sterile saline. These applications are known as "dry-to-dry" or "wet-to-dry," and the function is to help debride the wound and remove necrotic tissue and loose debris. Once wound-healing is under way, non-adherent dressings are chosen so as to minimize further damage to the wound when changing the dressing. These tend to disrupt the wound surface minimally and are recommended for wounds starting to granulate. Granulation tissue does not grow in the presence of infection and so its presence indicates that infection is controlled. Non-adherent dressings prevent wound desiccation and allow moisture to be retained at the wound site, promoting healing. Furthermore, traditional wound dressings can be classified as topical pharmaceutical formulations and traditional dressings.

Topical pharmaceutical formulations

These formulations can be liquids such as solutions and suspensions or semi-liquid materials such as ointments and creams. These formulations can be used in the initial stages of wound healing, for example as antibacterials [153].

Traditional dressings

These are, unlike topical formulations, dry materials such as cotton wool and natural or synthetic gauzes. These dressings are more used in chronic wounds and burn wounds because liquid and semi-liquid dressings do not remain on the wound over optimal time [153]. While some clinicians insist that gauzes are as effective as new dressings, some studies show that moisture-retentive dressings are associated with faster healing time. Gauzes have been one of the most popular wound dressings but there are several disadvantages with the use of these. They can promote desiccation of the wound base, they bind to the wound bed, which causes pain and trauma for patients during dressing changes. They do not provide a good barrier against bacterial growth because they are susceptible to full thickness saturation with wound fluid [161]. The main aim of modern wound dressing is to create a moist environment for the wound to faclitate the healing process. Modern wound dressings are often classified as hydrocolloid dressings, alginate dressings, hydrogel dressings, dressings in form of gels, foams and films, etc. Modern dressings do not enhance the reepithelialization, but stimulate collagen synthesis, promote angiogenesis, and can inhibit bacterial growth by maintaining a barrier against external contamination and some of them by decreasing pH at the wound surface. The dry wound healing process would not only delay the wound healing process, but can cause further tissue death [161]. Five key points that technology and methodology of biomaterial design should consider when formulating biomaterials are as follows:

1. The first key point is to produce cell scaffolds for enhanced repair, proliferation, and differentiation process. In a big area of tissue defect, both cells and ECM are missing. Therefore, it is important to make a three-dimensional scaffold "ECM" that provides a suitable environment for the healing process and helps the attachment of cells while the natural ECM is produced. ECM is not only a physical support for cells, but also provides a natural environment for cell proliferation and differentiation. This artificial scaffold should be porous, because cells can infiltrate into the scaffold. The porosity provides oxygen and nutrition for cells, and cell wastes can be washed out through the porosity. This artificial cell scaffold is a temporary ECM for the cells. Cells begin to produce ECM naturally after the tissue regeneration is initiated [162].

2. The second key is to provide the space for cell based tissue regeneration and supply nutrients and oxygen to cells by angiogenesis. When a defect happens in a body tissue, fibroblasts produce fibrous which occupies the defected area immediately. This is a process that fills and repairs the defected area. Once the fibroblasts have occupied the area, repair of the target tissue area will be hard. To enable the transplanted cells to survive, there should be a cell scaffold in the defected area that provides oxygen and supply for the transplanted cells. This artificial scaffold should be a barrier membrane that in addition to cells and space providing membranes, also contains signaling molecules. Signaling molecules can accelerate tissue regeneration [162].

3. The third key in the technology of tissue engineering is use of growth factor. In some cases growth factor is required to promote tissue regeneration. Growth factor cannot be injected as a solution to the regeneration site. It diffuses out from the defected area and becomes digested or deactivated fast. Therefore, it is necessary to incorporate the growth factor in a biomaterial carrier. It will then be possible to make a controlled release of growth factor at the regeneration site over a longer period of time. Presenting signaling cells as growth factor and vascular endothelial growth factor (VEGF) to the wound is important, since a wound whose matrix is lost does not have the ability to regenerate itself just by the presented cells [163]. Introducing VEGF to the wound can stimulate angiogenesis, collagen deposition, and

epithelialization. The molecule will particularly accelerate healing of chronic wounds. [164].

4. The fourth technology refers to finding the right type and quality of cell culture that has the ability to survive and regenerate in the defected site. For this purpose, isolation, induction and in vitro culture technologies are needed.

5. The fifth and last technological key to successful tissue engineering is to use genetic engineering to create cells that are suitable for the defected area. The stem cells should have properties that fit the base cells in that area. These cells should function in a same way as the base cells. This requires development of carriers for gene transfection and for efficient gene expression. The technology of providing carriers for gene expression is an important step in tissue engineering [162].

In spite of all the above-mentioned options there are many wounds resistant to treatment and a variety of new techniques are being researched. These include tissue engineering techniques like stem cells and gene therapy for achieving wound closure [165, 166]. Stem cells have the ability to migrate to the site of injury or inflammation, participate in regeneration of damaged tissue, stimulate proliferation and differentiation of resident progenitor cells, secrete growth factors, re-model matrix, inhibit scar formation, increase angiogenesis and improve tensile strength of the wound [167–170]. Recently, stem cell–based skin engineering along with gene recombination represents an alternative tool for regenerative strategies for wound therapy [171, 172]. Current drug delivery strategies cannot control the loss of drug activity due to physical inhibition and biological degradation. So, to optimize the delivery of factors for maximum efficacy, a molecular genetics approach is being researched in which genetically modified cells synthesize and deliver the desired growth factor in a time-regulated and locally restricted manner to the wound site to promote wound healing. If stem cells could be instructed to differentiate into one particular lineage and functionally integrate into injured tissue environment, they could replace cells that have been lost. The most cost-effective method of removing devitalized tissue is through autolytic debridement with wet dressings such as hydrogels. Hydrogel dressings are much easier to remove than other wound dressings, due to the gel nature. They are rarely painful and can usually be changed with no need for sedation. In conventional wound management products, no single product is suitable for all wound types or at all stages of healing so development of a new generation of wound care products that will help at various stages of wound healing is an evolving field of research today. Based on available evidence and clinical experience, modern dressings do appear to have roles at various stages of the wound healing process. Absorbable haemostats and hydrogel dressings are the part of modern wound management systems, which help in haemostasis and moist wound healing by enhancing cell proliferation.

9.7.3   Hydrogel Wound Dressings

Skin toxicity is a major problem in transdermal drug delivery systems. Drug molecules are required that are innocuous, creating neither irritation nor allergenicity, so in some circumstances using a larger patch area can alleviate the problem. Because of this issue, and taking into consideration one of the possible practical applications of "smart materials" in wound treatment, the idea of using sensitive hydrogels and microgels as transdermal drug-delivery systems was considered.

Hydrogel wound dressing is a type of synthetic dressing that is particularly good for wounds that need to be kept moist, such as burns or necrotic wounds. Necrotic wounds are wounds with dead and dying tissue that must be removed before healing can effectively take place. Some types of hydrogel wound dressing can clean wounds by removing necrotic and infected tissue from the wound bed. Hydrogel is also used as wound filler or as a flat dressing. It is available in sheet form, on pads in a variety of shapes, or simply as a gel that can be applied to wounds. Besides wound hydration, one of the main advantages of hydrogel wound dressing is that it does not stick to wounds which can mean much less pain during dressing changes. This means that it can be removed during dressing changes without damaging the wound further or disrupting healing. This is particularly useful for burns and other types of open or chronic wounds. Due to the high water content and gel-consistency, hydrogel also feels cooling and soothing on the wounded area. The hydrogels have also potential in drug release to the skin. They may be of particular importance where the skin barrier is compromised, as in a diseased state or in wound management. In this situation, controlled delivery to the skin of active materials can provide therapeutic levels where required and minimize systemic uptake. These gels could be used at higher temperatures and release more material, as can be anticipated within wound tissue. The use of hydrogels in the healing of wounds dates back to the late seventies or early eighties. Hydrogels act as a wet wound dressing material and have the ability to absorb and retain the wound exudates along with the foreign bodies, such as bacteria, within its network structure. In addition, hydrogels have been found to promote fibroblast proliferation by reducing the fluid loss from the wound surface and protecting the wound from external noxae necessary for rapid wound healing. Hydrogels help in maintaining a micro-climate for biosynthetic reactions on the wound surface necessary for cellular activities. Fibroblast proliferation is necessary for complete epithelialization of the wound, which starts from the edge of the wound. Since hydrogels help to keep the wound moist, keratinocytes can migrate on the surface. The process of angiogenesis can be initiated by using semi-occlusive hydrogel dressings, which is initiated due to temporary hypoxia. Angiogenesis of the wound ensures the growth of granulation tissue by maintaining adequate supply of oxygen and nutrients to the wound surface. Hydrogel sheets are also generally applied over the wound surface with backing of fabric or polymer film and are secured at the wound surface with adhesives or with bandages.

Hydrogels fit most criteria for a suitable wound dressing as they:

- Help the rehydration of dead tissues and increase the healing of debridement
- Are suitable for cleansing of dry, sloughy, or necrotic wounds
- Do not react with biological reacts
- Are permeable to metabolites
- Are nonirritant
- Promote moist healing
- Are non-adherent
- Cool the surface of the wound [153].

Other properties of hydrogels allow them to be utilized to deliver topical wound medications (e.g., metronidazole and silver sulfadiazine). The release mechanism resulting in the diffusion of the medication can be controlled by the degree of cross linkage in the gel. Both temperature and pH-sensitive gels have been the subject of investigation with the objective of developing new products. Despite their high water content, hydrogels are capable of additionally binding great volumes of liquid because of the presence of hydrophilic residues. Hydrogels swell extensively without changing their gelatinous structure and are available for use as amorphous (without shape) gels and in various types of application systems, e.g., flat sheet hydrogels and non-woven dressings impregnated with amorphous hydrogel solution. These products consist of hydrophilic homopolymers or copolymers, which interact with aqueous solutions, absorbing and retaining significant volumes of fluid. Flat sheet hydrogel dressings have a stable crosslinked macrostructure and therefore retain their physical form as they absorb fluid [173]. Additional advantages such as transparency, cushioning effect, cooling effects, etc., considerably increase the utility value of hydrogels, in particular concerning patient comfort and ease of application. Hydrogels may be transparent, depending on the nature of the polymers, and provide cushioning and cooling/soothing effects to the wound surface. The main advantage of the transparent hydrogels includes easily monitoring of the wound healing without removing the wound. There are some commercialized hydrogel wounds dressings under the trade names of Vigilon, Ivalon, Aqua gel, Kik gel, etc. [174]. However, some of the hydrogel dressings, due to their weak mechanical properties, for instance, low strength and elasticity or low fracture toughness, especially in the swollen state, did not satisfy the ideal dressing requirements, i.e., they might stick to the wound surface or be crushed under high stresses [175]. These disadvantages are mainly attributed to the constrained molecular motion of the polymer chains due to the large number of randomly arranged crosslinks in the conventional hydrogels [176]. Hence, a hydrogel having good strength and elasticity is expected to be better as a dressing material. In order to improve the physical and mechanical properties of hydrogels and to make them strong and elastic, small-scale inorganic particles are commonly used as reinforcing agents. Also, to improve the mechanical properties of hydrogels, several manufacturing methods were proposed, among which three resulted in significant improvements in the mechanical properties. Namely:

- Double network hydrogels;
- Hydrogels containing sliding crosslinking agents;
- Nanocomposite hydrogels [177].

Double network hydrogels

In this method, two hydrogels are combined together. One of them is a highly crosslinked polyelectrolyte and the other one is a loosely crosslinked or maybe uncrosslinked natural hydrogel. This combination will result in an effective relaxation of locally applied stress and dissipation of crack energy [177]. The double network (DN) gels possess IPN structure where the properties of two networks exist in sharp contrast such as network density, rigidity, molecular weight, crosslinking density, etc. They are generally synthesized via a two-step sequential free-radical polymerization process in which a high relative molecular mass neutral second polymer network is incorporated within a swollen heterogeneous polyelectrolyte first network [178]. The mechanical properties of DN gels prepared from many different polymer pairs were shown to be much better than that of the individual components. Under an optimized structure, the DN gels, containing about

90 wt% water, possess hardness, strength (failure tensile stress 1–10 MPa, strain 1000–2000%; failure compressive stress 20–60 MPa, strain 90–95%), and toughness (tearing fracture energy of 100–4400 J.m^{-2}) [179]. These excellent mechanical performances had never been realized before in synthetic hydrogels, and are comparable to and even exceed some soft load-bearing tissues.

Hydrogels containing sliding crosslinking agents

In this method two cyclodextrin molecules get crosslinked. These molecules will create double rings that can move slightly along the PEG chains. This will result in an excellent mechanical property for hydrogel. It will provide a hydrogel with a high degree of swelling and a high stretching ratio without fracture [177].

Nanocomposite hydrogels

These kinds of hybrid hydrogels can be synthesized through polymerization of hydrophilic monomers in the presence of nanoparticles. Different kinds of nanoparticles such as gold, silver, iron oxide, carbon nanotubes, and various types of clays, e.g., laponite, montmorillonite, hydrotalcite, and bentonite have been used successfully in production of nanocomposite hydrogels [180]. Since 2002, several attempts have been made to reinforce the structure of hydrogels by incorporating nanoparticles or nanostructures to obtain nanocomposite hydrogels with improved mechanical, physical, and chemical properties [181]. The recent research on nanocomposite materials has shown that some properties of polymers and gels significantly improve by adding organoclay into polymeric matrix. The nanocomposite hydrogel exhibited extraordinary mechanical, optical, and swelling/de-swelling properties, which could simultaneously overcome the limitations of conventional chemically crosslinked hydrogels. The construction of nanocomposite hydrogels was achieved, not by the mere incorporation of clay nanoparticles into a chemically crosslinked network, but by allowing the clay platelets to act as multifunctional crosslinkers in the formation of polymer/clay networks. Due to their superior properties, nanocomposite hydrogels have attracted much attention and are believed to be a revolutionary type of hydrogel. Typically, in this method polymer N-isopropylacrylamide (NIPAAm) that is clay-contained is combined with hectorite [$Mg_{5.34}Li_{0.66}Si_8O_{20}(OH)_4$]$Na_{0.66}$ as a multifunctional crosslinker. The mechanical property of the hydrogel was enhanced and the tensile module and strength were proportional with clay content [177]. An interesting result with respect to applications obtained with the

IPN hydrogels is that these are two-phase systems (two glass transition temperatures), with the hydrophilic domains behaving essentially like the pure hydrophilic component [182–184]. Thus, the two basic functions of these IPN hydrogels with respect to applications, namely hydrophilicity and mechanical stability, are separately taken over by the two IPN components, the hydrophilic and hydrophobic domains, respectively. Hydrogels are poor bacterial barriers and require a secondary dressing to secure them in place. Therefore, the hydrogel wound dressing may also contain additives such as antibiotics, anti-fungals, and/or other organic compounds to prevent infection and to control odor.

9.7.3.1 Preparation Methods of Hydrogel Wound Dressings

Current approaches that are used for making hydrogels with controlled features can be categorized based on the technologies used to crosslink the gel during fabrication, including approaches such as emulsification, micro-molding, photolithography, and micro-fluidic techniques. These techniques are shown schematically in Figure 9.8. Some other known methods of making hydrogel dressing using hydrophilic polymers include: physical method by repeated freezing and thawing, chemical method using chemicals like borax, boric acid, glutaraldehyde, formaldehyde, and irradiation. PVA is a synthetic hydrogel that has been used in biomedical applications especially in wound dressing due to its good chemical and mechanical stability, processability, biocompatibility, and biodegradability. This kind of synthetic hydrogel is prepared by crosslinking the linear chains of PVA [185]. Crosslinked PVA based hydrogels possess most of the aforementioned properties, which make them ideal candidates as wound dressing material. Additionally, these hydrogels also create moist wound environments, which further accelerate the healing process. PVA can be crosslinked by different physical or chemical methods such as electron beam, c-irradiation, repeated freeze-thaw cycles, photo-crosslinking, reacting with bi-functional reagents like boric acid, phenyl boronic acid, dialdehydes, dicarboxylic acids, dianhydrides, acid chlorides, epichlorohydrin, etc. [186–191].

Crosslinked networks of synthetic polymers such as PEO, polyvinyl pyrollidone (PVP), PLA, polyacrylic acid (PAA), polymethacrylate (PMA), PEG, or natural biopolymers such as alginate, chitosan, carrageenan, hyaluronan, and carboxymethyl cellulose (CMC) have been reported for this purpose. The various preparation techniques adopted are hydrogen bonding interactions, physical crosslinking, chemical crosslinking, grafting polymerization, and radiation crosslinking. Hydrogen bonding

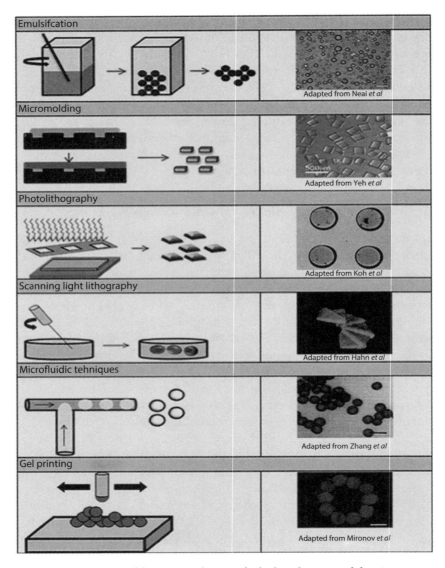

Figure 9.8 Various microfabrication techniques for hydrogels as wound dressing materials [24]. Reprinted from (*Adv. Mater*), Vol. 21 Copyright (2009) with permission from (WILEY-VCH Verlag GmbH & Co. KGaA, Weinheim).

interactions can be used to produce hydrogels in vitro by freezing-thawing, e.g., in the formulation of PVA-based hydrogels [192]. Physical crosslinking of polymer chains can be achieved using a variety of environmental triggers (pH, temperature, ionic strength) and a variety of physicochemical interactions (hydrophobic interactions, charge condensation, hydrogen bonding, stereocomplexation, or supramolecular chemistry). There has been an increased interest in physical or reversible gels due to relative ease of production and the advantage of not using crosslinking agents. These agents affect the integrity of substances to be entrapped (e.g., cells, proteins, etc.) as well as the need for their removal before application. Careful selection of hydrocolloid type, concentration and pH can lead to the formation of a broad range of gel textures and is currently an area receiving considerable attention, particularly in wound dressing application. The various methods reported in literature to obtain physically crosslinked hydrogels are:

- Freezing-thawing
- Heating/cooling a polymer solution
- Ionic interaction
- Complex coacervation
- H-bonding
- Maturation

Chemical crosslinking involves grafting of monomers on the backbone of the polymers or the use of a crosslinking agent to link two polymer chains. The crosslinking of natural and synthetic polymers can be achieved through the reaction of their functional groups (such as OH, COOH, and NH_2) with crosslinkers such as aldehyde (e.g., glutaraldehyde, adipic acid dihydrazide). Among several chemical crosslinking methods, IPN (polymerizing a monomer within another solid polymer to form interpenetrating network structure) and hydrophobic interactions [193] (incorporating a polar hydrophilic group by hydrolysis or oxidation followed by covalent crosslinking) are also used to obtain chemically crosslinked permanent hydrogels.

High-energy radiations like gamma and electron beams have been used to prepare hydrogel of unsaturated compound. The irradiation of aqueous polymer solution results in the formation of radical on the polymer chain.

PVP is another example of polymer applied for the synthesis of hydrogel to be used in different biomedical applications, especially wound dressing [194, 195]. PVP hydrogels can be obtained by gamma irradiation of PVP/water solutions. The physical and mechanical characteristics of the resultant gel depend on the radiation dose as well as the presence of additive in the solution. The irradiation causes crosslinking between the PVP chains and consequently results in the formation of a polymer network. In gel synthesis the presence of chemical substances different of PVP in the starting solution as well as the radiation dose influences the mechanical behavior of the resultant product, since it influences the network crosslinking density.

The network crosslinking density is one of the decisive parameters on the mechanical behavior of the gel. While copolymerization/crosslinking of reactive monomers and crosslinking of linear polymers are the main methods for producing hydrogels, there are other techniques that also deserve mention, particularly chemical conversion and the formation of interpenetrating networks. Production of hydrogels by chemical conversion uses chemical reactions to convert one type of gel to another. This can involve the conversion of a non-hydrogel to a hydrogel or the conversion of one type of hydrogel to another. Examples of the former include the hydrolysis of polyacrylonitrile to polyacrylamide or of poly(vinyl acetate) to PVA; an example of the latter includes the alkaline hydrolysis of polyacrylamide to PAA. As another example, PVA gels with superior mechanical properties can be produced by the bulk polymerization of vinyl tri-fluoroacetate (forming a non-hydrogel), followed by solvolysis of PVTFA to PVA [196]. It is well known that poly(hydroxyethyl methacrylate) (PHEMA) is one of the other most widely applied and important hydrogels. Since 1955 it has been modified with many natural and synthetic substances and by various methods, and has been applied in the production of contact lenses and dressings, and for drug delivery and tissue engineering purposes. The properties of PHEMA depend, among other things, on the method of synthesis, polymer content, degree of crosslinking, temperature and final application environment. The synthesis can be carried out with simultaneous crosslinking by UV-radiation. In one possible way of preparing PHEMA [197], 2-hydroxyethyl methacrylate (HEMA) as a monomer, polyethylene glycol dimethacrylate as the crosslinking agent and benzoin isobutyl ether (BIE) as the UV-sensitive initiator were used. De-ionized water (DI) in an amount appropriate to the desired concentration should then be added to the system prepared from the listed components. The final products are obtained in the form of films or membranes by treating them with UV radiation. Next, the film is immersed for 24 h in water until it is fully saturated in order to remove toxic or un-reacted substances that could damage living tissue. This kind of hydrogel can be applied in artificial skin manufacturing, and dressings, especially burn dressings, as it ensures good wound-healing conditions.

Also, PEG is one of the other most widely used hydrogels in medicine and biomedicine. Hydrogels based on its derivatives—polyethylene glycol methacrylate (PEGMA), polyethylene glycol dimethacrylate (PEGDMA), and polyethylene glycol diacrylate (PEGDA) are likewise widely applied. PEG-based hydrogels are characterized by their high biocompatibility, lack of toxic influence on surrounding tissue, and solubility in water, which makes them good candidates for wound dressing and drug delivery system applications [198]. Another group widely used in biomedical applications is hydrogels based on PVP. They are present in wound dressings and are usually obtained with the radiation technique, an apparently simple, efficient, clean and environmentally friendly process. The application of radiation in the formation of hydrogels for biomedical use offers a unique opportunity to combine the formation and sterilization of a product in one single technological step. PVP hydrogels can thus be regarded as advantageous owing to their simple formation and lower production costs, softness and elasticity; their ability to store large amounts of liquid while retaining quite good mechanical properties makes them optimal candidates for the manufacture of dressings. To improve its mechanical properties, PVP is usually blended with other polymers (mainly natural), which also enhances its biocompatibility and water uptake. Agar, cellulose and PEG are the usual polymers added [139, 199, 200]. Besides, polyurethane (PU) hydrogels are also applied in wound dressing manufacture. These gels are obtained in a one-step bulk synthesis using 1,6-hexamethylene diisocyanate (HDI) or methylene bis-(4-cyclohexy-lisocyanate) (HMDI) and 1,4-butanediol or maleic acid as a chain extender with the 0.02–0.05 wt % addition of a dibutyltin dilaurate or $FeCl_3$ catalyst. The reaction occurs within a few minutes, after which the mixture is placed in a propylene cube at 368 K for 20 h to finish the process. The final product can be also obtained in the form of films, using 5% solution of polymer in methanol [201].

9.7.3.2   Characterization of Hydrogel Wound Dressings

Since the presence of different functional groups plays an important role in the water holding capacity of the hydrogel, it becomes necessary to analyze the presence of different functional groups in a newly synthesized hydrogel wound dressing. Also, determination of the functional group can provide some information on the composition of the polymeric network. The various techniques used for characterization of hydrogels include UV-Vis spectroscopy, infrared (IR) spectroscopy, mass spectrometry and NMR. The chemical bonds in a molecule are always either in stretching or in bending motion. The IR spectroscopy involves excitation of the functional groups with IR irradiation of a particular wavelength, which results in the increase in the amplitude of the vibrations (bond stretching and bending) of the functional groups. The stretching vibrations can either be symmetric or asymmetric, while the bending vibrations can either be in-plane or out-of-plane. The change in the amplitude of the vibrations of the bonds is recorded by the IR instrument.

At present the most commonly used method to determine the mechanical properties of hydrogels are by tensile testing or strip extensiometry. These methods have been extensively used to study the mechanical behavior of various hydrogels [47, 202]. This destructive technique involves applying a tensile force to strips of material held between two grips. Alternatively, the force can be applied to a ring instead of a single strip. Applied force and the elongation of the material are used to obtain a stress-strain chart. This chart can be used to derive several mechanical properties of the hydrogel including Young's modulus, yield strength, and ultimate tensile strength. Tensile testing is performed to determine elastic modulus, ultimate stress, and ultimate strain for sheet hydrogels. In tensile testing, a "well sized" shaped sample is placed in the grips of movable and stationary fixtures in a screw driven device, which pulls the sample until it breaks and measures applied load versus elongation of the sample.

Another commonly used method to determine the mechanical properties of hydrogels are Dynamic Mechanical Analysis (DMA). This technique determines elastic modulus, loss modulus, and damping coefficient as a function of temperature, frequency, or time. The approach is often used to determine glass transition temperature as well. In this method, the sample is clamped into movable and stationary fixtures and then enclosed in a thermal chamber. The DMA applies torsional oscillation to the sample while slowly moving through the specified temperature range. Results are typically recorded as a graphical plot of elastic modulus, loss modulus, and damping coefficient versus temperature. DMA, like tensile testing, is a destructive technique.

Thermal Gravimetric Analysis (TGA) continuously measures the weight of a sample as a function of temperature and time. The sample is placed in a pan held in a microbalance. The pan and sample are heated in a controlled manner and weight is measured throughout the heating cycle. Changes in weight at specific temperatures correspond to reaction or changes in the sample such as decomposition.

Extensiometry can also be used to examine the viscoelastic characteristics of a hydrogel material by elongating the material strip to a particular length and examining the stress relaxation response over time at a constant strain. There are several shortcomings with using extensiometry, which include: only hydrogel strips or rings can be measured, measurements can only be performed once on each test piece, the potential misalignment of grips and the strain is limited to being uniaxial. The destructive nature of this type of test makes it difficult to monitor the change in mechanical properties of the hydrogels over time, an important parameter in engineering tissues.

Among the many parameters used to characterize hydrogels, the polymer volume fraction in the swollen state, the molecular weight of the polymer chain between crosslinks (M_c), and the mesh size of the gel (ζ) are among the most informative, especially for drug delivery and wound dressing applications (Figure 9.9). Also, the crosslinking density, the number of chemical or physical crosslinks in a given volume controls many fundamental hydrogel properties that are important in biomedical applications. These properties include the volumetric swelling ratio (Q), the shear modulus (G) and the diffusion coefficient (D) of entrapped molecules. As can be seen in Figure 9.9, two network structures representative of low and high crosslinking densities are depicted to demonstrate the relationship between crosslinking density and basic hydrogel properties. As crosslinking density increases, mesh size (ζ), which is a measure of the space that is available between macromolecular chains for the diffusion, decreases.

For more details about using the equilibrium swelling theory of rubber elasticity for determination of these three important parameters, see refs [34], [47], and [202–211].

The use of XRD for determining the crystalline nature of a substance has long been used. XRD throws light on the properties of the different phases (for example, structural make-up and crystallite dimension, orientation and strains) present in the polymer- polymer blend matrix. The XRD is based on the principle of diffraction of the X-rays from the atomic lattice of the specimen when the specimen is being irradiated with a beam of monochromatic x-rays. These diffracted rays are analyzed carefully to

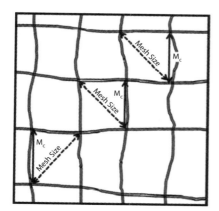
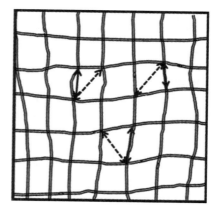

Figure 9.9 Schematic representation of a polymeric hydrogel where the molecular weight of the polymer chain between crosslinks (Mc), and the mesh size of the gel (ζ) are represented.

find out the angles of diffraction and the intensity of the diffracted rays. The angle of diffraction is useful in calculating interplanar atomic spacing (d-spacing). The d-spacing can reveal information, viz. unit cell size or lattice parameter, on the arrangement of atoms in a compound. The d-spacing and intensity information are unique for a particular material and act as a unique fingerprint for the material. Hence, the information can be used to characterize materials, both quantitatively and qualitatively, even in a mutli-component mixture. The width of the diffracted peaks can reveal information on the crystallite size and the micro-strain developed within a specimen. The main methods used to characterize and quantify the amount of free and bound water in hydrogels are DSC and NMR. The proton NMR gives information about the interchange of water molecules between the so-called free and bound states [212]. The use of DSC is based on the assumption that only the free water may be frozen, so it is assumed that the endotherm measured when warming the frozen gel represents the melting of the free water, and that value will yield the amount of free water in the hydrogel sample being tested. The bound water is then obtained by difference of the measured total water content of the hydrogel test specimen, and the calculated free water content [130]. Different behavior of water fractions in a hydrogel is most evident in differential thermal analysis and DSC. Upon slow heating of a frozen hydrogel (water content of about 30% and above), one often observes two endothermic peaks, a sharp one corresponding to the melting of ice (~273 K) and a broader one at 10–40 K below. They are usually associated with freezable free water and freezable bound water [213, 214]. It should be noted, however, that a different explanation has also been given for the case of poly(hydroxyethyl methacrylate) (pHEMA) hydrogels: the apparent two endothermic peaks are really only one peak which is interrupted by an exothermic peak caused by amorphous ice undergoing crystallization just below the melting transition [215, 216].

Raman measurements allow an estimate of defects in water-water hydrogen bonding in hydrogels as compared to liquid water or ice. They show no qualitative difference between pure water and hydrogels when one goes through the temperatures at which the DSC peaks occur [217]. In fact, Raman spectroscopy is very sensitive to other phase transitions such as the coil-to-globule transition in poly [N- (3-ethoxypropyl) acrylamide] hydrogels.

SEM can be used to provide information about the sample's surface topography, composition and other properties such as electrical conductivity. Magnification in SEM can be controlled over a range of up to 6 orders of magnitude from about 10 to 500,000 times. This is a powerful technique widely used to capture the characteristic "network" structure in hydrogels [218, 219].

Atomic forced microscopy (AFM) is commonly used for determining the surface properties of the hydrogels. AFM is a contact profilometer and can be operated either in contact mode or in tapping mode. In tapping mode, the tip of the cantilever has a piezoelectric element (PZTe). The cantilever is oscillated in the resonance frequency of the PZTe. As the tip approaches the surface of the material, there is an increased interaction between the surface and the tip, thereby resulting in a decrease in the amplitude of the oscillation. The decrease in the amplitude of oscillation is then recorded and compared with an external reference, which provides information on the surface characteristics of the material. While in the static mode, the tip of the AFM instrument drags over the sample surface. The surface properties are measured as a function of the tip deflection as it moves over the surface. The deflection of the tip is generally measured using a laser beam detector, which detects the reflected laser beam from the upper surface of the cantilever holding the tip.

A compression test is another technique that has previously been used to examine the mechanical properties of several different types of hydrogels. This technique involves placing the material between two plates and compressing it. The pressure applied to the surface of the hydrogel and distance the hydrogel is compressed can be used to calculate the mechanical properties of the hydrogels using a theoretical model. One of the advantages of the compression test over extensiometry is that it does not limit the hydrogel geometry to strips or rings although it does require a flat surface. This approach has several limitations including bulging of the hydrogel under compression and difficulty in applying pressure evenly. Bulging can be overcome by confining the hydrogel around its outer edge although this changes the nature of the measurements [48, 220]. However, non-destructive online characterization of the mechanical properties of hydrogels is not achievable with this technique.

9.8   Recent Developments on Hydrogels

A recent subject that has attracted much attention in the field of gel science is the fabrication of hydrogels with the rapid stimuli responsiveness and superior mechanical properties required for many applications of stimuli-responsive hydrogels. The creation of durable hydrogels that exhibit reversible and rapid changes in shape or volume, the likes of which are found in living organisms such as muscle, has been a challenge for materials chemists.

Hydrogels synthesized from monomer solutions by radical polymerization, a general-purpose method for making synthetic hydrogels, however, face several constraints. The inherently weak mechanical properties caused by underlying spatial inhomogeneity during polymerization and extremely slow responsiveness caused by critical slowing down and vitrification during the shrinking process restrict the widespread use of these hydrogels [221]. Although some approaches have been developed to improve the stimuli sensitivities and mechanical properties of hydrogels, it remains a challenge to design ideal gel networks with a combination of desired properties. In the past few years, there has been increased interest in the development and applications of hydrogel nanocomposites, specifically as a new class of biomaterials. Nanocomposite hydrogels have good potential to be used as functional soft materials in biomedical applications, due to their excellent mechanical properties, high water content, and good biocompatibility [222]. To date, there is growing interest on the research, synthesis, and the development of nanocomposite hydrogels for novel applications especially in wound dressing systems [223]. In some cases, the nanoparticles (e.g., gold, magnetic, carbon nanotubes) can absorb specific stimuli (e.g., alternating magnetic fields, near-IR light) and generate heat. This unique ability to remotely heat the nanocomposites allows for their remote controlled (RC) applications, including the ability to remotely drive the polymer through a transition event (e.g., swelling transition, glass transition). The nanocomposite hydrogels, referred to crosslinked polymer networks swollen with water in the presence of nanoparticles or nanostructures, are new generation materials that can be used in a wide variety of applications including stimuli-responsive sensors and actuators, pharmaceutical, and biomedical devices [224]. The most potential use of nanocomposite hydrogels is for novel biomaterials in tissue engineering, drug delivery, and hyperthermia treatment because they, in comparison with conventional hydrogels, can provide improved properties such as increased mechanical strength and ability for remote controlling [225]. Due to the excellent dispersion of cellulose nanocrystals (CNCs) in water, the fabrication, molding, and application of hydrogels containing CNCs without modification has many advantages compared with other nanofillers such as polymer and metal nanoparticles [226, 227]. Moreover, CNCs possess the long-term biocompatibility and controlled biodegradability, which is beneficial to further develop applications of nanocomposite hydrogels used as biomaterials. Nakayama et al. [228] for the first time reported cellulose-polymer nanocomposite hydrogels composed of bacterial cellulose (BC) and gelatin. Bacterial cellulose is biosynthesized by microorganisms, and displays unique properties, including high mechanical strength, high water absorption capacity, high crystallinity, and an ultra-fine and highly pure fiber (10–100 nm) network structure. By immersing BC gel in aqueous gelatin solution followed by crosslinking with N-(3-dimethylaminopropyl)-N'-ethylcarbodiimide hydrochloride, high mechanical strength DN nanocomposite hydrogels were prepared. In addition, these double network nanocomposite hydrogels exhibit not only a mechanical strength as high as several megapascals but also a low frictional coefficient of the order of 10^{-3}. These kinds of hydrogels not only exhibit superior mechanical properties (compression strength of up to 10 MPa) and withstand long-term cyclic stresses (up to 2000–6000 cycles) without substantial reduction of mechanical properties, but also show anisotropic behavior on both swelling and deformation. Cellulose nanocrytals or nanofibers isolated from plants have lower cost and higher price-performance ratio than BC, and their size can be facilely adjusted to meet the requirement of hydrogels properties for the various applications, e.g., pharmaceutical and biomedical devices and the potential impact for nanocomposite hydrogels to influence the lives of the general public continues to grow. The large amount of recent reports has explained some of the physics and chemistry behind the unique properties of these hydrogel materials. Responsive hydrogels that change properties and function as response to external stimuli such as artificial muscles are often inspired by natural systems. Desirable properties such as injectability and precise drug delivery that cannot be controlled with certain macroscopic hydrogels are achieved with nanocomposite micro- and nanogels. Future directions certainly include the rational design of biomedical nanocomposite hydrogels that require not only control of chemical and physical properties, but also the consideration of biological variables. While hydrogels can be used to simulate biological tissues, significant challenges arise when it comes to designing mechanically strong gels with long-term biocompatibility and controlled biodegradability. Reviewed literature suggests that the development of synthetic routes and fabrication technologies for fundamental understanding of nanocomposite hydrogels will continue, but with a greater emphasis on designing sophisticated multi-component and complex materials that can be tailored to very specific applications. Emerging new techniques strongly support the systematic characterization of nanocomposite gels, which, in return, drives research forward and impacts the rational design of materials. Although developing hydrogel systems that respond quickly to external stimuli is a challenge, films that have a fast, easily observable response to minimal stimuli have undergone tremendous improvement. Furthermore, developing a single system that can respond to multiple stimuli is often challenging and requires precise molecular engineering. Some of the recent studies in the development of the hydrogel nanocomposites are discussed here. An example of multi-responsive hydrogels have been developed by the Serpe group, including unique p(NIPAM-co-AAc) microgel film assemblies on gold substrates, called etalons, capable of color tunability over 300 nm in response to temperature and pH changes [229–231], which may find applications as microarray sensors or in display technologies. The recent research on nanocomposite materials has shown that some properties of polymers and gels significantly improve by adding organoclay into polymeric matrix. Recently, Kokabi *et al.* [232] reported a new nanocomposite hydrogel based on PVA and reinforcing agents, i.e., OMONT. In their research work, in order to obtain wound dressing with better properties, nanocomposite hydrogel wound dressing was prepared, using a combination of PVA hydrogel and organoclay, i.e., Na-montmorillonite, via the freezing-thawing method. Because of its desired water absorption, extensive swelling in water, and cation exchange capacity, montmorillonite is recognized as the main mineral clay used to prepare polymer nanocomposites [233, 234]. In another attempt, a new generation of nanocomposite hydrogel with excellent mechanical properties, good water-stability and unique water-activated shape memory behavior were prepared with *in situ* free-radical polymerization using titania nanoparticles as the crosslinking agent by Xu *et al.* [235]. Reviewing the literature on the nanocomposite hydrogels shows that numerous research studies have been performed on those nanocomposite hydrogels, which their crosslinking process is done mainly via using chemical crosslinking agents (so-called chemical nanocomposite hydrogels). Almost no essential attention has been paid to the nanocomposite hydrogels crosslinked physically in the absence of chemical crosslinking agents, like those which can be prepared via freezing-thawing technique. Recently, Sirousazar *et al.* [236] have prepared the PVA nanocomposite hydrogels via a physical crosslinking method (freezing-thawing cyclic technique) using OMONT as nanoparticles. Based on the obtained swelling characteristics of the PVA-OMONT nanocomposite hydrogels, the clay loading level in nanocomposite hydrogels was optimized and the best one to biomedical applications was chosen by them. It should be noted that, surely, imitating and/or designing of this new type of advanced materials will help aid in the development of more applications.

9.9 Conclusions

Hydrogels are crosslinked, hydrophilic polymeric networks with a broad range of materials and attractive diverse characteristics. The utilization of hydrogels has made great progress in the last 50 years. These compounds have played a very important role in various branches of science and industry especially in biomedical applications. Hydrogels can be synthesized by any of the techniques used to create crosslinked polymers. Covalently crosslinked networks can be created by copolymerization/ crosslinking of monomers and by linking linear polymers with chemical reagents or through ionizing radiation. Hydrogels can also be crosslinked by non-covalent interactions, including entanglements, electrostatics, hydrogen bonds, hydrophobic interactions, and crystallites. Hydrogels can be synthesized by any of the techniques used to create crosslinked polymers. Covalently crosslinked networks can be created by copolymerization/ crosslinking of monomers and by linking linear polymers with chemical reagents or through ionizing radiation. With increasing efforts devoted to controlled molecule release, the applications of hydrogels will continue to grow in the future. Proper network design and accurate mathematical modeling are keys to tuning the drug release rates as well as to modulating tissue regeneration. The future success of hydrogels as advanced materials relies on the development of novel materials that can address specific biological and medical challenges. Recent enhancements in the field of polymer science and technology have led to the development of various stimuli sensitive hydrogels. Either pH-sensitive and/or temperature-sensitive hydrogels can be used for site-specific controlled drug delivery and wound dressing applications. The development of synthetic occlusive wound dressing material is currently an area of great commercial interest. Crosslinked-based hydrogels possess most of the aforementioned properties, which make them an ideal candidate as wound dressing material.

References

1. J.M.G. Cowie, *Polymers: Chemistry and physics of modern materials*, 2nd ed., Stanley Thornes, Cheltenham, UK, 1991.
2. S.M. Gomes, H. Azevedo, P. Malafaya, S. Silva, J. Oliveira, G. Silva, R. Sousa, J. Mano and R. Reis, "Natural Polymers in Tissue Engineering Applications," In: C. V. Blitterswijk, Ed., *Tissue Engineering*, Academic Press, Waltham, pp. 146–191, 2008.
3. A.N. Peppas, A.R. Khare, *Adv. Drug. Deliv. Rev*, Vol. 11, pp. 1–35, 1993.
4. A.S. Hoffman, *Adv. Drug. Deliv. Rev*, Vol. 43, pp. 3–12. 2002.
5. P. Prang, P. Muller, A. Eljaouhari, K. Heckmann, W. Kunz, T. Weber, C. Faber, M. Vroemen, U. Bogdahn , N. Weidner, *Biomaterials*,Vol. 27, pp. 3560–3569, 2006.
6. D.M. Veríssimo, R.F.C. Leitão, R.A. Ribeiro, S.D. Figueiró, A.S.B. Sombra, J.C. Góes, and G.A.C. Brito, *Acta Biomaterialia*, Vol. 6, pp. 4011–4018, 2010.
7. M. Rani, A. Agarwal, Y.S. Negi, *BioResource*s, Vol. 5, pp. 2765–2807, 2010.

8. H. J. Van der Linden, S. Herber, W. Olthius, P. Bergveld, *Analyst*, Vol. 128, pp. 325–331, 2003.

9. C. Gao, M. Liu, J. Chen, X. Zhang, *Polym. Degrad. Stabil.*, Vol. 94, pp. 1405–1410, 2009.

10. K. Pal, A. K. Banthia and D. K. Majumdar, *Des. Monomers Polym*, Vol. 12, pp. 197–220, 2009.

11. S. Simões, A. Figueiras, F. Veiga, *J. Biomater. Nanobiotech*, Vol. 3, pp. 185–199, 2012.

12. G.A. Mun, I.E Suleimenov, R.B. Bakytbekov, E.S.M. Negim, N.V. Semenyakin, D.B. Shaltykova, *World Appl. Sci. J.* Vol. 17, pp. 1504–1509, 2012.

13. M.R. Kemp, P.J. Fryer, *Innov. Food Sci. Emerg*, Vol. 8, pp. 143–153, 2007.

14. T. Li, X.W. Shi, Y.M. Du and Y.F. Tang, *J. Biomed. Mater. Res. A*, Vol. 83, pp. 383–390, 2007.

15. A. Richter, G. Paschew, S. Klatt, J. Lienig, K. F. Arndt, H.J. P. Adler, *Sensor*, Vol. 8, pp. 561–581, 2008.

16. K.Y. Lee, D.J. Mooney, *Chem. Rev.*, Vol. 101, pp. 1869–1879, 2001.

17. A.M. Mathur, S.K. Moorjani, A.B. Scranton, *J. Macromol. Sci. Polymer Rev*, Vol. 36, pp. 405–430, 1996.

18. S. Dwivedi, P. Khatri, G. R. Mehra and V. Kumar, *Int. J. Pharm. Biol. Arch.*, Vol. 2, pp. 1588–1597, 2011.

19. F. Ganji, S. Vasheghani-Farahani, Ebrahim Vasheghani-Farahani, *Iran. Polym. J*, Vol. 19, pp. 375–398, 2010.

20. B. Obradović, *Cell and Tissue Engineering, - Biochemical engineering*, Springer.199. 3642219136, 9783642219139, p. 275, 2012.

21. T.R. Hoare, D.S. Kohane, *Polymer*, Vol. 49, pp. 1993–2007, 2008.

22. D. Campoccia, P. Doherty, M. Radice, P. Brun, G. Abatangelo and D.F. Williams, *Biomaterials* Vol. 19, pp. 2101–2127, 1998.

23. R.A. Gemienhart and Ch. Guo; Fast Swelling Hydrogel Systems. In: Yui N., Mrsny R.J., Park K. (eds), *Reflexive Polymers and Hydrogels*. CRC Press, New York, pp. 245–258, 2004.

24. B. V. Slaughter, S.S. Khurshid, O.Z. Fisher, A. Khademhosseini, N.A. Peppas, *Adv. Mater*, Vol. 21, pp. 3307–3329, 2009.

25. T. Katayama, M. Nakauma, S. Todoriki, G.O. Phillips, M. Tada, *Food Hydrocolloid.*, Vol. 20, pp. 983–989, 2006.

26. N.A. Peppas, P. Bures, W. Leobandung, H. Ichikawa, *Eur. J. Pharm. BioPharm*, Vol. 50, pp. 27–46. 2000.

27. S.K.S. Kushwaha, P. Saxena, A.K. Rai, *Int. J. Pharm. Investig*, Vol. 2, pp. 54–60, 2012.

28. M. Bassil, M. Ibrahim, M.El-Tahchi, *Soft Matter*, Vol. 7, pp. 4833–4838, 2011.

29. F. Ganji, E.Vasheghani-Farahani, *Iran. Polym. J*, Vol. 18, pp. 63–88, 2009.

30. M. Guenther, G. Gerlach, C. Corten, D. Kuckling, J. Sorber, K.F. Arndt. *Sensor Actuat. B-Chem*, Vol. 132, pp. 471–476, 2008.

31. Y.B. Schuetz, R. Gurny , O. Jordan, *Eur. J. Pharm. Bio. Pharm*, Vol. 68, pp. 19–25, 2008.

32. C.S. Satish, K.P. Satish and H.G Shivakumar, *Indian J. Pharm. Sci*, Vol. 68, pp. 133–140, 2006.

33. S.S. Bushetti, V. Singh, S.A. Raju, A. javed, V. ram, *Indian J. Pharm. Educ. Res*, Vol. 43, pp. 241–250, 2009.

34. C.C. Lin, A.T. Metters, *Adv. Drug. Deliv. Rev*, Vol. 58, pp. 1379–1408, 2006.

35. J.C. Leroux , E. Ruel-Gariepy, *Eur. J. Pharm. BioPharm,*Vol. 58, pp. 409–426, 2004.

36. Y.Qui, K. Park, *Adv. Drug. Deliv. Rev*, Vol. 53, pp. 321–339, 2001.

37. S. Murdan, *J. Control Release*, Vol. 92, pp. 1–17, 2003.

38. R. Masteikova, Z. Chalupova, Z. Sklubalova, *Medicina*, Vol. 39, pp. 19–24, 2003.

39. P. Bawa, V. Pillay, Y.E Choonara and L.C. Du-Toit, *Biomed. Mater*, Vol. 4, pp. 1–15, 2009.

40. J. Hur, K. Im, S. Hwang, B. Choi, S. Kim, S. Hwang, N. Park, K. Kim, *Sci. Rep*, Vol. 3, pp. 1282–1288, 2013.

41. C. Wang, B. Yu, B. Knudsen, J. Harmon, F. Moussy and Y. Moussy, *Biomacromolecules*, Vol. 9, pp. 561–567, 2008.

42. J.P. Gong, Y. Katsuyama, T. Kurokawa, Y. Osada, *Adv. Mater*, Vol. 15, pp. 1155–1158, 2003.

43. Y. Tanaka, J.P. Gong, Y. Osada, *Prog. Polym. Sci*, Vol. 30, pp. 1–9, 2005.

44. Y.H. Na, T. Kurokawa, Y. Katsuyama, H. Tsukeshiba, J.P. Gong, Y. Osada, S. Okabe, T. Karino, M. Shibayama , *Macromolecules*, Vol. 37, p. 5370–5374, 2004.

45. K. Haraguchi, T. Takehisa, S. Fan, *Macromolecules*, Vol. 35, pp. 10162–10171, 2002.

46. K. Haraguchi, T. Takehisa, *Adv. Mater*, Vol. 14, pp. 1120–1124, 2002.

47. L. Jeffrey, A. Hinkley, L.D. Morgret, S.H. Gehrke, *Polymer*, Vol. 45, pp. 8837–8843, 2004.

48. A. Svensson, E. Nicklasson, T. Harrah, B. Panilaitis, D.L. Kaplan, M. Brittberg, P. Gatenholm, *Biomaterials*, Vol. 26, pp. 419–431, 2005.

49. R.A. Mirshams and R.M. Pothapragada, *Acta. Mater*, Vol. 54, pp. 1123–1134, 2006.

50. R. Lapasin and S. Pricl, *Rheology of industrial polysaccharides: theory and application.* Cornwall, U.K.: Blackie Academic and Professional; 1995.

51. N. Sahiner, M. Singh, D. De-Kee, V.T. John, G.L. McPherson, *Polymer*, Vol. 47, pp. 1124–1131, 2006.

52. L.L. Hyland, M.B. Taraban, Y. Feng, B. Hammouda, Y.B. Yu, *Biopolymers*, Vol. 97, pp. 177–188, 2011.

53. S.K. De, N.R. Aluru, B. Johnson, W.C. Crone, D.J. Beebe, J. Moore, *J Microelectromech S.*, Vol. 11, pp. 544–555, 2002.

54. S. Gunasekaran, T. Wang, C. Chai, *J. Appl. Polym. Sci*, Vol. 102, pp. 4665–4671, 2006.

55. C. Ruana, K. Zenga, C.A. Grimes, *Anal. Chim. Acta*, Vol. 497, pp. 123–131, 2003.

56. H. Omidian, J.G. Rocca, K. Park, *J. Controlled Release*, Vol. 102, pp. 3–12, 2005.

57. F. Jianqi, G. Lixia, *Eur. Polym. J*, Vol. 38, pp.1653–1658, 2002.
58. P. Gupta, K. Vermani, S. Garg, *Drug Discovery Today*. Vol. 7, pp. 569–579, 2002.
59. K. Urayama, T. Takigawa, T. Masuda, *Macromolecules*, Vol. 26, pp. 3092–3096, 1993.
60. P.F. Kiser, G. Wilson, D. Needham, *J. Controlled Release*, Vol. 68, pp. 9–22, 2000.
61. M.Y. Kizilay, O. Okay, *Macromolecueles*, Vol. 36, pp. 6856–6862, 2003.
62. B. Amsden, *Macromolecules*, Vol. 31, pp. 8382–8395, 1998.
63. M.E. McNeill, N.B. Graham, *J. Biomater. Sci. Polymer Edn*, Vol. 4, pp. 305–322, 1993.
64. J. Crank and G.S. Park, Methods of Measurements. Ed. Crank J. Academic Press London and New York, pp. 1–39, 1968.
65. J. Chen, H. Park, K. Park, *J. Biomed. Mater. Res. A*, Vol. 44, pp. 53–62, 1999.
66. N.A. Peppas and J.J. Sahlin, *Int. J. Pharm*, 57(2), pp. 169–172, 1989.
67. E. Diez-Pena, I. Quijada-Garrido, J.M. Barrales-Rienda. *Macromolecules*, Vol. 35, pp. 8882–8888, 2002.
68. Y.K. Bhardwaj, V.Kumar, S. Sabharwal, *J. Appl. Polym. Sci.*, Vol. 88, pp. 730–742, 2003.
69. S. Wu, H. Li, J.P. Chen, *J. Macromol. Sci., Polym. Rev.*, Vol. 44, pp. 113–130, 2004.
70. M. Takigami, H. Amada, N. Nagasawa, T. Yagi, T. Kasahara, S. Takigami, M. Tamada, *Trans. Mater. Res. Soc. Japan*, Vol. 32, pp. 713–716, 2007.
71. W.E. Hennink and C.F. Van-Nostrum. *Adv. Drug Deliv. Rev*, Vol. 54, pp. 13–36, 2002.
72. H. Tan, A.J. Defail, J.P. Rubin, C.R. Chu and K.G. Marra, *J. Biomed. Mater. Res. A*, Vol. 92A, pp. 979–987, 2010.
73. A.K. Bajpai, S. K. Shukla, S. Bhanu, S. Kankane, *Prog. Polym. Sci*, Vol. 33, pp. 1088–1118, 2008.
74. Q.S. Zhao, Q.X. Ji, K. Xing, X.Y. Li, C.S. Liu, X.G. Chen, *Carbohydr. Polym*, Vol. 76, pp. 410–416, 2009.
75. C. Esteban and D. Severian, Polyionic hydrogels based on xanthan and chitosan for stabilising and controlled release of vitamins, *Vol. WO0004086 (A1)* (ed. U. United States Patent), Kemestrie Inc [CA], USA, 2000.
76. D. Magnin, J. Lefebvre, E. Chornet, S. Dumitriu, *Carbohydr. Polym,* 55, pp. 437–453, 2004.
77. D. Eagland, N.J. Crowther, C.J. Butler, *Eur. Polym. J*, Vol. 30, pp. 767–773, 1994.
78. J.C. Gayet, G. Fortier, *J. Control. Rel*, Vol. 38, pp. 177–184, 1996.
79. T. Pongjanyakul, S. Puttipipatkhachorn, *Int. J. Pharm.*, Vol. 331, pp. 61–71, 2007.
80. M. Himi and S.D. Maurya, *J. Drug Deliv. Therapeutics*, Vol. 3, pp. 131–140. 2013.
81. C.Sandeep, S.L. Harikumar, Kanupriya, *Int. J. Res. Pharm. Chem*, Vol. 2, pp. 603–614, 2012.

82. S. Gupta, T.J. Webster, A.Sinha, *J. Mater. Sci. Mater. Med*, Vol. 22, pp. 1763–1772, 2011.
83. C.M. Hassan, N. A. Peppas, *Macromolecules*, Vol. 33, pp. 2472–2479, 2000.
84. P. Giannouli, E.R. Morris, *Food Hydrocolloids*, Vol. 17, pp. 495–501, 2003.
85. H. Aoki, S. Al-Assaf, T. Katayama, G.O. Phillips, *Food Hydrocolloids*, Vol. 21, pp. 329–337, 2007.
86. C. Chang and L. Zhang, "Cellulose-based hydrogels: Present status and application prospects." *Carbohydr. Polym*,Vol. 84, pp. 40–53, 2011.
87. Syed K. H. Gulrez, S. Al-Assaf, G. O. Phillips, *Hydrogels: Methods of Preparation, Characterisation and Applications, Progress in Molecular and Environmental Bioengineering—From Analysis and Modeling to Technology Applications*, Prof. Angelo Carpi (Ed.), ISBN: 978-953-307-268-5. 2011.
88. N.A. Peppas, A.G. Mikos, In: Peppas, N.A., Eds,. Preparation Methods and Structure of Hydrogels, Hydrogels in Medicine and Pharmacy, Vol I, CRC Press, Boca Raton, FL, 986, 1.
89. M.B. Huglin, M.B. Zakaria, *J. Appl. Polym. Sci.*,Vol. 31, pp. 457–475, 1986.
90. D. Sarydin, E. Karadag, N. Sahiner, O. Guven, *J. Mater. Sci*, Vol. 37, pp. 3217–3223, 2002.
91. E. Jabbari, S. Nozari, *Eur. Polym. J*, Vol. 36, pp. 2685–2692, 2000.
92. K. Makuuchi, *Radiat. Phys. Chem*, Vol. 79, pp. 267–271, 2010.
93. L.S. Wang, C. Du, J.E. Chung and M. Kurisawa, *Acta. Biomater.*, Vol. 8, pp.1826–1837, 2012.
94. J.J. Sperinde, L.G. Griffith, *Macromolecules*, Vol. 33, pp. 5476–5480, 2000.
95. J. L. Koenig, *Spectroscopy of polymers*, Elsevier Science, New York, USA 1999.
96. J. M. Chalmers and P. R. Griffiths (Eds.) *Handbook of vibrational spectroscopy*, Vol. 2, John Wiley & Sons, Chichester, UK 2002.
97. L.L. Hench and J.R. Jones. Biomaterials, artificial organs and tissue engineering, Woodhead Publishing Limited, Cambridge, England, p.39. 2005
98. D.T.R. Austin, B.P. Hills, *Appl. Magn. Reson*, Vol. 35, pp. 581–591, 2009.
99. F.A. Dorkoosh, J. Brussee, J.C. Verhoef, G. Borchard, M. Rafiee-Tehrani, H.E. Junginer, *Polymer*, Vol. 41, pp. 8213–8220, 2000.
100. A. Teleman, J. Lundqvist, F. Tjerneld, H. Stalbrand, O. Dahlman, *Carbohy. Res*, Vol. 329, pp. 807–815, 2000.
101. F. Brandl, F, Kastner, R.M. Gschwind, T. Blunk, J. Tessmar and A. Gopferich, *J. Control. Release*, Vol. 142, pp. 221–228, 2010.
102. http://www.nanoscience.com/education/AFM.html
103. M. Caldorera-Moore, M.K. Kang, Z. Moore, V, Singh, S.V. Sreenivasan, L. Shi, R. Huang, K. Roy, *Soft Matter*, Vol. 7, pp. 2879–2887, 2011.
104. C. Maldonado-Codina and N. Efron, *Clin. Exp. Optom*, Vol. 86, pp. 396–404, 2005.
105. F. Castelli, M.G. Sarpietro, D. Micieli, S. Ottimo, G. Pitarresi, G. Tripodo, B. Carlisi, G. Giammona, *Eur J Pharm Sci*, Vol. 35, pp. 76–85, 2008.
106. G. Deepa, A.K. Thulasidasan, R.J. Anto, J.J. Pillai and G.S. Kumar, *Int. J. Nanomedicine*, Vol. 2012, pp. 4077–4088, 2012.
107. M. Guvendiren, H.D. Lu, J.A. Burdick, *Soft Matter*, Vol. 8, pp. 260–272, 2012.

108. A.R. Goodall, M.C. Wilkinson, J. Hearn, *J. Colloid Interface Sci*, Vol. 53, pp. 327–331, 1975.

109. M.J. Murray and M.J. Snowden, *Adv. Colloid Interface Sci*, Vol. 54, pp. 73–91, 1995.

110. A. Pourjavadi, M. Kurdtabar, *Eur. Polym. J.*, Vol. 43, pp. 877–889, 2007.

111. J. Silva-Correia, B. Zavan , V.Vindigni , T.H. Silva , J.M. Oliveira , G. Abatangelo, R.L. Reis, *Adv. Healthc. Mater.*,Vol. 2, pp. 568–575, 2013.

112. S. Al-Assaf, G.O. Phillips, H. Aoki, Y. Sasaki, *Food Hydrocolloids*, Vol. 21, pp. 319–328, 2007.

113. S. Al-Assaf, G.O. Phillips, P.A. Williams, *Food Hydrocolloids*, Vol. 20, pp. 369–377, 2006.

114. A. Patanarut, A. Luchini, P.J. Botterell, A. Mohan, C. Longo, P. Vorster, A.F. Petricoin, L.A. Liotta and B. Bishop, *Colloids Surf. A Physicochem. Eng. Asp*, Vol. 362, pp. 8–19, 2010.

115. S.B. Debord, A. Lyon, *J. Phys. Chem. B*, Vol.107, pp. 2927–2932, 2003.

116. E.A. Nieuwenhuis, C. Pathmamanoharan, A. Vrij, *J. Colloid Interface Sci*, Vol. 81, pp. 196–213, 1981.

117. M.C. Wilkinson, J. Hearnand, P.A. Steward, *Adv. Colloid Interface Sci*,Vol. 81,pp. 77–165,1999.

118. K. Laslo, A. Guillermo, A. Fluerasu, A. Maussaid, E. Geissler, *Langmuir*, Vol. 26, pp. 4415–4420, 2010.

119. S.S. Singh, V.K. Aswal, H.B. Bohidar, *Eur. Phys. J. E. Soft Matter*, Vol. 34, pp. 1–9, 2011.

120. S. Yang, S. Fu, X. Li, Y. Zhou, H. Zhan, *BioResources*, Vol. 5, pp. 1114–1125, 2010.

121. J. Mao, S. Kondu, H.F. Ji, M.J. Mcshane, *Biotechnol. Bioeng*, Vol. 95, pp. 333–341, 2006.

122. R. Flamia, A.M. Salvi, L. D'Alessio, J.E. Castle, A.M. Tamburro, *Biomacromolecules*, Vol. 8, pp. 128–138, 2007.

123. A. Chenite, M. Buschmann, D. Wang, C. Chaput, N. Kandani, *Carbohydr. Polym*, Vol. 46, pp. 39–47, 2001.

124. A.W. Watkins, S.L. Southard, K.S. Anseth, *Acta Biomater*, Vol. 3, pp. 439–448, 2007.

125. A. D'Emanuele, C. Gilpin, *Scanning*, Vol. 18, pp. 522–527, 1996.

126. Y. Hou, C.A. Schoener, K.R. Regan, D. Munoz-Pinto, M. S. Hahn, M.A. Grunlan, *Biomacromolecules*, Vol. 11, pp. 648–656, 2010.

127. N.A. Peppas, J.Z. Hilt, A. Khademhosseini, R. Langer, *Adv Mater*, Vol. 18, pp.1345–1360, 2006.

128. D.T. Eddington, D.J. Beebe, *Adv. Drug. Deliv. Rev*, Vol.56, pp. 199–210, 2004.

129. M. Sirousazar, M. Yari, *Chin. J. Polym. Sci*, Vol. 28, pp. 573–580, 2010.

130. M. Sirousazar, M. Kokabi, M. Yari, *Iran. J. Pharm. Sci*, Vol. 4, pp. 51–56, 2008.

131. N. Nakabayashi, D.F. Williams, *Biomaterials*, Vol. 24, pp. 2431–2435, 2003.

132. J. L. Drury, D.J. Mooney, *Biomaterials*, Vol. 24, pp. 4337– 4351, 2003.

133. S. Varghese, J.H. Elisseeff, *Adv. Polym. Sci*, Vol. 203, pp. 95–144, 2006.

134. S. Nayak, L.A. Lyon, *Angew. Chem. Int. Ed*, Vol. 44, pp. 7686–7708, 2005.

135. J. Kopecek, *Nature*, Vol. 417, pp. 388–391, 2002.

136. Z. Yang, Y. Zhang, P. Markland, V. C. Yang, *Wiley Periodicals, Inc. J Biomed Mater Res,* Vol. 62, pp. 14–21, 2002.

137. N.A. Peppas, J.J. Sahlin, *Biomaterials*, Vol.17, pp.1553–1561, 1996.

138. O.Wichterle, D. Lim, *Nature*, Vol. 185, pp. 117–118, 1960.

139. J.L. Panza, W.R. Wagner , H.L. Rilo , R.H. Rao , E.J. Beckman, A.J. Russell, *Biomaterials*, Vol. 21, pp. 1155–1164, 2000.

140. Z. Ajji, I. Othman, J.M. Rosiak, *Nucl Instrum Meth B*, Vol. 229, pp. 375–380, 2005.

141. S. Cohen , E. Lobel, A. Trevgoda, Y, Peled, *J. Control Release*, Vol. 44, pp. 201– 208, 1997.

142. S. H. Lee, H. Shin, *Adv Drug Deliv Rev*, Vol. 59, pp. 339–359, 2007.

143. R. Langer, J. P. Vacanti, *Science*, Vol. 260, pp. 920–926, 1993.

144. S. Sershen, G. Mensing, M. Ng, N. Halas, D. Beebe, J. West, *Adv. Mater,* Vol. 17, pp. 1366–1368, 2005.

145. S. W. Kim, Y. H. Bae, "Stimuli-Modulated Delivery Systems," In: G. L.Amidon, P. I. Lee and E. M. Topp, Eds., *Transport Processes in Pharmaceutical Systems*, Marcel Dekker, New York, pp. 547–573, 2000.

146. L. B. Peppas, "Polymers in Controlled Drug Delivery", *Medical Plastics and Biomaterials Magazine*, November, 1997.

147. L. Achar, N.A. Peppas, *J. Controlled Release*, Vol. 31, pp. 271– 276, 1994.

148. F.M. Veronese, C. Mammucari, P. Caliceti, O. Schiavon, S. Lora, *J. Bioact Compat Polym*, Vol. 14, pp. 315–330, 1999.

149. L. Klouda, A.G. Mikos, *Eur. J. Pharm. Biopharm*, Vol. 68, pp. 34–45, 2008.

150. P. Colombo, R. Bettini, P. Santi, A. De Ascentiis, N.A. Peppas, *J. Controlled. Release,* Vol. 39, pp. 231–237, 1996.

151. J. Siepmann, F. Siepmann, *Int. J. Pharm*, Vol. 364, pp. 328–343, 2008.

152. N.A. Peppas, P. Bures, W. Leobandung, H. Ichikawa, *Eur. J. Pharm. Biopharm*, Vol. 50, pp. 27–46, 2000.

153. S. J. Holland, B. J. Tighe, P. L. Gould, *J. Controlled Release*, Vol. 4, pp. 155–180, 1986.

154. J.S. Boateng, K.H. Matthews, H.N.E. Stevens, G.M. Eccleston, *J. Pharm. Sci,* Vol. 97, pp. 2892–2923, 2008.

155. G.D. Winter, *Nature*, Vol. 193, pp. 293–294, 1962.

156. S.K. Purna, M. Babu, *Burns*, Vol. 26, pp. 54–62, 2000.

157. S.Y. Lin, K.S. Chen, L. Run-Chu, *Biomaterials*, Vol.22, pp. 2999–3004, 2001.

158. G.A. Kannon, A.B. Garrett, *Dermatol. Surg*, Vol. 21, pp. 583–590, 1995.

159. R. Snyder, Proteases in wound healing: understanding how these chemical mediators work represents a paradigm shift in wound care. *Podiatry Management* (www.podiatrym.com - subscription required), 2004.

160. K.J. Quinn, J.M. Courtney, J.H. Evans, J.D.S. Gaylor, W.H. Reid, *Biomaterials*,Vol.6, pp. 369–377, 1985.

161. R. J. Morin, N. L. Tomaselli, *Clinics in Plastic Surgery*, vol. 34, pp. 643–658, 2007.

162. M.A. Fonder, G.S. Lazarus, D.A. Cowan, B. Aronson–Cook, A.R. Kohli, A.J. Mamelak, *J Am Acad Dermatol*, Vol. 58, pp. 185–206, 2008.

163. Y. Tabata, *Journal of the Royal Society, Interface,* Vol. 6, S311–S324, 2009.

164. S. Huang and X. Fu, *J. Control Release,* Vol. 142, pp. 149–159, 2010.

165. P. Bao, A. Kodra, M. Tomic-Canic, M.S. Golinko, H.P. Ehrlich, H. Berm, *Journal of Surgical Research*, Vol. 153, pp. 347–358, 2009.

166. S.E. Hanson, M.L. Bentz, P. Hematti, *Plast Reconstr Surg*, Vol. 125, pp. 510–516, 2010.

167. J. Cha, V. Falanga, *Clin Dermatol*, Vol. 25, pp. 73–78, 2007.

168. A. Stoff, A.A. Riviera, N.S. Banerjee, S.T. Moore, T.M. Numnum, A. Espinosa-de-los-Monteros, D.F. Richter, G.P. Siegal, L.T. Chow, D. Feldman, L.O. Vasconez, J.M. Mathis, M.A. Stoff-Khalili, D.T. Curiel, *Exp. Dermatol*, Vol 18, pp. 362–369, 2009.

169. T. Yoshikama, H. Mitsumo, I. Nonaka, Y. Sen, K. Kawanishi, Y. Inada, Y. Takakura, K. Okuchi, A. Nonomura, *Plast Reconstr Surg*, Vol. 121, pp. 860–877, 2008.

170. S.H. Ko, A. Nauta, V. Wong, J. Glotzbach, G.C. Gurtner, M.T. Longaker, *Plast Reconstr Surg.* Vol. 127(Suppl 1): pp.10S–20S, 2011.

171. M. Cherubino, J.P. Rubin, N. Miljkovic, A. Kelmendi–Doko, K.G. Marra, *Ann Plast Surg*, Vol. 66, pp. 210–215, 2011.

172. G.G. Gauglitz, M.G. Jeschke, *Mol. Pharma*, Vol. 8, pp. 1471–1479, 2011.

173. L.K. Branski, G.G. Gauglitz, D.N. Herndon, M.G. Jeschke, *Burns*, Vol. 35, pp.171–180, 2009.

174. T. Vernon, *Br J Community Nurs*, Vol. 5, pp. 511–516, 2000.

175. M.T. Razzak, D. Darwis, Z. Sukirno, *Radiat. Phys. Chem,* Vol. 62, pp. 107–113, 2001.

176. F. Yoshii, Y. Zhanshan, K. Isobe, K. Shiozaki, K. Makunchi, *Radiat. Phys. Chem*, Vol. 55, pp. 133–138, 1999.

177. M. Sirousazar, M. Kokabi, Z. M. Hassan, A. R. Bahramian, *J. Macromol. Sci., Phys*, Vol. 51, pp. 1335–1350, 2012.

178. J. Kopecek, *J. Polym. Sci. Pol. Chem*, Vol. 47, pp. 5929–5946, 2009.

179. J.P. Gong, *Soft Matter*, Vol. 6, pp. 2583–2590, 2010.

180. M. A. Haque, T. Kurokawa, J. P. Gong, *Polymer*, Vol. 53, pp. 1805–1822, 2012.

181. M. Sirousazar, M. Kokabi, Intelligent Nanocomposite Hydrogels, in: Ashutosh Tiwari, Ajay K. Mishra, Hisatoshi Kobayashi, Anthony P.F. Turner (Eds.), Intelligent Nanomaterials: Processes, Properties, and Applications , John Wiley & Sons Inc., Hoboken, NJ, USA , pp. 487–531, 2012.

182. M. Sirousazar, M. Kokabi, Z. M. Hassan, A. R. Bahramian, *J. Macromol. Sci., Phys*, Vol. 51, pp. 1583–1595, 2012.

183. G. G. Ferrer, M. M. Pradas, J. L. G. Ribelles, P. Pissis, *J. Non-Cryst. Solids*, Vols. 235–237, pp. 692–696, 1998.

184. J. L. G. Ribelles, M. M. Pradas, G. G. Ferrer, N. P. Torres, V. P. Gimenez, P. Pissis, A. Kyritsis, *J. Poly. Sci. B*, Vol. 37, pp. 1587–1599, 1999.

185. G. G. Ferrer, J. M. S. Melia, J. H. Canales, J. M. M. Duenas, F. R. Colomer, M. M. Pradas, J. L. G. Ribelles, P. Pissis, G. Polizos, *Colloid. Polym. Sci*, Vol. 283, pp. 681–690, 2005. 185. M. Sirousazar, M. Kokabi, Z. M. Hassan, A. R. Bahramian, *J. Appl. Polym. Sci,*Vol. 125, E122–E130, 2012.

186. S. L. Bourke, M. A. Khalili, T. Briggs, B.B. Michniak, J. Kohn, L.A.P. Warren, *AAPS Pharm Sci*, Vol. 5, pp. 101–111, 2003.

187. H. H. Wang, T. W. Shyr, M. S. Hu, *J. Appl. Polym. Sci*, Vol. 74, pp. 3046–3052, 1999.

188. P. R. Hari, K. Sreenivasan, *J. App.l Polym. Sci*, Vol. 82, pp. 143–149, 2001.

189. J. Varshosaz, N. Koopaie, *Iran. Polym. J*, Vol. 11, pp. 123–131, 2002.

190. W.S. Dai, T.A. Barbari, *J. Memb. Sci,*Vol. 156, pp. 67–79, 1999.

191. C. Xiao, G. Zhou, *Polym. Degrad. Stab*, Vol. 81, pp. 297–301, 2003.

192. R. Ricciardi, C. Gaillet, G. Ducouret, F. Lafuma, F. Laupretre, *Polymer*, Vol. 44, pp. 3375–3380, 2003.

193. W. E. Hennink, C. F. V. Nostrum, *Adv. Drug. Deliv. Rev*, Vol. 54, pp.13–36, 2002.

194. J. M. Rosiak, P. Ulanski, A. Rzeinicki, *Nucl. Instrum & Methods in Phys. Res. B*, Vol. 105, pp. 335–339, 1995.

195. A.B. Lugão, L.D.B. Machado, L.F. Miranda, M.R. Alvarez, J.M. Rosiak, *Radiat. Phys. Chem*, Vol. 52, pp. 319–322, 1998.

196. R.F. Ofstead, C.I. Posner, Semicrystalline poly (vinyl alcohol) hydrogels. In: J Glass, ed. Polymers in Aqueous Media. Washington, DC: *American Chemical Society*, pp. 61–72, 1989.

197. C. Young, J.R. Wu, T.L. Tsou, *J. Membrane Sci*, Vol. 146, pp.83–93, 1998.

198. G. Underhill, A. Chen, D. Albrecht, S. Bhatia, *Biomaterials*, Vol. 28, pp. 256–270, 2007.

199. S. Benamer, M. Mahlous, A. Boukrif, B. Mansouri, S. L. Youcef, *Nucl. Instrum & Methods in Phys. Res. B*, Vol. 248, pp. 248–290, 2006.

200. M. Wang, L. Xu, H. Hu, M. Zhai, J. Peng , Y. Nho, J. Li, G. Wei, *Nucl. Instrum & Methods in Phys. Res. B*, Vol. 265, pp. 385–389, 2007.

201. P. Petrini, M. Tanzi, *J. Mater. Sci.:Mater. in Medicine*, Vo. 10, pp. 635–639, 1999.

202. J. L. Drury, R.G. Dennis, D.J Mooney, *Biomaterials*, Vol. 25, pp.3187–3199, 2004.

203. P. J. Flory, *Polymer*, Vol. 20, pp. 1317–1320, 1979.

204. N.A. Peppas, E.W. Merrill, *J. Appl. Polym. Sci*, Vol. 21, pp.1763–1770, 1977.

205. M. Kim, S. Tang, B. D. Olsen. *J. Polym. Sci. Part B: Polym. Phys*, Vol. 51, pp. 587–601, 2013.

206. J. Rička, T. Tanaka, *Macromolecules,*Vol. 17, pp. 2916–2921, 1984.

207. A. Pourjavadi, H. Salimi, *J. Ind. Eng. Chem*, Vol. 47, pp. 9206– 9213, 2008.

208. M. G. Swann, W. Bras, P. D. Topham, J. R. Howse, A. J. Ryan. *Langmuir*, Vol. 26, pp. 10191–10197, 2010.

209. M. Gao, K. Gawel, B. T. Stokke, *Soft. Matter*, Vol. 7, pp. 1741–1746, 2011.

210. I. Donati, Y. A. Mørch, B. L. Strand, G. Skjåk-Bræk, S. Paoletti. *J. Phys. Chem. B*, Vol. 113, pp. 12916–12922, 2009.
211. S. E. Kudaibergenov, V. B. Sigitov, *Langmuir*, Vol. 15, pp. 4230–4235, 1999.
212. G. O. Phillips, T. A. D. Plessis,, S. Al-Assaf, P. A. Williams, Biopolymers obtained by solid state irradiation in an unsaturated gaseous atmosphere, Vol. *6,610,810*, (ed. U. S. Patent), Phillips Hydrocolloid Research limited, UK., 2003.
213. R.M. Hodge, G.H. Edward, G.P. Simon, *Polymer*, Vol. 37, pp. 1371–1376, 1996.
214. W. I. Cha, S. H. Hyon, Y. Ikada, *Macromol. Chem. Physic,* Vol. 194, pp. 2433–2441, 1993.
215. W.E. Roorda, J.A. Bouwstra, M.A. Salomons-de Vries, H.E. Junginger, *Biomaterials*, Vol. 9, pp. 494–499, 1988.
216. J.A. Bouwstra, M.A. Salomons-de Vries, J.C. van Miltenburg, *Thermochim. Acta*, Vol. 248, pp. 319–327, 1995.
217. Y. Maeda, N. Tsukida, H. Kitano, T. Terada, J. Yamanaka, *J. Phys. Chem*, Vol. 97, pp. 13903–13906, 1993.
218. F. A. Aouada, M. R. de Moura, P. R. G. Fernandes, A. F. Rubira, E. C. Muniz, (2005) *Eur. Polym. J.* Vol. 41, pp. 2134–2141, 2005.
219. M. El Fray, A Pilaszkiewicz, W. Swieszkowski, K. J. Kurzydlowski, *Eur. Polym. J.* Vol. 43, pp. 2035–2040, 2007.
220. H.A. Awad, M. Q. Wickham, H.A. Leddy, J.M. Gimble, F. Guilak, *Biomaterials*, Vol. 25, pp. 3211–3222, 2004.
221. A. B. Imran, T. Seki, Y. Takeoka, *Polym. J*, Vol. 42, pp. 839–851, 2010.
222. M. Sirousazar, M. Kokabi, Z.M. Hassan, A.R. Bahramian, *Scientia Iranica, Transaction F: Nanotechnology*, Vol. 18, pp. 780–784, 2011.
223. M. Sirousazar, M. Kokabi, Z.M. Hassan, *J. Biomater. Sci, Polym. Ed*, Vol. 22, pp. 1023–1033, 2011.
224. P. Schexnailder, G.Schmidt, *Colloid. Polym. Sci*, Vol.287, pp. 1–11, 2009.
225. A. Samantha, K. W. A, Meenach, J. Z. Hilt, Hydrogel Nanocomposites: Biomedical Applications, *Biocompatibility, and Toxicity Analysis*. New York: Springer, 2009.
226. P. Saravanan, M. P. Raju, S. Alam, *Mater. Chem. Phys*, Vol.103, pp. 278–282, 2007.
227. Y. T. Wu, Z. Zhou, Q. Q.Fan, L. Chen, M. F. Zhu, *J. Mater. Chem*, Vol.19, pp. 7340–7346, 2009.
228. A. Nakayama, A. Kakugo, J. P. Gong, Y. Osada, M. Takai, T. Erata, S. Kawano, *Adv. Funct. Mater,* Vol. 14, pp.1124–1128, 2004.
229. C. D. Sorrell, M. C. D. Carter, M. J. Serpe, *Adv. Funct. Mater*, Vol. 21, pp. 425–433, 2011.
230. L. Hu, M. J. Serpe. *J. Mater. Chem*, Vol. 22, pp. 8199–8202, 2012.
231. K. C. C. Johnson, F. Mendez, M. J. Serpe, *Anal. Chim. Acta*, Vol. 739, pp. 739, 83–88, 2012.

232. M. Kokabi, M. Sirousazar, Z. M. Hassan, *Eur. Polym. J*, Vol. 43, pp. 773–781, 2007.

233. M. Sirousazar, M. Yari, B.F. Achachlouei, J. Arsalani, and Y. Mansoori, e-polymers, no. 027, pp. 1–9, 2007.

234. Y. Mansoori, S.V. Atghia, M.R. Zamanloo, Gh. Imanzadeh, and M. Sirousazar, *Eur. Polym. J*, Vol. 46, pp. 1844–1853, 2010.

235. B. Xu, H. Li, Y. Wang, G. Zhang, Q. Zhang, *RSC. Adv*, Vol. 3, pp. 7233–7236, 2013.

236. M. Sirousazar, M. Kokabi, Z. M. Hassan, *J. Appl. Polym. Sci*, Vol. 123, pp. 50–58, 2012.

Part 4

EMERGING BIO-ENGINEERING DEVICES

10

Modified Natural Zeolites—Functional Characterization and Biomedical Application

Jela Milić[1]*, Aleksandra Daković[2], Danina Krajišnik[1] and George E. Rottinghaus[3]

[1]*Department of Pharmaceutical Technology and Cosmetology, University of Belgrade–Faculty of Pharmacy, Belgrade, Serbia*
[2]*Institute for Technology of Nuclear and Other Mineral Raw Materials, Belgrade, Serbia*
[3]*Veterinary Medical Diagnostic Laboratory, College of Veterinary Medicine, University of Missouri, Columbia, MO, USA*

Abstract

Natural and synthetic zeolites have emerged as potential materials for biomedical application in recent years. Zeolites are hydrated microporous tektoaluminosilicates consisting of three-dimensional frameworks of SiO_4 and AlO_4 tetrahedra linked through shared oxygen atoms. Clinoptilolite, a mineral from the heulandite group of zeolites, $((Na,K)_6(Al_6Si_3O)O_{72} \cdot nH_2O)$, is the most abundant sedimentary zeolite in nature. In this chapter an overview of surface modification of clinoptilolite by interaction with cationic surfactants is given alongside with the methods used for characterization of the surfactant modified zeolites (organozeolites/composites). Different organozeolites were tested under conditions for adsorption of several mycotoxins commonly found in animal feed. Study of *in vitro* surfactant desorption *in vitro* followed by *in vivo* acute toxicity testing was used to demonstrate the nontoxic nature of these improved mineral materials. Functionality related characteristics of modified natural zeolites were analyzed for investigation of their potential use as pharmaceutical excipients for modified release of active pharmaceutical ingredients.

*Corresponding author: jela@pharmacy.bg.ac.rs

Ashutosh Tiwari (ed.) Advanced Healthcare Materials, (359–404) 2014 © Scrivener Publishing LLC

Keywords: Clinoptilolite, cationic surfactant, mycotoxins, drugs, adsorption, organozeolites/composites, pharmaceutical excipients, functionality related characteristics

10.1   Introduction

Zeolites were discovered in the 18th century by the Swedish mineralogist Axel Fredrik Cronstedt. Natural zeolites are safe, environmentally friendly, naturally occurring minerals. Briefly, they are crystalline, hydrated tectoaluminosilicates of alkali and alkaline earth cations having an infinite, open, three-dimensional cage-like structure. Zeolites are able to lose and gain water reversibly and to exchange extraframework cations, both without change of crystal structure. The large structural cavities and the entry channels leading into them contain water molecules, which form hydration spheres around exchangeable cations. Some of the more common natural zeolites are: analcime, chabazite, heulandite, natrolite, phillipsite, and stilbite [1, 2].

Most zeolites in volcanogenic sedimentary rocks were formed by the dissolution of volcanic glass (ash) and later precipitation of micrometer-size crystals, which mimic the shape and morphology of their basalt counterparts. Sedimentary zeolitic tuffs are generally soft, friable, lightweight and commonly contain between 50 and 90% of a single zeolite. As accessory minerals, zeolitic tuffs usually contain unreacted volcanic glass, quartz, K-feldspar, montmorillonite, calcite, gypsum, and cristobalite/tridymite [2].

Generally, zeolites consist of frameworks of SiO_4^{4-} tetrahedra wherein all oxygens of each tetrahedron are shared with adjacent tetrahedra. In their structures, some of the quadrivalent Si is replaced by trivalent Al, giving rise to a deficiency of positive charge in the framework. This negative charge is balanced by extraframework positive ions, usually sodium, calcium, magnesium, and potassium. The general formula for a zeolite is:

$$(M^+_x, M^{2+}_y) [(Al_{(x+2y)} Si_{n-(x+2y)} O_{2n}] mH_2O$$

where M^+ and M^{2+} within the first set of parentheses are monovalent and divalent extraframework cations, those within the second set are the structural cations, while m is the number of molecules of water. No zeolites have been found that contain more Al than Si, hence, the Si/Al atomic ratio of zeolite is $\geq$ 1:1 [3].

The structure of natural zeolite is characterized in the following way: the primary building units (PBU) of zeolites are the SiO_4 and AlO_4 tetrahedra linked through oxygen into simple geometrical forms—secondary building units (SBU). Secondary building units may be connected together in a variety of ways. This specific arrangement of SBUs directly influences the internal pores or cages and connecting channels in the zeolite. Loosely held water molecules within cavities and channels of the framework surround the exchangeable cations. These water molecules can be easily removed by continuously heating the zeolite from room temperature to about 350 °C. Once water is removed, the cations move back to positions on the inner surface of the pore and channels close to the seat of charge (i.e., where Al^{3+} substitutes for Si^{4+} in the tetrahedron) [3, 4].

The extraframework exchangeable cations are usually weakly bound in the tetrahedral framework and can be removed or exchanged readily by washing with a strong solution of another cation. This property is called cation exchange capacity (CEC) of the zeolite and is basically a function of the amount of Al that is substituted for Si in the framework tetrahedra. Thus, the greater the Al content, the more extraframework cations are needed to balance the negative charge of framework. Natural zeolites have CECs from 2 to 4 milliequivalents per g (meq/g).

10.1.1   Clinoptilolite

Among the known natural zeolites, clinoptilolite is the most common, occurring in large amounts (millions of tons) in altered volcanic tuffs and saline, alkaline-lake deposits. Clinoptilolite and heulandite are isostructural and belong to the same group (HEU) of zeolites.

HEU type zeolites have a two-dimensional channel system that allows the mineral to act as a molecular sieve. In clinoptilolite, the 10-ring and 8-ring channels parallel to the c-axis of the structure have free dimensions of 0.44 by 0.72 nm and 0.41 by 0.47 nm, respectively. A third 8-ring channel system parallels the a-axis and has free dimensions of 0.40 by 0.55 nm [3].

Characterization of the starting zeolite is crucial for potential practical application of the material. It usually includes chemical and instrumental analyses of the samples. The chemical composition is generally determined by several different methods: classical chemical analysis—gravimetric method, atomic absorption spectrometry, or X-ray fluorescence spectrometry. It may provide insight into the main basic oxide components (SiO_2 and Al_2O_3), exchangeable cations, and other elements present in smaller concentrations [4].

There are several deposits of HEU type zeolites in Serbia. The most commonly studied zeolitic tuff is from the Zlatokop deposit, near Vranjska Banja (southern Serbia). The zeolitic tuff (ZVB) contains approximately

70% clinoptilolite while accessory minerals are quartz, feldspar and pyrite [5]. Some basic mineralogical and physochemical characteristics of this zeolitic tuff are described below. The surface morphology of ZVB was investigated using a scanning electron microscope (SEM). Fig. 10.1 presents the SEM micrograph of the raw zeolitic tuff. Field observation and microscopic study of the sample revealed that the natural zeolitic mineral occurs predominantly as well-formed fine-sized crystals. Many of the well defined plates display tabular morphology characteristic of a monoclinic crystal system of clinoptilolite. The plates are commonly 0.025 to 0.050 μm in thickness and 0.2 to 1 μm in length [6].

Heulandite and clinoptilolite cannot be easily distinguished solely on the basis of X-ray powder diffraction data. Thus, these two species, with a common HEU framework topology, are distinguished based on the Si/Al ratio, exchangeable-cation composition and behavior upon heat treatment (thermal stability) or framework chemistry. Other methods that may be used for differentiation include optical properties, thermogravimetric analysis or differential scanning calorimetry and proton nuclear magnetic resonance (1H NMR) [7, 8].

The Zeolite Subcommittee of the Commission on New Minerals and Mineral Names of the International Mineralogical Association (IMA) proposed that zeolite mineral species should not be distinguished solely on the basis of the framework Si : Al ratio. However, an exception is made in the case of heulandite and clinoptilolite; heulandite is defined as the zeolite

Figure 10.1 SEM micrograph of the natural zeolite ZVB [6].

mineral series having the distinctive framework topology of heulandite and the ratio Si/Al < 4, while clinoptilolite is defined as the series with the same framework topology and Si/Al > 4 [7]. The chemical composition of ZVB is presented in Table 10.1.

From the results in Table 10.1, the calculated Si/Al ratio is over 4, indicating that ZVB is rich in clinoptilolite. Additionally, clinoptilolite is thermally stable in air to nearly 700 °C, whereas heulandite transforms into heulandite-B phase near 250 °C and becomes amorphous to X-rays near 350 °C [3].

From previous thermal stability investigations it was shown that ZVB is thermally stable near 600 °C, providing evidence that ZVB is rich in clinoptilolite [9]. The DTA/TGA curve of zeolite ZVB is presented in Fig. 10.2.

Table 10.1 Chemical composition of the natural zeolite ZVB.

Natural zeolite	Content, wt%							
	SiO_2	Al_2O_3	Fe_2O_3	CaO	MgO	K_2O	Na_2O	I.L.*
	66.57	13.13	2.30	3.85	0.56	1.17	1.27	11.05

I.L.*- ignition loss

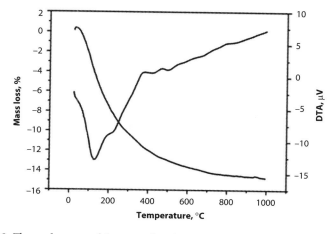

Figure 10.2 Thermal curves of the natural zeolite ZVB.

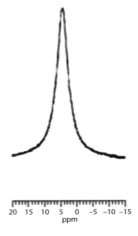

Figure 10.3 Proton NMR spectra of the natural zeolite ZVB [10].

Ward and McKague [8] and Daković *et al.* [10] demonstrated that clinoptilolite and heulandite can be differentiated based on their ¹H solid state NMR spectra, independent of the origin of the zeolite. Randomly dispersed water molecules give rise to a Gaussian peak shape for clinoptilolite. For heulandite, a Pake doublet was observed, indicating an ordered arrangement of the crystal water in the lattice. Proton NMR spectra of ZVB [10] is presented in Fig. 10.3. A Lorentzian line shape was observed, with a chemical shift of 4.4 ppm, and a line width of about 4 ppm (1.6 kHz). Both line width and shape indicate motional narrowing, and the signal is most likely caused by physisorbed water on the exterior and interior surface. The much narrower line width than that observed by Ward and McKague [8] is due to additional water adsorbed from the environment. The absence of any splitting suggests that ZVB is rich in clinoptilite, confirming the results from the thermal stability study.

10.1.2   Biomedical Application of Natural Zeolites

Biomedical application of natural zeolites is based on its ion-exchange or adsorption capability and it is strongly connected to the integrity of its structure. Having an alkaline nature, zeolites are unstable in an acid environment. Zeolite breakdown, due to long exposure to low pHs, is usually preceded by partial replacement of indigenous cations with hydronium ions followed by dealumination. Resistance to acids strongly depends on the Si/Al ratio, which increases with increasing Si content. The high-silica clinoptilolite is particularly acid resistant: it resists acid attack for several months at pH even lower than zero [11]. Considering that the pH of the stomach is very low, acid resistance of zeolite is an imperative pre-requisite, especially in oral administrations. This is the main reason for the preference of siliceous zeolites, such as clinoptilolite and mordenite, in biomedicine compared to the more aluminous phillipsite and chabazite [2, 12].

The application of clinoptilolite and mordenite-rich tuffs for animal nutrition began in Japan in the mid-1960s with their successful use in slowing down the passage of nutrients in the digestive system and therefore resulting in better nutrient assimilation and calorific efficiency [12]. Clinoptilolite has been used in human medicine as antacid, haemostatic and wound-healing accelerator and antihyperglycemic agent [13–18]. Furthermore, antidiarrhoeic, hypocholesterolemic, antiviral, immunomodulatory and antitumor activity of clinoptilolite has been reported [19–28].

10.2   Surfactant Modified Zeolites (SMZs)

The natural zeolites, due their net negative charge and hydrophilic surface, have no ability to adsorb nonpolar organic molecules as well as inorganic anions. It has been demonstrated that the long chain cationic surfactants due to a great affinity to negative charge of the zeolitic surface are commonly used for preparation of SMZs (organozeolites/composites). The adsorption of surfactants at the solid–liquid interface may modify the properties of the solid surface and favor the uptake of molecules from solution that do not adsorb onto the solid in the absence of surfactants. This phenomenon is usually described as surface solubilization, adsolubilization or co-adsorption [29]. The cationic surfactants that are most commonly used to modify natural zeolites are usually long alkyl chains with a quaternary ammonium head group at one end of the chain such as hexadecyltrimethylammonium (HDTMA) bromide.

The HDTMA molecule (Fig. 10.4) has the following properties: it consists of a 16-carbon chain tail group attached to a 3-methyl quaternary

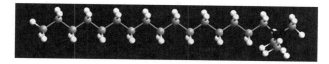

Figure 10.4 Molecular model of HDTMA ion (blue-nitrogen; grey-carbon; white-oxygen).

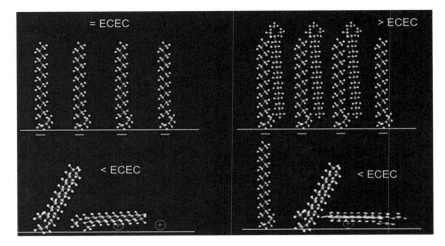

Figure 10.5 Concept of surfactant adsorption at the zeolitic surface (below and above ECEC).

amine head group with the permanent +1 charge; it is water soluble and exists as monomers in solution below the critical micelle concentration (CMC) or as micelles above the CMC of 9×10^{-4} M; summation of the van der Waals packing radii for a 3-methyl ammonium head group yields a diameter of 0.694 nm; a fully extended HDTMA chain length is 3.5 nm; a typical carbon chain diameter is 0.4 nm [30].

Cationic surfactants, such as HDTMA, may exchange native inorganic cations (sodium, calcium, potassium, and magnesium) at the clinoptilolite surface producing an organic carbon enriched surface. Based on the dimensions of commonly studied surfactant HDTMA and the dimensions of clinoptilolite channels, it is obvious that these cations are excluded from the interior channels of clinoptilolite and their adsorption is limited to the external surface of the zeolitic mineral. The external cation exchange capacity (ECEC) characterizes the exchange capacity of the mineral surfaces for surfactants. The adsorption of cationic surfactants onto a negatively charged surface involves both cation exchange and hydrophobic binding. When the amount of surfactants is below the ECEC value, surfactant monomers are adsorbed, eventually developing a monolayer at the zeolitic surface. Monomers adsorb individually by coulombic interactions to available exchange sites and also by hydrophobic tail group interactions. When the amount of surfactant is $\geq$ ECEC a bilayer exists at the mineral surface. The bilayer formation results in charge reversal on the external surface, providing sites where anions will be retained, cations repelled, while neutral species can partition into the hydrophobic core. The possible arrangement of surfactants at the zeolitic surface is presented in Fig. 10.5.

Initial studies using surfactant-modified clays in environmental remediation were focused on removal of hydrophobic organic contaminants from water due to the increase in total organic carbon content of the sorbents after modification. However, once the surfactant loading on zeolite exceeded their ECEC, the modified mineral will reverse their surface charge and show strong affinities for anions such as chromate, nitrate and sulfate [31–34]. The sorption and retention of anions by surfactant-modified zeolite is attributed to surfactant-bilayer formation on which positive charges will develop and surface anion exchange is the main adsorption mechanism. The adsorption ability of SMZs for mycotoxins, metabolites, pollutants and/or drugs [10, 35–38] will be described in the following sections.

10.2.1   Application of SMZs as Sorbents of Mycotoxins

Mycotoxins are a group of structurally diverse secondary fungal metabolites that occur as contaminants of grain worldwide. It is estimated that mycotoxin contamination may affect as much as 25% of the world's food crops each year. Many of these mycotoxins can cause serious problems in livestock resulting in substantial economic losses [39]. The most common mycotoxins found in animal feed are the aflatoxins, ochratoxins, trichothecenes, fumonisins, zearalenone, and ergot alkaloids.

The most promising and economical approach for detoxifying mycotoxin-contaminated feedstuffs is the addition of nutritionally inert mineral adsorbents to the diet to decrease the bioavailability of the mycotoxins during absorption in the gastrointestinal tract. Aluminosilicates (natural zeolite–clinoptilolite and natural bentonite–montmorillonite) are the most widely utilized adsorbents [40].

The unmodified surfaces of aluminosilicates are very effective in binding aflatoxin B1; however, their negatively charged surfaces are ineffective in binding other fairly nonpolar mycotoxins [41, 42]. The permanent negative charge in their crystal structure makes them suitable for modification using long-chain organic cations (surfactants), which results in an increased hydrophobicity of the mineral surface and a high affinity for hydrophobic mycotoxins [43]. The long chain organic cations exchange with inorganic cations only on the external surfaces of zeolites, whereas in montmorillonite, all exchangeable positions are equally available, therefore, for zeolites, less organic phase is needed for modification [44].

The binding efficacy of adsorbents for mycotoxins is dependent on their crystal structures and physical properties (the total charge and charge distribution, the size of pores, accessible surface area, etc.) as well as the physicochemical properties of the mycotoxins (polarity, solubility, shape, charge distribution, dissociation constants, etc.) [39]. Adsorption of two mycotoxins zearalenon and fumonisin B_1 by organozeolites obtained by treatment of ZVB with octadecyldimethyl benzyl (ODMBA) ions is presented. Three different levels 25%, 50%, and 100% of ODMBA ions were used for modification. Samples were denoted as OZ-2, OZ-5 and OZ-10. Measurements of inorganic cations released from the surface, unreacted ODMBA content in the supernatants after ODMBA adsorption, as well as through the determination of the CEC of the modified organozeolites confirmed quantitative ion exchange has occurred [45].

Zearalenone (ZEN) is a hydrophobic molecule, only slightly soluble in water. ZEN has two phenolic hydroxyl groups (Fig. 10.6) with an estimated $pK_a1 = 7.62$ [43], suggesting that at pH 3, it is mainly in neutral form, while at pH 7, the phenolate anion is present in water solution. Adsorption of ZEN by organozeolites was studied at pH 3 and 7.

Natural zeolite - clinoptilolite, due to its net negative charge on the surface and the presence of exchangeable metal cations (Ca^{2+}, Mg^{2+}, Na^+, and K^+) at the hydrophilic surface was ineffective in adsorbing ZEN at pH 3 and 7. However, the presence of the hydrophobic ODMBA at the surface of each organozeolite significantly increased ZEN adsorption, at both pH values indicating that ODMBA ions may be the active site relevant for ZEN adsorption on the organozeolites [45].

In order to investigate the mechanism of ZEN sorption by the three organozeolites, the adsorption isotherms were determined at pH 3 and 7 (Figs. 10.7–8). These isotherms were obtained by plotting the concentration of ZEN in solution after equilibrium against the amount of ZEN adsorbed per unit of weight of each adsorbent [10]. All three organozeolites showed increasing adsorption of ZEN with increasing ZEN concentration. The lowest adsorption was observed for organozeolite OZ-2 at all

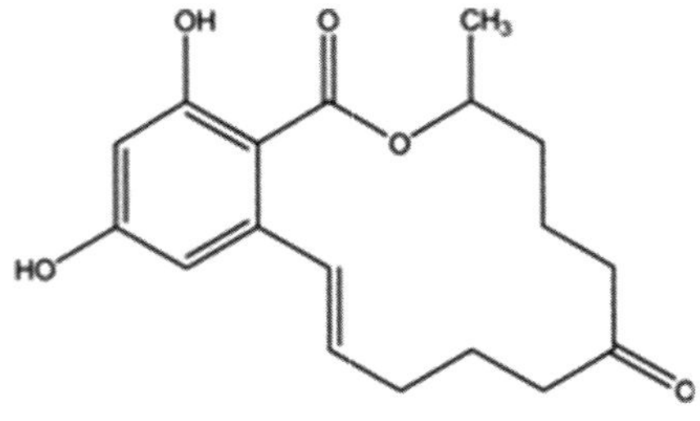

Figure 10.6 Chemical structure of zearalenone.

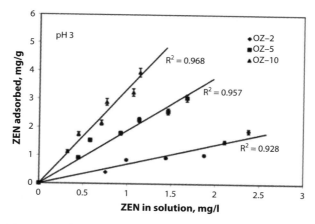

Figure 10.7 Adsorption of ZEN by organozeolites at pH 3 [10].

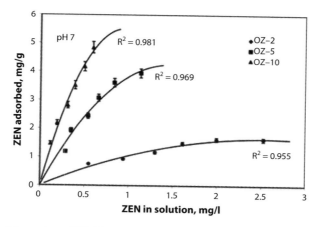

Figure 10.8 Adsorption of ZEN by organozeolites at pH 7 [10].

pH values. Adsorption of ZEN by organozeolites was best represented by a linear type of isotherm at pH 3 (Fig. 10.7), while at pH 7 adsorption of ZEN by organozeolites followed a non linear type of isotherm (Fig. 10.8).

Groisman et al. [46] reported that short-chain organoclays are better sorbents for non ionic compounds (NOC) than the long-chain organoclays only in the case of low molecular weight compounds of relatively low hydrophobicity. Namely, they studied the sorption of six compounds—atrazine, ametryn, prometryn, terbutryn, pyrene, and permnethrin, with the range of log K_{ow} (K_{ow}, the octanol-water partition coefficient, is the measure of hydrophobicity) values from 2.5 to 6 by TMA- and octadecyltrimethyl

(ODTMA)-bentonite from water solution. They reported that compounds with low or medium hydrophobicities (log K_{ow} 2.5–3.8) were more strongly sorbed on the short-chain organoclay, whereas the more hydrophobic compounds (log K_{ow} 5.2–6.1) were better sorbed on the long-chain organoclay. Isotherms of the ODTMA-clay were linear for all the compounds, whereas those of the TMA-clay were non linear.

The data for ZEN adsorption on the organozeolites with different levels of ODMBA at the zeolitic surface at different pH values suggest that the adsorption mechanism is dependent on the form of ZEN in solution. At pH 3, ZEN exists in solution as the neutral form, thus the linear isotherms at pH 3 suggest that hydrophobic interactions are probably responsible for adsorption of neutral, hydrophobic ZEN on the hydrophobic surface of the organozeolites. At pH 7, the phenolate anion is present in water solution, thus data for ZEN adsorption by the three organozeolites at this pH suggest that sorption appears to be the result of the adsorption process as well as partitioning.

Fumonisin B_1 contains carboxylic, hydroxyl, and amino functional groups (Fig. 10.9) [47, 48]. Compared to ZEN, FB_1 is more polar and water soluble [49, 50]. The carboxylic and amino functional groups suggest that FB_1 exists in solution either in cationic or anionic form depending on pH. Adsorption of FB_1 on ZVB (C_{0FB1} = 2 ppm, and $C_{suspension}$ = 10 g/L) showed that its adsorption was 90.3 and 2.0% at pH 3 and 7, respectively [6]. Based on the literature, the unmodified zeolitic surface with its negative charge has little or no affinity for anionic species [31] suggesting that FB_1 may exists in the anionic form at pH 7.

The sorption isotherms (Figs. 10.10–11) of fumonisin B_1 by the three organozeolites OZ-2, OZ-5 and OZ-10 were studied [6] and it was found that the presence of ODMBA ions at the zeolitic surface greatly improved

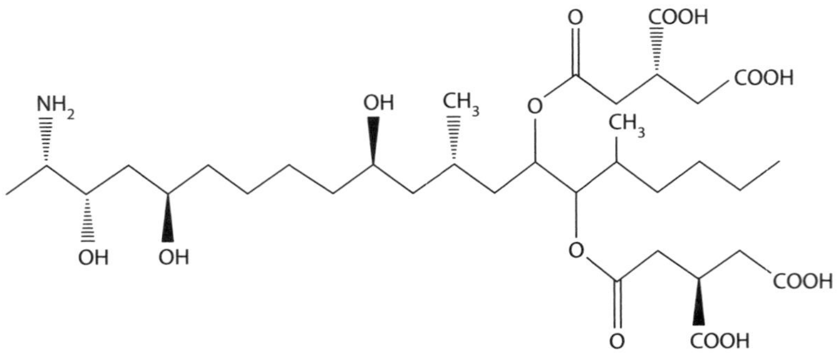

Figure 10.9 Chemical structure of fumonisin B_1.

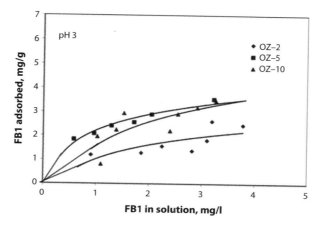

Figure 10.10 Fumonisin B_1 adsorption by organozeolites at pH 3 [10].

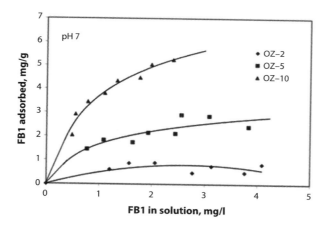

Figure 10.11 Fumonisin B_1 adsorption by organozeolites at pH 7 [10].

adsorption of FB_1 at different pHs, suggesting that ODMBA ions at the surface may provide active sites onto which FB_1 is adsorbed. Furthermore, FB_1 adsorption by all the three organozeolites showed a nonlinear type of isotherm at pH 3 and 7 and the highest FB_1 sorption was achieved when the zeolitic surface was completely covered with ODMBA (OZ-10).

Avantaggiato *et al.* [48] reported nonlinear isotherms for FB_1 sorption on activated carbon and cholestyramine at pH 7, indicating that binding of FB_1 is a saturable process. Usually, sorption of nonionic organic substances by smectites modified with large organic cations (HDTMA) is due to linear solute partitioning into the hydrophobic phase formed by the large alkyl chains of the HDTMA. Nonlinear isotherms, indicative of adsorption or co-adsorption, were observed when smectite was modified with smaller organic cations (tetramethylammonium – TMA) [51, 52]. Li *et al.* [53] reported the nonlinear isotherms for sorption of polar phenol and p-nitrophenol on HDTMA-bentonite, indicating that sorption appeared to result from adsorption in addition to partition. The presumed low partition was attributed to the large disparity of the (polar) phenols and the (nonpolar) HDTMA medium. Also, Rawajfih and Nsour [54] studied the sorption of phenol, p-chlorophenol, an 2,4-dichlorophenol by HDTMA-bentonite and reported nonlinear isotherms suggesting that sorption occured via an adsorption/partition mechanism, in which partitioning of organic compounds into organobentonite and adsorption take place simultaneously. Fumonisin B_1 is a larger molecule than phenol and it contains one NH_2-group, three hydroxyl, and four carboxylic groups (Fig. 10.9), although it also has hydrophobic parts of the molecule and can partition into hydrophobic phase created by ODMBA alkyl chains. The interactions between carboxylic groups of FB_1 and nitrogen of head groups of the surfactant may be involved in the sorption process. Once FB_1 is attached to the head groups of surfactant, then hydrophobic interactions between hydrophobic areas of FB_1 and alkyl chains of ODMBA could enhance the sorption. The nonlinear isotherms that were obtained for sorption of polar FB_1 to the three organozeolites suggested a sorption/partition mechanism.

Effective adsorbents for either ZEN or FB_1 were obtained by modification of ZVB with ODMBA ions. Adsorption of both mycotoxins increased with increasing the amounts of ODMBA at the zeolitic surface at pH 3 and 7. Results on adsorption of these two mycotoxins, which have different polarity, by organozeolites indicated that different adsorption of ZEN and FB_1 is related to different functional groups and different shapes of these molecules. Knowledge of the physical and chemical properties of mycotoxins and the properties of the modified minerals together with the determination of the adsorption isotherms may indicate the mechanism mycotoxins are adsorbed by organozeolites.

10.3   Minerals as Pharmaceutical Excipients

Almost all therapeutic products (for human and veterinary use) contain both pharmacologically active compounds and excipients added to aid the formulation and the manufacture of the subsequent dosage form for administration to patients. According to pharmacopoeias, an excipient (auxiliary substance) is any constituent of a medicinal product that is not an active substance. Excipients are used in pharmaceutical preparations to: (a) enhance their organoleptic characteristics, such as flavor (flavor correctors) and color (pigments); (b) improve their physicochemical properties, such as viscosity of the active ingredient (emulsifying, thickening, and anticaking agents), (c) facilitate their elaboration (lubricants, diluents, binders, isotonic agents) or conservation (desiccants, opacifiers), and (d) facilitate liberation of the active ingredient within the organism (disintegrants, carrier-releasers). In accordance with the above roles, excipients are indispensable components of medicinal products and, in most cases, comprise the greatest proportion of the dosage unit.

The properties of the final dosage form (i.e., its bioavailability and stability) are, for the most part, highly dependent on the excipients chosen, their concentration, and interaction with both the active compound and each other. Excipients can be defined according to their functional roles as solubility or bioavailability modifiers, stability enhancers, crystal form stabilizers, buffers and pH-adjusting agents, propellants, tablet binders, dispersing agents and the like. Many studies have been focused on providing information about this rapidly evolving area, for which regulatory guidance is only developing [55–59].

In earlier days, excipients were considered pharmacologically inactive and indifferent ingredients. Over time, pharmaceutical scientists learned that excipients are not inert (or slightly active) and frequently have substantial impact on the manufacture, quality, safety, and efficacy of the drug substance(s) in a dosage form [57]. The types of excipients that can, for example, delay or extend the release of a drug substance are very important to achieving the desired bioavailability of the active ingredient (without consideration of its solubility) at the required location in the body [60].

A wide range and variety of minerals have been used as excipients in pharmaceutical preparations because they have certain desirable physical and physicochemical properties, such as high adsorption capacity, specific surface area, swelling capacity and reactivity to acids. Other important properties are water solubility and dispersivity, hygroscopicity, unctuosity, thixotropy, slightly alkaline reaction (pH), plasticity, opacity, and color. Clearly such minerals must not be toxic to humans [59].

The following minerals are commonly used as excipients in pharmaceutical preparations: oxides (rutile, zincite, periclase, hematite, maghemite, magnetite), hydroxides (goethite) carbonates (calcite, magnesite), sulfates (gypsum, anhydrite), chlorides (halite, sylvite), phosphates (hydroxyapatite) and phyllosilicates/clay minerals (smectites, palygorskite, sepiolite, kaolinite and talc). Within the smectite group, montmorillonite, saponite, and hectorite are the most widely used species. Colloidal anhydrous silica and various aluminosilicates are used as excipients in pharmaceuticals, cosmetics, and food products because of their adsorbent, anticaking, bulking, binding, stabilizing, and other similar properties. More recently, some tectosilicates (zeolites) also feature in pharmaceutical preparations [55, 61, 62].

10.3.1   Minerals and Modified Drug Delivery

In addition to aforementioned classical pharmaceutical uses, some minerals (so far, mostly clays) have been used effectively in development of new drug delivery systems with modified drug release.

To overcome disadvantages of conventional drug therapy, such as short duration of drug action due to inability to control temporal drug delivery, fluctuation in drug level, potentiation or reduction in drug activity during chronic dosing, appearance of side effects and problems with patient compliance during chronic use, development of modified drug delivery dosage forms has been, and continues to be, the focus of a great deal of research.

The mineral–organic interaction can be used to control the release of active ingredients (drugs) with improved therapeutic properties. Here the minerals first serve as a carrier, and then as a releaser of the active ingredient. Because of their large specific surface area and high adsorption capacity, clays (smectites, palygorskite, sepiolite, kaolinite) are well suited to act as drug carriers and have been proposed as fundamental constituents for obtaining the modified drug release, with different purposes and acting through various mechanisms [63–69].

10.3.2   Clinoptilolite as Potential Pharmaceutical Excipient/ Drug Delivery

Several toxicological studies can be found in the literature that demonstrate certain zeolites (e.g. clinoptilolite) are non-toxic and safe for use in human and veterinary medicine. In addition, clinoptilolite crystals, having a non-fibrous morphology, contrary to erionite, are completely safe as regards the possible induction of mesothelioma [23, 70].

Clinoptilolite (from Vranje deposit, southern Serbia) was investigated in a preclinical toxicology trial by setting the "limit" test [25]. This test refers to administering high doses of clinoptilolite (2 × 200 and 2 × 500 mg/mouse per day orally by gavage) for 6, 14, and 30 days. Since the clinoptilolite did not cause death of mice in this limit test, an "up-and-down" test on mice was performed, with daily doses ranging from 60 to 4000 mg/ per mouse, in which no toxicity was observed. Therefore, classical acute, subacute, and chronic toxicity studies on mice and rats of both sexes (separately) were conducted. The duration of the study was as follows: acute, 1 month; subacute, up to 3 months; chronic toxicity, up to 6 months. Animals were monitored for phenotypic changes, changes in behavior and survival, changes in body weight, amount of food and water consumed, changes in hematological and serum clinical chemistry parameters. Pathohistological analysis of liver, spleen, kidney, brain, lung, testes, ovary, duodenum, eye, stomach, large and small intestine, muscles, myocardium, pancreas, thymus, and axillary lymph nodes was carried out on sacrificed experimental and control mice. The results of all of these studies were that oral (in diet) administration of clinoptilolite to mice and rats for 6 and 12 months, respectively, caused no change that could be considered a toxic effect of treatment.

Local tolerance of the fine powder of natural clinoptilolite, obtained by tribomechanical micronization, was evaluated to ascertain whether the test substance is tolerated at the sites in the body which may come into contact with the product as a result of its administration [23]. Repeated-dose dermal tolerance testing of these small-sized particles (MZ) was performed on male Wistar rats and male BALB/c mice. MZ was applied to the shaved skin of the whole dorsal region of animals in three ways: (a) as original powder; (b) mixed with neutral creme at the ratio of 1:1; (c) mixed with paraffin oil at the ratio of 1:1. The animals were treated twice a day for 28 days. Macroscopic changes in the treated skin were examined daily. The left dorsal region of the animal was used as control. For microscopic analysis of the possible changes, skin samples were collected 1 day after the last treatment. The results revealed that MZ was not toxic or allergenic for the skin.

The Cosmetic Ingredient Review (CIR) Expert Panel [71] reviewed the safety of claylike ingredients (Aluminum Silicate, Calcium Silicate, Magnesium Aluminum Silicate, Magnesium Silicate, Magnesium Trisilicate, Sodium Magnesium Silicate, Zirconium Silicate, Attapulgite, Bentonite, Fuller's Earth, Hectorite, Kaolin, Lithium Magnesium Silicate, Lithium Magnesium Sodium Silicate, Montmorillonite, Pyrophyllite and Zeolite) used in cosmetic formulations. The Panel considered that most of formulations are not respirable, and of the preparations that are respirable, the concentration of the ingredient is very low. Even so, the Panel considered that any spray containing these solids should be formulated to minimize their inhalation. With this admonition to the cosmetics industry, the CIR Expert Panel concluded that these ingredients (i.e., clinoptilolite as representative of zeolites) are safe as currently used in cosmetic formulations.

The fully comprehensive review by Colella [12] gives a critical reconsideration of biomedical and veterinary applications of natural zeolites.

Namely, as claimed by the author, studies of natural zeolites in life sciences should involve a preliminary characterization of the material to evaluate the specificity of its use and to plan, when possible, suitable modifications. Unfortunately, most studies do not pay attention to this point; thereby the zeolitic material is often used in its original form irreproducible if other similar materials are used. In a few studies and/or applications, attention is paid, however, to the characterization and standardization of the material. A protocol is usually set up to obtain, through suitable treatment (e.g., through ion exchange, exhaustive grinding, etc), a product, which can be used safely and with the certainty of results.

Commercial products presenting a higher grade of standardization, e.g., tribomechanically activated zeolites are sometimes used for experiments [27]. For some of the commercial products such as clinoptilolite for the natural detoxification, suppliers state that it complies with the quality requirements of the European Pharmacopoeia. Since the clinoptilolite (or any other zeolite) is not included in the European Pharmacopoeia, this statement is probably related to the *European Pharmacopoeia requirements* for heavy metals limitation for substances of mineral origin such as bentonite, kaolin or talk. Namely, the limit for extractable heavy metals for these substances is a maximum 50 ppm, i.e., maximum 25 ppm if intended for internal use.

Narin *et al.* [72] reported that the zeolitic tuff (a 67 wt% of clinoptilolite) was successfully loaded with NO after thermal activation. The irreversibly adsorbed NO could be stored under dry atmosphere and was released when the NO-loaded natural zeolite came in contact with aqueous medium under physiological conditions. The released NO exhibited antibacterial and bactericidal activities against both Gram-positive and Gram-negative bacteria. Hence, the local clinoptilolite-rich natural zeolite seems a promising adsorbent for NO storage and release for medical applications.

A clinoptilolite-rich rock was evaluated as inorganic Zn^{2+} releasing carrier for antibiotic erythromycin in the topical treatment of acne, given the efficacy of zinc-erythromycin combination against resistant *Propionibacterium* strains [62]. A 66 wt% of clinoptilolite content was determined by means of XRD analysis while its CEC was 1.45±0.08 mEq/g. Using a specific exchange method, the material was previously Na-conditioned then Zn-conditioned. A substantially complete Zn-form was obtained, as demonstrated by AAS analyses. The Zn-conditioned powder was then micronized to achieve a volume/surface ratio suitable for a topical therapy. After micronization, the specific surface area, determined by BET gas adsorption, was 30.2 m^2/g, and 92% of the powdered rock was lower than 30 μm. The Zn^{2+} release from the micronized rock was measured at 37 °C both in physiologic solution (9 g/l NaCl) and in 0.05 M

KH_2PO_4/Na_2HPO_4 buffer. Under these conditions, a prompt and significant zinc release was recorded: after 30 min, 68% and 60%, respectively. Erythromycin was charged onto the micronized material using a solvent evaporation method. HPLC determinations showed that 85% of the drug in contact with the carrier was loaded. The simultaneous release of zinc and erythromycin were evaluated in the phosphate buffer and it was determined that 82% of the loaded antibiotic was released after 30 min. It was also interesting that zinc and erythromycin had similar release patterns demonstrating that the presence of cation does not significantly interfere with the release of the antibiotic. According to the authors, under these circumstances, the two components of the preparation (zinc and antibiotic) can jointly develop their effects against the *P. acnes*.

Investigations into the effectiveness of clinoptilolite as a drug carrier have been done mainly by means of its active external surface interaction with drug molecules. Lam *et al.* [73] reported a theoretical study of physical adsorption of aspirin on the natural zeolite clinoptilolite. It was demonstrated through a theoretical approach that the natural zeolite is able to bind aspirin (acetylsalicylic acid) molecules on its surface, allowing a modulation of drug activity by its slow availability. In the following papers regarding investigations from the same research team, theoretical studies of metronidazole adsorption [74] in addition to the physicochemical interactions of metronidazole and sulfamethoxazole [75] (Table 10.2) with the purified natural zeolite were presented. The natural zeolite used in these studies was from Tasajeras deposit (Cuba) and it met the requirements for the use in the pharmaceutical industry, which includes the fulfillment of the physical, chemical, and technological tests, as well as toxicological, pharmacological, stability, and clinical studies [19, 76]. In view of the fact that the drugs used in these investigations after oral administration may cause a group of side effects mainly associated with gastrointestinal disturbances, it was interesting to evaluate the chemical behavior of both organic molecules in the presence of natural zeolite, which is used itself as an antacid. As expected, the polar zeolite surface interacted preferably with the more hydrophilic metronidazole than with the more hydrophobic sulfamethoxazole, i.e., the difference in the adsorptive behavior is fundamentally related with the polarity of the molecules. It was demonstrated that the drugs do not show signs of degradation after the contact with the zeolite at any pH values, therefore simultaneous administration to a patient without any loss of the individual pharmaceutical effect is possible. To enhance the clinoptilolite interaction with these drugs, research was directed towards preparation and characterization of clinoptilolite-surfactant drug composites and investigation of their use as drug carriers.

10.4 SMZs for Pharmaceutical Application

As previously mentioned, the chemical modification of zeolite with long chain organic cations results in an increased hydrophobicity of the mineral surface, providing a high affinity for organic, i.e., drug molecules. Nowadays it is also important to investigate adsorption from aqueous solutions as useful and safe alternative to already reported organic media such as n-hexane, dimethyl sulfoxide, dimethylformamide and diethylether, employed for drugs adsorption on both natural and synthetic aluminosilicates [77–79].

The usefulness of SMZs reflects not only upon their efficiency of diverse contaminants (from water or soil) removal but also upon their long-term chemical and biological stability. In the study of Li *et al.* [80] the chemical (over a wide range of pH values and ionic strengths) and biological (under aerobic, anaerobic, saturated, and unsaturated conditions) stability of clinoptilolite modified with HDTMA–Br or –Cl was investigated. SMZs were stable in high ionic strength and low- and high- pH environments, with more than 90% of HDTMA remaining bound to the zeolite surface after washing with 100 pore volumes. HDTMA bound to the zeolite surface was resistant to microbial degradation, with more than 98% of the original HDTMA remaining after 12–17 weeks of incubation under aerobic or anaerobic conditions. While aqueous HDTMA was biocidal, HDTMA bound on SMZs did not inhibit microbial growth. On the basis of the results of this study, SMZs appear suitable as a sorbent for long term *in situ* applications and as a substrate for enhanced bioremediation.

Interactions of cationic surfactant with natural zeolites have been extensively studied since prepared modified natural zeolites proved to be excellent adsorbents for various drug molecules.

10.4.1 Preparation and Characterization of SMZs for Drug Delivery

In the study of Rivera *et al.* [81] cationic surfactant benzalkonium chloride (BC) (Table 10.3) was used for modification of the pure natural zeolite from Tasajeras deposit (Cuba) (see above) and preliminary characterization of drug support systems based on natural clinoptilolite was performed. The modification process took place at room temperature and under agitation. The model drugs used for investigation were sulfamethoxazole, acetylsalicylic acid, and metronidazole (Table 10.2).

Table 10.2 Overview of the structural formulas and physicochemical properties of the model drugs.

Compound	Structural formula	MWt	pKa	Solubility in water[a]
Acetylsalicylic acid *2-(Acetyloxy)benzoic acid*		180.2	3.5	Slightly soluble
Diclofenac diethylamine *Diethylammonium 2-[(2,6-dichloroanilino) phenyl] acetate*		369.3	4.87	Sparingly soluble
Diclofenac sodium *2-[(2,6-dichlorophenyl)amino]phenyl]acetate*		318.1	4.0	Sparingly soluble
Ibuprofen *(2RS)-2-[4-(2-methylpropyl)phenyl]propanoic acid*		206.3	4.55	Practically insoluble

(Continued)

Table 10.2 (*Cont.*)

Compound	Structural formula	MWt	pKa	Solubility in water[a]
Metronidazole *2-(2-Methyl-5-nitro-1H-imidazol-1-yl)ethanol*		171.2	2.62	Slightly soluble
Sulfamethoxazole *4-Amino-N-(5-methylisoxazol-3-yl)* *benzenesulphonamide*		253.3	5.6	Practically insoluble

[a] Approximate volume of solvent in millilitres per gram of solute (Ph. Eur. 7):

- sparingly soluble: from 30 to 100
- slightly soluble: from 100 to 1000
- practically insoluble: more than 10 000

The drug/modified zeolites composites were characterized by 27Al and 29Si MAS NMR and 13C CPMAS NMR spectroscopy, and nitrogen adsorption isotherm measurements. The overall results suggested that the presence of surfactant and drug molecules on the zeolitic surface does not produce structural changes, resulting in a strong decrease of the specific surface area, allowing the zeolitic materials to support drugs of very different nature.

In the following investigation the same research team used a zeolitic sample from a national deposit (a mixture of about 70% clinoptilolite-heulandite, 5% mordenite, 15% anortite and 10% of quartz)[35]. The natural zeolitic sample (NZ) was a powder of 37–90 μm particle size, which has a Si/Al ratio ≈ 5.3. The adsorption of sulfamethoxazole and metronidazole as guest compounds were evaluated on clinoptilolite-surfactant (BC) composites. The drug adsorption assays showed that the NZ–BC composite adsorbs a considerable amount of drug, in particular sulfamethoxazole, as a less polar drug. It was proven by means of TG thermogravimetric analysis that the composites are more stable thermally when the admicelles contain sulfamethoxazole in their interior. The studies by IR spectroscopy indicated that the zeolite structure remained unchanged after the modification with the surfactant and the drug. The results of the drug release revealed that approximately 80% of sulfamethoxazole was released from the NZ–BC–drug composite in about 24 h.

The reported results of investigation of SMZs in drug delivery [35, 36, 81, 82] revealed the possibility of their potential use as prospective excipient for pharmaceutical application. Nevertheless, it should be emphasized that the formulation of a final dosage form includes further investigation of its overall composition and possible interactions of SMZs (as one of its components) with other auxiliary substances.

Latest investigations on cationic surfactants-modified natural zeolites were directed toward analysis of their potential use as drug formulation excipients with improved functionality [83–85]. In recent years, increased attention is being given to functional roles of excipients in pharmaceutical dosage forms. In terms of functionality is the ability of an excipient to interact with the active ingredient in the formulated dosage form and/or provide a matrix that can affect critical quality attributes of the drug substance, including stability and bioavailability [57]. Since the 5th Edition of the European Pharmacopoeia, a number of excipient monographs have contained a non-mandatory section on functionality-related characteristics (FRCs) [86]. The aim of this section is to provide users with a list of physical and physicochemical characteristics that are critical to the typical uses of the concerned excipient, and to provide the general methods required to assess these characteristics.

The starting material for these experiments was, as previously mentioned, ZVB. Three cationic surfactants of different chain length and polar head groups: benzalkonium chloride, hexadecyltrimethylammonium bromide and cetylpyridinium chloride were used for preparation of organozeolites (Table 10.3).

The modified zeolites with the different surfactant contents were used for additional adsorption investigations of three model drugs: diclofenac diethylamine (DDEA), diclofenac sodium (DS) and ibuprofen (IB) (Table 10.2) from a buffer solution at pH 7.4, by means of adsorption isotherm analysis.

The model drugs used in the investigations are representatives of nonsteroidal anti-inflammatory drugs (NSAIDs), a group of structurally unrelated organic acids that have analgesic, anti-inflammatory, and antipyretic properties.

10.4.1.1 Physicochemical Analysis of ZCPCs

Comprehensive physiochemical analysis of the organozeolites/composites obtained by the modification of ZVB with CPC at three different levels, i.e., 10, 20, and 30 mmol/100g (denoted as ZCPC-10, ZCPC-20 and ZCPC-30) was performed by Fourier transform infrared spectroscopy (FTIR), electrokinetic (zeta) potential measurements and thermal analysis (DTA/TGA/DTG) [84].

The IR spectra for clinoptilolite ZVB and ZCPC 10–30 composites are shown in Fig. 10.12. Bands positioned at 3630, 3420, and 1640 cm^{-1} (dashed line) are characteristic bands of the clinoptilolite connected to acidic hydroxyls Si-O(H)-Al, hydrogen-bonding hydroxyl groups and deformation vibration of absorbed water, respectively [87]. However, two new bands appear (solid line), which correspond to the CPC present on the composite, associated with the C-H stretching vibrations of the hydrocarbon chain −2917 and 2847 cm^{-1} [88]. The relative intensity of these bands increases with increasing amounts of adsorbed CPC. No relevant variations in the frequency of the bands assigned to the clinoptilolite after the treatment with the CPC were observed, which indicates that the zeolite structure remains unaltered after the modification, and that surfactant is present only at the zeolite surface.

Measurements of zeta potential, as a reflection of surface potential, showed that the initial zeta potential of ZVB (−27.4 mV) changed in correlation with the initial surfactants loadings (Fig. 10.13). When the

Table 10.3 The molecular structure, CMC, and molecular weight of cationic surfactants used for modification of natural zeolites.

Surfactant	Structural formula	MWt	CMC (mmol/l)
Benzalkonium chloride (BC)[a]	$R = C_8H_{17}$ to $C_{18}H_{37}$	354.0	5.0
Hexadecyltrimethylammonium bromide (HDTMA-Br)		364.5	0.9
Cetylpyridinium chloride (CPC)		358.0	0.9

[a] Benzalkonium chloride is a mixture of alkylbenzyldimethylammonium chlorides, the alkyl groups having chain lengths of C_8 to C_{18}. It contains not less than 95.0% and not more than the equivalent of 104.0% of alkylbenzyldimethylammonium chlorides, calculated as $C_{22}H_{40}ClN$ (Mr 354.0) with reference to the anhydrous substance.

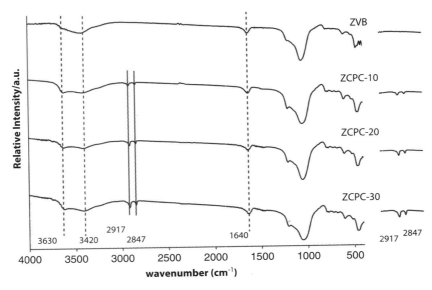

Figure 10.12 IR spectra for ZVB and ZCPC 10–30 composites [84].

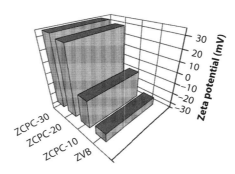

Figure 10.13 Zeta potentials of ZVB and ZCPC 10–30 composites.

concentration of surfactant was equivalent to 100% of ECEC of starting zeolite, zeta potential of composite ZCPC-10 was –8.4 mV. Increasing the surfactant loading to 200% of ECEC influenced the more pronounced alteration of the zeta potential for sample ZCPC-20 (+29.7 mV), while after modification with surfactant amount equivalent to 300% of ECEC zeta potential of sample ZCPC-30 slightly increased to + 32.2 mV (Fig. 10.13).

It is well known that when the concentration of surfactant is ≤ ECEC of the initial zeolite, a general model of sorption of ionic surfactants on a solid surface is the formation of monolayer or hemimicelle at the solid-aqueous interface via strong coulomb (ionic) bonds. The sorbed surfactant creates an organic-rich layer on the previously negative zeolite surface, and the charge on the surface is reversed from more to less negative or even positive value with increasing surfactant concentration.

If the initial surfactant concentration exceeds the CMC, according to experimental conditions in this and previously reported studies [37, 89, 90], the hydrophobic tails of the surfactant molecules will associate to form a bilayer or aggregates, i.e., admicelles. As a result, the formation of surface aggregates can be envisaged as a superficial analogue of micellization in the bulk solution. Furthermore, as with the micelles in solution, the aggregates may have different shapes and sizes such as patchy bilayer, bilayer, small spherical micelle, and multilayer. A desired surfactant configuration on the surface can be achieved by controlling the initial and final surfactant concentrations in the system.

Since the initial concentration of CPC in our study was above its CMC, when the amount of CPC was equal to ECEC value (ZCPC-10), zeta potential increased and approached zero confirming nearly monolayer formation and almost complete hydrophobicity of the zeolitic surface. At CPC amount equal to 200% and 300% ECEC zeta potential became positive pointing charge reversal and bilayer formation at the zeolitic surface. It is possible that at the initial CPC concentration of 200% of ECEC, micelle produced less extensive bilayer at the zeolitic surface. A slight increase in zeta potential of sample ZCPC-30 suggests that when the initial concentration of CPC was equal to 300% of ECEC, possibly, the micelle systems formed complete bilayer or extensive admicelles at the zeolitic surface.

The thermogravimetric (TG), differential thermogravimetric (DTG) and differential thermal analysis (DTA) curves of ZVB and ZCPC 10–30 are given in Fig. 10.14, while the mass loss for all the samples in different temperature regions are presented in Table 10.4. Thermal curves of the pure CPC are presented in Fig. 10.15.

Thermal analysis of organoclay complexes supplies information on the thermal reactions, properties and stability of the complexes, and the amount and properties of adsorbed water in the organoclay and on the bonding between organic species and the clay [91, 92].

Since in our study, CPC initial concentrations were above CMC and surfactant was added in amounts ≥ ECEC of the starting zeolite, from the TG curves of the three organozeolites (Fig. 10.14a), it is expected that an initial mass loss ~ 230 °C is associated with dehydratation, as well as with elimination of non- or weakly adsorbed organic molecules from the zeolitic surface.

Zeta potential measurements indicated almost monolayer formation for ZCPC-10, less extensive bilayer for ZCPC-20 and complete bilayer or

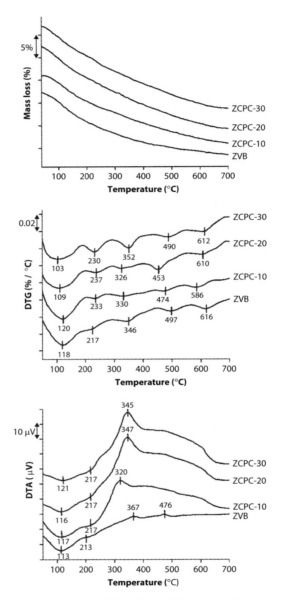

Figure 10.14 Thermal: TG (a), DTG (b), and DTA (c) curves of the starting zeolitic tuff (ZVB) and ZCPC 10–30 composites [84].

extensive admicelles for ZCPC-30. Compared to the mass loss of the start-ing zeolite, in the first temperature region (Table 10.4) (7.36%), the mass loss for ZCPC-10 decreased −6.49% confirming increased hydrophobic-ity. The increase of mass loss of ZCPC-20 (8.01%) indicates elimination of

Table 10.4 Mass loss for ZVB and ZCPC 10-30 samples in different temperature regions [84].

Sample	Mass loss, %			
	20–230 °C	230–500 °C	500–700 °C	20–700 °C
ZVB	7.36	4.93	1.64	13.83
ZCPC-10	6.49	5.80	2.34	14.99
ZCPC-20	8.01	7.25	2.90	18.09
ZCPC-30	7.53	7.65	3.00	18.33

weakly bound CPC from less extended bilayer, while slightly lower mass loss for ZCPC-30 (7.53%) indicates an elimination of part of CPC from ordered complete bilayer or admicelle. The mass loss between 230 °C and 700 °C corresponds to the oxidation of organic material present at the zeolitic surface, as well as to the water coordinated to the cations remaining in the zeolite channels.

The DTA curve (Fig. 10.14c) of the initial zeolitic tuff had an endothermic peak at 113 °C and a shoulder at 213 °C, characterictic of Ca-clinoptilolite from the Zlatokop deposit [93]. Usually, because some water is replaced by the adsorbed organic matter, the size of this peak, in the DTA curves of the organoclay complexes, is smaller than that in the DTA curve of the untreated clay [92]. At DTA curves of organozeolites, in first temperature region, the intensity of endothermic peak decreases with increasing the amount of CPC used for modification, and also the peak is slightly shifted toward higher temperatures with increasing the amount of CPC at the zeolitic surface (from 117 °C for ZCP-10 to 121 °C for ZCPC-30). Besides dehydratation, in the first temperature region, melting and boiling of non-adsorbed organic compounds are identified by their endothermic peaks, thus, broad endothermic peak observed around 217 °C for organozeolites may additionally suggest eliminaton of weakly bound CPC.

Additonally, at DTG curves of organozeolites (Fig. 10.14b), in same temperature region, shift of the first DTG peak toward lower temperatures was observed with increasing amounts of CPC (from 120 °C for ZCPC-10 to 103 °C for ZCPC-30). Consequently, increasing in intensity of the second DTG peak (around 230 °C) on the DTG curves with increasing amounts of CPC is observed. Thus, desorption of water and elimination of weakly bound organic matter from zeolitic surface are the thermal reactions in first region.

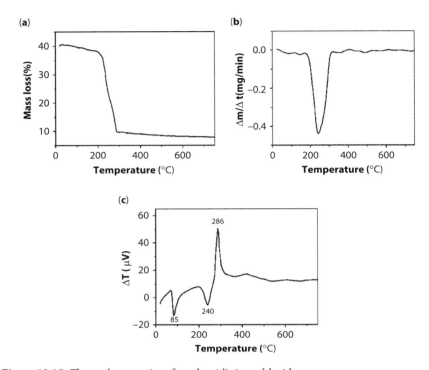

Figure 10.15 Thermal properties of cetylpyridinium chloride.

In the second temperature region (230–700 °C), the sharp exothermic peak at 476 °C on the DTA curve of the natural zeolitic tuff demonstrates the oxidation of pyrite that is present in the raw material [93]. In the DTA curves of the organozeolites, an intensive exothermic peak at temperatures >300 °C was observed representing oxidation of the organic matter present at the zeolitic surface. The relative intensity of this exothermic peak increases with increasing amounts of organic phase at the zeolitic surface. It is observed that exothermic peak shifts toward higher temperatures with increasing amounts of CPC at the zeolitic surface (from 320 °C for ZCPC-10 to 345 °C for ZCPC-30).

The DTA curve of pure cetylpyridinium chloride (Fig. 10.15c) has one exothermic peak visible at 286 °C. Compared to the temperature of the exothermic peak on DTA curve of pure surfactant, the higher temperature of the exothermic peak observed for organozeolites (320 °C for ZCPC-10, 347 °C for ZCPC-20, and 345 °C for ZCPC-30) suggests CPC chemically bonded to the zeolitic surface and more heat is required to remove the surfactant molecules from the mineral surface. Since CPC was added in amounts above ECEC of the zeolitic tuff, the broad and not well defined exothermic peak at temperatures >400 °C may characterize the extent of organic matter at the zeolitic surface whose oxidation has not been completed at lower temperatures. In the DTA curves of the organozeolites, pyrite oxidation peak is not visible due to overlapping with the peak that originated from the extent of the surfactant.

Thus, the results of thermal analysis confirmed that within the CPC amounts evaluated, the organic molecules are both chemisorbed and physisorbed at the zeolitic surface. Chemisorption is mainly occurred up to complete hydrophobicity of the zeolitic surface, while elimination of weakly bound CPC together with chemisorption takes place when the bilayer is formed.

10.4.1.2   Evaluation of Possible SMZs Pharmaceutical Application

Evaluation of ZCPCs composites as possible pharmaceutical excipient was performed through *in vivo* acute toxicity testing, drug adsorption study, *in vitro* drug release and surfactant desorption testing. Additionally, pharmaceutical technical characterization of the composites was used for assessment of their FRCs relevant for pharmaceutical application.

The dose of 2000 mg/kg of ZVB and ZCPC-30 was administered *via* oral route according to FDA guidance for *Potential excipient intended for short term use* [94]. During the observation period (72 h) treated animals were observed for mortality or symptoms of toxicity. The results of *in vivo* experiments demonstrated that neither natural clinoptilolite nor organozeolite with the highest surfactant amount did cause death or any kind of toxicological reaction during the period of observation. These results along with previous findings of non-toxic clinoptilolite effects make this mineral material a suitable applicant as excipient in pharmaceutical formulations. The non-toxic nature of the organozeolites modified with the highest amount (30 mmol/100g) of BC and HDTMA-Br was also proven [95, 96].

Diclofenac sodium is hydrophobic organic molecule (Table 10.2), sparingly soluble in water, thus it is assumed that it will sorb at the hydrophobic phase created by surfactant tail groups at the zeolitic surface. It is also a weak organic acid and at pH 7.4 (phosphate buffer), exists in ionized (more 99%) anionic form. Diclofenac is used mainly as the sodium salt for the relief of pain and inflammation in various conditions. The usual oral or rectal dose of diclofenac sodium is 75 to 150 mg daily in divided doses [97]. Its use is associated with gastrointestinal-intestinal disturbances including discomfort, nausea, diarrhoea, and occasionally bleeding or ulceration. Due to these side effects and its short elimination half time (1–2 h) it is an ideal candidate for modified release preparations with the aim to maintain therapeutic activity, reduce side effects and to improve patient compliance.

It was already mentioned that the natural zeolite due to its net negative charge and presence of hydrated inorganic cations (Ca^{2+}, Mg^{2+}, Na^+, and K^+) at the hydrophilic surface, has no affinity for neither hydrophobic nor anionic species, and in preliminary experiments it was confirmed that ZVB was ineffective in DS adsorption. The drug adsorption by ZCPCs composites was studied through the evaluation of the adsorption isotherms. The DS adsorption isotherms by ZCPCs composites are presented in Fig. 10.16.

Adsorption of DS by all three modified zeolites followed nonlinear isotherms. The Langmuir and Freundlich models as a good approximation of adsorption behavior were used to fit the equilibrium data. These models are suitable for isotherm evaluation in system comprising inorganic or functionalized sorbents. The better fits of the experimental data were obtained using the Langmuir model (Table 10.5).

The presence of CPC at the zeolitic surface significantly increased DS sorption and the drug sorbed amounts increased with increasing the initial DS concentrations in solution (Fig. 10.16). Also, it was observed that DS adsorbed amounts increased with increasing the amount of surfactant at the surface of ZCPCs composites. These results suggest that organic phase at zeolitic surface is most likely relevant for DS sorption by the organozeolites.

The DS release profiles from the DS/ZCPCs composites are shown in Fig. 10.17. As can be seen from Fig. 10.17, prolonged release of DS from

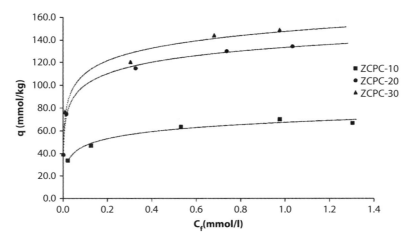

Figure 10.16 Adsorption of DS by ZCPC 10–30 composites [84].

Table 10.5 Fitted Langmuir parameters for DS adsorption by different composites [84].

Sample	Q_m (mmol/kg)	K (l/mmmol)	R^2
ZCPC-10	70.17	24.20	0.997
ZCPC-20	134.93	42.65	0.998
ZCPC-30	159.57	46.82	0.995

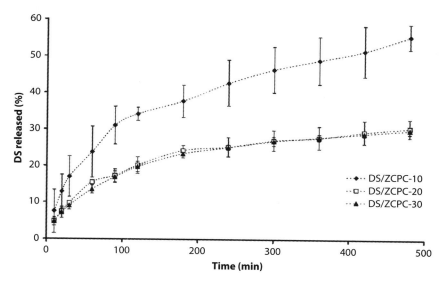

Figure 10.17 *In vitro* drug release profiles of DS from the ZCPC 10-30 composites [85].

all the three composites over 8 h was achieved. For the DS/ZCPC-10 composite, 55% of the drug was released, while a lower amount (30%) of DS was released from the DS/ZCPC-20 and DS/ZCPC-30 composites. These results indicated that the drug molecules were released slowly from the materials.

It was suggested that mainly a monolayer of CPC ions exists at the surface of ZCPC-10, thus alkyl ammonium ions and hydrophobic alkyl chains are the relevant sites for DS adsorption from buffer solution. During release of DS from DS/ZCPC-10, desorption from these sites occurred. However, a complete bilayer was formed on the samples ZCPC-20 and ZCPC-30, thus beside the above-mentioned active sites for DS adsorption, counter

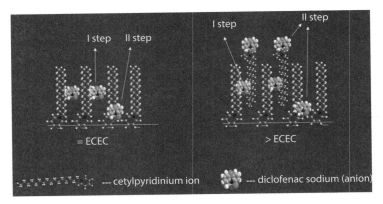

Figure 10.18 Schematic representation of the possible DS release from the ZCPC 10–30 composites (blue-nitrogen; grey-carbon; white-oxygen; green-chlorine; red-sodium) [85].

ions from the bilayer can interact with the anionic form of DS. After drug adsorption, the anionic form of DS in the bilayer of the surfactant may interact with phosphates from the buffer solution; thus for drug release from DS/ZCPC-20 and DS/ZCPC-30, anion exchange together with DS desorption from hydrophobic alkyl chains occurred. This behavior is schematically presented in Fig. 10.18.

Hydrophobic interactions are much weaker than electrostatic interactions, thus DS desorbed more easily and faster from the surface of ZCPC-10 than from ZCPC-20 and ZCPC-30. This may be an explanation for the higher amount of DS released from ZCPC-10 than from ZCPC-20 and ZCPC-30 during the test period.

The results of the surfactant desorption (after ultrasonication) and in a buffer solution at pH 6.8 (under same conditions as in the drug release testing) are presented in Table 10.6 [96].

It was found that for ZCPC-10, no desorption occurred, while for ZCPC-20 and ZCPC-30, the amount of CPC desorbed was 9% and 5%, respectively. These results showed that at initial concentrations of CPC equal to and above 200% of ECEC, only small amounts of the CPC was desorbed from the surface of the composites, confirming the high stability of the ZCPCs, in both *in vitro* and *in vivo* conditions.

To study whether a similar and prolonged release of the drug could be achieved from a sample obtained by mixing the composite and DS, release experiments were performed from a physical mixture containing ZCPC-10 and DS. This composite was chosen because it showed the highest percentage of released drug in addition to its lack of surfactant desorption. For comparison, the drug release from a physical mixture containing NZ

Table 10.6 Results of *in vitro* surfactant desorption for ZCPC 10-30 composites.

Sample	Medium	
	Water	Phosphate buffer (pH 6.8)
ZCPC-10	/	/
ZCPC-20	2*	9
ZCPC-30	4	5

* (%) of the surfactant desorbed

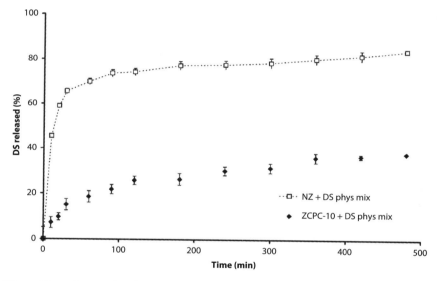

Figure 10.19 *In vitro* dissolution profiles of DS from DS/ZCPC-10 and DS/NZ physical mixtures [85].

and DS (DS/NZ) was also investigated. The quantity of DS in both physical mixtures was the same as DS content in DS/ZCPC-10 composite ($\approx$ 22 mg DS/g). The drug release profiles from the physical mixtures of ZCPC-10 and DS and of DS/NZ are presented in Fig 10.19.

DS release from both physical mixtures and the composite was permantent over a period of 8 h. For the physical mixture of DS/natural zeolite (NZ), maximum amount of released drug was over 85% compared to the DS release from the sample DS/ZCPC-10 (max 55%, see Fig. 10.17), a similar prolonged, but slightly lower release of DS (max 38%) was observed from the physical mixture containing ZCPC-10 and DS. A much faster release of DS from a physical mixture Mg–Al–hydrotalcite (HTIc) than from intercalated composite HTIc–DS was observed by Ambrogi *et al.* [98].

In their study, the release of DS from the physical mixture was already completed after 15 min, whereas the release of DS composite HTIc–DS was 38% after 15 min, 60% after 90 min and 90% after 9 h. Since CPC species at the zeolite surface has a high affinity for DS, the prolonged release of DS from the mixture containing ZCPC-10 and DS suggested that the drug molecules in the buffer solution first dissociated and adsorb at the CPC active sites at the surface of ZCPC-10 creating DS/ZCPC-10. After completion of this process, slow release of DS from the composite occurred. Furthermore, from the slope of the curves presented in Figs. 17 and 19, it can be seen that the drug release kinetics for DS/ZCPC-10 and physical mixture containing ZCPC-10 and DS were similar, but a much faster release of DS was achieved from the physical mixture DS/NZ (70% after 1 h). This may be explained by the fact that the negative surface of NZ had no affinity to adsorb the hydrophobic anionic form of DS from the buffer solution [83]; thus, only a more rapid and more complete release of DS from the physical mixture DS/NZ occurred.

FRCs of the ZCPC 10–30 composite was evaluated by determination of apparent density and flowability according to procedures of Ph. Eur. 7 [99, 100]. The apparent density (also called *bulk density*) is determined by measuring the volume of a known mass of powder that has been passed through a screen into a graduated cylinder. The test for flowability is intended to determine the ability of divided solids (for example, powders and granules) to flow vertically under defined conditions.

The results of these investigations (Table 10.7) revealed that modification of the zeolitic surface by treatment with cationic surfactant induced enlargement of apparent density probably caused by changes of hydrophilic/hydrophobic properties (noticed during modification procedure) of the modified zeolite surfaces. Furthermore, the results of mannitol (an excipient often used in formulation of solid dosage forms) characterization, were in accordance with the literature data (bulk density 0.430 g/cm^3 for powder [55].

The starting ZVB sample showed worse flowability of mannitol, however the modification procedure induced improvement of the powders flow properties, especially for the sample ZCPC-10. It was demonstrated previously that particular pharmaceutical characteristics of a solid dosage form could be affected by hydrophobicity/hydrophilicity of the pharmaceutical excipient [101]. The overall understanding of excipients' role in

Table 10.7 Results of apparent density and flowability determination.

Sample	Apparent density (g/ml)	Flowability (s)*
ZVB	0.36	81.4
ZCPC-10	0.56	32.4
ZCPC-20	0.67	22.6
ZCPC-30	0.76	25.1
Mannitol	0.52	29.6

*result related to 100 g of the sample

pharmaceutical dosage forms contributed to their alteration from "inter ingredient" to "functional component" of the formulation [102]. The improved functionality can be in terms of improved process ability such as flow properties, compressibility, content uniformity, dilution potential, or improved performance such as disintegration and dissolution profile [103]. The presented results exposed the influence of the modification procedure on the particular pharmaceutical characteristics of the obtained modified zeolites, which could be attributed as their FRCs.

10.5   Conclusions

A wide range of minerals have been used for biomedical/pharmaceutical applications, because they have desirable physical and physicochemical properties, such as high adsorption capacity, specific surface area, swelling capacity and reactivity to acids. Aluminosilicates (natural zeolite–clinoptilolite and natural bentonite–montmorillonite) and surfactant-modified zeolites (SMZs) are the most widely utilized as mycotoxins adsorbents. Because of their large specific surface area and high adsorption capacity, clinoptilolite and surfactants-modified clinoptilolite (surfactants - clinoptilolite composite) are well suited to act as drug carriers. The latest investigations on cationic surfactants-modified natural zeolites were directed toward analysis of their potential use as drug formulation excipients with improved functionality. The improved functionality can be in terms of improved process ability such as flow properties, compressibility, content uniformity, dilution potential, or improved performance such as disintegration and dissolution profile.

The presented results discuss the influence of the modification procedure on the particular pharmaceutical characteristics (functionality-related characteristics/FRCs) of the obtained modified zeolites.

Acknowledgement

This work was realized within the framework of the projects TR 34031 and ON 172018 supported by the Ministry of Education, Science and Technological Development of Republic of Serbia. The company "Nemetali" in Vranjska Banja, Serbia, kindly provided the natural zeolite sample for the researches.

References

1. D. Fuoco, *Nanomaterials*, Vol. 2, pp. 79–91, 2012.
2. F.A. Mumpton, *Proc. Natl. Acad. Sci. USA*, Vol. 96, 3463–3470, 1999.
3. D.W. Ming and F.A. Mumpton, "Zeolites in Soils," in J.B. Dixon and S.B. Weed, eds., *Minerals in Soil Environment*, Soil Science Society of America, Madison, Wisconsin, USA, pp. 873–911, 1989.
4. K. Margeta, N. Zabukovec Logar, M. Šiljeg and A. Farkaš, "Natural Zeolites in Water Treatment – How Effective is their Use" in W. Elshorbagy and R. K. Chowdhury, eds., *Water treatment*, InTech, pp. 81–112, 2013.
5. M. Kragović, A. Daković, Ž. Sekulić, M. Trgo, M. Ugrina, J. Perić and G.D. Gatta, *Appl. Surf. Sci.*, Vol. 258, pp. 3667–3673, 2012.
6. A. Daković, M. Tomašević-Čanović, G. E. Rottinghaus, S. Matijašević and Ž. Sekulić, *Micropor. Mesopor. Mat.*, Vol. 105, pp. 285–290, 2007.
7. D.L. Bish and J.M. Boak, "Clinoptilolite-Heulandite Nomenclature," in D.L. Bish and D.W. Ming, eds., *Reviews in Mineralogy and Geochemistry, Natural Zeolites: Occurrence, Properties, Applications*, Mineralogical Society of America, Vol 45, pp. 207–216, 2001.
8. R.L. Ward and H.L. McKague, *J. Phys. Chem.*, Vol. 98, pp. 1232–1237, 1994.
9. A. Radosaljević-Mihajlović, Characterization and stability of dealuminated clinoptilolite rich tuffs from different deposits, Master Thesis, Faculty of Physical Chemistry, University of Belgrade, 2003 (in Serbian).
10. A. Daković, S. Matijašević, G.E. Rottinghaus, V. Dondur, T. Pietrass and C.F.M. Clewett, *J. Colloid Interface Sci.*, Vol. 311, pp. 8–13, 2007.
11. A. Dyer, *An Introduction to Zeolite Molecular Sieves*, Chichester, John Wiley & Sons, UK, pp. 113–115, 1988.
12. C. Colella, *Clay Miner.*, Vol. 46, pp. 295–309, 2011.
13. A. Rivera, G. Rodríguez-Fuentes and E. Altshuler, *Micropor. Mesopor. Mat.*, Vol. 24, pp. 51–58, 1998.

14. G. Rodríguez-Fuentes, A.R. Denis, M.A. Barrios Álvarez and A.I. Colarte, *Micropor. Mesopor. Mat.*, Vol. 94, pp. 200–207, 2006.
15. T.A. Ostomel, Q. Shi, P.K. Stoimenov and G.D. Stucky, *Langmuir*, Vol. 23, pp. 11233–11238, 2007.
16. P. Rhee, C. Brown, M. Martin, A. Salim, D. Plurad, D. Green, L. Chambers, D. Demetriades, G. Velmahos and H. Alam, *J. Trauma*, Vol. 64, pp. 1093–1099, 2008.
17. B. Concepción-Rosabal, G. Rodríguez-Fuentes and R. Simón-Carballo, *Zeolites*, Vol. 19, pp. 47–50, 1997.
18. B. Concepción-Rosabal, J. Balmaceda-Era and G. Rodríguez-Fuentes, *Micropor. Mesopor. Mat.*, Vol. 38, pp. 161–166, 2000.
19. G. Rodríguez-Fuentes, M.A. Barrios, A. Iraizos, I. Perdomo and B. Cedré, *Zeolites*, Vol. 19, pp. 441–448, 1997.
20. J. Ramu, K. Clark, G.N. Woode, A.B. Sarr and T.D. Phillips, *J. Food Protect.*, Vol. 60, pp. 358–362, 1997.
21. A.A. Sadeghi and P. Shawrang, *Livest. Sci.*, Vol. 113, pp. 307–310, 2008.
22. R. Simón Carballo, G. Rodríguez-Fuentes, C. Urbina and A. Fleitas, Paper 32-O-03 (CD-ROM), in A. Galarneau, F. Di Renzo, F. Fajula and J. Vedrine, eds., *Zeolites and Mesoporous Materials at the Dawn of the 21st Century Studies in Surface Science and Catalysis*, No. 135, Elsevier, Amsterdam, The Netherlands, 2001.
23. K. Pavelić, M. Hadžija, L. Bedrica, J. Pavelić, I. Dikić, M. Katić, M. Kralj, M.H. Bosnar, S. Kapitanović, M. Poljak-Blaži, S. Križanac, R. Stojković, M. Jurin, B. Subotić and M. Čolić, *J. Mol. Med.*, Vol. 78, 708–720, 2001.
24. K. Pavelić, M. Katić, V. Sverko, T. Marotti, B. Bošnjak, T. Balog, R. Stojković, M. Radačić, M. Čolić and M. Poljak-Blaži, *J. Cancer. Res. Clin.*, Vol. 128, pp. 37–44, 2002.
25. K. Pavelić and M. Hadžija, "Biomedical application of zeolites," in S.M. Auerbach, K.A. Carrado and P.K. Dutta, eds., *Handbook of Zeolite Science and Technology*, Marcel Dekker, New York, pp. 1143–1174, 2003.
26. M. Grce and K. Pavelić, *Microp. Mesop. Mat.*, Vol. 79, pp. 165–169, 2005.
27. S. Ivkovic, U. Deutsch, A. Silberbach, E. Walraph and M. Mannel, *Adv. Ther.*, Vol. 21, pp. 135–147, 2004.
28. L. Munjas Jurkić, I. Cepanec, S. Kraljević Pavelić and K. Pavelić, *Nutr. Metab.*, Vol. 10, art. No 2, 2013.
29. R. Sharma, ed., *Surfactant Adsorption and Surface Solubilization*, American Chemical Society, Washington DC, 1999.
30. E. J. Sullivan, D. B. Hunter and R. S. Bowman, *Clays Clay Miner.*, Vol. 45, pp. 42–53, 1997.
31. G. M. Haggerty and R. S. Bowman, *Environ. Sci. Technol.*, Vol. 28, pp. 452–458, 1994.
32. R.S. Bowman, G. M. Haggerty, R. G. Huddleston, D. Neel, and M. M. Flynn, "Sorption of nonpolar organic compounds, inorganic cations, and inorganic oxyanions by surfactant-modified zeolites," in D. A. Sabatini, R. C. Knox, and

J. H. Harwell, eds., *Surfactant-Enhanced Subsurface Remediation*, vol. 594 of ACS Symposium Series, American Chemical Society, Washington, DC, USA, pp. 54–64, 1995.

33. A. D. Vujaković, M. R. Tomašević-Čanović, A. S. Daković and V. T. Dondur, *Appl. Clay Sci.*, Vol. 17, pp. 265–277, 2000.

34. Z. Li and R. S. Bowman, *Environ. Sci. Technol.*, Vol. 33, pp. 2407–2412, 1997.

35. A. Rivera, T. Farías, *Micropor. Mesopor. Mat.*, Vol. 80, pp. 337–346, 2005.

36. T. Farías, L.C. de Ménorval, J. Zajacb and A. Rivera, *Colloids Surf. B*, Vol. 76, pp. 421–426, 2010.

37. S. Wang, W. Gong, X. Liu, B. Gao and Q.Yue, *Sep. Purif. Technol.*, Vol. 51, pp. 367–373, 2006.

38. U. Wingenfelder, G. Furrer and R. Schulin, *Micropor. Mesopor. Mat.*, Vol. 95, pp. 265–271, 2006.

39. A. Huwig, S. Freimund, O. Käppeli and H. Dutler, *Toxicol. Lett.*, Vol. 122, pp. 179–188, 2001.

40. D. Papaioannou, P.D. Katsoulos, N. Panousis and H. Karatzias, *Micropor. Mesopor. Mat.*, Vol. 84, pp. 161–170, 2005.

41. T. D. Phillips, B. Sarr and P. Grant, *Nat. Toxins*, Vol. 3, pp. 204–213, 1995.

42. M. Tomašević–Čanović, M. Dumić, O. Vukićević, P. Radošević, I. Rajić and T. Palić, *Acta Vet. Beograd*, Vol. 44, pp. 309–318, 1994.

43. S.L. Lemke, P.G. Grant and T.D. Phillips, *J. Agr. Food Chem.*, Vol. 46, pp. 3789–3796, 1998.

44. A. Daković, M. Tomašević-Čanović, V. Dondur, D. Stojšić and G. Rottinghaus, "Zeolites and mesoporous materials at the down of the 21st century," in: A. Galarneau, F. Di Renzo, F. Fajula, J. Vedrina, eds., Proceedings of the 13th International Conference, Montpellier, France, *Stud. Surf. Sci. Catal.*, vol. 135, pp. 5276–5283, 2001.

45. A. Daković, M. Tomašević-Čanović. V. Dondur, G. E. Rottinghaus, V. Medaković and S. Zarić, *Colloids Surf. B*, Vol. 46, 20–25, 2005.

46. L. Groisman, C. Rav-Acha, Z. Gerstl, U. Mingelgrin, *Appl. Clay Sci.*, Vol. 24, 159–166, 2004.

47. IARC Monographs on the Evaluation of Carcinogenic Risk to Humans. Some Naturally Occurring Substances: Food Items and Constituents, Heterocyclic Aromatic Amines and Mycotoxins, Vol. 56 (Lyon, France 9–16 June 1992), WHO, International Agency for Research on Cancer, 1993.

48. G. Avantaggiato, M. Solfrizzo and A. Visconti, *Food Addit. Contam.*, Vol. 22, pp. 379–388, 2005.

49. S. Hendrich, K. A. Miller, T. M. Wilson and P. A. Murphy, *J. Agric. Food Chem.*, Vol. 41 pp. 1649–1654, 1993.

50. A. E. Pohland, "Occurrence of fumonisins in the U.S. food supply," in L. S. Jackson, J. W. DeVries, L. B. Bullerman, eds., *Fumonisins in Food*, Plenum Press, New York, pp. 19–26, 1996.

51. L. Zhu, B. Chen and X. Shen, *Environ. Sci. Technol.*, Vol. 34, pp. 468–475, 2000.

52. D.I. Song and S.S. Won, *Environ. Sci. Technol.*, Vol. 39, pp. 1138–1143, 2005.
53. Z. Li, T. Burt and R. S. Bowman, *Environ. Sci. Technol.*, Vol. 34, pp. 3756–3760, 2000.
54. Z. Rawajfih and N. Nsour, *J. Colloid Interface Sci.*, Vol. 298, pp. 39–49, 2006.
55. R.C. Rowe, P.J. Sheskey, and S.C. Owen, eds., *Handbook of pharmaceutical excipients*, Pharmaceutical Press and American Pharmacists Association, London & Washington DC, 2006.
56. European Pharmacopoeia 7th ed., Directorate for the Quality of Medicines of the Council of Europe, Strasbourg (F), 2010.
57. L. Bhattacharyya, S. Schuber, C. Sheehan, and R. William, "Excipients: Background/Introduction," in A. Katdare and M.V. Chaubal, eds., *Excipient Development for Pharmaceutical, Biotechnology and Drug Delivery Systems*, Informa Healthcare USA Inc., New York, pp 1–2, 2006.
58. P.J. Crowley and L.G. Martini, "Excipients for Pharmaceutical Dosage Forms." in J. Swarbrick, ed., *Encyclopedia of Pharmaceutical Technology*, 3rd ed., Informa Healthcare USA, Inc., New York, pp. 1609–1621, 2007.
59. M.I. Carretero and M. Pozo, *Appl. Clay Sci.*, Vol. 46, pp. 73–80, 2009.
60. A. Parker, *Chim. Oggi.*, Vol. 27, pp. 5–7, 2009.
61. M.I. Carretero and M. Pozo, *Appl. Clay Sci.*, Vol. 47, pp. 171–181, 2010.
62. G. Cerri, M. de' Gennaro, M.C. Bonferoni and C. Caramella, *Appl. Clay Sci.*, Vol. 27, pp. 141–150, 2004.
63. C. Aguzzi, P. Cerezo, C. Viseras, and C. Caramella. *Appl. Clay Sci.*, Vol. 36, pp. 22–36, 2007.
64. J.P. Zheng, L. Luan, H.Y. Wang, L.F. Xi, and K.D. Yao, *Appl. Clay Sci.*, Vol. 36, pp. 297–301, 2007.
65. J.K. Park, Y.B. Choy, J.-M. Oh, J.Y. Kima, S.-J. Hwanga and J.-H. Choy, *Int. J. Pharm.* Vol. 359, pp. 198–204, 2008.
66. G.V. Joshi, B.D. Kevadiyaa, H.A. Patel, H.C. Bajaj and R.V. Jasrab, *Int. J. Pharm.* Vol. 374, pp. 53–57. 2009.
67. S. Rojtanatanya and T. Pongjanyakul, *Int. J. Pharm.*, Vol. 383, pp. 106–115, 2008.
68. S.L. Madurai, S.W. Joseph, A.B. Mandal, J. Tsibouklis and B.S.R. Reddy, Nanoscale Res. Lett., Vol. 6, pp. 1–8, 2011.
69. L.A.D.S. Rodrigues, A. Figueiras, F. Veiga, R.M. de Freitas, L.C.C. Nunes, E.C. da Silva Filho and C.M. da Silva Leite, *Colloids Surf. B*, Vol. 103, pp. 642–651, 2013.
70. Z. Adamis, E. Tátrai, K.É.S. Honma and G. Ungváry, *Ann. Occup. Hyg.*, Vol. 44, pp. 67–74, 2000.
71. F.A. Andersen, *Int. J. Toxicol.*, Vol. 22, pp. 37–102, 2003.
72. G. Narin, B.Ç. Albayrak, and S. Ülkü, *Appl. Clay Sci.*, Vol. 50, pp. 560–568, 2010.
73. A. Lam, L.R. Sierra, G. Rojas, A. Rivera, G. Rodríuez-Fuentes and L.A. Montero, *Micropor. Mesopor. Mat.*, Vol. 23, 247–252, 1998.
74. A. Lam, A. Rivera and G. Rodrídguez-Fuentes, *Micropor. Mesopor. Mat.*, Vol. 49, pp. 157–162, 2001.

75. T. Farías, A.R. Ruiz–Salvador and A. Rivera, *Micropor. Mesopor. Mat.,* Vol. 61, pp. 117–125, 2003.
76. NRIB, 1152: Quality requirements, Natural Zeolites for Pharmaceutical Industry, Drug Quality Control of Cuba, 1992.
77. C. Charnay, S. Bégu, C. Tourné-Péteilh, L. Nicole, D.A. Lerner and J.M. Devoisselle, *Eur. J. Pharm. Biopharm.,* Vol. 57, pp. 533–540, 2004.
78. P. Horcajada, C. Márquez-Alvarez, A. Rámila, J. Pérez-Pariente and M. Vallet-Regí, *Solid State Sci.,* 8, pp. 1459–1465, 2006.
79. M.G. Rimoli, M.R. Rabaioli, D. Melisi, A. Cucio, S. Mondello, R. Mirabelli and E. Sbignete, *J. Biomed. Mater. Res. A,* Vol. 87A, pp. 156–164, 2007.
80. Z. Li, S.J. Roy, Y. Zou and R. S. Bowman, *Environ. Sci. Technol.,* Vol. 32, pp. 2628–2632, 1998.
81. A. Rivera, T. Farías, A.R. Ruiz-Salvador and L.C. de Menorval, *Micropor. Mesopor. Mat.,* Vol. 61, pp. 249–259, 2003.
82. T. Farías, L. C. de Ménorval, J. Zajac and A. Rivera, *J. Colloid Interf. Sci.,* Vol. 363, pp. 465–475, 2011.
83. D. Krajišnik, M. Milojević, A. Malenović, A. Daković, S. Ibrić, S. Savić, V. Dondur, S. Matijašević, A. Radulović, R. Daniels and Milić J, *Drug Dev. Ind. Pharm.,* Vol. 36, pp. 1215–1224, 2010.
84. D. Krajišnik, A. Daković, M. Milojević, A. Malenović, M. Kragović, D. Bajuk Bogdanović, V. Dondur and J. Milić, *Colloids Surf. B,* Vol. 83, pp. 165–172, 2011.
85. D. Krajišnik, A. Daković, A. Malenović, Lj. Djekić, M. Kragović, V. Dobričić and J. Milić. *Micropor. Mesopor. Mat.,* Vol. 167, pp. 94–101, 2013.
86. European Pharmacopoeia 5th ed., Directorate for the Quality of Medicines of the Council of Europe, Strasbourg (F), 2004.
87. O. Korkuna, R. Leboda, Z.J. Skubiszewska, T. Vrublevska, V.M. Gunko and J. Ryczkowski, *Micropor. Mesopor. Mat.,* Vol. 87, pp. 243–254, 2006.
88. T. Kawai, Y. Yamada and T. Kondo, *J. Phys. Chem. C,* Vol. 112, pp. 2040–2044, 2008.
89. Z. Li and R. Bowman, *Environ. Sci.Technol.,* Vol. 32, pp. 2278–2282, 1998.
90. E.J. Sullivan, J.W. Carey and R.S. Bowman, *J. Colloid Interf. Sci.,* Vol. 206, pp. 369–380, 1998.
91. S. Yariv, *Appl. Clay Sci.,* Vol. 24, 225–236, 2004.
92. Y. Xi, W. Martens, H. He and R. L. Frost, *J. Therm. Anal. Calorim.,* Vol. 81. 91–97, 2005.
93. A.D. Vujaković, M.A. Đuričić and M.R. Tomašević-Čanović, *J. Therm. Anal. Calorim.,* Vol. 63, 161–172, 2001.
94. Center for Drug Evaluation and Research (CDER). Guidance for Industry: Nonclinical Studies for Development of Pharmaceutical Excipients. Final guidance issued by FDA CDER, 2005.
95. D. Krajišnik, A. Daković, M. Kragović, A. Malenović, M. Milojević, V. Dondur, S. Ristić and J. Milić, Proceedings of the 8th International Conference of the Occurrence, Properties, and Utilization of Natural Zeolites, Sofia, Bulgaria: Book of abstracts, pp. 151–152, 2010.

96. D. Krajišnik, A. Daković, A. Malenović, S. Ristić and J. Milić, Proceedings of the CESIO Austria, Vienna; Book of abstracts CD ROM, P19, 2011.

97. S.C. Sweetman, ed., *Martindale: The Complete Drug Reference,* 36th ed., Pharmaceutical Press–London, 2009.

98. V. Ambrogi, G. Fardella, G. Grandolini and L. Perioli, *Int. J. Pharm.,* Vol. 220, pp. 23–32, 2001.

99. D. Krajišnik, Preformulation investigations of modified alumosilicates as potential pharmaceutical excipients, PhD Thesis, Faculty of Pharmacy, University of Belgrade, 2011 (in Serbian).

100. D. Krajišnik, A. Daković, J. Janićijević and J. Milić, Proceedings of the 50th Meeting of the Serbian Chemical Society, Book of abstracts, p. 61, 2012.

101. S. Jonat, S. Hasenzahl, A. Gray and P.C. Schmidt, *Drug Dev. Ind. Pharm.,* Vol. 31, 687–696, 2005.

102. G. Pifferi, P. Santoro and M. Pedrani, *Farmaco,* Vol. 54, pp. 1–14, 1999.

103. P. Gupta, S.K. Nachaegari and A.K. Bansal, "Improved Excipient Functionality by Coprocessing," in A. Katdare and M.V. Chaubal, eds., *Excipient Development for Pharmaceutical, Biotechnology, and Drug Delivery Systems,* Informa Healthcare, Inc., New York, pp. 109–129, 2006.

11

Supramolecular Hydrogels Based on Cyclodextrin Poly(Pseudo)Rotaxane for New and Emerging Biomedical Applications

Jin Huang[1,*], Jing Hao[1], Debbie P. Anderson[2] and Peter R. Chang[2,*]

[1]College of Chemical Engineering, Wuhan University of Technology, Wuhan, China
[2]Bioproducts and Bioprocesses National Science Program, Agriculture and Agri-Food Canada, Government of Canada, Saskatoon, SK, Canada

Abstract

This chapter introduces/presents an interesting new advanced material based on the supramolecular network formed between a spontaneous aggregation of poly(pseudo)rotaxanes (which provides columnar crystalline domains) and its absorbed water (which serves as the hydrophilic segment). Fabrication mechanisms and formation conditions of cyclodextrin poly(pseudo)rotaxane supramolecular hydrogels will be reviewed and updated. The intrinsic properties of such hydrogels, including shear-thinning and temperature-sensitivity, as well as smart properties contributed by the uncovered guest molecules or nanoparticles, will be depicted. Due to their biocompatible nature and stimulus-response properties and functions, these hydrogels show great potential in the biomedical field. Recently, attempts have been made to develop hydrogels as injectable biomaterials for use as drug carriers, scaffolds, and gene carriers.

Keywords: Biomedical applications, cyclodextrin, host-guest interactions, inclusion complexes, nanocomposite, poly(pseudo)rotaxane, stimulus-response properties, supramolecular hydrogels

*Corresponding authors: peter.chang@agr.gc.ca; huangjin@iccas.ac.cn

Ashutosh Tiwari (ed.) Advanced Healthcare Materials, (405–438) 2014 © Scrivener Publishing LLC

405

11.1 Introduction

Cyclodextrins (CDs) are a series of cyclic oligosaccharides with the most common being α-, β-, and γ-CD, which consist of 6, 7, and 8 D(+)-glucose units, respectively [1, 2, 3]. The conformation and orientation of hydroxyl groups in the cyclic molecular structure of CDs produce a hydrophobic inner cavity with a depth of ca. 7.0 Å, an internal diameter of ca. 4.5, 7.0, or 8.5 Å (α-, β-, and γ-CD, respectively), and a hydrophilic outer surface (Figure 11.1) [4, 5, 6, 7]. CDs can act as host molecules to include a great variety of molecular guests in their cavities as a result of geometric compatibility, van der Waals forces, and host-guest hydrophobic interactions between the CDs. Based on host-guest interactions, suitable linear polymer chain guest molecules penetrate the cavities of CDs to form inclusion complexes (ICs) and a poly(pseudo)rotaxane is

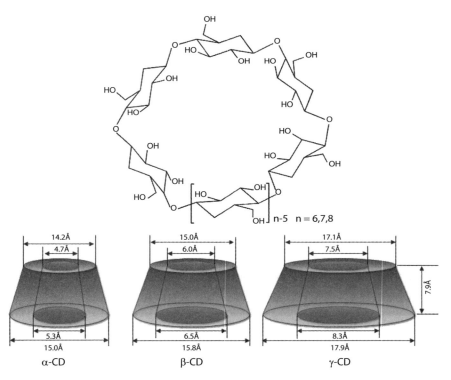

Figure 11.1 Molecular structure of cyclodextrins (CDs) and their corresponding dimensions.

thus formed with or without bulky groups (or stoppers) on both terminal ends [7, 8, 9, 10, 11]. Once threaded onto the guest polymer chains, CD molecules become hydrophobic because their hydroxyl groups are consumed by hydrogen bonding [12]. In detail, two stages are involved with the formation of supramolecular hydrogels: first, the CD host threads onto the guest polymer chains to form individual poly(pseudo)rotaxanes as a result of host-guest hydrophobic interactions. Subsequently, due to strong intermolecular hydrogen bonding between neighboring CDs, the cyclodextrin poly(pseudo)rotaxanes aggregate into columnar α-CD crystalline domains between the neighboring threaded α-CDs, acting as physical crosslinks that give rise to a polymer network crystallite structure [2, 13]. The remaining hydrophilic segments, such as partially unthreaded poly(ethylene glycol) (PEG) blocks or hydrophilic blocks introduced via grafting or blocking, function as the water absorbing portion and form large amorphous domains that are essential for the retention of a large volume of water within the supramolecular hydrogel structure [14, 15, 16]. Figure 11.2 depicts the process of CD poly(pseudo)

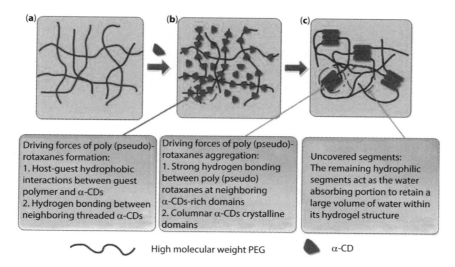

Figure 11.2 Formation of poly(pseudo)rotaxane hydrogels from α-cyclodextrin (α-CD) and high molecular weight PEG. A clear PEG aqueous solution without α-CD (a); α-CD is added and threads onto the polymer to form poly(pseudo)rotaxanes (b); and aggregation of poly(pseudo)rotaxanes and uncovered hydrophilic PEG segments gives rise to hydrogels (c).

rotaxane supramolecular hydrogel formation using unmodified CDs and homopolymers of high molecular mass PEG as an example.

In some cases the hydrophobic chains, such as poly(ε-caprolactone) [PCL] and poly(propylene-oxide) [PPO], remain unthreaded and may self-aggregate due to hydrophobic interactions that contribute to the physical crosslink network and improve the strength of the hydrogels [17, 18]. Supramolecular hybrid and nanocomposite hydrogels can also be established on CD–poly(pseudo)rotaxane inclusions with carbon nanomaterials, such as grapheme [19], single-walled carbon nanotube (SWNT) [20, 21], and multi-walled carbon nanotube (MWNT) [22]; with metal-based nanoparticles of magnetite [23] and silver [24]; with silica nanoparticles [25]; or with polysaccharide nanocrystals. In all these examples, multiple PEG or PEG-containing copolymers were first attached to the nanomaterials through chemical bonding or physical interactions and then α-CDs were threaded onto the PEG-decorated materials. Remarkably, the functions of the nanoparticles were inherited and new network junctions increased.

Supramolecular hydrogels based on the aggregation of cyclodextrin poly(pseudo)rotaxanes in aqueous solution without chemical cross-linking reagents were found to be thixotropic, reversible, and injectable through needles, which was characteristic of the supramolecular self-assembly system [14, 26]. The intrinsic properties of supramolecular hydrogels, such as shear-thinning and temperature-sensitivity, were due mainly to the aggregation of poly(pseudo)rotaxanes; while properties such as pH-sensitivity [12], reduction-sensitivity [27], or temperature and pH dual-responsivity [28, 29] resulted from the unthreaded guest polymers or from introduced nanomaterials. The stimulus-response function and reversible thixotropy, as well as the biocompatibility of the compositions, allow them to be used in a unique injectable hydrogel drug delivery system using a fine needle [29, 17], which has great potential in the injectable biomaterials field. Hydrogels have also shown great potential as drug carriers (due to its sustained and controlled released properties), gene carriers [30], and cell-adhesive scaffolds [31]. Table 11.1 summarizes the properties of polymer guest molecules and applications for poly(pseudo)rotaxane supramolecular hydrogels.

11.2   Fabrication of Cyclodextrin Poly(pseudo)rotaxane-Based Hydrogels

11.2.1   Homopolymers as Guest Molecules

There are two kinds of homopolymers used as guest molecules: high molecular mass PEG and linear polyethylenimines (LPEIs, molecular mass

Table 11.1 Polymer Guest Molecules, Properties and Applications of Poly(pseudo)rotaxane Supramolecular Hydrogels.

Type of CD	Threaded Chains	Structure of Guest Polymer	Unthreaded Chains	Properties	Application
α-CD	PEG	Homopolymer	Hydrated PEG	Shear-thinning	Injectable dextran-FITC carrier[14]
α-CD	PEG	Diblock copolymer PEG-b-PCL	Hydrated PEG, PCL	Shear-thinning	Injectable dextran-FITC carrier with long-term sustained release[17]
α-CD	PEG	Diblock copolymer PEG-b-PLG	Hydrated PEG, PLG	Temperature and pH dual-responsivity	Injectable DOX carrier with sustained release[28]
α-CD	PEG	Triblock copolymer PEG-b-PPO-b-PEG	Hydrated PEG, PPO	Shear-thinning temperature- responsivity	Injectable BSA-FITC carrier with long-term sustained release[4]
α-CD α-CD	PEG MPEG	Triblock copolymer PEG-b-PHB-b-PEG Grafted copolymer mPEG-g-polyphosphazenes	Hydrated PEG, PHB Hydrated PEG, polyphosphazenes	Shear-thinning Shear-thinning temperature-responsivity	Injectable dextran-FITC carrier with long-term sustained release[26] Injectable BSA carrier with controlled sustained release[32]
α-CD	MPEG	Graft copolymer COS-g-PCL-b-MPEG	Hydrated COS, PCL	Shear-thinning	Injectable BSA carrier with controlled release[33]
α-CD	MPEG	Graft copolymer mPEG-g-SS-PAA	Hydrated mPEG, PAA	Reduction-sensitivity	Injectable BSA carrier with controlled release[27]

(Continued)

Table 11.1 (*Cont.*)

Type of CD	Threaded Chains	Structure of Guest Polymer	Unthreaded Chains	Properties	Application
α-CD	PPEGMA	Graft copolymer PPEGMA-co-PDMA	Hydrated PPEOMA, PDMA	Temperature and pH dual-responsivity	Injectable BSA-DTAF carrier with controlled release[29]
α-CD	PEG	Polysaccharide nanocrystal-doped nanocomposite hydrogels	Hydrated PEG, PPO	Shear-thinning reinforcement	Injectable BSA carrier with sustained release[61]
α-CD	PEG	Modified CN-doped nanocomposite hydrogel	Hydrated PEG, PPO	Shear-thinning reinforcement	Injectable DOX carrier with sustained release[34]
α-CD	PEG	PEG-b-PCL and PEG-b-PAA	Hydrated PEG, PCL, PAA	Shear-thinning	Injectable dual DOX/cisplatin carrier with respective sustained release[35]
α-CD	PEG	Heparin–F-127	Hydrated PEG, Heparin, PPO	Shear-thinning	Injectable dual G-CSF/CPT Carrier with respective sustained release[36]
α-CD	MPEG	MPEG-b-PCL-b-PDMAEMA	Hydrated MPEG, PCL PDMAEMA	Shear-thinning	Injectable gene carrier with long-term sustained release[30]
α-CD	PEG	PRxRGD-CHPA	CHPA, RGD	Shear-thinning	Injectable cell-adhesive scaffold[31]

greater than 3.5 kDa). Homopolymers used in cyclodextrin poly(pseudo) rotaxane hydrogels must have the following characteristics: first, the molecular mass of linear chains must be greater than 3.5 kDa, otherwise the normally fast α-CD threading will cause shorter chains (less than 2 kDa) to be fully covered and precipitate[37]; and second, the uncovered chain units must be long enough to remain unthreaded and water soluble in order to form large amorphous domains, which are essential for the retention of a large volume of water within the polymer structure. Without this water absorbing portion, the hydrogen bond interaction with water diminishes during the growth of columnar α-CD domains, leading to precipitation from aqueous solution rather than hydrogel formation. The aggregation of cyclodextrin poly(pseudo)rotaxanes forms physical crosslinks that induce a supramolecular polymer network and provide the primary driving force for hydrogelation. The uncovered hydrophilic segments allow water to fill the network during supramolecular self-assembly [13, 38, 14].

The first report of cyclodextrin poly(pseudo)rotaxane supramolecular hydrogels dates back to 1994 [37]. Water-soluble linear high molecular mass PEG polymers partially penetrate the inner cavity of α-CD from both terminals to form cyclodextrin poly(pseudo)rotaxanes with a necklace-like structure [13]. Due to hydrogen bonding and crystallite formation, the poly(pseudo)rotaxane chains rapidly aggregate into bundles containing nanosized columnar α-CD domains between neighboring CD-rich segments. These domains continue to assemble into larger particles leading to a turbid sol and subsequent increase in viscosity. Eventually, with many of these α-CD domains acting as physical crosslinks, hydrogelation occurs [39, 37, 8]. Kinetically, hydrogelation is favored by low temperature and by a high α-CD to PEG ratio. The properties and structures of supramolecular hydrogels can be fine-tuned by adjusting the composition ratio between α-CD and PEG, the molecular mass, and the chemical structure of the polymer [14]. Recently, based on the Poisson distribution of α-CD/PEG, Sabadini *et al.* [40] found the complexation of PEG and α-CD had a small dependence on the polymer molecular mass or structure (linear or star-like) while the kinetics of complexation were strongly dependent on the PEG structure.

Similar to α-CD/PEG-based hydrogels, linear polyethylenimines (LPEIs) of various molecular masses were used to prepare CD poly(pseudo) rotaxane hydrogels under specific conditions (pH and/or temperature dependent) [41]. In this work, the high molecular mass LPEIs partially penetrated the inner cavity of α-CD from both terminals to form cyclodextrin poly(pseudo)rotaxanes, which then aggregated into columnar α-CD crystalline domains. Hydrogelation, however, required a high gelation temperature and a relatively long time to achieve critical threading of α-CDs. The unthreaded LPEI chains acted as the water absorbing portion to retain water within the hydrogel structure. In the case of low molecular mass LPEI (2.2 kDa), all the LPEI chains easily penetrated the inner cavity of α-CD and the uncovered or unoccupied EI units were too short to form a hydrogel. It is worth noting that the physical state of the hydrogels spontaneously turned into a crystalline precipitate during repeated heating and cooling with continuous stirring. The intermolecular hydrogen bonds between the hydrated LPEI chains and/or water played a major role in the stabilization of the hydrogels as a result of deprotonation of almost all the secondary amines (pKa = 8.9) at pH 11.0.

11.2.2 Block-Copolymers as Guest Molecules

To avoid using PEG or LPEI as the high molecular mass homopolymer as described above, an alternative is to utilize a flanking PEG block or threadable segments. Copolymers that could be used to induce cyclodextrin poly(pseudo)rotaxane hydrogels include block, graft, and other branched structures either with or without PEG. The combination of partial inclusion complexation between PEG blocks and CDs and hydrophobic interaction between hydrophobic blocks in the copolymers induced the formation of stable macromolecular supramolecular networks [17, 42].

11.2.2.1 Diblock Copolymer

While α-CD and high molecular mass PEG can form supramolecular hydrogels based on inclusion complexation, it is also possible to form a hydrogel with α-CD and a diblock copolymer comprising a PEG block and a hydrophobic block. A series of diblock copolymers such as PEG-b-PCL [17], PEG-b-PLL [12] and PEG-b-PLG [28] have been reported to form cyclodextrin poly(pseudo)rotaxane hydrogels. In these systems, the introduction of hydrophobic blocks decreases the molecular mass of the threadable PEG block.

Biodegradable amphiphilic poly(ethylene glycol)-b-poly(ε-caprolactone) (PEG -b-PCL) diblock copolymer has been used to form novel supramolecular hydrogels [17]. In this work, PEG with a low molecular mass of 5 kDa was used to obtain a viscous mixture after adding α-CD aqueous solution to a diblock copolymer aqueous solution. A homogeneous hydrogel formed within a few minutes. In the system, α-CD preferred to thread onto PEG blocks to partially form cyclodextrin poly(pseudo)rotaxane ICs [17, 43]. These ICs subsequently self-assembled into columnar α-CD crystalline domains, which act as physical crosslinks.The unthreaded PEG segments remained water soluble and may have formed large water-soluble amorphous domains; however, whole PCL blocks were left unthreaded, particularly when the ratio of α-CD to PEG-*b*-PCL copolymer was low, which facilitated hydrogel formation and improved stability via hydrophobic interactions between uncovered PCL blocks. As a result, a phase-separated structure consisting of uncovered hydrophobic PCL blocks, crystalline α-CD/PEG-*b*-PCL poly(pseudo)rotaxane ICs, and unthreaded hydrated PEG blocks formed. The combination of the aggregated cyclodextrin poly(pseudo)rotaxane ICs between PEG blocks of PEG-*b*-PCL diblock copolymer and α-CD and the hydrophobic interaction between the uncovered PCL blocks provided the primary driving force for hydrogel formation during supramolecular self-assembly. Based on a similar principle, poly(ethylene glycol)-*b*-poly(L-lysine) (PEG-*b*-PLL) diblock copolymer was used to form cyclodextrin poly(pseudo)rotaxane supramolecular hydrogels under alkaline conditions (pH 10) in which the PLL blocks were hydrophobic as a result of the amino end groups being neutral and insoluble -NH$_2$ groups [12]. Interestingly, poly(ethylene glycol)-*b*-poly(L-glutamic acid) (PEG-*b*-PLG) has also been used to fabricate supramolecular polypeptide-based hydrogels in two forms: normal micellar hydrogels and reverse micellar hydrogels [28]. The normal micellar hydrogels existed in the form of polypeptide-cored micelles with a PEG corona, which was mediated through hydrogen bonding interactions between PLG chains, followed by inclusion complexation between the PEG corona and α-CDs acting as physical cross-linking points. In a completely different process, reverse micellar hydrogels existed in the form of α-CD/PEG poly(pseudo)rotaxane-cored micelles with a polypeptide corona, with inclusion complexation occurring first, followed by hydrogen bonding interactions among the polypeptide coronas.

11.2.2.2   Triblock Copolymer

A series of triblock copolymers comprising PEG blocks such as PEG-*b*-PPO-*b*-PEG [18], PEG-*b*-PHB-*b*-PEG [44], PHB-*b*-PEG-*b*-PHB [45], PCL-*b*-PEG-*b*-PCL [42], and PLA-*b*-PEG-*b*-PLA [46] have been used to construct cyclodextrin poly(pseudo)rotaxane hydrogels. These systems used self-assembly based on CDs and triblock copolymer inclusion complexes, which aggregated into microcrystals that acted as physical crosslinks and led to the formation of a supramolecular polymer network. At the same time, hydrophobic interaction between the unthreaded blocks facilitated hydrogel formation and improved stability. Supramolecular hydrogels based on triblock amphipathic copolymers poly(ethylene glycol)-*b*-poly[(R)-3-hydroxybutyrate]-*b*-poly(ethylene glycol)) (PEG-*b*-PHB-*b*-PEG) and α-CD have been reported [26, 44, 47]. In the α-CD/PEG-*b*-PHB-*b*-PEG system, low molecular mass PEG blocks (5 kDa) partially penetrated the inner cavity of α-CD from both terminals to form cyclodextrin poly(pseudo)rotaxane ICs, which then aggregated into microcrystals and acted as physical cross-links leading to the formation of a supramolecular polymer network. The uncovered PEG remained water soluble and retained water within the supramolecular structure. The PHB chains could not penetrate the inner cavity of α-CDs to form ICs, but facilitated the formation of a polymer network due to hydrophobic interaction between the middle PHB blocks, making the macromolecular network stronger. Similar results were reported when the middle PHB blocks were replaced by poly(propylene-oxide) (PPO) in PEG-*b*-PPO-*b*-PEG triblock copolymers (also known as Pluronics) [18, 48].

Based on a similar principle, some triblock amphipathic copolymers have threadable PEG blocks in the middle in which α-CD molecules can partially thread past the telechelic segments with a larger cross-sectional area than their cavity size, and preferentially settle on the relatively more hydrophilic middle PEG segments forming block-selected cyclodextrin poly(pseudo)rotaxanes with necklace-like structures[49]. The remaining unthreaded blocks could decrease the molecular mass to build an extensive and strong hydrogel network through hydrophobic interaction, such as PHB-*b*-PEG-*b*-PHB [45] and PCL-*b*-PEG-*b*-PCL [42] triblock copolymers being used to form cyclodextrin poly(pseudo)rotaxanes.

UV irradiation with a photoinitiator caused the physically cross-linked hydrogel precursor to rapidly turn into a chemically cross-linked hydrogel when the cyclodextrin poly(pseudo)rotaxane was end-capped with photocurable methacryloyl groups or acryloyl groups. The system of amphiphilic poly(L-lactide)-*b*-poly(ethylene glycol)-*b*-poly(L-lactide) is an example of this process. PLA-*b*-PEG-*b*-PLA copolymers were end-capped with methacryloyl groups [46, 50], in which the polymethacrylates formed crosslink junctions as topological stoppers for the polyrotaxane chains of the hydrogel network. Similar results can be achieved based on the system of Pluronic F68/PCL block copolymer end-capped with acryloyl groups and α-CD [51] and Pluronic F68/PCL block copolymer end-capped with acryloyl groups and β-CD, in which β-CD can selectively thread onto the middle PPO block [52].

11.2.3   Graft-Copolymers as Guests

Different from block copolymers with linear guest polymers, graft copolymer-based cyclodextrin poly(pseudo)rotaxane supramolecular hydrogels could lead to a higher degree of cross-linking than a linear guest polymer. PEG (with a molecular mass of less than 5 kDa)-grafted high molecular mass biodegradable backbone (molecular mass greater than 10 kDa) copolymers, for example PEG-grafted dextrans [16], PEG-grafted chitosans [15], PEG-grafted hyaluronic acid (HA) [53], and PPEGMA-grafted- poly[2-(dimethylamino)ethyl methacrylate] (PDMA) [29], have been used as guest polymers to form α-CDs poly(pseudo)rotaxane hydrogels. Guest polymers such as poly(ε-lysine) (PL)-grafted dextran and α-CDs [54], and poly(propylene glycol) (PPG)-grafted dextran and β-CDs [55] have also been used to prepare cyclodextrin poly(pseudo)rotaxane supramolecular hydrogels. In these systems, the PEG, PL, and PPG side-chains grafted to the hydrated backbones thread the cavities of CDs to form cyclodextrin poly(pseudo)rotaxane ICs. They have a channel-type structure, which aggregates into columnar crystalline domains as physical crosslink junctions. The backbones, such as dextran, chitosan, hyaluronic acid and PDMA, did not form ICs under certain conditions and therefore remained hydrophilic in the hydrated state and acted as the water absorbing portion. The supramolecular hydrogels that formed had a phase-separated structure that consisted of the hydrophobic crystalline IC domains and the hydrophilic domain of hydrated dextran chains. If the backbone was an ionic polymer with pH-dependent water-solubility, such as chitosan, hyaluronic acid, or PDMA, the hydrogel formation showed pH-dependent properties. Poly(organophosphazenes) were synthesized through grafting poly(ethylene glycol) methyl ether (mPEG) and about 20% glycine ethyl ester to the phosphazene backbone to prepare novel supramolecular hydrogels with α-CDs [32]. The aggregation of ICs between the α-CD and mPEG side chains served as physical cross-links, while the uncovered mPEG side chains acted as a water absorbing reagent.

Amphiphilic MPEG-*b*-PCL-grafted chitooligosaccharide (COS-*g*-PCL-*b*-MPEG) copolymers have been used to fabricate supramolecular hydrogels [33]. In this work, an amphiphilic MPEG-*b*-PCL diblock copolymer (MPEG of 2 kDa) side chain was grafted onto a chitooligosaccharide backbone to produce a novel ternary amphiphilic graft copolymer (COS-*g*-PCL-*b*-MPEG). α-CD can thread the PEG and PCL chains to form cyclodextrin poly(pseudo)rotaxane ICs, but the rapid gelation and strong hydrophobic interaction between PCL blocks may have hindered α-CD molecules sliding over PEG blocks to cover PCL blocks; consequently not all PCL blocks were occupied. The unthreaded COS backbone remained hydrophilic and retained a large volume of water in its supramolecular structure.

11.2.4 Other Branched Polymers as Guests

A double-grafted polymer brush with a high density of PEG grafts, poly(2- (2-bromoisobutyryloxy)ethyl methacrylate) (PBIEM) *graft* poly(poly(ethylene glycol)methyl ether methacrylate) (PEGMA) (PBIEM-*g*-P(PEGMA)) was used to fabricate supramolecular hydrogels with α-CDs [56]. The high density PEG chains located on the outer end of the side chains penetrated the inner cavity of α-CD to easily form cyclodextrin poly(pseudo)rotaxane ICs. In the case of P(PEGMA)-based hydrogels, the column domains were perpendicular to the main chain, while in the PBIEM-*g*-P(PEGMA) brush the column domains were parallel to the main chain. The orientation of the column domains for the polymer brush may have facilitated physical cross-linking. On the other hand, steric hindrance prevented CD threading of the PEG segments attached close to the backbone, so those segments remained unthreaded and hydrophilic, allowing for retention of a large volume of water in the supramolecular structure. These two factors contributed to the formation of supramolecular hydrogels. A novel reduction-sensitive supramolecular hydrogel based on [poly(ethylene oxide) monomethyl ether]-*graft*-[disulfide-linked poly(amido amine)] (mPEG-*g*-SS–PAA) and α-CDs was introduced in which the reduction-sensitivity was ascribed to the disulfide linkage in the unthreaded SS–PAA main chain [27]. Another interesting type of cyclodextrin poly(pseudo)rotaxane hydrogel in which "multi-arm" physical crosslinkers, i.e., well-defined biodegradable triblock copolymer methoxy-poly(ethylene glycol)-*b*-poly(ε-caprolactone)-*b*-poly[2-(dimethylamino)ethylmethacrylate] (MPEG-*b*-PCL-*b*-PDMAEMA) (denoted as ECD, where E represents MPEG, C represents PCL, and D represents PDMAEMA), were used to condense plasmid DNA (pDNA) into polyplexes (ECD/pDNA polyplexes) has been reported[30]. By adding a low solids content of PEG (1%) and α-CD (8%) to ECD/pDNA polyplexes, poly(pseudo)rotaxane hydrogels formed. The MPEG-*b*-PCL-*b*-PDMAEMA copolymers condensed pDNA into 275–405 nm polyplexes with hydrophilic MPEG in the outer corona. Due to its amphiphilicity, the copolymer self-assembled into micelles with a hydrophobic PCL core and a hydrophilic MPEG/PDMAEMA corona at a copolymer concentration above the critical micellization concentration (CMC). The pDNA condensed into ECD/pDNA polyplexes through electrostatic attraction interaction to form a multi-arm structure. The multiple MPEG in the outer corona and the added PEG partially penetrated the inner cavity of α-CD to form cyclodextrin poly(pseudo)rotaxane ICs that could act as physical crosslinks. The multiple α-CD/ECD/pDNA and α-CD/PEG complexes formed poly(pseudo)rotaxane columnar crystalline domains due to strong intermolecular hydrogen bonding between CDs, which then formed the supramolecular hydrogel, as depicted in Figure 11.3. These ECD/pDNA polyplexes acted as "multi-arm" crosslinkers in the hydrogel formation, which facilitated the gelation process and improved the hydrogel

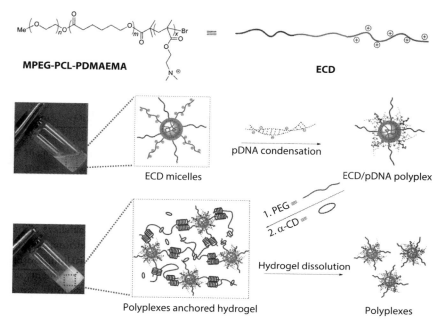

Figure 11.3 Design of ECD/pDNA anchored α-CD/PEG supramolecular hydrogels. Amphiphilic ECD forms micelles with a PCL core and MPEG/PDMAEMA corona in an aqueous environment. PDMAEMA at the corona is responsible for pDNA binding while MPEG serves both as a stabilizing moiety for the resultant ECD/pDNA polyplex and as a hydrogel anchoring segment. Sustained release of pDNA polyplexes is achieved via hydrogel dissolution over time. Reprinted with permission from Ref 30.

performance. The unthreaded PCL chains made the hydrogel stronger and more stable due to hydrophobic interaction.

11.3 Stimulus-Response Properties of Cyclodextrin Poly(pseudo)rotaxane Based Hydrogels

As previously mentioned, supramolecular hydrogels that are based on the aggregation of cyclodextrin poly(pseudo)rotaxane ICs in aqueous solutions, without the use of chemical cross-linking reagents, are thixotropic and could be injectable through needles [14, 57]. Although α-CD/PEG poly(pseudo)rotaxane hydrogels are already stimuli-responsive as a result of their thixotropic nature and their thermo-reversible sol-gel transition, it is nevertheless of great interest to incorporate other stimuli-responsive properties into this kind of hydrogel to obtain multi-responsive "smart" materials. Stimulus-response properties, including shear-thinning, pH, temperature, reduction response, and so on, may result from the aggregation of poly(pseudo)rotaxane, from uncovered polymer segments, or both.

11.3.1   Stimulus-Response Properties Derived from Cyclodextrin Poly(pseudo)rotaxanes and their Aggregates

The major properties that result from the aggregation of poly(pseudo) rotaxanes include reversible shear-thinning and thermosensitive sol-gel transition.

11.3.1.1   Shear-Thinning

Reversibility of the hydrogels is a basic and crucial feature of supramolecular systems [57]. Rheological studies on supramolecular-structured hydrogels formed by α-CD and high molecular mass PEG homopolymer indicated that they were thixotropic and reversible, and that the viscosity of the hydrogel was sensitive to the shear rate, i.e., the viscosity decreased as the shear rate increased[14]. In Figure 11.4, we can see that the viscosity of the hydrogel significantly decreased when agitated (Figure 11.4(a)), but once the agitation disappeared the viscosity was eventually restored toward the original values, in most cases within hours (Figure 11.4(b)). It is this property that makes hydrogels injectable even through a fine needle where viscosity decreases markedly after injection. The finer needle leads to a higher shear rate causing the hydrogel viscosity to decrease more deeply and steeply.

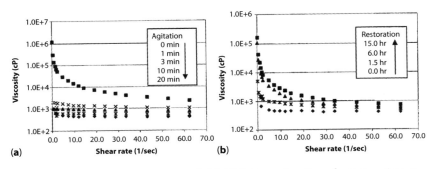

Figure 11.4 Changes in the viscosities of Gel-20K-60 as a function of agitation time at a shear rate of 120 s⁻¹(a); and (b) restoration of the gel viscosities after 20 of agitation at a shear rate of 120 s⁻¹. Reprinted with permission from Ref. 14.

11.3.1.2   Temperature-sensitivity

In the grafted dextran-based supramolecular-structured poly(pseudo) rotaxane hydrogels, the sol-gel transition was based on supramolecular assembly and dissociation, corresponding to a process of threading-dethreading the α-CD molecules along the PEG grafts [55, 16]. The gelation temperature ($T_{gelation}$) was defined as the point where the polymer solutions no longer flowed when the vial was inverted. The gels became mobile as the temperature increased and the point where the gels started to flow was taken as the gel-melting temperature ($T_{gel-melting}$). The hydrogels experienced three different phases. For example, in the case of IC hydrogels of PEG grafted dextrans (PEG of 750) [16], when the temperature was below $T_{gelation}$ (the temperature at which the ICs formed) they exhibited an opaque gel phase and were not mobile. With increasing temperature they remained opaque and mobility increased until the temperature was greater than $T_{gel-melting}$, at which point they became clear solutions. There are several parameters that can be used to control the properties of the hydrogel, including variation of the polymer concentration and feed ratio of the guest [PEG]/host [CD] as well as the PEG content in the graft copolymers. When the polymer concentration was high, the PEG chains formed more stable cyclodextrin poly(pseudo)rotaxane ICs with a high $T_{gel-melting}$. The stoichiometry of [PEG]/[CD] is known to be 2 [58]. When the feed ratio of [PEG]/[CD] increased from 1 to 2, both $T_{gel-melting}$ and $T_{gelation}$ increased. This could be explained by the fact that with a high feed ratio, that is to say with more guests, the increased guest moieties have more opportunities to form more stable poly(pseudo)rotaxane ICs with the host CD, leading to an enhanced physical crosslinking network. However, a decreased concentration of CD moieties leads to fewer opportunities for disturbing the network formation induced by the hydrophobic interactions of ICs. The number of grafted chains per backbone must be sufficient to provide enough hydrophobic moieties to associate with CDs, which may be more advantageous to host–guest interactions. On the other hand, the formation of poly(pseudo)rotaxane ICs is entropically unfavorable because it is hindered by more favorable interactions such as hydrogen bonding and hydrophobic interactions. As the temperature increases, the polymer chains dethread from the CDs, restoring their intrinsic entropy from random conformations, and leading to dissociation of the poly(pseudo)rotaxane ICs[16]. With the repeated heating-cooling that accompanied the sol-gel transitions that corresponded to threaded-dethreaded CDs, the supramolecular hydrogel showed a thermally reversible sol-gel transition [55]. A thermosensitive hydrogel based on an alkylpyridinium and α-CD poly(pseudo)rotaxane, which showed sol-gel transitions between 7–67°C (depending on the type and amount of the guest compound) has also been reported [59]. Another unique thermoreversible sol-gel transition has been seen in hydrogels based on mPEG modified biodegradable polyphosphazenes and α-CD [32].

Rapid response thermosensitive supramolecular hydrogels, such as Pluronic F68/poly(ε-caprolactone)/α-CD [51] and Pluronic F68/poly(ε-caprolactone)/β-CD [52] based hydrogels have also been reported. In these systems the Pluronic F68/poly(ε-caprolactone) acts as a thermosensitive macromer. In addition to the aggregation of cyclodextrin poly(pseudo)rotaxane ICs, which mainly provides temperature-sensitivity, hydrophobic aggregation between the unthreaded hydrophobic blocks also facilitates the network. These two factors contribute to rapid thermosensitivity.

11.3.2   Stimulus-Response Properties Derived from Uncovered Segments

11.3.2.1   pH Sensitivity

In PEG-*b*-PLL/α-CD-based hydrogels, the α-CD cannot thread the PLL chains to form stable cyclodextrin poly(pseudo)rotaxane ICs, so pH-sensitive gelation was induced by the uncovered PLL segments. The pH-inducible hydrogel from supramolecular micelles was found to be completely reversible upon change in pH [12]. The hydrogels were formed under alkaline conditions at pH 10, but the gel could be readily reversed to sol by acidification. When protonated in a neutral or acidic solution the PLL chains remained water soluble, adopting a random coil conformation due to electrostatic repulsion between the side chains, which consist of charged amino end groups that prefer to stay outside of the hydrophobic threaded CD cavities. Once the pH increased to greater than 10, the amine salt ($-NH_3^+$) in the branched chains of the PLL blocks was deprotonated to form neutral insoluble $-NH_2$ groups, causing the PLL chains to be more hydrophobic. Aggregation of PLL chains due to hydrophobic effects facilitated the polymer network. The $-NH_2$ groups were protonated again by acid causing hydrogel dissociation, and then deprotonated again with alkali to form hydrogels. This repeated protonation-deprotonation by changing pH showed the hydrogels had a pH reversible gel–sol/sol–gel transition.

It is worth mentioning that some temperature/pH dual-responsive hydrogels have been reported. The copolymers used in dual-responsive hydrogels contain two kinds of blocks: one that forms cyclodextrin poly(pseudo)rotaxane ICs, such as PEG and PL; and another that contains ionizable chains with carboxyl or amino groups, such as PEG side-chains on chitosan backbones [15], PEG side-chains on PDMA backbones [29], and diblock copolymers (PEG-*b*-PDMAEMA) [60] (all containing amino-groups), as well as (PEG)-grafted hyaluronic acid (HA) [53], which contains carboxyl groups. The pH-responsivity resulted from pH-dependent water-solubility where the hydrogels showed pH reversible gel–sol/sol–gel transitions. At the same time, high temperature disrupts weak interactions, such as hydrogen bonding and hydrophobic interaction, which may cause CDs to dethread from the chains and thereby weaken the hydrophobic interactions between unthreaded hydrophobic blocks, leading to dissociation of the hydrogels. Therefore, it is the aggregation of poly(pseudo) rotaxane ICs and the uncovered blocks that contributes to temperature and pH dual-responsivity.

11.3.2.2 Reduction Sensitivity

A novel reduction-sensitive supramolecular hydrogel based on mPEG-*g*-SS–PAA and α-CD was fabricated [27]. The reduction-sensitivity was ascribed to the disulfide linkage in the unthreaded SS–PAA main chain. The disulfide bond is sensitive to reduction conditions in the human body due to the thiol–disulfide exchange reaction. This means that disulfide bonds are stable at low concentrations of glutathione tripeptide (GSH), ca. 2–20 μM, in the body and extracellular milieu; however, they may be quickly cleaved at high GSH concentrations, ca. 0.5–10 mM. In this work, dithiothreitol (DTT), a reducing agent model and a substitute for GSH, cleaved the disulfide bonds of the SS–PAA main chain because a more stable six-membered cyclic disulfide formed after the oxidation of DTT. The mPEG-*g*-SS–PAA polymer could be cleaved into a short segment, oligomer, or small molecule under high reductant concentration. As the DTT concentration increased, the mPEG-*g*-SS–PAA degradation rate increased; this was attributed to the incorporation of more DTT, which provided more opportunities to cleave the disulfide bonds of the main chain mPEG-*g*-SS–PAA. When the mPEG-*g*-SS–PAA main chain broke, followed by breakage of the cyclodextrin poly(pseudo)rotaxane ICs, the hydrogel turned into a turbid or transparent solution. It was concluded that reduction-sensitivity resulted from the unthreaded segments and was not caused by the aggregation of poly(pseudo)rotaxanes.

11.4 Nanocomposite Supramolecular Hydrogels

Based on a principle similar to that discussed in the approach using PEG-based biodegradable polymers, carbon nanomaterials, metal nanoparticles, and polysaccharide nanocrystals have been used to construct nanocomposite or hybrid supramolecular hydrogels. In all these systems, multiple PEG or PEG-containing copolymers were first attached to the nanomaterials through chemical bonding or physical interactions, followed by threading of α-CD onto the PEG-decorated materials. The incorporation of nanoparticles such as SWNTs, graphene, and silica may accelerate hydrogel formation.

11.4.1   Nanocomposite Hydrogel Filled with Carbon Nanoparticles

In the carbon-based nanocomposite hydrogel system based on the non-ionic commercially available Pluronic copolymer (PEG-*b*-PPO-*b*-PEG) and α-cyclodextrin, the Pluronic copolymer has a dual role. Due to its hydrophobic effect in water, the copolymer disperses the pristine carbon nanomaterial to form a stable and homogeneous copolymer-coated solution, in which the PPO blocks are bonded onto the hydrophobic surface of the carbon nanomaterial via hydrophobic interactions. The PEG chains partially penetrate the inner cavity of α-CD to form cyclodextrin poly(pseudo)rotaxanes, which aggregate into columnar α-CD crystallite domains as physical crosslinks. The introduction of carbon nanomaterials such as grapheme [19] and single-walled carbon nanotubes (SWNTs) [20, 21], may influence the morphology of the hydrogel. In SWNT nanocomposite hydrogels [20] the hybrid hydrogel was relatively smooth while the native hydrogel appeared coarse with undulant pimples, which may result from the hydrophobic interaction of the PPO block with SWNTs. Such hydrophobic interaction could limit the movement and micellization of the PPO block and result in the changed morphology. The SWNT-hybridized hydrogel [21] inherited the loose sponge-like structure of the native hydrogels, but had many agglomerates and a crystalline character, which may be attributed to nucleation of SWNTs. Graphene-hybridized supramolecular hydrogels, on the other hand, had a lamellar structure [19]. The introduction of graphene changed the geometric/spatial state of the block copolymer, making the PEG chains more ordered, and thus accelerated formation of the hydrogel.

There is another kind of carbon-based supramolecular nanocomposite hydrogel based on poly(ethylene glycol)-grafted-multiwalled carbon nanotube (MWNT-*g*-PEG) [22]. Similar to PEG-*b*-PCL/α-CD, it is a combination of the aggregation of cyclodextrin poly(pseudo)rotaxane ICs based on selective inclusion complexation between the PEG block and α-CD and the strong hydrophobic interaction that provides the primary driving force for hydrogel formation. In a comparative study, three different polymers were used: MWNT-*g*-PEG (2000, 4000, and 6000). In the case of MWNT-*g*-PEG$_{2000}$/α-CDs, there was only precipitate formation, no hydrogel, because one end of the uncomplexed PEG chain was anchored to a MWNT and the other end of the uncomplexed segment was not long enough to form a network due to the low molecular mass. In the case of MWNT-*g*-PEG$_{6000}$/α-CDs, the grafted PEG$_{6000}$ chain was so long that the MWNT hindrance was almost negligible and the distance between the PEG$_{6000}$ chains was too large for the cyclodextrin poly(pseudo)rotaxanes to aggregate and form physical cross-links. The MWNT-*g*- PEG$_{4000}$/α-CDs had a high molecular mass and a high graft density of 590; therefore the PEG chains were long enough to thread the α-CDs and the high graft density provided enough physical crosslinking. The uncomplexed PEG was also long enough to absorb a large volume of water to fill in the network (Figure 11.5).

11.4.2 Nanocomposite Hydrogels Filled with Metal-Based Nanoparticles

Similar to SWNT-based nanocomposite hydrogels, magnetic supramolecular hydrogels based on PEG-*b*-PCL/α-CD stabilized magnetic

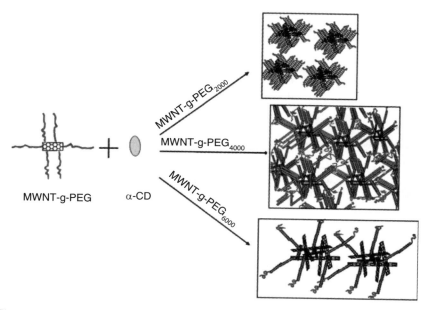

Figure 11.5 Schematic representation of the hybrid hydrogels made of MWNT-g-PEO/α-CDs. Reprinted with permission from Ref. 22.

iron oxide nanoparticles (Fe$_3$O$_4$) and α-CD have been reported [23]. And a novel strategy for *in situ* incorporation of silver nanoparticles into the supramolecular hydrogel networks was reported by Ma *et al.* [24]. In the silver nanoparticle-based hybrid hydrogels the amphiphilic block copolymer Pluronic F-68 (PEG-*b*-PPO-*b*-PEG) not only forms cyclodextrin poly(pseudo)rotaxane ICs to give rise to a network, but also absorbs silver ions to form colloidal silver hydrosols at the micelle-water interface. These hydrosols were the result of silver ions forming complexes with oxygen atoms in the presence of Pluronic F-68 chains. The silver ions were reduced at the interface leading to the incorporation of silver nanoparticles in the micellar aggregates. As the concentration of Pluronic F-68 increased more micelles aggregated, depositing more silver nanoparticles onto the micellar surface, preventing the silver nanoparticles from gathering together and hence enhanced colloidal stability. In this case, Pluronic F-68 functioned as a stabilizing and capturing agent for the formation of colloidally stable hydrosols containing silver nanoparticles. Consequently, the Pluronic F-68-based hydrogels contained colloidally stable silver nanoparticles. It is worthwhile noting that the incorporation of silver nanoparticles prolonged the gelation process, which could be explained by the formation of silver nanoparticles in the mixed system that weakened the ICs between α-CD and PEG blocks, resulting in an increased gelation time.

Another supramolecular hydrogel was prepared based on adamantane monoend-functionalized low molecular mass PEG (Ada-PEG of 1.1 or 2K) and α-CD and their hybridization with β-CD surface-functionalized silica nanoparticles (β-CD-SiO$_2$) through host-guest interaction (Figure 11.6) [25]. In this work, the Ada-PEG penetrated the inner cavity of α-CD to form cyclodextrin poly(pseudo)rotaxane ICs to give rise to

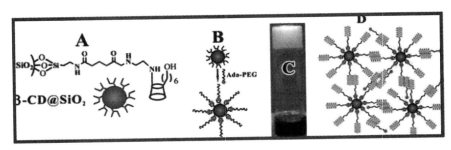

Figure 11.6 (A) Chemical structure and (B) schematic representation of β-CD -SiO$_2$ and its inclusion complex with Ada-PEG. (C) Optical photo and (D) schematic representation of the supramolecular hydrogels made of Ada-PEG2K, α-CD, and β-CD- SiO$_2$. Reprinted with permission from Ref. 25.

supramolecular hydrogels rather than crystalline precipitates. The hydrophobic Ada group is an important element in the formation of a homogeneous hydrogel. Because the bulky Ada group is larger than the α-CD cavities it can partially prevent the threading of CD, thereby decreasing the relative amount of threaded CD. The remaining uncovered PEG is then long enough to absorb water within the network. The Ada group provides additional physical crosslinks via its hydrophobic aggregation which enhances the network. What's more, based on the fact that β-CD and Ada can form ICs (with a stability constant as high as 5×10^5 M^{-1}), β-CD-SiO$_2$ nanoparticles were used as supra-cross-links that attached to PEG chains to form nanoparticle-hybridized supramolecular hydrogels. The incorporated inorganic β-CD-SiO$_2$ nanoparticles were fixed within the network through inclusion complex interaction between the β-CD groups and the Ada groups, which not only accelerated the gelation process but also strengthened the hydrogels. The combination of the aggregation of partial cyclodextrin poly(pseudo)rotaxane ICs with one-end low-molecular mass PEG chains and α-CD, as well as the ICs formed between the Ada group and β-CD-SiO$_2$ nanoparticles, cooperatively leads to a strong network structure and gives rise to novel nanoparticle-hybridized supramolecular hydrogels.

11.4.3 Nanocomposite Hydrogel Filled with Polysaccharide Nanoparticles

It is worth noting that polysaccharide nanocrystal-doped supramolecular hydrogels based on PEG$_{137}$-b-PPO$_{44}$-b-PEG$_{137}$/α-CD inclusion complexation have been constructed [61]. In this work, a combination of the aggregation of cyclodextrin poly(pseudo)rotaxanes based on selective inclusion complexation between PEG blocks and α-CD and micellization between PPO blocks due to strong hydrophobic interaction provided the primary driving force for hydrogel formation. The introduction of cellulose whisker (CW) and crab shell chitin whisker (CHW) to the nanocomposite hydrogels improved the gelation process. Recently, β-CD modified cellulose nanocrystals have been reported to construct supramolecular hydrogels based on Pluronic polymers and α-CD [34]. The PEG blocks first facilitate the dispersion and compatibility of nanocrystals to a high loading level, and then form ICs with α-CD to give rise to in situ hydrogels.

11.4.4 Role of Nanoparticles

In the broader context of materials design, incorporating functional nanoparticles into three dimensional cyclodextrin poly(pseudo)rotaxane hydrogel networks allows for the design of hybrid hydrogels where properties of nanomaterials could be exploited. In many cases, the incorporation of nanoparticles not only accelerated gelation, but also regulated the mechanical performance of the hydrogels, for example the superparamagnetic effect from incorporating Fe_3O_4 nanoparticles and the catalytic activity of incorporated Ag nanoparticles.

11.4.4.1   Reinforcement

In polysaccharide nanocrystal-doped supramolecular hydrogels the incorporation of polysaccharide nanocrystals, such as the rod-like whiskers of cellulose (CW) and chitin (CHW), and the platelet-like starch nanocrystal (StN) [61], into supramolecular hydrogels speeded gelation, enhanced mechanical strength, improved erosion resistance of solution, and facilitated long-term sustained release of drugs. Figure 11.7 depicts the dynamic

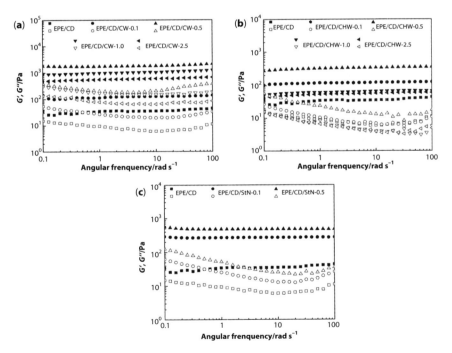

Figure 11.7 Dynamic rheological behavior of nanocomposite hydrogels containing various loading levels of CW (A), CHW (B), and StN (C), as well as the native EPE/CD hydrogel for comparison. (Solid symbols for storage modulus, G', and hollow symbols for loss modulus, G''). Reprinted with permission from Ref. 61.

rheological behaviors of the native PEG-*b*-PPO-*b*-PEG (EPE)/CD hydrogel and of the nanocomposite hydrogels with different polysaccharide nanocrystals and loading levels. As shown, the storage modulus (G') of both hydrogels was greater than the loss modulus (G") over the entire frequency range, exhibiting a substantial elastic response with strength and rigidity. Both G' and G" increased due to the nucleation effect of polysaccharide nanocrystals, which improved the crystallization and physical cross-linking caused by hydrogen bonding between the incorporated polysaccharide nanocrystals and the hydrogel matrix and resulted in enhanced mechanical strength. Both the type and loading level of polysaccharide nanocrystals regulated the G' and G" values of the nanocomposite hydrogels. The G' values initially increased with an increase in the loading level of polysaccharide nanocrystals, but only up to 0.5 wt%, which produced the highest mechanical strength. When the loading level was 1.0 wt% and greater, the G' values initially decreased due to increased aggregation of the polysaccharide nanocrystals. The G' values were in the order of EPE/CD/CW-0.5 > EPE/CD/CHW-0.5 > EPE/CD/StN-0.5 at the same 0.5 wt% loading level. Cellulose whisker at a 0.5 wt% loading level gave the highest storage modulus. Its maximum G' value of ca. 1700 Pa×s was approximately 50 times higher than that of the native EPE/CD hydrogel. This was attributed to the cellulose whisker's ability to avoid protonation of amino groups on the CHW surface and the self-aggregation of StN, which inhibited hydrogen bonding between the polysaccharide nanocrystals and hydrogel matrix. The introduction of polysaccharide nanocrystals positively affected reinforcement because nucleation of the polysaccharide nanocrystals induced crystallization of ICs, and hydrogen bonding between the polysaccharide nanocrystals and the hydrogel matrix acted as physical crosslinks which improved the stability of the hydrogel framework. Similar to the polysaccharide nanocrystal-doped supramolecular hydrogels, β-CD modified cellulose nanocrystals have been used to construct nanocomposite hydrogels with higher loading levels (> 5 wt %), in which both thermal stability and mechanical properties were improved [34].

Some carbon-based nanocomposite supramolecular hydrogel systems also showed reinforcement, such as the incorporation of multiwalled carbon nanotube (MWNT-*g*-PEG) that promoted the strength and viscosity of hydrogels[22]. The storage modulus (G') of these hydrogels was almost two times higher than the native hydrogels, showing an obviously improved strength. At the same time, the thermal stability was also greatly improved as the decomposition temperatures of nanocomposite hydrogels were nearly 100 °C higher than that of native hydrogel [22]. In the SWNT-incorporated nanocomposite hydrogels the higher strength and rigidness were due to mechanical reinforcement of its distinct nanostructure and hydrophobic interaction between SWNT and PPO blocks [21]. The viscosities and ultimate strengths of SWNT-based [20] and graphene sheet-based[19] hybrid hydrogel systems decreased as compared to native hydrogels. This phenomenon was attributed to the adsorption and binding of PPO chains onto the hydrophobic surface of the carbon material, which restrained the micellization and entanglement of PPO blocks with each other and decreased the density of crosslinking points. However, the mechanical strength and the viscosity of the SiO_2 nanoparticle hybrid hydrogel were greatly improved [25]. Compared to the native hydrogel, the storage modulus (G') of the hybrid hydrogel was about four times higher over a broad frequency range; the viscosity at a low shear rate (up to ⊠1300 Pa×s) was approximately eight times higher.

11.4.4.2   Other Functions

The *in situ* incorporation of silver nanoparticles into hybrid hydrogel[24] decreased the G'and G″ and increased the gelation time due to weakened inclusion complexation resulting from the formation of silver nanoparticles by the reduction reaction of $AgNO_3$ with the PEG blocks. The hydrogel showed good catalytic activity for the reduction of methylene blue dye (MB) by sodium borohydride $NaBH_4$.

The SWNT-hybridized supramolecular hydrogel exhibited better antimicrobial activity against *Escherichia coli* and *Staphylococcus aureus* [21] than the native hydrogel and shows promise as a wound dressing.

The incorporation of iron oxide (Fe_3O_4) nanoparticles in the nanocomposite magnetic supramolecular hybrid hydrogels [23] accelerated the gelation process by shortening gelation time and enhanced mechanical strength due to the interaction of PEG-*b*-PCL with Fe_3O_4 nanoparticles, which is favorable for the formation of subsequent cyclodextrin poly(pseudo)rotaxane ICs. More importantly, these hydrogels were superparamagnetic and the saturation magnetization was lower with a higher PEG-*b*-PCL concentration.

11.5   Biomedical Application of Cyclodextrin Poly(pseudo)rotaxane-Based Hydrogels

11.5.1   Drug Carriers

It was previously mentioned that supramolecular hydrogels are thixotropic, reversible, and injectable through needles. The properties of thixotropy and reversibility allow hydrogels to be used as a unique injectable drug delivery system using a fine needle. Drugs can be encapsulated directly into the hydrogels *in situ* at room temperature without organic solvents or chemical crosslinking required during the gelation process [14, 35]. Encapsulation not only facilitates enhanced drug loading levels, but also avoids structural changes to the drug and the interference of cell activity caused by chemical crosslinking. With gelation restored spontaneously *in situ* after injection, the hydrogel serves as a depot for controlled release [14]. In addition, in aqueous environments, hydrogels gradually disintegrated and dissolved in the release medium during the release process; therefore, the release of encapsulated drugs is controlled by dissolution, dissociation, and erosion of the hydrogel that corresponds to dethreading of the cyclodextrin poly(pseudo)rotaxanes rather than by diffusion of the drugs through the hydrogel. Because of this, the drug release kinetics were less dependent on the properties of the drugs, leading to a potential advantage for rendering the delivery system more widely available for different drugs [14, 17, 29]. Supramolecular hydrogels based on α-CD/PEG homopolymer have been used to study release and drug-delivery properties using fluorescein isothiocyanate-labeled dextran (dextran-FITC) as a model drug [14]. However, the rapid dissociation and release kinetics upon contact with the incubation solution make such hydrogels unsuitable for long-term drug release (most entrapped dextran-FITC was released within 5 days). By increasing the molecular mass of PEG, the release rate decreased markedly, with the hydrogels formed by PEG 100 kDa showing the most sustained release kinetics. The high molecular mass PEG (PEG > 10 kDa), however, is difficult to filter through the human kidney [17], so it was important to either find a biodegradable high molecular mass replacement or look for other associative forces to strengthen the hydrogel network. To this end, the use of PEG-block biodegradable amphiphilic copolymers or nanocomposite hydrogels has been explored. The molecular mass of these copolymers can be greater than 10 kDa as long as the non-degradable PEG portion is less than the renal clearance limit. This allows for continued exploitation of polymer entanglement to strengthen the hydrogel network. Because these polyesters are hydrophobic and semicrystalline, they provide additional physical crosslinking via hydrophobic interaction and segmental crystallization to strengthen the hydrogel network, resulting in more stable hydrogels for long-term controlled and sustained release [62]. For example, when dextran-FITC was used as a model drug in α-CD/PEG-*b*-PCL diblock based supramolecular hydrogels [17] the sustained release increased significantly over the pure α-CD/PEG hydrogels, and almost 20% of the encapsulated dextran-FITC was retained by the hydrogel for up to 1 month, even if the molecular mass of the PEG block was only 5 kDa. Another supramolecular hydrogel based on the amphiphilic copolymer PEG-*b*-PHB-*b*-PEG drug showed excellent controlled release properties using dextran-FITC as a model by sustaining the drug's release for more than 1 month [26]. These results further support the idea that cooperation of the aggregation of cyclodextrin poly(pseudo) rotaxane ICs between α-CD and PEG segments and the hydrophobic interaction between unthreaded blocks leads to the formation of hydrogels with strong supramolecular networks, as well as the properties for long-term sustained release. These hydrogels gradually dissolved during the release of dextran-FITC and disappeared when all dextran-FITC was released. Based on a similar principle, another supramolecular hydrogel based on the COS-*g*-PCL-*b*-MPEG copolymer with α-cyclodextrin showed sustained release of bovine serum albumin (BSA) [33].

Some responsive hydrogels used as sustained release injectable carriers are triggered by changes in pH, temperature, or other conditions of the environment. For example, in temperature and pH dual-responsive hydrogels [29] the release kinetics of 5-[(4,6-dichlorotriazin-2-yl)amino] fluorescein-labeled BSA (BSA-DTAF) could be controlled by regulating temperature and pH. In work on reduction-sensitive injectable supramolecular hydrogels [27] the release rate of BSA as a model protein drug could be regulated via the reduction conditions, including concentration of the reducing agent dithiothreitol (DTT) in the reduction media (phosphate buffer solution with pH of 7.4). In the absence of DTT, the release profile showed a prominent sustained release characteristic with an accumulative release ratio of ca. 70% for up to 30 days. The release rate obviously increased as the DTT concentration increased due to degradation of mPEG-*g*-SS–PAA in the presence of DTT. Remarkably, a higher molar ratio of DTT vs. disulfide bond, corresponding to higher DTT concentrations, led to a higher release rate of BSA, accompanied by rapid erosion of the supramolecular hydrogel due to severe cleavage of mPEG-*g*-SS–PAA.

In other work, polysaccharide nanocrystal-doped nanocomposite supramolecular hydrogels exhibited potential for application as injectable biomaterials in which the release profiles of BSA exhibited sustained release [61]. Compared to the native hydrogel that released encapsulated BSA in 18 days, the cumulative release rate of the nanocomposite hydrogel was only ca. 80% after 20 days. A loading level of 0.5 wt% showed the most sustained release in the order of EPE/CD/CW-0.5 > EPE/CD/CHW-0.5 >EPE/CD/StN-0.5, which is the same order as was found for G' values. These sustained effects could be due to drug release being controlled by erosion of the hydrogel at the interface between the hydrogel exterior and solution, and by drug diffusion. Nucleation of polysaccharide nanocrystals to induce crystallization and hydrogen bonding between the polysaccharide nanocrystals and hydrogel matrix resisted erosion of the hydrogel exterior that was immersed in solution and diffusion of external solution into the hydrogel. Also, the polar surface of the polysaccharide nanocrystals immobilized encapsulated BSA through intermolecular interaction and hence slowed the drug's sustained release. Similar nanocomposite hydrogels based on β-CD modified cellulose nanocrystals exhibited promising controlled and sustained release behavior about three times higher than the native hydrogels [34].

Recently, cyclodextrin poly(pseudo)rotaxane dual-drug loaded supramolecular hydrogels have created great interest. Two drugs could be co-encapsulated *in situ* forming a hydrogel matrix and showing both long-term and sustained release behaviors. For example, the in vitro release profiles of supramolecular hydrogel/micelle composites for co-encapsulation and prolonged release of camptothecin (CPT) and granulocyte colony-stimulating factor (G-CSF) showed that loaded CPT and G-CSF could be released simultaneously and have dual independent release behaviors [36]. A doxorubicin (DOX)/cisplatin dual-drug loaded hydrogel has been reported that showed higher cytotoxicity than single drug loaded hydrogels [35], providing a convenient and effective way to formulate cocktail-drug loaded hydrogel systems for cancer therapy.

11.5.2   Gene Carriers

The cyclodextrin poly(pseudo)rotaxane supramolecular hydrogels also showed immense potential as injectable carriers for sustained gene delivery based on thixotropy and *in situ* gelling[30]. In this work, active cationic copolymer MPEG-PCL-PDMAEMA (denoted as ECD) condensed plasmid pDNA into ECD/pDNA polyplexes with hydrophilic MPEG in the outer corona that anchored the polyplexes within the poly(pseudo)rotaxane hydrogel and acted as a sustained gene delivery carrier. The results showed the sustained release of DNA was up to 6 days, which was longer than in α-CD/PEG hydrogel in its naked form (up to 3 days) with the same amount of pDNA (70 μg) encapsulated. In all cases, there was no significant burst release and the hydrogels were completely eroded after 100% DNA release and showed a sustained release profile. This indicated the DNA release was not diffusion controlled, but controlled mainly by the gel disassociation mechanism. The gradual erosion of the hydrogels that resulted from dethreading of the α-CD from the PEG and MPEG chains contributed to surface erosion. Regions that were lower in ECD/pDNA polyplexes resulted in areas of poor dispersibility and compositional inhomogeneity of corresponding polyplexes during hydrogel formation. These regions were more susceptible to erosion and thus weakened the network links resulting in faster hydrogel erosion than the more homogeneous hydrogel. In addition, released pDNA was found in intact nanoparticle ECD/pDNA polyplexes as a result of electrostatic bonding between the cationic segment of the MPEG-PCL-PDMAEMA and pDNA. It is worthwhile noting that the bioactivity of released pDNA polyplexes did not decrease during encapsulation by in vitro gene transfection. Also, 3-(4,5-dimethylthiazol-2-yl)-2,5-diphenylte-trazolium bromide (MTT) measurements showed that released ECD/pDNA polyplexes had no significant cytotoxicity.

11.5.3   Cell-Adhesive Scaffold

Cyclodextrin poly(pseudo)rotaxane supramolecular hydrogels based on chitosan derivative hydroxyphenyl acetamide chitosan (CHPA) and Arg-Gly-Asp (RGD)-conjugated poly(pseudo)rotaxane (PRxRGD-CHPA) have been used as cell-adhesive injectable hydrogel scaffolds[31]. It is known that RGD is the cell adhesion motif for integrin receptors and this peptide promotes proliferation and differentiation of a variety of cells. Figure 11.8

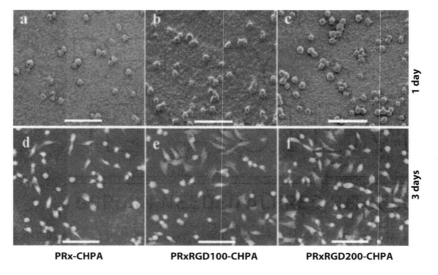

Figure 11.8 Scanning electron microscope images showing the morphology of fibroblasts adhered to PRx-CHPA and PRxRGD-CHPA hydrogels after incubation times of a-c: 1 day and d-f: 3 days (scale bar = 100 μm). Reprinted with permission from Ref. 31.

shows the morphology of the cells that adhered to the hydrogels. In the presence of RGD more cells adhered to the hydrogels, and the number of adhered cells increased after 3 days of incubation. Cells adhering to the hydrogels were spread out and adopted a spindle-shape. It was concluded that the RGD-containing hydrogels more effectively promoted cell adhesion and enhanced spreading and proliferation, and that all hydrogels were cyto-compatible and therefore show potential for use as injectable biomimetic hydrogel scaffold in tissue engineering applications.

11.6 Conclusions and Prospects

Supramolecular hydrogels based on host-guest interactions and aggregation of cyclodextrin poly(pseudo)rotaxane ICf materials design, various kinds and shapes of copolymers containing threadable chains as guest molecules could be used in poly(pseudo)rotaxane systems to construct novel supramolecular hydrogels with special properties and functions, such as stimuli-sensitivity. The introduction of hydrophobic blocks or nanomaterials is expected to reinforce and stabilize the polymer network. Also, hybrid and nanocomposite hydrogels inherited the properties of the nanomaterials resulting in smart functions. Based on the intrinsic shear-thinning thixotropic and thermo-reversible properties, as well as biocompatibility, these supramolecular hydrogels have great potential as injectable biomaterials and break through the limits of conventional hydrogels. Supramolecular hydrogels with drugs physically encapsulated *in situ* showed long-term controlled and sustained release properties. In addition, the supramolecular hydrogels also showed potential for application in sustained gene delivery systems and in injectable cell-adhesive scaffolds. The upcoming research and development on designing smart hydrogel biomaterials with specified properties to broaden their biological and practical applications is worth anticipating.

References

1. Y. He, P. Fu, X. Shen, and H. Gao, "Cyclodextrin-based aggregates and characterization by microscopy," *Micron*, Vol. 39 (5), pp. 495–516, 2009.
2. A. Harada, J. Li, and M. Kamachi, "The molecular necklace: a rotaxane containing many threaded α-cyclodextrins," *Nature*, Vol. 356 (6367), pp. 325–327, 1992.
3. E. Sabadini, T. Cosgrove, "Inclusion complex formed between star-poly(ethylene glycol) and cyclodextrins," *Langmuir*, Vol. 19 (23), pp. 9680–9683, 2003.

4. J. Li, "Self-assembled supramolecular hydrogels based on polymer–cyclodextrin inclusion complexes for drug delivery," *NPG Asia Materials*, Vol. 2 (30, pp. 112–118, **2010.**

5. S. Das, M.T. Joseph, D. Sarkar, "Hydrogen bonding interpolymer complex formation and study of its host-guest interaction with cyclodextrin and its application as an active delivery vehicle," *Langmuir*, Vol. 29 (6), pp. 1818–1830, 2013.

6. J. Li, "Cyclodextrin inclusion polymers forming hydrogels." In *Inclusion Polymers (Advances in Polymer Science)*, G. Wenz, Ed. Springer-Verlag, Berlin, Vol. 222, pp. 79–112, 2009.

7. J. Zhao, S.J. Yao, "Functional and physicochemical properties of cyclodextrins," In *Starch-Based Polymeric Materials and Nanocomposites*, Ahmed, J.; Tiwari B. K.; Imam, S. H.; Rao, M. A., Eds. CRC Press, Boca Raton, pp. 183–230, 2012.

8. P. Lo Nostro, L. Giustini, E. Fratini, B.W. Ninham, F. Ridi, P. Baglioni, "Threading, growth, and aggregation of pseudopolyrotaxanes" *The Journal of Physical Chemistry B*, Vol. 112 (4), pp. 1071–1081, 2008.

9. J. Wang, L. Li, Y. Zhu, P. Liu, X. Guo, "Hydrogels assembled by inclusion complexation of poly(ethylene glycol) with α-cyclodextrin," *Asia-Pacific Journal of Chemical Engineering*, Vol. 4 (5), pp. 544–550, 2009.

10. F. Yuen, K.C. Tam, "α-Cyclodextrin assisted self-assembly of poly(ethylene glycol)-block-poly(N-isopropylacrylamide) in aqueous media" *Journal of Applied Polymer Science*, Vol. 127 (6), pp. 4785–4794, 2013.

11. B.-H. Han, M. Antonietti, "Cyclodextrin-based pseudopolyrotaxanes as templates for the generation of porous silica materials," *Chemistry of Materials* Vol. 14 (8), pp. 3477–3485, 2002.

12. R.X. Yuan, X.T. Shuai, "Supramolecular micellization and pH-inducible gelation of a hydrophilic block copolymer by block-specific threading of α-cyclodextrin" *Journal of Polymer Science Part B-Polymer Physics* Vol. 46 (8), pp. 782–790, 2008.

13. A. Harada, J. Li, M. Kamachi, "Double-stranded inclusion complexes of cyclodextrin threaded on poly(ethylene glycol)," *Nature*, Vol. 370 (6485), pp. 126–128, 1994.

14. J. Li, X.P. Ni, K.W. Leong, "Injectable drug-delivery systems based on supramolecular hydrogels formed by poly(ethylene oxide)s and α-cyclodextrin," *Journal of Biomedical Materials Research Part A* Vol. 65A (2), pp. 196–202, 2003.

15. K.M. Huh, Y.W. Cho, H. Chung, I.C. Kwon, S.Y. Jeong, T. Ooya, W.K. Lee, S. Sasaki, N. Yui, "Supramolecular hydrogel formation based on inclusion complexation between poly(ethylene glycol)-modified chitosan and α-cyclodextrin," *Macromolecular Bioscience* Vol. 4 (2), pp. 92–99, 2004.

16. K.M. Huh, T. Ooya, W.K. Lee, S. Sasaki, I.C. Kwon, S.Y. Jeong, N. Yui, "Supramolecular-structured hydrogels showing a reversible phase transition by inclusion complexation between poly(ethylene glycol) grafted dextran and α-cyclodextrin," *Macromolecules*, Vol. 34 (25), pp. 8657–8662, 2001.

17. X. Li, J. Li, "Supramolecular hydrogels based on inclusion complexation between poly(ethylene oxide)-b-poly (ε -caprolactone) diblock copolymer and α-cyclodextrin and their controlled release property," *Journal of Biomedical Materials Research. Part A*, Vol. 86A (4), pp. 1055–1061, 2008.

18. J. Li, X. Li, Z. Zhou, X. Ni, K.W. Leong, "Formation of supramolecular hydrogels induced by inclusion complexation between Pluronics and α-cyclodextrin," *Macromolecules*, Vol. 34 (21) pp. 7236–7237, 2001.

19. S.-Z. Zu, B.-H. Han, "Aqueous dispersion of graphene sheets stabilized by Pluronic copolymers: formation of supramolecular hydrogel," *The Journal of Physical Chemistry C*, Vol. 113 (31), pp. 13651–13657, 2009.

20. Z. Wang, Y. Chen, "Supramolecular hydrogels hybridized with single-walled carbon nanotubes," *Macromolecules*, Vol. 40 (9), pp. 3402–3407, 2007.

21. Z. Hui, X. Zhang, J. Yu, J. Huang, Z. Liang, D. Wang, H. Huang, P. Xu, "Carbon nanotube-hybridized supramolecular hydrogel based on PEO-b-PPO-b-PEO/α-cyclodextrin as a potential biomaterial," *Journal of Applied Polymer Science*, Vol. 116 (4), pp. 1894–1901, 2010.

22. K.Y. Sui, S. Gao, W.W. Wu, Y.Z. Xia, "Injectable supramolecular hybrid hydrogels formed by MWNT-grafted-poly(ethylene glycol) and α-cyclodextrin," *Journal of Polymer Science Part A-Polymer Chemistry*, Vol. 48 (14), pp. 3145–3151, 2010.

23. D. Ma, L.M. Zhang, "Fabrication and modulation of magnetically supramolecular hydrogels," *Journal of Physical Chemistry B*, Vol. 112, (20), pp. 6315–6321, 2008.

24. D. Ma, X. Xie, L.-M. Zhang, "A novel route to in-situ incorporation of silver nanoparticles into supramolecular hydrogel networks," *Journal of Polymer Science Part B: Polymer Physics*, Vol. 47 (7), pp. 740–749, 2009.

25. M.Y. Guo, M. Jiang, S. Pispas, W. Yu, C.X. Zhou, "Supramolecular hydrogels made of end-functionalized low-molecular-weight PEG and α-cyclodextrin and their hybridization with SiO_2 nanoparticles through host-guest interaction," *Macromolecules*, Vol. 41 (24), pp. 9744–9749, 2008.

26. J. Li, X. Li, X.P. Ni, X. Wang, H.Z. Li, K.W. Leong, "Self-assembled supramolecular hydrogels formed by biodegradable PEO-PHB-PEO triblock copolymers and α-cyclodextrin for controlled drug delivery," *Biomaterials*, Vol. 27 (22) pp. 4132–4140, 2006.

27. J. Yu, H. Fan, J. Huang, J. Chen, "Fabrication and evaluation of reduction-sensitive supramolecular hydrogel based on cyclodextrin/polymer inclusion for injectable drug-carrier application," *Soft Matter*, Vol. 7 (16), pp. 7386–7394, 2011.

28. Y. Chen, X.H. Pang, C.M. Dong, "Dual stimuli-responsive supramolecular polypeptide-based hydrogel and reverse micellar hydrogel mediated by host-guest chemistry," *Advanced Functional Materials*, Vol. 20 (4), pp. 579–586, 2010.

29. L. Ren, L. He, T. Sun, X. Dong, Y. Chen, J. Huang, C. Wang, "Dual-responsive supramolecular hydrogels from water-soluble PEG-grafted copolymers and cyclodextrin," *Macromolecular Bioscience*, Vol. 9 (9), pp. 902–910, 2009.

30. Z.B. Li, H. Yin, Z.X. Zhang, K.L. Liu, J. Li, "Supramolecular anchoring of DNA polyplexes in cyclodextrin-based polypseudorotaxane hydrogels for sustained gene delivery," *Biomacromolecules*, Vol. 13 (10), pp. 3162–3172, 2012.

31. N.Q. Tran, Y.K. Joung, E. Lih, K.M. Park, K.D. Park, "RGD-conjugated in situ forming hydrogels as cell-adhesive injectable scaffolds," *Macromolecular Research*, Vol. 19 (3), pp. 300–306, 2011.

32. Z. Tian, C. Chen, H.R. Allcock, "Injectable and biodegradable supramolecular hydrogels by inclusion complexation between poly(organophosphazenes) and α-cyclodextrin," *Macromolecules*, Vol. 46 (7), pp. 2715–2724, 2013.

33. S. Zhao, J. Lee, W. Xu, "Supramolecular hydrogels formed from biodegradable ternary COS-g-PCL-b-MPEG copolymer with α-cyclodextrin and their drug release," *Carbohydrate Research*, Vol. 344 (16), pp. 2201–2208, 2009.

34. N. Lin, A. Dufresne, "Supramolecular hydrogels from in situ host-guest inclusion between chemically modified cellulose nanocrystals and cyclodextrin," *Biomacromolecules*, Vol. 14 (3), pp. 871–880, 2013.

35. W. Zhu, Y.L. Li, L.X. Liu, Y.M. Chen, F. Xi, "Supramolecular hydrogels as a universal scaffold for stepwise delivering Dox and Dox/cisplatin loaded block copolymer micelles," *International Journal of Pharmaceutics*, Vol. 437 (1–2), pp. 11–19, 2012.

36. D. Ma, H.-B. Zhang, K. Tu, L.-M. Zhang, "Novel supramolecular hydrogel/micelle composite for co-delivery of anticancer drug and growth factor," *Soft Matter*, Vol. 8 (13), pp. 3665–3672, 2012.

37. J. Li, A. Harada, M. Kamachi, "Sol-gel transition during inclusion complex formation between α-cyclodextrin and high molecular weight poly(ethylene glycol)s in aqueous solution," *Polymer Journal*, Vol. 26 (9), pp. 1019–1026, 1994.

38. A. Harada, M. Kamachi, "Complex formation between poly(ethylene glycol) and α-cyclodextrin," *Macromolecules*, Vol. 23 (10), pp. 2821–2823, 1990.

39. C. Travelet, G. Schlatter, P. Hébraud, C. Brochon, A. Lapp, G. Hadziioannou, "Formation and self-organization kinetics of α-CD/PEO-based pseudo-polyrotaxanes in water. A specific behavior at 30 °C," *Langmuir*, Vol. 25 (15), pp. 8723–8734, 2009.

40. E. Sabadini, F.C. Egídio, T. Cosgrove, "More on polypseudorotaxanes formed between poly(ethylene glycol) and α-cyclodextrin," *Langmuir*, Vol. 29 (15), pp. 4664–4669, 2013.

41. H.S. Choi, T. Ooya, S. Sasaki, N. Yui, M. Kurisawa, H. Uyama, S. Kobayashi, "Spontaneous change of physical state from hydrogels to crystalline precipitates during poly-pseudorotaxane formation," *Chemphyschem*, Vol. 5 (9), pp. 1431–1434, 2004.

42. S.-P. Zhao, L.-M. Zhang, D. Ma, "Supramolecular hydrogels induced rapidly by inclusion complexation of poly(ε-caprolactone)-poly(ethylene glycol)-poly(ε-caprolactone) block copolymers with α-cyclodextrin in aqueous solutions," *The Journal of Physical Chemistry B*, Vol. 110 (25), pp. 12225–12229, 2006.

43. Z. Gan, T.F. Jim, M. Li, Z. Yuer, S. Wang, C. Wu, "Enzymatic biodegradation of poly(ethylene oxide-b-ε-caprolactone) diblock copolymer and its potential biomedical applications," *Macromolecules*, Vol. 32 (3), pp. 590–594, 1999.

44. X. Li, J. Li, K.W. Leong, "Preparation and characterization of inclusion complexes of biodegradable amphiphilic poly(ethylene oxide)-poly (R)-3-hydroxybutyrate-poly(ethylene oxide) triblock copolymers with cyclodextrins," *Macromolecules*, Vol. 36 (4), pp. 1209–1214, 2003.

45. K.L. Liu, J.-L. Zhu, J. Li, "Elucidating rheological property enhancements in supramolecular hydrogels of short poly[(R,S)-3-hydroxybutyrate]-based amphiphilic triblock copolymer and α-cyclodextrin for injectable hydrogel applications," *Soft Matter*, Vol. 6 (10), pp. 2300–2311, 2010.

46. H Wei, J. He, L.-G. Sun, K. Zhu, Z.-G. Feng, "Gel formation and photopolymerization during supramolecular self-assemblies of α-CDs with LA–PEG–LA copolymer end-capped with methacryloyl groups," *European Polymer Journal*, Vol. 41 (5), pp. 948–957, 2005.

47. X. Li, J. Li, K.W. Leong, "Role of intermolecular interaction between hydrophobic blocks in block-selected inclusion complexation of amphiphilic poly(ethylene oxide)-poly (R)-3-hydroxybutyrate-poly(ethylene oxide) triblock copolymers with cyclodextrins," *Polymer*, Vol. 45 (20), pp. 6845–6851, 2004.

48. X. Ni, A. Cheng, J. Li, "Supramolecular hydrogels based on self-assembly between PEO-PPO-PEO triblock copolymers and α-cyclodextrin," *Journal of Biomedical Materials Research. Part A*, Vol. 88 (4), pp. 1031–1036, 2009.

49. K.L. Liu, S.H. Goh, J. Li, "Threading α-cyclodextrin through poly[(R,S)-3-hydroxybutyrate] in poly[(R,S)-3-hydroxybutyrate]-poly(ethylene glycol)-Poly[(R,S)-3-hydroxybutyrate] triblock copolymers: formation of block-selected polypseudorotaxanes," *Macromolecules*, Vol. 41 (16), pp. 6027–6034, 2008.

50. D. Hou, X. Tong, H. Yu, A.Y. Zhang, Z.G. Feng, "A kind of novel biodegradable hydrogel made from copolymerization of gelatin with polypseudorotaxanes based on α-CDs," *Biomedical Naterials*, Vol. 2 (3), pp. S147–152, 2007.

51. S.P. Zhao, L.M. Zhang, D. Ma, C. Yang, L. Yan, "Fabrication of novel supramolecular hydrogels with high mechanical strength and adjustable thermosensitivity," *Journal of Physical Chemistry B*, Vol. 110 (33), pp. 16503–16507, 2006.

52. S.-P. Zhao, W.-L. Xu, "Thermo-sensitive hydrogels formed from the photocrosslinkable polypseudorotaxanes consisting of β-cyclodextrin and Pluronic F68/PCL macromer," *Journal of Polymer Research*, Vol. 17 (4), pp. 503–510, 2010.

53. T. Nakama, T. Ooya, N. Yui, "Temperature- and pH-controlled hydrogelation of poly(ethylene glycol)-grafted hyaluronic acid by inclusion complexation with α-cyclodextrin," *Polymer Journal*, Vol. 36 (4), pp. 338–344, 2004.

54. H.S. Choi, K. Yamamoto, T. Ooya, N. Yui, "Synthesis of poly(ε-lysine)-grafted dextrans and their pH- and thermosensitive hydrogelation with cyclodextrins," *Chemphyschem*, Vol. 6 (6), pp. 1081–1086, 2005.

55. H.S. Choi, K. Kontani, K.M. Huh, S. Sasaki, T. Ooya, W.K. Lee, N. Yui, "Rapid induction of thermoreversible hydrogel formation based on poly(propylene glycol)-grafted dextran inclusion complexes," *Macromolecular Bioscience*, Vol. 2 (6) pp. 298–303, 2002.

56. L. He, J. Huang, Y. Chen, X. Xu, L. Liu, "Inclusion interaction of highly densely PEO grafted polymer brush and α-cyclodextrin," *Macromolecules*, Vol. 38 (9), pp. 3845–3851, 2005.

57. G. Chen, M. Jiang, "Cyclodextrin-based inclusion complexation bridging supramolecular chemistry and macromolecular self-assembly," *Chemical Society Reviews*, Vol. 40 (5), pp. 2254–2266, 2011.

58. A. Harada, J. Li, M. Kamachi, "Preparation and properties of inclusion complexes of polyethylene glycol with α-cyclodextrin," *Macromolecules*, Vol. 26 (21), pp. 5698–5703, 1993.

59. T. Taira, Y. Suzaki, K. Osakada, "Thermosensitive hydrogels composed of cyclodextrin pseudorotaxanes. Role of [3]pseudorotaxane in the gel formation,". *Chemical Communications*, Vol. (45), pp. 7027–7029, 2009..

60. K. Sui, X. Shan, S. Gao, Y. Xia, Q. Zheng, D. Xie, "Dual-responsive supramolecular inclusion complexes of block copolymer poly(ethylene glycol)-block-poly[(2-dimethylamino)ethyl methacrylate] with α-cyclodextrin," *Journal of Polymer Science Part A: Polymer Chemistry*, Vol. 48 (10), pp. 2143–2153 2010.

61. X. Zhang, J. Huang, P.R. Chang, J. Li, Y. Chen, D. Wang, J. Yu, J. Chen, "Structure and properties of polysaccharide nanocrystal-doped supramolecular hydrogels based on cyclodextrin inclusion," *Polymer* , Vol. 51 (9), pp. 4398–4407, 2010.

62. K.L. Liu, Z. Zhang, J. Li, "Supramolecular hydrogels based on cyclodextrin–polymer polypseudorotaxanes: materials design and hydrogel properties," *Soft Matter*, Vol. 7 (24), pp. 11290–11297, 2011.

12

Polyhydroxyalkanoate-Based Biomaterials for Applications in Biomedical Engineering

Chenghao Zhu and Qizhi Chen*

Department of Materials Engineering, Moansh University, Clayton, Australia

Abstract

Bacterially synthesized polyhydroxyalkanoate (PHA) polymers have attracted much attention because of their tuneable biodegradability and superb biocompatibility. Poly(3-hydroxybutyrate) (P3HB), for example, has been intensively studied for applications in tissue engineering. However, PHA polymers also have disadvantages, including being mechanically rigid due to crystallization and limited elastic stretchability. In contrast, chemically crosslinked biodegradable elastomers, poly(polyolsebacate) (PPS) for example, exhibit a large elastic elongation at break but tend to show cytotoxicity. Hence, it has been envisaged that the blending of PHA with the crosslinked elastomers is a potential approach to maximizing the advantages and minimizing the disadvantages of these polymers, achieving a satisfactory combination of the mechanical properties of elastomeric polymers and biocompatibility/degradation kinetics of PHA. This chapter provides a review of PHAs and PHA-based polymer blend biomaterials that are developed for applications in tissue engineering, focusing on the biocompatibility, biodegradability, and mechanical properties of these materials, which are three essential features of biomaterials in most tissue engineering applications. At the end of the chapter, the major achievements on the development of PHA-based polymer blends are summarized, and future directions are highlighted.

Keywords: Polyhydroxyalkanoate (PHA), PHA-based polymer blends, biocompatibility, biodegradability, mechanical properties

*Corresponding author: qizhi.chen@monash.edu

Ashutosh Tiwari (ed.) Advanced Healthcare Materials, (439–464) 2014 © Scrivener Publishing LLC

12.1 Introduction

From the viewpoint of terminology, a polyhydroxyalkanoate (PHA) is the polymer formed from a hydroxyl-alkanoic acid (i.e., a carboxylic acid), which can be defined as any acid with the structure HO-R-COOH, where R is an alkyl unit having the composition of C_nH_{2n}. Based on this definition, polymers such as poly(lactic acid) (O-CH(CH$_3$)-COO), poly(6-hydroxy-hexanoic acid) (otherwise known as polycaprolactone, (O-(CH$_2$)$_5$-COO)), and poly(4-hydroxybutyrate) (O-(CH$_2$)$_3$-COO) are all PHAs. However, in the biomaterials field, PHA is usually restricted to those members that are derived from bacterial sources [1–4]. Hence, poly(lactic acid) (PLA), which is formed by conventional synthesis from bio-derived monomers, and polycaprolactone, which is derived from synthetic monomers, are usually excluded from the PHA class. The general structure of PHA is shown below in Fig.12.1.

PHAs are a relatively old family of biomaterials. The discovery of its first member, P3HB, can be traced back to the 1920s. The potential of P3HB for biomedical applications was first recognized in 1962, but it was not until the 1980s that P3HB again became of interest in biomedical research, when P3HB tablets for sustained drug delivery were studied [6]. PHAs have since attracted increasing attention as tissue engineering materials due to their biocompatibility, general biodegradability and tuneable mechanical properties. This review chapter provides a review on PHAs and related elastomeric biomaterials used in tissue engineering, including PHAs, PHA-co-polymers, and PHA-based polymer blends. Each section includes a description of the synthesis of the material that is critical to the understanding of its unique properties, followed by a detailed discussion of its biocompatibility, biodegradability, and mechanical properties. Lastly,

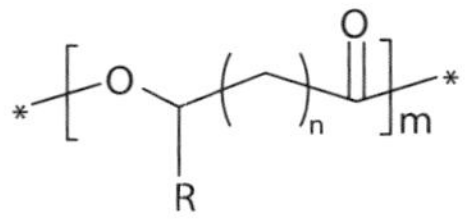

Figure 12.1 General structure of poly(hydroxyl alkanoates) [5].
n = 1 R = hydrogen poly(3-hydroxypropionate)
R = methyl poly(3-hydroxybutyrate)
R = ethyl poly(3-hydroxyvalerate)
R = propyl poly(3-hydroxyhexanoate)
R = pentyl poly(3-hydroxyoctanoate)
R = nonyl poly(3-hydroxydodecanoate)
n = 2 R = hydrogen poly(4-hydroxybutyrate)
n = 3 R = hydrogen poly(5-hydroxyvalerate)

the major achievements and remaining challenges for elastomeric biomaterials are summarized.

12.2   Synthesis of PHAs

In the 1920s, Lemoigne reported 3-hydroxybutyric acid, a granular component produced by the bacteria Bacillus Megaterum, and characterized
the solid material in the bacterial cell as a polymer of 3-hydroxybutyric
acid [7, 8]. To date, over 150 different aliphatic polyesters of the same general structure with different R-pendant groups have been discovered [6, 9,
10]. Some PHAs are listed in Table 12.1.

Although PHAs can be man-made, these polymers are primarily
produced by microorganisms. Many bacteria can synsesize polyesters
of hydroxyalkanoic acids and can accumulate water-insoluble PHAs
in the cytoplasm as inclusions and as storage compounds for energy
and carbon. These inclusions are referred to as PHA granules [12].
Accumulation of PHAs in the bacterial cell usually occurs under conditions of depleted nutrients (e.g., ammonium, potassium, sulphate, magnesium, and phosphate, which is essential for growth) but with an excess
carbon source [6, 9, 12–14]. Under these conditions, PHAs are synthesized by diverting either central intermediates of the carbon metabolism or derivatives from precursor substrates, which are provided as
carbon sources for the growth of the bacteria, to hydroxyacyl-CoA thio-
esters. The latter are then polymerized to PHAs, which become bound
to the surface of PHA granules together with other proteins. PHAs produced by bacteria can be homo-polymers or copolymers in which two
or more hydroxyalkanoates are randomly arranged or occur in blocks.
Alcaligenes eutrophus (A. eutrophus) and Pseudomonas oleovorans (P.
oleovorans) are the two types of PHA synthesis bacteria that are the
most widely reported.

Chemically synthesized PHAs are less competitive than biosynthesized
PHAs due to the relatively high cost of the former. However, PHAs of low
molecular weight with single composition have not been achieved by biosynthesis, which limits the possible modifications on the PHAs. Hirt and
co-workers [15] produced low molecular weight P3HB and P(3HB-co-
3HV) by trans-esterification. By dissolving P3HB and P(3HB-co-3HV)
in diglyme and adding ethylene glycol and dibutyltindilaurate as catalysts, low molecular weight P3HB and P(3HB-co-3HV) can be produced
after the above reactant mixture is held at 140 ºC for 7.5 hours [15]. The

Table 12.1 Names and structures of some common PHA units [11]

Full name of PHA	Abbreviation	m	R	structure
poly(3-hydroxybu-tyrate)	P3HB	1	$-CH_3$	
poly(4-hydroxybu-tyrate)	P4HB	2	H	
poly(3-hydroxyval-erate)	P3HV	1	$-CH_2CH_3$	
poly(3-hydroxyhex-anoate)	P3HH	1	$-(CH_2)_2CH_3$	
poly(3-hydroxyoc-tanoate)	P3HO	1	$-(CH_2)_4CH_3$	
poly(3-hydroxydecano-ate)	P3HD	1	$-(CH_2)_6CH_3$	

molecular weight drops from over 30000 in the raw material to approximately 2000in the final products. Other low molecular weight PHAs could also be obtained by the same method [16]. With low molecular weight PHAs, block copolymers based on PHAs can be produced, such as P3HB-co-polycaprolactone(PCL) and P3HB-co-poly(ethylene glycol)(PEG) diblock polymers [17, 18].

12.3   Processing and its Influence on the Mechanical Properties of PHAs

The final products of PHAs are typically processed in the shape of thin sheets/films or fibers, which sensitively affect the mechanical properties of the polymers. PHAs are thermoplastic rubbers. In principle, PHAs have good processability, which is characteristic of all thermoplastics. In these polymers, there are two separated micro phases, crystalline rigid segments and amorphous flexible segments. The flexible polymer chains are held together by crystalline regions. The rigid segments function as crosslinkers that provide mechanical strength, whereas amorphous segments provide flexibility [19].

The difficulty in processing P3HB, however, is caused by its rigidity and brittleness. When the spherulites of P3HB homo-polymer are grown from the melt, large cracks are often observed in the spherulites. The formation of such cracks in P3HB spherulites is one of the causes of embrittlement [20]. The sealing of cracks can be achieved by rolling the spherulitic films. Spherulites without cracks can be made by increasing the nucleation density until the spherulites become sufficiently small to prevent cracks. The P3HBsheets/films without cracks can be obtained using the self-seeding procedure with a spherulite size of under 20μm [20]. De Koning and Lestra reported that the embrittlement of P3HB materials occurs due to the progress of secondary crystallization during storage at room temperature [21]. This secondary crystallization has been argued to be involved with the reorganization of lamellar crystals formed at the initial crystallization, which tightly contains the amorphous chains between the crystals. It has been demonstrated that molded P3HB materials can be toughened by annealing after their initial crystallization.

The elongation at break of the stretched P3HB film was markedly improved relative to that of non-stretched film (Fig. 12.2), and the improvement in the tensile strength was due to the orientation of the P3HB molecular chains, which was facilitated through the increasing draw ratio. Further improvement in the mechanical properties of P3HB films was achieved by subsequent annealing against the nine or ten times hot drawn UHMW P3HB film. It was also found that the mechanical properties of the stretched P3HB film did not deteriorate during storage at room temperature for 6 months (Fig. 12.2).

The X-ray fiber pattern obtained from the two-step-drawn P3HB films showed a new equatorial reflection derived from the planar zigzag conformation (β-form), together with a 2_1 helix conformation(α-form)

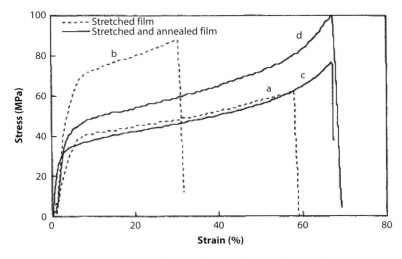

Figure 12.2 Stress-strain curves of ultra-high-molecular-weight P3HB films after different storage times at room temperature. (a) stretched film stored for 7 days (b) stretched film stored for 190 days (c) stretched-annealed film stored for 7 days and (d) stretched-annealed film stored for 180 days [22].

(Fig. 12.3).It has been considered that the β-form be introduced by the orientation of free molecular chains in amorphous regions between α-form lamellar crystals [23]. Specifically, tie molecules between the lamellar crystals are important in the generation of high mechanical properties. By a two-step drawing at room temperature, the tie molecules are strongly extended and, as a result, the planar zigzag conformation is generated, which is responsible for the good mechanical properties.

12.4 Mechanical Properties of PHA Sheets/Films

The PHA family of polyesters offers a wide variety of material properties, from hard crystalline plastics to elastic rubbers. For P(3HB-co-3HV), the tensile strength and Young's modulus of the films decreased from 45 to 18 MPa and from 3.8 to 1.2 GPa, respectively, as the 3HV fraction was increased from 0 to 34 mol%. The elongation at break increased from 4 to 970% as the 3HV fraction was increased from 0 to 34 mol% [24]. As mentioned above, P(3HB-co-3HV) copolymers show a high degree of crystallinity due to co-crystallization. Therefore, the mechanical characteristics of P(3HB-co-3HV) with (R)-3HV content up to 20% are not significantly

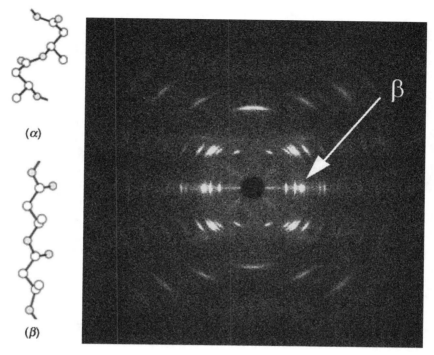

(α)

(β)

Figure 12.3 A wide-angle X-ray diffraction pattern of two-step-drawn films of ultrahigh molecular weight P3HB (draw-ratio = 15) [23].

improved in comparison with the P3HB homopolymer. However, if sufficient -3HV units are incorporated into P3HB molecules, the films show high elongation at break, for example, that of P(3HB-co-28 mol%3HV) and P(3HB-co-34 mol%3HV) reached up to 700 and 970%, respectively.

In contrast to the above-discussed P3HB-co-HV polymers, other P3HB-based copolymers with longer side chains on the 3rd carbon atom do not display co-crystallization. The crystallinity of P(3HB-co-3HHx) is rapidly reduced by the introduction of 3HHx units (Fig. 12.4). The tensile strength of the films decreased from 43 to 20 MPa as the 3HHx fraction was increased from 0 to 17 mol%. The elongation at break dramatically increased from 6 to 850%. Thus, the P(3HB-co-3HHx) becomes a soft and flexible material by copolymerizing with a small amount of the 3HHx unit [25].

If the side chain is on the 4th carbon atom and the side chain is also short, the polymer still does not display co-crystallization. For example, the mechanical properties of the P(3HB-co-4HB) copolymer have been

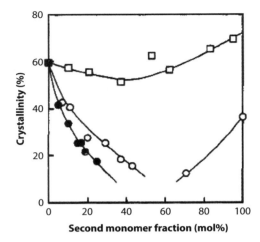

Figure 12.4 Effects of co-polyester composition on the degrees of crystallinity of different microbial co-polyesters [25]. (●) P(3HBco-3HH); (○) P(3HB-co-3HP);(□) P(3HB-co-3HV); P(3HB-co-3HP) is a random copolymer of (R)-3-hydroxybutyric and 3-hydroxypropionic acids.

determined using solution-cast films that have a wide range of compositions from 0 to 100 mol% 4HB. The tensile strength of P(3HB-co-4HB) films decreased from 43 to 26MPa with an increase in the4HB fraction from 0 to 16mol%, whereas the elongation to break increased from 5 to 444%. The tensile strength of the films with compositions of 64–100 mol% 4HB increased from 17 to 104 MPa with an increasing 4HB fraction [26, 27].

Even though P3HB-based copolymers have slow crystallization rates and exhibit soft and flexible material properties, secondary crystallization occurs during storage at room temperature and results in the embrittlement of materials that are similar to the P3HB homopolymer. In contrast to the secondary crystallization of the P3HB homopolymer that involves the reorganization of lamellar crystals, the small and thin crystallites produced at the interlamellar amorphous regions reduce the mobility of molecular segments with the progress of secondary crystallization.

Similar treatments of cold-drawing and hot-drawing are also applied on P3HB copolymers. Iwata and co-workers applied a cold drawing technique for copolymer samples of P(3HB-co-8 mol%3HV) ($M_w=1.0\times10^6$) and P(3HB-co-5 mol%3HHx) ($M_w=0.8\times10^6$). Independent of the co-monomer structure, cold drawing of melt-quenched amorphous films of P(3HB-co-8 mol%3HV) ($M_w=1.0\times10^6$) and P(3HB-co-5 mol%3HHx) ($M_w=0.8\times10^6$) succeeded easily and reproducibly at the temperature of ice water. The tensile strength of the P(3HB-co-8 mol%3HV) films was drastically increased from 19 to 117 MPa. In the case of P(3HB-co-5 mol% 3HHx) films, melt-quenched films in the rubber state could be stretched reproducibly. The 5-times cold–drawn films of P(3HB-co-5 mol%3HHx) had a tensile strength of 80 MPa, with a high elongation to break of 258%. By using two-step drawing for the cold drawn films, the tensile strength of P(3HB-co-8 mol%3HV) (M_w=1.0×10^6) and P(3HB-co-5 mol%3HHx) (M_w=0.8×10^6) increased from 117 to 185 MPa and from 80 to 140 MPa, respectively [28, 29].

In summary, the important mechanical properties, except for Young's modulus, of PHA sheets/films are usually improved by drawing, including cold drawing, hot drawing and two-step drawing. The higher the drawn ratio is, the higher the tensile strength and elongation at break that can be achieved. However, Young's modulus remains unchanged, especially after one-step cold drawing. According to previous studies [29, 30], the mechanical properties of PHA sheets can be improved with the two-step drawing method.

12.5 PHA-Based Polymer Blends

Besides direct biosynthesis of PHAs from carbon sources in bacteria, blending PHA with other polymers is an alternative way to tune the physical properties. Polymer blends are physical mixtures of chemically different polymers, and the mixture of two polymers forms a homogeneous or heterogeneous phase in the amorphous region on a microscopic scale at equilibrium. When a mixture of two polymers in the amorphous phase exists as a single phase, the blend is considered to be miscible in a thermodynamic sense. In contrast, if a mixture of two polymers separates into two distinct phases that are primarily composed of individual components, then the two components are considered to be immiscible in the thermodynamic sense. The physical properties of a blend polymer are strongly dependent on the phase structures [35].

12.5.1 Miscibility of PHAs with Other Polymers

Blends with other polyesters

The miscibility of component polymers is the critical property for successful synthesis of their polymer blends. Hence, many studies performed on PHA-containing blends have focused on the miscibility of the polymers. Being polyesters, PHAs are in general miscible with other polyesters.

Table 12.2 Mechanical properties of P3HB and copolymer sheets/films.

Materials	Drawing method	Draw ratio	Tensile strength (MPa)	Elongation at break (%)	Young's modulus(GPa)	Refs
P3HB	Not-drawn	1	45	4	3.8	(Avella, Martuscelli et al. 2000)
	Not-drawn	1	14	9	0.9	(Iwata, Tsunoda et al. 2003)
	Not-drawn	1	36	2.5	2.5	(Engelberg and Kohn 1991)
	Cold drawing	10	195	100	0.9	(Iwata, Tsunoda et al. 2003)
Ultra high molecular weight (UHMW) P3HB	Not-drawn	1	36	14	1.3	(Aoyagi, Yamashita et al. 2002)
	Hot drawing	10	277	84	2.3	
	Two-step hot drawing (10×1.5)	15	388	25	2.9	
	Cold drawing	10	237	112	1.5	
	Two-step cold drawing (10×1.5)	15	287	53	1.8	
P(3HB-co-7 mol% 3HV)	Not-drawn	1	24	2.8	1.4	(Engelberg and Kohn 1991)
P(3HB-co-8 mol% 3HV)	Not-drawn	1	19	35	0.1	(Iwata and Doi 2005)

Materials	Drawing method	Draw ratio	Tensile strength (MPa)	Elongation at break (%)	Young's modulus(GPa)	Refs
P(3HB-co-11 mol% 3HV)	Not-drawn	1	20	17	1.1	(Engelberg and Kohn 1991)
P(3HB-co-19 mol% 3HV)	Not-drawn	1	18	25	1.5	(Ramsay, Langlade et al. 1993)
P(3HB-co-22 mol% 3HV)	Not-drawn	1	16	36	0.62	(Engelberg and Kohn 1991)
P(3HB-co-28 mol% 3HV)	Not-drawn	1	21	700	1.5	(Avella, Martuscelli et al. 2000)
P(3HB-co-34 mol% 3HV)	Not-drawn	1	18	970	1.2	
P(3HB-co-8 mol% 3HV)	Cold drawing	10	117	109	0.5	(Iwata and Doi 2005)
	Two-step cold drawing (10×1.4)	14	185	63	1.4	
P(3HB-co-2.5 mol% 3HH)	Not-drawn	1	26	7	0.63	(Asrar, Valentin et al. 2002)
P(3HB-co-7 mol% 3HH)	Not-drawn	1	17	24	0.29	
P(3HB-co-9.5 mol% 3HH)	Not-drawn	1	8.8	43	0.155	
P(3HB-co-5 mol% 3HH)	Not-drawn	1	32	267	0.5	(Josefine Fischer, Aoyagi et al. 2004)
	Two-step cold drawing (4×2.5)	10	140	116	1.5	

According to Blümm and Owen's work [36], low-molecular-weight poly(L-lactic acid)(PLLA) (M_n = 1759) is miscible with P3HB in melt over the whole composition range, whereas a blend of high-molecular-weight PLLA (M_n = 159400) with P3HB exhibits phase separation, which is in agreement with the prediction of Flory-Huggins theory [36]. DSC analysis revealed that the structure of solid P3HB/P(CL-co-LA) blends was strongly dependent on the copolymer composition of the P(CL-co-LA) component. The miscible blends of P3HB have been prepared with amorphous P(CL-co-LA) that ranges from 30 to 100 mol % LA [37].

Blends with non-polyester polymers

It has been shown that PHA polymers exhibit possibilities for interaction with a number of other non-polyester polymers. This finding has been demonstrated in blends with poly(ethylene oxide) (PEO)and poly(vinyl alcohol) (PVA). Differential scanning calorimetry (DSC) and optical microscopy were used to determine the miscibility behaviour of P3HB and PEO mixtures. It was found that P3HB and PEO are miscible in the melt [38]. Yoichiro Azuma and co-workers [39] studied thermal behaviour and miscibility of P3HB and PVA blends. The results show that the increase in PVA content in a blend enhances the miscibility, and the crystallinity of P3HB in the blends also decreases with an increment in the PVA content.

When P3HB was blended with poly(butylene succinate-co-butylene adipate) (PBSA) or poly(butylene succinate-co-ε-caprolactone) (PBSC), the DSC thermogram analysis of each blend system revealed two distinct glass transitions that were independent of the blend composition, indicating that P3HB is immiscible with either PBSA or PBSC in the amorphous state [40]. Melt-quenched P3HB/poly(methyl methacrylate) (PMMA) blends containing up to 20 wt% P3HB are single-phase amorphous materials with a composition-dependent T_g. At a higher P3HB content, PMMA molecules are fully engaged in a 20/80 P3HB/PMMA miscible phase, while excessive P3HB segregates and forms a partially crystalline phase [41].

Ethylene-vinyl acetate (EVA) is a softer, more rubbery material. P(3HB-co-15%HV)/EVA (28 mol% VA) blends are immiscible. The partially crystalline solid blends of P(3HB-co-3HV) and EVA showed two glass transitions and melting regions, which relate to the two separate components. The dynamic modulus and failure strength were strongly affected by the blend composition. In the region above 70% P(3HB-co-3HV), the samples were stiff and relatively brittle, and the behaviour was dominated by the P(3HB-co-3HV) component. Below 70% P(3HB-co-3HV), the behaviour was dominated by the softer EVA component, which appears to form a continuous matrix in which BIOPOL domains are embedded [42]. Holmes and co-workers reported the partial miscibility of PHAs with polymers that contain chlorine or nitrile groups. Blends of PHAs and several chlorinated or nitrile-containing polymers showed improvements in the mechanical properties. It was suggested that the chlorine and nitrile groups promote hydrogen bonding with the carbonyl groups of the PHA polymers [43].

Blends with natural rubbers

A number of rubbers, including natural rubber, nitrile rubber and butadiene rubber, were blended with a PHA that has a medium chain length (mcl-PHA) at room temperature using a solution with the blending technique by Rachana Bhatt and co-workers [44]. Blends of 5, 10 and 15 % of mcl-PHA with various rubbers were produced (the author did not specify which mcl-PHA they used). Thermo-gravimetric analysis of mcl-PHA showed that the melting temperature of the polymer was approximately 50 °C. Thermal properties of the synthesized blend were studied by DSC, which confirmed effective blending between the polymers.

Imam and co-workers prepared injection-moulded specimens by blending poly (hydroxybutyrate-co-valerate) (PHBV) with corn starch [45]. Blended formulations incorporated 30 or 50% starch in the presence or absence of PEO, which enhances the adherence of starch granules to PHBV.

12.5.2   Degradability of PHA-Based Polymer Blends

Biomedical degradation

The biodegradability of binary blends of P3HB with PCL, poly(1,4-butylene adipate) (PBA) and PVA has been studied [46]. The enzymatic degradation of the P3HB-based blend films was conducted at 37°C and pH 7.4 in an aqueous solution of an extracellular P3HB depolymerized from Alcaligenesfaecalis T1. The profiles of enzymatic degradation of the P3HB-based blends were strongly dependent on the polymer component blended with P3HB. In the case of the P3HB/PCL blend film, a complicated dependence of the PCL weight fraction on the rate of enzymatic degradation was observed. In contrast, the weight loss of the P3HB/PBA or P3HB/PVA blend film by the P3HB depolymerization decreased monotonically with an increase in the weight fraction of PBA or PVA. The enzymatic degradation data are shown in Fig. 12.5 [46].

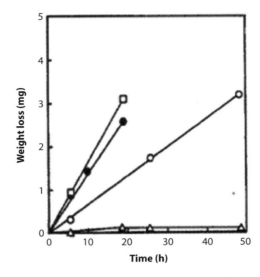

Figure 12.5 Enzymatic degradation (erosion) profiles of films of P3HB-based blends in an aqueous solution (pH 7.4) of P3HB depolymerase (8 μg) at 37°C. ●, P3HB; □, P3HB/PCL [77/23(w/w)]; O, P3HB/PBA [75/25(w/w)]; △, P3HB/PVA[74/26(w/w)] [46].

Films of P3HB blended with PEG in various proportions of P3HB/PEG (100/0, 98/2, 95/5, 90/10, 80/20 and 60/40 (wt. %)) were studied for their degradation. Samples that contained 20and 40% PEG lost 50% of their mass after 401 and 431 hr of aging with amylase. The same amount of mass loss was seen after 853, 650 and 784 hr for 2, 5 and 10 % P3HB, respectively. The addition of larger amounts of PEG (20 and 40 %) increased the number of polar groups in the samples and promoted the interaction of these groups with water molecules in the hydrolysis that preceded biodegradation [47].

Environmental degradation

Degradation studies on the blends of 5, 10 and 15 % of mcl-PHA with various natural rubbers were conducted using a soil isolate, Pseudomonas sp. 202, for 30 days. Imam and co-workers investigated the degradation of PHBV blended with corn starch[45]. Blended formulations incorporated 30 or 50% starch in the presence or absence of PEO. Starch in blends degraded faster than PHBV and accelerated the PHBV degradation. Additionally, PHBV did not retard the starch degradation. Gel permeation chromatography (GPC) analyses revealed that, while the number of average molecular weight (M_n) PHBVs in all of the exposed samples decreased, there was no significant difference in this decrease between neat PHBV as opposed to PHBV blended with starch. SEM showed homogeneously distributed starch granules embedded in a PHBV matrix, which is typical of a filler material. Starch granules were rapidly depleted during exposure to compost, increasing the surface eroding area of the PHBV matrix.

In short, the blending of PHAs with other polymers influences the overall rate of biodegradation. Immiscible blends frequently show enhanced degradation rates, but the homogeneity of these blends must be well controlled to obtain good reproducibility [48].

12.5.3   Biocompatibility of PHA-Based Polymer Blends

Research into the biocompatibility of PHA-based blends has been focused on the blends of the members of the PHA family. Blends of polyhydroxyalkanoates containing polyhydroxybutyrate (PHB) and poly(hydroxybutyrate-co-hydroxyhexanoate) (PHBHHx) were fabricated into films and scaffolds [49]. Scaffolds made of PHBHHx/P3HB consisting of 60 wt% PHBHHx showed strong growth and proliferation of chondrocytes on the blending materials under a scanning electron microscope. The chondrocyte cells grown on PHBHHx/P3HB scaffolds exhibited effective physiological functions for the generation of cartilage [49].

Similar blending is also used for rabbit articular cartilage chondrocytes [50]. It was shown that the blend with a 1:1 ratio of P3HB/PHBHHx possessed the highest surface free energy, which was the most optimal material for chondrocyte adhesion. After 24 h, the amount of chondrocytes adhered on films of P3HB/PHBHHx (1:1) was 2.1×10^4 cells/cm^2, 5 times more compared with that on the P3HB films (0.4×10^4 cells/cm^2). The polarity of the blends increased with decreasing crystallinity. After 8 days of cultivation, the chondrocytes attached on P3HB films were surrounded by both collagen II and collagen X, and the amount of extracellular collagen X decreased with increasing polarity contributed by increasing PHBHHx content in the blend, while chondrocytes changed shapes from spherical to flat with increasing polarity [50].

Poly(3-hydroxybutyrate-co-3-hydroxyhexanoate)   (PHBHHx)   and poly(propylene carbonate) (PPC) were blended for the fibronectin (FN) and rabbit aorta smooth muscle cells adsorption test [51]. The amounts of FN absorbed on 7:3, 3:7 PHBHHx/PPC blend and PPC films were not significantly different from that on PHBHHx films. The amount on 5:5 PHBHHx/PPC blend film was higher than that on PHBHHx films. For rabbit aorta smooth muscle cells, the adhesion to 7:3 and 3:7 PHBHHx/PPC blend films showed no significant difference compared with that of PHBHHx films. This finding was consistent with the results of FN adsorption, except for that on PPC films.

12.5.4   Mechanical Properties of PHA-Based Polymer Blends

Blends with other thermoplastic polyesters

The mechanical properties of the films of P3HB/PCL and P3HB/PBA blends have been studied by Yoshiharu Kumagai & Yoshiharu Doi [46]. The Young's modulus and tensile strength of P3HB film at 23°C were 1560 and 38MPa, respectively. When P3HB was blended with PCL, both of the values of the mechanical properties of the blend films markedly decreased with an increase in the weight fraction of the PCL component and reached minimum values (110 and 4MPa) at approximately 50 wt% of the PCL component. The Young's modulus and tensile strength of the P3HB/PBA blend films decreased linearly with an increase in the weight fraction of PBA in the film, from 1050 MPa (Young's modulus) and 32MPa (tensile strength) to 480 and 10MPa [46].

Blends of P3HB with PLLA were prepared at various compositions, and their drawing behaviours were investigated in terms of mechanical properties [52]. The mechanical property was enhanced in proportion to the PLLA content due to a good interfacial adhesion between two polymers and the reinforcing role of the PLLA components. More specifically, the ultra-high molecular weight (UHMW) P3HB/PLLA system showed remarkable improvement due to the simultaneous orientation of the PLLA domains.

Blends with other elastomeric polyesters (poly polyol sebacate)

The present authors have very recently developed a new family of PHA-containing cross-linked elastomeric blends that is biocompatible and elastic, with broad tuneability on mechanical properties that are suitable for a range of tissue repair applications. In this study, one of the poly(polyol sebacate) (PPS) members, poly(glycerol sebacate) (PGS), was used as the crosslink polymer component. Homogenous blends of PGS-P34HB have been achieved at a content of P34HB lower than 50 wt %, beyond which phase separation occurred.

Compared with pure P34HB polymers, the PGS-P34HB blends of P34HB (up to 50 wt % of P34HB) show the typical elastomeric behaviour of cross-linked elastomers, with trajectory stress-strain curves (Fig. 12.6). With P34HB content further increasing ($\geq$ 70wt % P34HB), however, the material becomes as brittle as P34HB.

Compared with the mechanical properties of pure PGS, the Young's modulus, UTS and elongation at break of PGS-P34HB blends increase simultaneously with the increment in the P34HB component up to 50 wt% (Fig. 12.7). The above phenomenon can be attributed to the two opposing

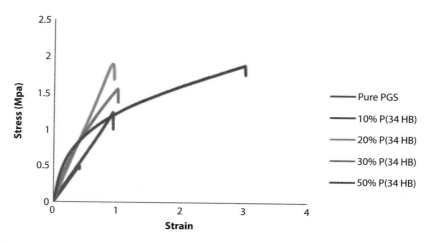

Figure 12.6 Stress strain curves of P34HB-PGS blending after crosslink treatment.

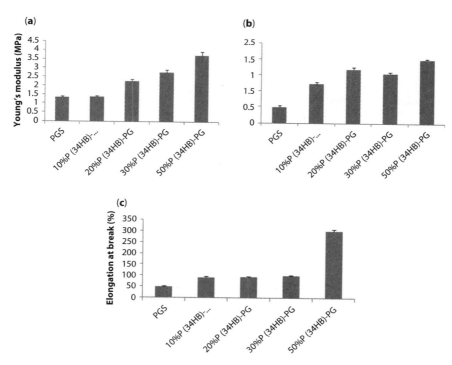

Figure 12.7 Mechanical properties of pure PGS and PGS-P34HB blends. (a) Young's modulus, (b) ultimate stress, and (c) elongation at break.

effects of P34HB on the mechanical properties of the network. First, the addition of P34HB introduced a physical barrier to the crosslinking reaction and, thus, reduced the mechanical strength of the network. Second, the relatively long and side chained P34HB had the effect of hardening the network. The mechanical properties of PGS-P34HB depends on the competition of the above two antagonistic effects. When the content of P34HB is low, the hydroxyl group of glycerol and the carboxyl groups of sebasic acid still have a sufficient possibility to meet and react to crosslink polymer chains. The increased Young's modulus can be attributed to the relatively long and side-chained P34HB molecules. In addition, the well dispersed P34HB could form a microcrystal structure that could contribute to the mechanical properties. At the 50 wt% P34HB, the elongation at break increases to 300%, which could be due to small crosslinks in the blends that are caused by the immiscibility that occurs at such a high proportion of P34HB.

The cyclic tests (Fig. 12.8) show that the resilience of the PGS-P34HB blend is satisfactorily higher than that of pure P34HB (65%), although it drops from 99 (pure PGS) to 80% when the P34HB proportion increases to 50 wt%.

Blends with non-polyester polymers

To obtain binary P3HB blends by melt-mixing, the following minor components with functional ester or anhydride groups were used: (1) ethylene propylene rubber (EPR), (2) EPR grafted with dibutyl maleate (EPR-g-DBM), (3) EPR grafted with succinic anhydride (EPR-g-SA), (4) ethylene-vinyl acetate (EVA), and (5) EVA polymer containing OH groups (EVAL) [53]. The addition of 20% rubber to P3HB results in a decrease in the tensile strength σ_R and the modulus of elasticity E, but it results in an improvement in the elongation to break ε_R (Table 12.3). Blends of P3HB/poly(methylene oxide) (POM) have been prepared by melt mixing and subsequent compression moulding [54]. The immiscibility of the two polymers in the liquid state is reported. The tensile test results are summarised in Table 12.4 [54].

Poly(3-hydroxybutyrate-co-3-hydroxyhexanoate) (PHBHHx) and poly(propylene carbonate) (PPC) were blended by a solvent casting method into films at various weight ratios in order to obtain materials with properties that were more suitable for blood vessel tissue engineering than pure PHBHHx alone [51]. With increasing PPC content in the PHBHHx/PPC blends, the elastic modulus and tensile strength decreased from 93 to 43MPaand from 10 to 6MPa, respectively, while the elongation at break increased from 16 to 1456 % [51].

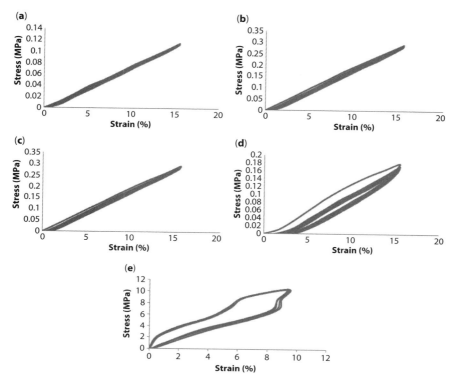

Figure 12.8 The cyclic stress-strain curves of pure P34HB and PGS-P34HB blending. (a) PGS-10 wt% P34HB, (b) PGS-20 wt% P34HB, (c) PGS-30 wt% P34HB, (d) PGS-50 wt% P34HB, and (e) P34HB.

Table 12.3 Modulus E, stress σ_R, and elongation ε_R at rupture for P3HB homo-polymer and for PHB/rubber blends.

Polymer	E (10^{-3}kg/cm^{-2})	σ_R (kg/cm^{-2})	ε_R (%)
P3HB	2.1	290	1.5
P3HB-EPR	1.5	170	2.0
P3HB-EVA	1.6	175	2.0
P3HB-EVAL	1.7	185	3.0
P3HB-EPR-g-DBM	1.6	175	4.0
P3HB-EPR-g-SA	1.6	182	6.5

Table 12.4 Tensile properties of PHB/POM blends.

Polymer	Young's modulus (GPa)	Stress at break (MPa)	Strain at break (%)
P3HB	2	28	2.6
P3HB 80	1.9	18	1.0
P3HB 60	1.9	16	0.9
P3HB 40	1.7	13	1.1
P3HB 20	1.9	34	4.4
POM	1.9	58	10.7

Table 12.5 Tensile mechanical properties of the various P3HB/PEG blends.

Polymer	Tensile strength at maximum (MPa)	Elongation at break (%)
P3HB/PEG 100/0	28±3	9±3
P3HB/PEG 98/2	26±8	25±6
P3HB/PEG 95/5	11±3	25±10
P3HB/PEG 90/10	12±3	25±8
P3HB/PEG 80/20	13±8	32±8
P3HB/PEG 60/40	13±4	31±10

Blends of P3HB with PEG (P3HB/PEG) in different proportions (100/0, 98/2, 95/5, 90/10, 80/20 and 60/40 wt %) were investigated for their mechanical properties. Compared to pure P3HB film, the tensile strength was reduced by the addition of PEG (Table 12.5) [47].

Atactic poly(methyl methacrylate)(aPMMA) was blended with P3HB up to a maximum composition of 25% of polyester, at 190°C in a Brabender-like apparatus [55]. Tensile experiments showed that, at room temperature, the 10/90 and 20/80 P3HB/aPMMA blends exhibited higher values of strain at break, and there were slight decreases in the modulus and stress at break compared to neat aPMMA (Table 12.6) [55].

Table 12.6 Tensile properties of P3HB/aPMMA blends at 25°C.

P3HB/ aMMA(wt%)	E (MPa)	σ_R (MPa)	Elongation at break ε_R (%)	Energy at break ($\times 10^6$J/m^3)
0/100	2600±150	53±8	2.6±0.5	71±8
10/90	2300±200	55±5	3.8±0.4	133±15
20/80	2100±180	49±5	3.7±0.4	112±13
25/75	2000±200	45±7	3.5±0.6	103±10

12.6 Summary

PHAs, which are highly biocompatible and have good processability, have been extensively investigated for applications in tissue engineering [56–67]. The major drawbacks of PHAs include their mechanical rigidity and heterogeneous degradation properties [68, 69], which are caused by the crystalline or rigid amorphous domains that contribute to strength but resist degradation. To minimize the above-mentioned drawbacks, PHA-based polymer blends have been developed [54]. In general, PHA-based blends with other thermoplastics frequently show little-improved or even compromised mechanical properties and accelerated degradation rates. In contrast, blends of PHA with chemically crosslinkable polyesters, such as PGS, show significantly improved elastic properties, with Young's modulus, UTS and elongation at break increasing simultaneously. Further work is needed: 1) to expand the family of PHA-PPS elastomers, 2) to control the homogeneity of these blends to obtain good reproducibility, 3) to investigate the degradation kinetics of this new family of polymer blends, 4) to evaluate the toxicity of these elastomeric blends in vitro, and 5) to conduct in vivo animal trials.

References

1. T. Freier "Biopolyesters in tissue engineering applications," *Polymers for Regenerative Medicine*, pp. 1–61, 2006. 2006:1–61.
2. C.J. Spaans, V.W. Belgraver, O. Rienstra J.H. De Groot, A.J. Pennings, "Biomedical polyurethane-amide, its preparation and use," Google Patents; 2001.
3. J.E. Dumas, T. Davis, G.E. Holt, T. Yoshii, D.S. Perrien, J.S. Nyman, *et al.*, "Synthesis, characterization, and remodeling of weight-bearing allograft

bone/polyurethane composites in the rabbit," *Acta Biomaterialia*, Vol. 6, pp. 2394–406, 2010.

4.   T. Yoshii, A.E. Hafeman, J.S. Nyman, J.M. Esparza, K. Shinomiya, D.M. Spengler. *et al.* "A sustained release of lovastatin from biodegradable, elastomeric polyurethane scaffolds for enhanced bone regeneration," *Tissue Eng Part A*, Vol. 16, pp. 2369–79, 2010.

5.   D. Byrom, Polyhydroxyalkanoates. Plastic from microbes: microbial synthesis of polymers and polymer precursors" Hanser Munich. 1994:5–33.

6.   T. Freier, "Biopolyesters in tissue engineering applications," *Polymers for Regenerative Medicine*, pp. 1–61, 2006.

7.   M. Lemoigne, "Production d'acide β-oxybutyrique par certaines bactéeries du groupe du Bacillus subtilis," *CR Hebd Seances Acad Sci.*, Vol. 176, p. 1761, 1923..

8.   M. Lemoigne, "Etudes sur l-autolyse microbienne origigine de l'acide beta-oxybutyrique forme par autolyse," *Ann Inst Pasteur*, Vol. 41, pp. 148–65, 1927.

9.   A. Steinbuchel, "Perspectives for biotechnological production and utilization of biopolymers: Metabolic engineering of polyhydroxyalkanoate biosynthesis pathways as a successful example," *Macromolecular Bioscience*, Vol. 1, pp. 1–24, 2001.

10. R.W. Lenz, R.H. Marchessault, "Bacterial polyesters: biosynthesis, biodegradable plastics and biotechnology," *Biomacromolecules*, Vol. 6, pp. 1–8, 2005.

11. Q.Z. Chen, S.L. Liang, G.A. Thouas, "Elastomeric biomaterials for soft tissue engineering," Progress in Polymer Science, 2012; doi.org/10.1016/j.progpolymsci.2012.05.003.

12. L.L. Madison, G.W. Huisman"Metabolic engineering of poly(3-hydroxyalkanoates): From DNA to plastic," *Microbiology and Molecular Biology Reviews*, Vol. 63, pp. 21–+, 1999.

13. M. Zinn, B. Witholt, T. Egli, "Occurrence, synthesis and medical application of bacterial polyhydroxyalkanoate," *Advanced Drug Delivery Reviews*, Vol. 53, pp. 5–21, 2001.

14. K. Sudesh, H. Abe, Y. Doi, "Synthesis, structure and properties of polyhydroxyalkanoates: biological polyesters," *Progress in Polymer Science*, Vol. 25, pp. 1503–55, 2000.

15. T.D. Hirt, P. Neuenschwander, U.W. Suter, "Telechelic diols from poly [(R)-3-hydroxybutyric acid] and poly {[(R)-3-hydroxybutyric acid]-co-[(R)-3-hydroxyvaleric acid]}," *Macromolecular Chemistry and Physics*, Vol. 197, pp. 1609–14, 2003.

16. T.D. Hirt, P. Neuenschwander, U.W. Suter, "Synthesis of degradable, biocompatible, and tough block-copolyesterurethanes," *Macromolecular Chemistry and Physics*, Vol. 197, pp. 4253–68, 2003.

17. L. Timbart, E. Renard, V. Langlois, P. Guerin, "Novel Biodegradable Copolyesters Containing Blocks of Poly (3-hydroxyoctanoate) and Poly

(ε-caprolactone): Synthesis and Characterization," *Macromolecular bioscience*, Vol. 4, pp. 1014–20, 2004.

18. F. Ravenelle, R.H. Marchessault, "One-step synthesis of amphiphilic diblock copolymers from bacterial poly ([R]-3-hydroxybutyric acid)," *Biomacromolecules*, Vol. 3, pp. 1057–64, 2002.

19. S. Hiki, M. Miyamoto, Y. Kimura, "Synthesis and characterization of hydroxy-terminated [RS]-poly (3-hydroxybutyrate) and its utilization to block copolymerization with l-lactide to obtain a biodegradable thermoplastic elastomer," *Polymer*, Vol. 41, pp. 7369–79, 2000.

20. J. Martinez-Salazar, M. Sanchez-Cuesta, P. Barham, A. Keller, "Thermal expansion and spherulite cracking in 3-hydroxybutyrate/3-hydroxyvalerate copolymers," *Journal of materials science letters*, Vol. 8, pp. 490–2, 1989.

21. G. De Koning, P. Lemstra, "Crystallization phenomena in bacterial poly [(R)-3-hydroxybutyrate]: 2. Embrittlement and rejuvenation," *Polymer*, Vol. 34, pp. 4089–94, 1993.

22. S. Kusaka, T. Iwata, Y. Doi, "Microbial synthesis and physical properties of ultra-high-molecular-weight poly [(R)-3-hydroxybutyrate]," *Journal of Macromolecular Science, Part A Pure and Applied Chemistry*, Vol. 35, pp. 319–35, 1998.

23. Y. Aoyagi, Y. Doi, T. Iwata. "Mechanical properties and highly ordered structure of ultra-high-molecular-weight poly [(R)-3-hydroxybutyrate] films: Effects of annealing and two-step drawing," *Polymer Degradation and Stability*, Vol. 79, pp. 209–16, 2003.

24. M. Avella, E. Martuscelli, M. Raimo, "Review Properties of blends and composites based on poly (3-hydroxy) butyrate (PHB) and poly (3-hydroxybutyrate-hydroxyvalerate)(PHBV) copolymers," *Journal of Materials Science*, Vol. 35, pp. 523–45, 2000.

25. Y. Doi, S. Kitamura, H. Abe, "Microbial synthesis and characterization of poly (3-hydroxybutyrate-co-3-hydroxyhexanoate)," *Macromolecules*, Vol. 28, pp. 4822–8, 1995.

26. Y. Saito, Y. Doi, "Microbial synthesis and properties of poly (3-hydroxybutyrate-co-4-hydroxybutyrate) in *Comamonas acidovorans*," *International journal of biological macromolecules*, Vol. 16, pp. 99–104, 1994.

27. Y. Saito, S. Nakamura. M. Hiramitsu, "Microbial synthesis and properties of poly (3-hydroxybutyrate-co-4-hydroxybutyrate)," *Polymer International*, Vol. 39, pp. 169–74, 1999.

28. T. Iwata, Y. Doi, Mechanical Properties of Uniaxially Cold-Drawn Films of Poly [(R)-3-hydroxybutyrate] and Its Copolymers. Macromolecular Symposia: Wiley Online Library, pp. 11–20. 2005.

29. J. Josefine Fischer, Y. Aoyagi, M. Enoki, Y. Doi, T. Iwata, "Mechanical properties and enzymatic degradation of poly ([R]-3-*hydroxybutyrate-co-*[R]-3-hydroxyhexanoate) uniaxially cold-drawn films," *Polymer Degradation and Stability*, Vol. 83, pp. 453–60, 2004.

30. Y. Aoyagi, K. Yamashita, Y. Doi, "Thermal degradation of poly [(*R*)-3-hydroxybutyrate], poly [ε-caprolactone], and poly [(*S*)-lactide]," *Polymer Degradation and Stability*, Vol 76, pp. 53–9, 2002.

31. T. Iwata, K. Tsunoda, Y. Aoyagi, S. Kusaka, N. Yonezawa, Y. Doi, "Mechanical properties of uniaxially cold-drawn films of poly ([*R*]-3-hydroxybutyrate)," *Polymer Degradation and Stability.*, Vol. 79, pp. 217–24, 2003.

32. I. Engelberg, J. Kohn, "Physico-mechanical properties of degradable polymers used in medical applications: a comparative study," *Biomaterials*, Vol. 12, pp. 292–304, 1991.

33. B. Ramsay, V. Langlade, P. Carreau, J. Ramsay, "Biodegradability and mechanical properties of poly-(beta-hydroxybutyrate-co-beta-hydroxyvalerate)-starch blends," *Applied and environmental microbiology*, Vol. 59, pp. 1242–6, 1993.

34. J. Asrar, H.E. Valentin, P.A. Berger, M. Tran, S.R. Padgette, J.R. Garbow, "Biosynthesis and properties of poly (3-hydroxybutyrate-co-3-hydroxyhexanoate) polymers," *Biomacromolecules*, Vol. 3, pp. 1006–12, 2002.

35. K. Sudesh, H. Abe, Practical guide to microbial polyhydroxyalkanoates: ISmithers; 2010.

36. E. Blümm, O. Owen, "Miscibility, crystallization and melting of poly (3-hydroxybutyrate)/poly (L-lactide) blends," *Polymer*, Vol. 36, pp. 4077–81, 1995.

37. N. Koyama, Y. Doi, "Miscibility, thermal properties, and enzymatic degradability of binary blends of poly [(R)-3-hydroxybutyric acid] with poly (ε-caprolactone-co-lactide)," *Macromolecules*, Vol. 29, pp. 5843–51, 1996.

38. M. Avella, E. Martuscelli, "Poly-d-(−)(3-hydroxybutyrate)/poly (ethylene oxide) blends: phase diagram, thermal and crystallization behaviour," *Polymer*, Vol. 29, pp. 1731–7, 1988.

39. Y. Azuma, N. Yoshie, M. Sakurai, Y. Inoue, R. Chûjô, "Thermal behaviour and miscibility of poly (3-hydroxybutyrate)/poly (vinyl alcohol) blends," *Polymer*, Vol. 33, pp. 4763–7, 1992.

40. Y. He, T. Masuda, A. Cao, N. Yoshie, Y. Doi, Y. Inoue, "Thermal, Crystallization, and Biodegradation Behavior of Poly (3-hydroxybutyrate) Blends with Poly (butylene succinate-co-butylene adipate) and Poly (butylene succinate-co-. EPSILON.-caprolactone)," *Polymer journal*, Vol. 31, pp. 184–92, 1999.

41. N. Lotti, M. Pizzoli, G. Ceccorulli, M. Scandola, "Binary blends of microbial poly (3-hydroxybutyrate) with polymethacrylates," *Polymer*, Vol. 34, pp. 4935–40, 1993.

42. F. Gassner, A. Owen, "On the physical properties of BIOPOL/ethylene-vinyl acetate blends," *Polymer*, Vol. 33, pp. 2508–12, 1992.

43. P.A. Holmes, F.M. Willmouth, A.B. Newton, Polymer blends containing polymer of. beta.-hydroxybutyric acid and chlorine or nitrile group containing polymer. Google Patents; 1983.

44. R. Bhatt, D. Shah, K.C. Patel, U. Trivedi, "PHA-rubber blends: Synthesis, characterization and biodegradation," *Bioresource Technology*, Vol. 99, pp. 4615–20, 2008.

45. S.H. Imam, L. Chen, S.H. Gorden, R.L. Shogren, D. Weisleder, R.V. Greene, "Biodegradation of injection molded starch-poly(3-hydroxybutyrate-co-3-hydroxyvalerate) blends in a natural compost environment," *Journal of Environmental Polymer Degradation*, Vol. 6, pp. 91–8, 1998..

46. Y. Kumagai, Y. Doi, "Enzymatic degradation and morphologies of binary blends of microbial poly (3-hydroxy butyrate) with poly (ε-caprolactone), poly (1, 4-butylene adipate and poly (vinyl acetate)," *Polymer Degradation and Stability*, Vol. 36, pp. 241–8, 1992.

47. D. Parra, J. Fusaro. F. Gaboardi, D. Rosa, "Influence of poly (ethylene glycol) on the thermal, mechanical, morphological, physical–chemical and biodegradation properties of poly (3-hydroxybutyrate)" *Polymer Degradation and Stability*, Vol. 91, pp. 1954–9, 2006.

48. P. Coussot-Rico, G. Clarotti, A. Ait Ben Aoumar, A. Najimi, J. Sledz, F. Schue, *et al.* "Relation entre l'energie de surface de membranes a base de polyhydroxyalkanoates et l'adsorption des proteines sur ces memes membranes," *European polymer journal*. Vol. 30, pp. 1327–33, 1994.

49. K. Zhao, Y. Deng, J.C. Chen, G.Q. Chen, "Polyhydroxyalkanoate (PHA) scaffolds with good mechanical properties and biocompatibility," *Biomaterials*, Vol. 24, pp. 1041–5, 2003.

50. Z. Zheng, F.F. Bei, H.L. Tian, G.Q. Chen, "Effects of crystallization of polyhydroxyalkanoate blend on surface physicochemical properties and interactions with rabbit articular cartilage chondrocytes," *Biomaterials*, Vol. 26, pp. 3537–48, 2005.

51. L. Zhang, Z. Zheng, J. Xi, Y. Gao, Q. Ao, Y. Gong, *et al.* "Improved mechanical property and biocompatibility of poly (3-hydroxybutyrate-co-3-hydroxyhexanoate) for blood vessel tissue engineering by blending with poly (propylene carbonate)," *European polymer journal*, Vol. 43, pp. 2975–86, 2007..

52. J.W. Park, Y. Doi, T. Iwata, "Uniaxial drawing and mechanical properties of poly [(R)-3-hydroxybutyrate]/poly (L-lactic acid) blends," *Biomacromolecules*, Vol. 5, pp. 1557–66, 2004.

53. M. Abbate, E. Martuscelli, G. Ragosta, G. Scarinzi, "Tensile properties and impact behaviour of poly (D (−) 3-hydroxybutyrate)/rubber blends," *Journal of Materials Science*, Vol. 26, pp. 1119–25, 1991.

54. M. Avella, E. Martuscelli, G. Orsello, M. Raimo, B. Pascucci, "Poly (3-hydroxybutyrate)/poly (methyleneoxide) blends: thermal, crystallization and mechanical behaviour," *Polymer*, Vol. 38, pp. 6135–43, 1997.

55. S. Cimmino, P. Iodice, C. Silvestre, F. Karasz, "Atactic poly (methyl methacrylate) blended with poly (3-D (−) hydroxybutyrate): Miscibility and mechanical properties," *Journal of Applied Polymer Science*, Vol. 75, pp. 746–53, 2000.

56. G.A. Skarja, K.A. Woodhouse, "Synthesis and characterization of degradable polyurethane elastomers containing an amino acid-based chain extender," *Journal of Biomaterials Science-Polymer Edition*, Vol. 9, pp. 271–95, 1998.

57. D. Barrett, M. Yousaf, "Design and Applications of Biodegradable Polyester Tissue Scaffolds Based on Endogenous Monomers Found in Human Metabolism," *Molecules*, Vol. 14, pp. 4022–50, 2009.

58. D.W. Grijpma, G.J. Zondervan, A.J. Pennings, "High molecular weight copolymers of l-lactide and ε-caprolactone as biodegradable elastomeric implant materials," *Polymer Bulletin*, Vol. 25, pp. 327–33, 1991.

59. L. Sipos, M. Zsuga, G. Deák, "Synthesis of poly(L-lactide)-block-polyisobutylene-block-poly(L-lactide), a new biodegradable thermoplastic elastomer," *Macromolecular Rapid Communications*, Vol. 16, pp. 935–40, 1995.

60. H-J Tao, W.J. MacKnight, K.D. Gagnon, R.W. Lenz, S.L. Hsu , "Spectroscopic Analysis of Chain Conformation Distribution in a Biodegradable Polyester Elastomer, Poly(.beta.-hydroxyoctanoate)," *Macromolecules*, Vol. 28, pp. 2016–22, 1995.

61. J. Kylmä, M. Hiljanen-Vainio, J. Seppälä, "Miscibility, morphology and mechanical properties of rubber-modified biodegradable poly(ester-urethanes)," *Journal of Applied Polymer Science*, Vol. 76, pp. 1074–84, 2000.

62. S. Hiki, M. Miyamoto, Y. Kimura, "Synthesis and characterization of hydroxy-terminated [RS]-poly(3-hydroxybutyrate) and its utilization to block copolymerization with -lactide to obtain a biodegradable thermoplastic elastomer," *Polymer*, Vol. 41, pp. 7369–79, 2000.

63. Z. Zhang, D.W. Grijpma, J. Feijen, "Triblock Copolymers Based on 1,3-Trimethylene Carbonate and Lactide as Biodegradable Thermoplastic Elastomers," *Macromolecular Chemistry and Physics*, Vol. 205, pp. 867–75, 2004.

64. S.I. Jeong, B-S Kim, S.W. Kang, J.H. Kwon, Y.M. Lee, S.H. Kim, *et al.* "In vivo biocompatibilty and degradation behavior of elastic poly(-lactide-co-[var epsilon]-caprolactone) scaffolds," *Biomaterials*, Vol. 25, pp. 5939–46, 2004.

65. K. Odelius, P. Plikk, A-C Albertsson, "Elastomeric Hydrolyzable Porous Scaffolds: copolymers of Aliphatic Polyesters and a Polyether-ester," *Biomacromolecules*, Vol. 6, pp. 2718–25, 2005.

66. M.K. Hassan, K.A. Mauritz, R.F. Storey, J.S. Wiggins, "Biodegradable aliphatic thermoplastic polyurethane based on poly(ε-caprolactone) and L-lysine diisocyanate," *Journal of Polymer Science Part A: Polymer Chemistry*, Vol. 44, pp. 2990–3000, 2006.

67. R.G. Sinclair, In: Pat. U, editor.1977.

68. G.G. Pitt, M.M. Gratzl, G.L. Kimmel, J. Surles, A. Sohindler, "Aliphatic polyesters II. The degradation of poly (DL-lactide), poly ([epsilon]-caprolactone), and their copolymers in vivo," *Biomaterials*, Vol. 2, pp. 215–20, 1981.

69. R.F. Storey, T.P. Hickey "Degradable polyurethane networks based on d,l-lactide, glycolide, ε-caprolactone, and trimethylene carbonate homopolyester and copolyester triols," *Polymer*, Vol. 35, pp. 830–8, 1994.

70. D.J. Stuckey, H. Ishii, Q.Z. Chen, A.R. Boccaccini, U. Ha nsen, C.A. Carr, *et al.* "Magnetic Resonance Imaging Evaluation of Remodeling by Cardiac Elastomeric Tissue Scaffold Biomaterials in a Rat Model of Myocardial Infarction," *Tissue Engineering Part A*, Vol. 16, pp. 3395–402, 2010.

71. B. Amsden, "Curable, biodegradable elastomers: emerging biomaterials for drug delivery and tissue engineering," *Soft Matter*, Vol. 3, pp. 1335–48, 2007.

13

Biomimetic Molecularly Imprinted Polymers as Smart Materials and Future Perspective in Health Care

Mohammad Reza Ganjali*, Farnoush Faridbod and Parviz Norouzi

Center of Excellence in Electrochemistry, University of Tehran, Tehran, Iran
Biosensor Research Center, Endocrinology and Metabolism Molecular-Cellular Sciences Institute, Tehran University of Medical Sciences, Tehran, Iran

Abstract

Molecularly imprinted polymers (MIPs) are smart synthetic receptors, characterized by a high selectivity for a selected template. They are known as biomimetic molecules because of their ability to mimic the behavior of a bio-recognition element, such as antibodies or biological receptors. MIPs can recognize both biological and chemical molecules including amino acids and proteins, nucleotides and their derivatives, and many drugs and poisons. MIPs have a wide range of applications and are used in separation and purification sciences, chemical sensors/biosensors, antibodies and receptors systems, and especially in drug delivery. In comparison with bio-molecules, MIPs are cheaper materials, more robust with longer shelf life, more stable in high temperatures and pressures and are chemically inert towards acids, bases, metal ions and organic solvents.

From the time of their discovery up to now, many various biological applications have been widely developed, especially in pharmacology and clinical chemistry and medicine. Even many efforts have been performed for miniaturized the size of MIPs to nano-scales (Nano-MIPs). However, in spite of these progresses in MIPs technology, several aspects have to be further studied to let this advanced materials are used in vivo or in point-of-care devices in the near future.

Hence, in this chapter, recent achievements in design, synthesis, and development of MIPs for use in drug delivery or in healthcare devices based on sensor/biosensor systems will be discussed.

*Corresponding author: ganjali@khayam.ut.ac.ir, ganjali@gmail.com

Ashutosh Tiwari (ed.) Advanced Healthcare Materials, (465–492) 2014 © Scrivener Publishing LLC

Keywords: Molecularly imprinted polymers, biomimetic molecules, drug delivery, sensors, healthcare devices

13.1   Molecularly Imprinted Polymer Technology

Molecularly imprinted polymer (MIP) is a smart artificial receptor, which has a high affinity to a special molecule. MIP technology deals with designing and synthesis of a polymeric supra-molecule according to a template molecule.

Perhaps the concept of using a molecular template to generate recognition sites for selective reaction within a polymeric network has come from Polyakov's studies on silica matrixes in 1931 [1].

Molecularly imprinted polymers (MIPs) are highly stable synthetic polymers that have some recognition sites in a specific three-dimensional (3-D) shapes for target molecules [2].

This selective recognition of molecularly imprinted polymers depends on the morphology and the number of functional groups of the template [3]. Increasing the number of interacting functional groups in a molecule will increase the selectivity of MIP. Accessibility of the interaction sites depends on the morphology of the MIP.

MIPs have also the ability to recognize one or a group of analytes depending on the choice of template. The selectivity of the imprinted polymer depends on the applications for which they are used. For example in separation of chiral molecule, synthesis of well-defined sites with high selectivity is required. But in drug delivery, other factors such as the ease of diffusion of analytes in and out of the polymer network are important [4, 5].

In comparison with natural receptors, molecularly imprinted polymers offer some advantages such as cost effectiveness of the preparation process, ability to design and synthesize most compounds, possessing similar affinity to natural biomolecules but often with better specificity, ability to work in organic solvents, physical robustness, stability at low/high pHs, pressure and temperature, compatibility with micro-fabrication.

13.2   Synthesis of MIPs

From the time MIPs discovered up to now, their synthesis have been discussed several times in reviews, chapters and books in detail [6–9].

Molecular imprinted polymer is synthesized by a reaction of a template molecule, functional monomers, a cross-linking monomer (or two), a polymerization initiator in a porogenic solvent. During polymerization

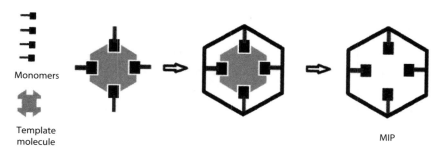

Monomers

Template
molecule

MIP

Figure 13.1 Schematic diagram of synthesis of MIPs.

process, a complex forms between the template molecule and the functional monomer, and the complex is surrounded by addition of cross-linker, and yielding a three-dimensional polymer network where the template molecule is trapped after completion of polymerization. By washing, the template molecule is eliminated to create a cavity matching to the template size, shape, and molecular interactions. Figure 13.1 shows a schematic diagram of synthesis of MIPs.

To obtain MIPs possessing a particular selectivity or specific morphology, the selection of the reagents (monomers or functional monomers, cross-linker and solvents or porogens), which are used in imprinting is very important. Also, the conditions used in the preparation and polymerization process should be controlled carefully.

General monomers which are used as building block of the polymer network are acrylic acid, methacrylic acid, methylmethacrylic acid, acryl amide, methacrylamide, styrene, 4-ethystyrene, 4-vinylpridine, p-vinylbenzoic acid.

Common crosslinkers are ethylene glycol dimethacrylate (EGDMA), divinylbenzene, N,N-methylenediacrylamide, N,N-Phenylenebisacrylamide, 2,6-Bisacrylamidopyridine, Trimethylolpropane trimethacrylate, and tetramethylene dimethacrylate. The amount of a crosslinking agent affects the rigidity of MIPs, while the nature of this agent significantly affects the physicochemical properties of a polymer matrix. Azobisisobutyronitrile (AIBN) is the most efficient initiator.

The nature of the solvent has a great effect on the stage of MIP synthesis and on the stage of the repeated binding of target molecules. As a rule, MIPs exhibit a better capability for molecular recognition in solvents with low permittivity. In these solvents, the monomer–template non-covalent interactions are stronger than in polar solvents.

MIPs were mostly prepared by free-radical polymerization, resulting in non-conducting polymers. This polymerization requires the presence of a polymerization initiator and light or heat to induce it. MIPs are generally prepared by thermal bulk polymerization. Free-radical polymerization is inhibited in the presence of oxygen; the reaction mixture is purged with nitrogen or argon in order to remove oxygen. The rate of radical polymerization depends on the nature and concentration of the initiator.

The synthesis of MIPs can also be performed at reduced temperatures (from 15 to $-20°C$) while initiating the polymerization reaction by UV irradiation (366 nm). In this case, MIPs with a greater capacity for molecular recognition were obtained. This was explained by the fact that the complex of a monomer and a template in the prepolymerization mixture is more stable at low temperatures [10].

Another method for preparation of conducting MIPs is electrochemical synthesis. Sharma *et al.* have discussed on this method extensively in a review article [11]. In this way, MIP film is prepared by electropolymerization of conducting monomers on the surface of a transducer (e.g., a gold electrode or glassy carbon electrode). MIP in this way has superior properties which is suitable for the use in sensors. Simplicity and speed of preparation, easy control of the film thickness and morphology, high reproducibility and the possibility of polymer preparation and operation in aqueous solutions are the advantages of this method. Figure 13.2 shows a schematic of electropolymerization synthesis of MIP.

For many biomolecules and biomacromolecules such as glucose, fructose, neomycine, kanamycine, lysozyme, histamine, atropine, dopamine,

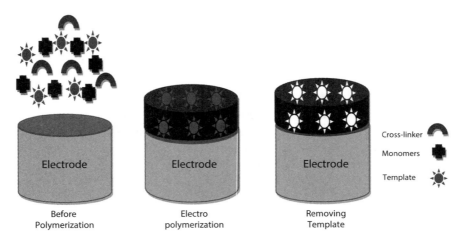

Figure 13.2 Synthesis of conducting MIPs by electropolymerization on the transducer surface.

morphine, folic acid electropolymerized conducting MIPs on Au, Pt, Glassy carbon electrodes have been reported [11].

In a recent review by Cheong *et al.*, synthesis of the MIPs for enantio-selective recognition has been discussed [12].

MIP can be prepared as monoliths, particles, nanoparticles, in the membrane or on the surface of a solid.

In general, MIPs can be synthesized in a covalent or a non-covalent way.

13.2.1   Molecular Covalent Imprinting Polymer

In this way the template molecule reacts chemically with the building blocks of the polymer (monomer). After the polymerization process is complete, the chemical bond is cleaved to obtain free selective binding sites. In this method a reversible covalent linkage occurs (Figure 13.3). The ability to fix template in place during polymerization, lower dispersity in binding sites, ability to perform in any solvent are the advantages of this method. However, due to the low kinetic rate of re-binding, covalent binding is not preferred for molecular recognitions.

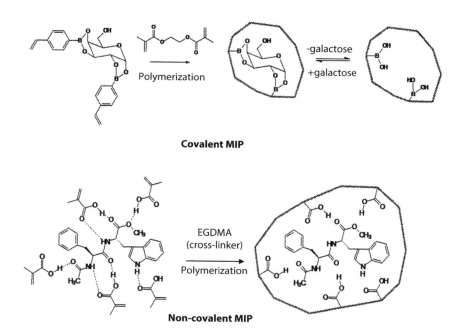

Figure 13.3 Comparison between covalent and non-covalent MIPs [13,14].

13.2.2   Molecular Non-Covalent Imprinting Polymer

Non-covalent imprinting polymer is based on self-assembly of the template molecule to the functional groups of monomers (complexation between monomer and template occurs). Then, the polymerization process is done in presence of a cross-linker. After that, the template molecule is removed from the polymer and the resulting polymer will contain a recognition site with functional groups in a defined 3-D array. Easy to remove template from polymer, good recovery of valuable templates and accessible binding sites, very large number of templates amenable to non-covalent imprinting, rapid kinetics of re-binding is advantages of this method.

The drawbacks of this method are the inability to fix the template in place during polymerization, polydispersity in binding sites, poor definition, the method generally requires low-polarity aprotic solvents, and it is incompatible with aqueous polymerizations.

13.2.3   Nano-Molecularly Imprinted Polymers (Nano-MIPs)

Most cases of the imprinted polymeric matrix is synthesized by polymerization of methacrylic acid monomers in bulk materials. Thus, a porous solid or monolith MIP is formed. MIPs made in this way need to grind to prepare size fractions suitable for separation and extraction applications. Other preparation methods such as suspension, emulsion, and precipitation polymerization for synthesis of MIPs with particle sizes of about 200 nm–100 μm have been used. Despite the fact that MIPs have molecular recognition ability similar to that of biological receptors, traditional bulky MIP materials usually exhibit a low binding capacity and slow binding kinetics to the target species. Moreover, the MIP materials lack the signal-output response to analyte binding events, when used as recognition elements in chemo/biosensors or bioassays [15]. The molecular imprinting nanotechnologies are expected to considerably increase the molecular affinity of MIP materials, and thus provide a wider range of applications approaching to biological receptors [16]. In the nano-structured, imprinted materials most of imprinted sites are situated at the surface or in the proximity of surface. Therefore, the forms of imprinted materials are expected to greatly improve the binding capacity, kinetics and site accessibility of imprinted materials. Compared with the imprinted films and surface-imprinted materials, the imprinted nanomaterials have shown a higher affinity and sensitivity to target analyte and a more homogeneous distribution of recognition sites [17].

However, recently researchers tried to prepare MIP particles less than 100 nm in size [18, 19]. MIP nanoparticles, which are called "Nano-MIPs," are compatible with a range of solvents and soluble as colloids, compatible with biological environments, and have a efficient diffusion path for binding and release of the template molecules. Moreover, using Nano-MIPs as direct replacements for antibodies and other natural binders in biosensors and immunoassays becomes a real possibility [20].

There are different ways to prepare materials at the nanoscale; these can be described either as top-down or bottom-up approaches [21]. The top-down method involves grinding or other means of reducing the size of the system (such as mini-emulsion polymerization). More control, however, can be exercized by using a bottom-up approach, assembling smaller units, but in a way that only produces nano-objects.

Molecularly imprinted polymer nanoparticles can be prepared by microemulsion polymerization according to the work reported by A.S. Belmont and coworkers [22].

Piletsky *et al.* reported the MIP nanoparticles as receptors, which are synthesized using the second approach [20, 21].

13.3   Application of MIPs

Molecularly imprinted polymer is a general technology which has a wide range of applications (Figure 13.4).

They are used in an increasing number of applications categorized as follows:

- As adsorbent in separation and purification sciences (e.g., chromatography even in chiral separation; selective permeable membranes; in selective solid phase extraction as adsorbents for separation of drugs, biomolecules,amino acids, poisons, pesticides, and many other organic compound in various clinical and environmental applications)
- As receptor or plastic antibody (e.g. antibody–receptor binding site mimics in recognition and assay systems or in microfluidic devices)
- As recognition elements in chemical sensors/biosensors
- As catalysts (enzyme mimics in catalytic applications or for facilitation of chemical synthesis
- As drug delivery systems (as drug/fragrance release matrices)

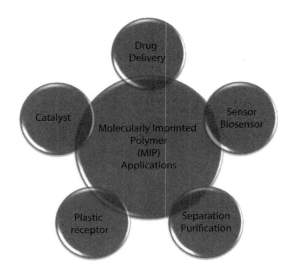

Figure 13.4 Schematic diagram of MIPs various field of applications.

13.4   Biomimetic Molecules

There is a rich and long history of gaining inspiration from nature for the design of practical materials and systems. It can be stated that the greatest impact of the biological sciences on technology may prove to be through the lessons learned by mimicking biological function.

Molecular biomimetics is an emerging field in which hybrid technologies are developed by using the tools of molecular biology and chemistry. Taking lessons from biology, polypeptides can now be genetically engineered to specifically bind to selected inorganic compounds for applications in nano- and biotechnology.

MIPs are one of the most promising areas of biomimetics. They are able to mimic the behavior of a bio-recognition element, such as antibodies or biological receptors. MIPs can recognize both biological and chemical molecules including amino acids and proteins, nucleotides and their derivatives, many drugs, and poisons.

In comparison with bio-molecules, MIPs are cheaper materials, more robust with longer shelf life, more stable in high temperatures and pressures, and are chemically inert towards acids, bases, metal ions, and organic solvents.

This technique is based on the system used by enzymes for substrate recognition, which is called the "lock and key" model (Figure 13.5).

Figure 13.5 "Lock and key" model of MIPs.

13.5   MIPs as Receptors in Bio-Molecular Recognition

Molecular recognition is the specific interaction between two or more molecules through non-covalent bonding such as metal coordination, hydrogen bonding, hydrophobic forces, van der Waals forces, π-π interactions, charge-dipole interaction, halogen bonding, electrostatic and electromagnetic interactions [23]. Besides these interactions, solvents have a dominant role in molecular recognitions [24, 25] Molecular recognition is also called "host-guest" interaction [26].

Chemists have demonstrated that artificial supramolecular systems can be designed that exhibit molecular recognition. One of the earliest examples of such systems is crown ethers which are able to bind selectively with specific cations. Cyclodextrines, calixarens, podanta, cryptands, dendrimers, fullerenes, and many other molecules have shown the recognition effect. Nowadays, a number of artificial systems have since been established. Molecularly imprinted polymers, a class of synthetic supramolecules, are an excellent choice for molecular recognition.

In general, molecular recognition can be divided into static molecular recognition and dynamic ones. In the static molecular recognition a kind of 1:1 complexation interaction occurs between host molecule and its guest like a key and a keyhole. To achieve advanced static molecular recognition, it is necessary to make recognition sites that are specific for guest molecules.

In dynamic molecular recognition, there are two or even more binding sites on the host molecule and two or more guest molecules can interact with the host. However, the first guest which binds to the first active site of a host affects the binding of a second guest with a second binding site. It can cause an increase (positive effect) or decrease (negative effect) in the association constant of second interaction. The dynamic molecular recognition can be seen in many biological systems, e.g., the interaction of albumin with some organic and inorganic molecules.

The most important interaction that occurs in biological systems is molecular recognition. This phenomenon can be observed where a ligand binds to a receptor or an antibody interacts with its antigen or even between DNA/RNA with proteins.

Smart materials like MIPs with biorecognition functions have enormous potential in the development of a new generation of stable biomimetic sensors, affinity separation matrices and in drug delivery systems. Replacement of antibodies with biomimetic MIPs as the binding component can be useful in immunoassays.

A comparison of the properties of MIPs with antibodies shows the clear advantages of MIPs. No need to the animals for producing, resistance to microorganism, reusability, longer lifetime, better shelf life, easier storage and the less production price are the most important superiority of MIPs to antibodies. Also, artificial binding sites of MIPs have the same features as the antibody binding sites, showing binding reversibility, enhanced selectivity, high affinity constant and a significant polyclonality (non-covalent MIPs) or monoclonality (covalent MIPs).

Piletsky in Cranfield University synthesized MIP nanoparticles and called them "plastic antibodies." They used "living" initiators, called photoiniferters, which allow the polymerization to be switched on and off by the presence or absence of UV light, as well as precipitation polymerization from diluted aqueous solution [20, 21].

Baggiani et al. reviewed the molecular imprinted polymers used as synthetic receptors for the analysis of some important toxins in human healthcare (myco- and phyco-toxins) [27]. As shown in this review, molecular imprinting can be successfully used to prepare intelligent materials for detection, clean-up and preconcentration of natural toxins in complex samples.

13.6 MIPs as Sensing Elements in Sensors/Biosensors

A sensor/biosensor is a simple device for fast and selective analysis of various species (called analyte) ranges from ions to organic molecules and biomacromolecules. Each sensor or biosensor is composed of three important parts: a sensing matrix, a transducer, and a recorder. These parts change a chemically important signal to a readable (electronical) signal. Sensor/biosensors are divided by two ways. One is according to the kind of transduction of chemical signal (electrochemical, optical, mass, and thermal sensors/biosensors) and the other one is based on the kind of sensing elements used in sensing layers (enzymatic biosensors, immuno-sensors, DNA sensors, MIP sensors, etc.).

The element which causes a selective, even a specific response, is a sensing material. Many supra-molecules and bio-molecules possessing recognition properties have been applied as sensing elements in designing many sensors/biosensors. MIPs are excellent sensing materials for use in sensor/biosensor systems.

To date, molecularly imprinted polymers have been successfully used with most types of transduction platforms (Table 13.1) and various methods have been used the platform with the polymer.

MIPs also can be used for optical sensing. A number of fluorescent monomers have been used to prepare fluorescent MIPs. In another approach, a fluorophore linked to the MIP structure.

Wang *et al.* developed a system that responded to the binding event with a significant fluorescence intensity change without the use of an external quencher [28]. Anthracene-containing monomer that was substituted with a boronic acid-containing group was used. When the template, d-fructose, was re-introduced into the system, a large change in fluorescence was observed. This was attributed to the reformation of the boronic ester with the cis-diol of the fructose (Figure 13.6). Other optical MIPs with different detection techniques have been also reported [29].

Piletsky *et al.* used a porous polymer, computationally designed, to interact with homoserine lactones. They prepared a "switch off" bioluminescence sensory system [30].

Table 13.1 summarizes the work of some researchers who used MIPs as sensing elements in sensory/biosensory systems.

13.7   MIPs as Drug Delivery Systems

One of the important applications of molecularly imprinted polymers is in modern drug delivery systems. MIPs are able to load a therapeutic drug and intelligently release it in response to the specific environment.

Drug delivery systems should be able to release the drug into the specific location and also control the rate of drug release. It means they cannot have any delay or speedup in the release. Hence, efficient drug delivery systems should provide a desired rate of delivery of the therapeutic dose at the most appropriate place in the living system. Drug delivery systems are usually used in case of some pharmaceutics which their consumption is difficult for the patients. By this way, the duration of pharmacological

Table 13.1 Some recent reports on MIPs for use as sensing elements in sensory/biosensory systems.

No	Template	The method of using MIP	Detection Technique	Detection Limit or linear range	[Ref.]
1	Cholesterol	A monolithic molecular imprinting sensor based on ceramic carbon electrode	Cyclic voltammetry	Detection limit: 1 nmol/L Linear range: 10–300 nmol/L	[31]
2	Metronidazole	Core-shell metronidazole-magnetic molecularly imprinted polymer (MMIP) was synthesized and then attached to the surface of magnetic glassy carbon electrode	Cyclic voltammetry	Detection limit: 1.6×10^{-8} mol/L Linear range: 5.0×10^{-8} -1.0×10^{-6} mol/L	[32]
3	Tramadol	Nano-molecularly imprinted polymer at $SiO_2@Fe_3O_4$ as the core and the supporting material	Square wave voltammetry	Linear range: 0.004- 20 µmol/ L	[33]
4	Chiral amino acides	Incorporation of molecularly imprinted materials into photonic crystals	Colorimetry	Linear range: 0.01–0.50 mmol/L.	[34]
5	Memantine	MIP was synthesized by precipitation polymerization, using memantine hydrochloride as a template molecule, methacrylic acid as a functional monomer, and ethylene glycol dimethacrylate as a cross-linker. The sensor was developed by dispersing the memantine imprinted polymer particles in dibutyl sebacate plasticizer and embedding in poly(vinyl chloride) matrix.	PVC membrane potentiometry	Detection Limit: 6.0×10^{-6} M Linear range: $10^{-5}–10^{-1}$ M	[35]
6	Heparin	Modification of heparin-imprinted polymer film onto a glassy carbon	Potentiometry	Detection limit: 0.001 µmol/L Linear range: 0.003–0.7 µmol/L	[36]

No	Template	The method of using MIP	Detection Technique	Detection Limit or linear range	[Ref.]
7	Dipyridamole	Molecularly imprinted polymer modified carbon paste electrode	Differential pulse adsorptive stripping voltammetry	Detection limit: 0.05 ng/ mL Linear range: 1.0–110 ng /mL	[37]
8	Glycoprotein	MIP was electropolymerized by o-phenylenediamine and 3-amino-phenylboronic acid monohydrate in the presence of template molecules (bovine serum albumin (BSA) on graphen- Au modified electrode	Voltammetry	Detection limit: 7.5×10^{-12} g /mL Linear range: 1.0×10^{-11} to 1.0×10^{-5} g/mL	[38]
9	Promethazine	Nano-imprinted polymers were synthesized by the ultrasonic assisted suspension polymerization in silicon oil. The MIP nanoparticles were then embedded in a carbon paste electrode	Voltammetry	Detection limit: 2.8×10^{-12} mol/L Linear ranges: 4×10^{-12} –1×10^{-10} and 1×10^{-9} –1×10^{-7} mol/L	[18]
10	Caffeine	Multiwalled carbon nanotubes and gold nanoparticles were first modified onto the glassy carbon electrode surface by potentiostatic deposition method. Then, o-aminothiophenol was assembled on the surface of the above electrode through Au-S bond. Then electropolymerization was done. During the assembled and electropolymerization processes, caffeine was embedded into the poly(o-aminothiophenol) (ATP) film through hydrogen bonding interaction between caffeine and ATP, forming an MIP electrochemical sensor.	Cyclic voltammetry, and differential pulse voltammetry	Detection Limit: 9.0×10^{-11} mol/ L Linear range: 5.0×10^{-10} to 1.6×10^{-7} mol/L	[39]

(Continued)

Table 13.1 (*Cont.*)

No	Template	The method of using MIP	Detection Technique	Detection Limit or linear range	[Ref.]
11	Epinephrine	Using the system of luminol-NaOH-H_2O_2 based on a graphene oxide-magnetite-molecularly imprinted polymer	Chemiluminescence	Detection limit: 1.09×10^{-9} mol/L Linear range: 1.04×10^{-7} – 7.06×10^{-3} mol/L	[40]
12	Dapsone	The molecularly imprinted polymers were coated at the surface of modified SiO_2 by the graft copolymerization	Flow injection chemi-luminescence	Detection limit: 5.27×10^{-7} mol/L Linear range: 1.0×10^{-6} to 1.0×10^{-4} mol/L	[41]
13	Clenbuterol	MIP was used in a PVC membrane	Potentiometry	detection limit: 7.0×10^{-8} mol/L Linear range: 1.0×10^{-7} to 1.0×10^{-4} mol/L	[42]
14	Chloramphenicol	A colloidal crystal template was first prepared from monodisperse SiO_2 nanospheres, then molecularly imprinted photonic polymer with numerous nano-cavities derived from the SiO_2	An inverse opal photonic crystal sensor	Linear range: 1 ng/mL to 1 mg/mL	[43]
15	Cocaine	A molecularly imprinted polymer containing a fluo-rescent moiety as the signalling group was formed and covalently attached to the distal end of an optical fiber	Fluorescence fiber-optic chemical sensor	Linear range: 0 - 500 µmol/L	[44]

No	Template	The method of using MIP	Detection Technique	Detection Limit or linear range	[Ref.]
16	Urea and Creatinine	Solvent evaporation processing of poly(ethylene-co-vinyl alcohol) is used to form molecularly imprinted polymers	Electrochemical impedance	Detection limit: 10 ng/mL Linear range: 0.02μg/mL to 3 μg/mL for urea and detection limit: 40 ng/mL with linear range from 0.05 mu g/mL to 2 μg/mL for creatinine	[45]
17	Acetylsalicylic acid	By co-polymerization of p-aminothiophenol (p-ATP) and HAuCl(4) on the Au electrode surface	Differential pulse voltamme-try, and cyclic voltammetry	Linear range: 1 nmol/L to 0.1 μmol L and 0.7 μmol/L to 0.1 mmol/L	[46]
18	Chlorpromazine	Methacrylic acid, 2-vinyl pyridine and 2-acrylamido-2-methyl-1-propanesulfonic acid based polymers. Molecularly imprinted particles were dispersed in 2-nitrophenyloctyl ether and entrapped in a poly(vinyl chloride) matrix	Potentiometry	Concentration range 1.0×10^{-4} to 1.0×10^{-2} mol/L	[47]
19	Salbutamol	poly(o-phenylenediamine) (POPD) film was pre-pared by the cyclic voltammetric deposition of o-phenylenediamine(OPD) with and without a template molecule, salbutamol, on the single-walled carbon nanotubes(SWNTs) modified electrodes	Amperometry	Detection limit: 6.08×10^{-8} mol/L Linear range: 7.94×10^{-8} – 1.36×10^{-5} mol/L	[48]

(Continued)

Table 13.1 (Cont.)

No	Template	The method of using MIP	Detection Technique	Detection Limit or linear range	[Ref.]
20	Diazepam	Mmolecularly imprinted films prepared on screen printed electrodes	Conductimetry	Detection limit: 0.008 mg/L Linear range: 0.04 to 0.62 mg/L	[49]
21	Caffeine	The MIP, embedded in the carbon paste electrode, functioned as a selective recognition element and a pre-concentrator agent for caffeine determination.	Voltammetry	Detection limit of 0.18 μg/mL	[50]
22	Glucose	An adduct of glucose and 4-vinylphenylboronic acid (VPBA) was synthesized by esterification and was then purified. The copolymer of the glucose/VPBA adduct and methylene bisacrylamide was grafted onto an indium tin oxide electrode surface. Glucose was washed out from the copolymer to obtain an MIP layer.	Electrochemical Enzyme free glucose sensor	Linear range: 0–900 mg/dl	[51]
23	Digoxin	Digoxin-specific bulk polymer was obtained by the UV initiated co-polymerisation of methacrylic acid and ethylene glycol dimethacrylate in acetonitrile as porogen. After extracting the template analyte, the ground polymer particles were mixed with plasti-cizer polyvinyl chloride to form a MIP membrane.	Optical sensor	Detection limit: 3.17×10^{-5} mg/L	[52]

No	Template	The method of using MIP	Detection Technique	Detection Limit or linear range	[Ref.]
24	Fenfluramine	The fenfluramine-imprinted polymer was prepared with acrylamide (AM) as functional monomer and ethylene glycol dimethacrylate (EGDMA) as cross-linker. Methyl and sulfonic group were introduced to rhodanine matrix, and a novel rhodanine ramification 3MORASP was synthesized and used as chemiluminescence reagent.	Chemiluminescence	Detection limit: 9.48×10^{-9} g/mL Linear range: 1.0×10^{-7} to 5.0×10^{-6} g/mL	[53]
25	Salicylic Acid	Electropolymerizing o-phenylenediamine on glassy carbon electrode in presence of template molecule (salicylic acid)	Square wave voltammetry	Detection limit 2×10^{-5} mol/L Linear range: 6×10^{-5}– 1×10^{-4} mol/L	[54]
26	Uracil and 5-Fluorouracil	The same MIP motif prepared from melamine and chloranil precursors was used for both analytes as a coating material for modification of a hanging mercury drop electrode (HMDE) by a drop-coating method using dimethylformamide casting Solution.	Voltammetry	Detection limit: for uracil 0.34 and 5-fluorouracil 0.26 ng/mL	[55]
27	Cetirizine	The MIP was synthesized by precipitation polymerization, using cetirizine dihydrochloride as a template molecule, methacrylic acid as a functional monomer and ethylene glycol dimethacrylate as a cross-linker.	Potentiometry	Detection limit: 7.0×10^{-7} mol/L Linear range: 1.0×10^{-6} to 1.0×10^{-2} mol/L	[56]

(Continued)

Table 13.1 (*Cont.*)

No	Template	The method of using MIP	Detection Technique	Detection Limit or linear range	[Ref.]
28	Creatinine	Based on a molecularly imprinted polymer-modified sol-gel film on graphite electrode	Differential pulse, cathodic stripping voltammetry	Detection limit: 0.37µg/mL Linear range 1.23 to 100 µg/mL	[57]
29	Hydroxyzine	MIP was synthesized using hydroxyzine dihydrochloride as a template molecule, methacrylic acid as a functional monomer and ethylene glycol dimethacrylat as a cross-linker.	Potentiometry	detection limit: 7.0×10^{-7} mol/L Linear range 1.0×10^{-6} to 1.0×10^{-1} mol/L	[58]
30	Terbutaline	Using terbutaline as the template, methacrylic acid as the functional monomer, ethylene glycol dimethacrylate as the cross-linker, and acetonitrile as the solvent The on line adsorbed terbutaline by the MIP can enhance the CL intensity of the reaction of luminol with ferricyanide.	Chemiluminescence (Micro flow sensor on a chip)	Detection limit: 4.0 ng/mL Linear range: 8.0 to 100 ng/mL	[59]
31	Uric acid	Uric acid-imprinted polymer was prepared from melamine and chloranil and coated directly onto the surface of a hanging mercury drop electrode	Differential pulse, cathodic stripping voltammetry	Detection limit: 0.024 µg/mL	[60]

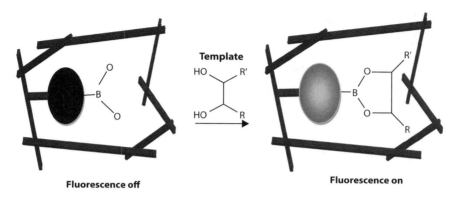

Figure 13.6 A fluorescent MIP sensor

action is prolonged and the side effects are reduced, the dosing frequency is minimized and finally patient compliance is enhanced [5].

MIPs as drug delivery system can sustain the plasma concentration of the drug in therapeutic window (below toxic levels and above the minimum effective level). They can also protect the drug from degradation by enzymes or other proteins. Hoshino *et al.* in 2010 reported the in vivo application of MIPs [61].

The main limitations in use of traditional MIPs in biomedical field are their toxicities, biocompatibilities and their best performance in hydrophobic organic solvents. Apolar solvents reduce the non-specific hydrophobic interactions and create the best environment for the interactions involved in the molecular recognition mechanism [62].

Puoci *et al.* in a recent review discuss more and complete about the MIPs for drug delivery system [5]. Table 13.2 summarizes the work of some researchers who used MIPs as drug delivery systems.

13.8   MIPs as Sorbent Materials in Separation Science

One of the wide applications of imprinted polymers is in separation and purification science. In a recent review by Cheong in journal of separation science this application of MIPs discussed completely [70]. Here, we just pointed to some of biological and clinical applications.

MIPs can be used in selective solid phase extraction techniques via on-line or off-line procedure, in the formats of mini-columns, knotted reactors, disks, membranes, renewable beads, or cartridges. They are powerful sample pretreatment/enrichment tools.

Table 13.2 Some recent reports on MIPs for use as drug delivery systems.

No	Drug name	method of using MIP	Release by	[Ref.]
1	Naltrexone	Nanoparticles of molecular imprinted polymers of acryl amide and ethylene glycol dimethacrylate (EGDMA) as cross-linker	Non-Fickian type diffusion mechanism	[63]
2	BSA	A novel molecularly imprinted polymer matrix, namely protein-imprinted N-maleoylchitosan-grafted-2-acrylamido-2-methylpropanesulfonic acid polymer matrix was prepared by using bovine serum albumin (BSA) as a template.	UV light	[64]
3	(S)-omeprazole (for enantioselective selective release)	Methacryloyl quinine and methacryloyl quinidine as monomers by suspension polymerization, using Ethylene glycol dimethacrylate as a cross-linker in chloroform	pH changes	[65]
4	Glycyrrhizic acid	Acidic (MAA), neutral (HEMA) and basic [2-(dimethylamino) ethyl methacrylate-DMAEMA] as functional monomers and EGDMA as cross-linker in DMF	pH changes	[66]
5	S-isomer of racemic propranolol	The transdermal patch based on the MIP thin-layer composited cellulose membrane-controlled release system	Complex formation with the selective receptor sites	[67]
6	a-Tocopherol	Polymers were synthesized using methacrylic acid as functional monomer and ethylene glycol dimethacrylate as cross-linker	pH changes	[68]
7	5-Fluorouracil	Polymers were synthesized using methacrylic acid as functional monomer and ethylene glycol dimethacrylate as cross-linker	pH changes	[69]

MIP membranes have been increasing used in the last few years. One of the recent examples is the work of Yusof *et al* in 2013. They used a MIP membrane as a sorbent for the removal of 2,4-dinitrophenol (2,4-DNP) from aqueous solution. They embedded MIP particles on cellulose acetate (CA) and polysulfone (PSf) to prepare the membrane. [71]. Several other articles have been published on using MIPs as sorbents, showing specific permeability or separation for template molecules such as cholesterol [72], amino acids [73] and various drugs [74–76] even enantiomeres [77].

Also, MIPs can be used as stationary phase in chromatography techniques [78]. There have been many reviews on capillary electrochromatography(CEC) application of MIPs too [79–87].

In a review published in 2007, Maier and Lindner [88] discussed on chiral separation applications of MIPs. Slow mass transfer characteristics and binding site heterogeneity are suggested as intrinsic limitations of MIPs in enantioselective separation. Non-covalent MIPs also can be used in affinity separations [89, 90]. Table 13.3 summarizes some of the recent reported MIPs used as sorbents.

Another application of MIPs as separation tools can be seen in separation of bacteria. Separation of compounds out of complex mixtures is a key issue that has been solved for small molecules. However, general methods for the separation of large bio-particles, such as cells, are still challenging. MIPs for small molecules separation are usually polymerized as bulk porous materials, which can be used as solid phase for example in chromatography, but for peptides and bigger biological objects and bacteria a surface molecular imprinting of polymers methodology has to be used. SMIPs technology was used for obtaining of the polymeric surface imprinting against various microorganisms: yeast, Saccharomyces, cerevisiae [101, 102].

Schirhagl et al. showed that imprinted polymeric films into a microfluidic chip, can preferentially capture cells matching an imprint template, and separate strains of cyanobacteria with 80–90% efficiency, despite a minimal difference in morphology and fluorescence, demonstrating its general nature. Capture specificity and separation can be further enhanced by orienting the imprints parallel to the flow vector and tuning the pH to a lower range [103].

In a study, N-Acyl homoserine lactone-binding polymers were used for the control of quorum sensing (the process of sensing bacterial numbers through signal molecule concentration in Gram negative bacteria) [104, 105]. In this study, molecularly imprinted polymers were used to bind the signal molecules (N-acyl homoserine lactones) in order to control the cells growth. The porous polymer, computationally designed to interact with homoserine lactones can switch off the bioluminescence of signaling molecule without affecting cell growth.

Table 13.3 Some reports on MIPs used as sorbents.

No	Template molecule	Method of using MIP	Techniques	[Ref.]
1	Glibenclamide	The novel surface molecularly imprinted polymers based on dendritic-grafting magnetic nanoparticles were developed	Solid phase extraction	[91]
2	Tetracycline	Magnetic molecularly imprinted polymers were prepared using bilayer modified Fe_3O_4 magnetite as the magnetically susceptible component, oxytetracycline and chlortetracycline as the mixed-template molecules, and methacrylic acid as the functional monomer	Solid phase extraction	[92]
3	Some estradiols (estriol, estradiol, estrone and diethylstilbestrol)	A molecularly imprinted polymer was synthesized by ultrasonic irradiation, with attapulgite as matrix using beta-naphthol as the template molecule, acryloyl-beta-cyclodextrin as the functional monomer, and N,N-methylenebiacrylamide as the cross-linker	Simultaneous on-line solid phase extraction and high performance liquid chromatography determination	[93]
4	Amiodarone	Water-compatible molecularly imprinted polymer	Solid phase extraction	[94]
5	Ephedrine and pseudoephedrine	Monolithic molecular imprinted polymer fiber by *in situ* polymerization in a silica capillary mold was produced	Solid phase micro extraction	[95]

No	Template molecule	Method of using MIP	Techniques	[Ref.]
6	Citaloperam	The molecularly imprinted polymers were prepared using methacrylic acid as functional monomer, ethylene glycol dimethacrylate as cross-linker, chloroform as porogen and citalopram hydrobromide as the template molecule.	Solid phase extraction	[96]
7	Neurotransmitters	A monolithic molecularly imprinted polymer column was prepared as the stationary phase for the capillary electrochromatographic separation. MIP composed of methacrylic acid (MAA) or itaconic acid (IA) together with a mixture of EDMA and AIBN in N,N-dimethylformamide	Chiral separation by capillary electrochromatography	[97]
8	Sulfachloropyridazine and sulfadiazine	MIP as the stationary phase	Chromatographic separation	[98]
9	D- and L-thyroxine	The imprinted polymer was coated on a vinyl functionalized self-assembled monolayer modified silver wire	Micro-solid phase extraction	[99]
10	Tramadol	The water-compatible molecularly imprinted polymers (MIPs) were prepared using methacrylic acid as functional monomer, ethylene glycol dimethacrylate as cross-linker, chloroform as porogen and tramadol as template molecule.	Solid-phase extraction as the sample clean-up technique combined with high-performance liquid chromatography	[100]

13.9   Future Perspective of MIP Technologies

From the time of their discovery up to now, various biological applications have been widely developed especially in pharmacology and clinical chemistry and medicine. Many efforts have been performed for miniaturized the size of MIPs to nano-scales (Nano-MIPs) or for synthesis of biocompatible MIPs. However, in spite of these progresses in MIPs technology, several aspects should be further studied to allow these advanced materials are used in vivo or in point-of-care devices for critical care applications or for direct use by patients in the near future.

Large-scale separation by MIPs is also another area of great potential market especially in chiral separation of pharmaceutical biotechnology industries or for separation of the bacterial from fermentation media.

In the future it is likely that a number of key developments in therapeutic monitoring and intelligent drug delivery will rely on MIPs technology.

13.10   Conclusion

In this chapter, recent achievements in design, synthesis and development of biocompatible MIPs for use in drug delivery or in healthcare devices work based on sensor/biosensor systems have been discussed. In spite of these progresses in MIPs technology, several aspects should be further studied to allow these advanced materials to be used in vivo or in point-of-care devices for critical care applications or for direct use by patients in the near future.

References

1. M. V. Polyakov, *Zhur Fiz Khim*, Vol. 2, p. 799, 1931.
2. O. Ramstrom, and K. Mosbach, *Curr. Opin. Chem. Biol.* Vol. 3, p. 759, 1999.
3. B. Sellergren, M. Lepistö M, and K. Mosbach, *J. Am. Chem. Soc.*, Vol. 110, p. 5853, 1988.
4. C. Alvarez-Lorenzo, and A. Concheiro, *J. Chromatogr. B*, Vol. 804, p. 231, 2004.
5. F. Puoci, G. Cirillo, M. Curcio, O.I. Parisi, F. Iemma and N. Picci, *Expert Opin. Drug Deliv.*, Vol. 8, p. 1379, 2011.
6. J. Steinke, D.C. Sherrington, and I.R. Dunkin, "Imprinting of synthetic polymers using molecular templates," in *Synthesis and Photosynthesis*, Springer-Verlag Berlin, Berlin, Vol. 123, p. 81, 1995.

7. K. Haupt, A.V. Linares, M. Bompart, and T.S.B. Bernadette, "Molecularly Imprinted Polymers," in *Molecular Imprinting*, edited by K. Haupt Vol. 325, p. 1, 2012.
8. G. Wulff, "Molecular imprinting - a way to prepare effective mimics of natural antibodies and enzymes," in *Nanoporous Materials*, edited by A. Sayari and M. Jaroniec, Elsevier Science Bv, Amsterdam, Vol. 141, p. 35, 2002.
9. P.A.G. Cormack and A.Z. Elorza, *J. Chromatogr. B*, Vol. 804, p. 173, 2004.
10. S.G. Dmitrienko, V.V. Irkha, A. Yu. Kuznetsova, and Yu. A. Zolotov, *J. Anal. Chem.*, Vol. 59, p. 808, 2004.
11. P.S. Sharma, A. Pietrzyk-Le, F. D'Souza and W. Kutner, *Anal. Bioanal. Chem.*, Vol. 402, p. 3177, 2012.
12. W.J. Cheong, F. Ali, J.H. Choi, J. O. K. Lee, and K. Y. Sung, *Talanta* Vol. 106, p. 45, 2013.
13. G. Wulff and S. Schauhoff, *J. Org. Chem.* Vol. 56, p. 395, 1991.
14. O. Ramström, I.A. Nicholls, and K. Mosbach, *Tetrahedron: Asymmetry*, Vol. 5, p. 649, 1994.
15. G. Guan, B. Liu, and Z. W. Z. Zhang, *Sensors*, Vol. 8, p. 8291, 2008.
16. N.A. Connor, D.A. Paisner, D. Huryn, and K.J. Shea, *J. Am. Chem. Soc.*, Vol. 130, p. 1680, 2008.
17. C. Xie, B. Liu, Z. Wang, D. Gao, G. Guan, and Z. Zhang, *Anal. Chem.*, 80 (2008) 437.
18. T. Alizadeh, M.R. Ganjali, and M. Akhoundian, *Int. J. Electrochem. Sci.*, Vol. 7, p. 10427, 2012.
19. T. Alizadeh, M.R. Ganjali, and M. Akhoundian, *Int. J. Electrochem. Sci.*, Vol. 7, p. 7655, 2012.
20. A.R. Guerreiro, I. Chianella, E. Piletska, M.J. Whitcombe, and S.A. Piletsky, *Biosens. Bioelectron.*, Vol. 24, p. 2740, 2009.
21. A. Poma, A.P.F. Turner, and S.A. Piletsky *Trends in Biotech.*, Vol. 28, p. 629, 2010.
22. A. S. Belmont, S. Jaeger, D. Knopp, R. Niessner, G. Gauglitz, and K. Haupt, *Biosens. Bioelectron.*, Vol. 22, p. 3267, 2007
23. I. Cosic, *IEEE transactions on bio-medical Engin.*, Vol. 41, p. 1101, 1994.
24. R. Baron, P. Setny, and J. A. McCammon, *J. American Chem. Soc.*, Vol. 132, p. 12091, 2010.
25. R. Baron, and J.A. McCammon, *Annual Rev. in Physical Chem.*, Vol. 64, p.151, 2013.
26. S.H. Gellman, *Chem. Rev.*, 97, p. 1231, 1997.
27. C. Baggiani, L. Anfossi and C. Giovannoli, *Analyst*, Vol. 133, p. 719, 2008.
28. S.H. Gao, W. Wang, and B.H. Wang, *Bioorg. Chem.*, Vol. 29, p. 308, 2001.
29. A.L. Hillberg, K.R. Brain, and C.J. Allender, *Adv. Drug Delivery Rev.*, Vol. 57, 1875, 2005.
30. E.V. Piletska, G. Stavroulakis, K. Karim, M.J. Whitcombe, I. Chianella, A. Sharma, K.E. Eboigbodin, G.K. Robinson, and S.A. Piletsky, *Biomacromolecules*, Vol. 11, p. 975, 2010.

31. Y.J. Tong, H.D. Li, H.M. Guan, J.M. Zhao, S. Majeed, S. Anjum, F. Liang, and G.B. Xu, *Biosens. Bioelectron.*, Vol.47, p. 553, 2013.

32. D. Chen, J. Deng, J. Liang, J. Xie, C.H. Hu, and K.H. Huang, *Sens. Actuators B*, Vol.183, p. 594, 2013.

33. A. Afkhami, H. Ghaedi, T. Madrakian, M. Ahmadi, H. Mahmood-Kashani, *Biosen. & Bioelectron.*, Vol. 44, p. 34, 2013.

34. Y. X. Zhang, P. Y. Zhao, and L. P. Yu, *Sens. Actuators B*, Vol. 181, p. 850, 2013.

35. M. Arvand, H.A. Samie, *Drug Testing Anal.*, Vol. 5, p. 461, 2013.

36. L. Li, Y. Liang, and Y. Liu, *Anal. Biochem.*, Vol. 434, p. 242, 2013.

37. M. Javanbakht, F. Fathollahi, F. Divsar, M.R. Ganjali, and P. Norouzi, *Sens. Actuators B*, Vol. 182, p. 362, 2013.

38. X.D. Wang, J. Dong, H.M. Ming, and S.Y. Ai, *Analyst*, Vol. 138, p. 1219, 2013.

39. X.W. Kan, T.T. Liu, C. Li, H. Zhou, Z.L. Xing, A.H. Zhu, *J. Solid State Electrochem.*, Vol. 16, p. 3207, 2012.

40. H.M. Qiu, C.N. Luo, M. Sun, F.G. Lu, L.L. Fan, and X.J. Li, *Carbon*, Vol. 50, p. 4052, 2012.

41. F.G. Lu, J.L. Yang, M. Sun, L.L. Fan, H.M. Qiu, X.J. Li, and C.N. Luo, *Anal. Bioanal. Chem.*, Vol. 404, p. 79, 2012.

42. R.N. Liang, Q. Gao, and W. Qin, *Chinese J. Anal. Chem.*, Vol. 40, p. 354, 2012.

43. C.H. Zhou, T.T. Wang, J.Q. Liu, C. Guo, Y. Peng, J.L. Bai, M. Liu, J.W. Dong, N. Gao, B.A. Ning, and Z.X. Gao, *Analyst*, Vol. 137, p. 4469, 2012.

44. T.H. Nguyen, S.A. Hardwick, T. Sun, and K.T.V. Grattan, *IEEE Sensors J.*, Vol. 12, p. 255, 2012.

45. B. Khadro, A. Betatache, C. Sanglar, A. Bonhomme, A. Errachid, and N. Jaffrezic-Renault, *Sensor Lett.*, Vol. 9, p. 2261, 2011.

46. Z.H. Wang, H. Li, J. Chen, Z.H. Xue, B.W. Wu, and X.Q. Lu, *Talanta*, Vol. 85, p. 1672, 2011.

47. F.T.C. Moreira, and M.G.F. Sales, *Mater. Sci. Eng. C*, Vol. 31, p. 1121, 2011.

48. Q.Y. Bing, Y. Liu, and Q.J. Song, *Chinese J. Anal. Chem.*, Vol. 39, p. 1053, 2011.

49. X.F. Liu, F. Li, B. Yao, L. Wang, G.Y. Liu, and C.Y. Chai, *Spectroscopy and Spectral Anal.*, Vol. 30, p. 2228, 2010.

50. T. Alizadeh, M.R. Ganjali, M. Zare, and P. Norouzi, *Electrochim. Acta*, Vol. 55, p. 1568, 2010.

51. Y. Yoshimi, A. Narimatsu, K. Nakayama, S. Sekine, K. Hattori, K. Sakai, *J. Artificial Organs*, Vol. 12, p. 264, 2009.

52. G. Paniagua Gonzalez, Pl Fernandez Hernando, J.S. Durand Alegria, *Anal. Chim. Acta*, Vol. 638, p. 209, 2009.

53. J.H. Yu, F.W. Wan, P. Dai, S.G. Ge, B. Li, and J.D. Huang, *Anal. Lett.*, Vol. 42, p. 746, 2009.

54. J.W. Kang, H.N. Zhang, Z.H. Wang, G.F. Wu, and X.Q. Lu, *Polymer-Plastics Tech. Eng.*, Vol. 48, p. 639, 2009.

55. B.B. Prasad, S. Srivastava, K. Tiwari, and P.S. Sharma, *Sens. Mater.*, Vol. 21, p. 291, 2009.

56. M. Javanbakht, S.E. Fard, M. Abdouss, A. Mohammadi, M.R. Ganjali, P. Norouzi, and L Safaraliee, *Electroanalysis*, Vol. 20, p. 2023, 2008.

57. A.K. Patel, P.S. Sharma, and B.B. Prasad, *Electroanalysis*, Vol. 20, p.2102, 2008.
58. M. Javanbakht, S.E. Fard, A. Mohammadi, M. Abdouss, M.R. Ganjali, P. Norouzi, and L. Safaraliee, *Anal. Chim. Acta*, Vol. 612, p. 65, 2008.
59. D.Y. He, Z.J. Zhang, H.J. Zhou, and Y. Huang, *Talanta*, Vol. 69, p. 1215, 2006.
60. D. Lakshmi, P.S. Sharma, and B.B. Prasad, *Electroanalysis*, Vol. 18, p. 918, 2006.
61. Y. Hoshino, H. Koide, and T. Urakami, *J. Am. Chem. Soc.*, Vol. 132, p. 6644, 2010.
62. E.V. Piletska, A.R. Guerreiro, and M. Romero-Guerra, *Anal. Chim. Acta*, Vol. 607, p. 54, 2008.
63. K. Rostamizadeh, M. Vahedpour, and S. Bozorgi, *Int. J. Pharm.*, Vol. 424, p. 67, 2012.
64. T. S. Anirudhan and S. Sandeep, *Polymer Chem.*, Vol. 2, p. 2052, 2011.
65. R. Suedee, C. Jantarat, W. Lindner, H. Viernstein, S. Songkro, and T. Srichana, *J. Control Release*, Vol. 142, p. 122, 2010.
66. G. Cirillo, O.I. Parisi, M. Curcio, F. Puoci, F. Iemma, and U. G. Spizzirri, *et al.* *J. Pharm. Pharmacol.*, Vol. 62, p.577, 2010.
67. R. Suedee, C. Bodhibukkana, N. Tangthong, C. Amnuaikit, S. Kaewnopparat, and T. Srichana, *J. Controlled Release*, Vol. 129, p. 170, 2008.
68. Puoci F, Cirillo G, Curcio M, F. Iemma, O. I. Parisi, M. Castiglione, and N. Picci, *Drug Delivery*, Vol. 15, p. 253, 2008.
69. F. Puoci, F. Iemma, G. Cirillo, *et al. Molecules,* Vol. 12, p. 805, 2007.
70. W. J. Cheong, S. H. Yang, and F. Ali, *J. Sep. Sci.*, Vol. 36, p. 609, 2013.
71. N. A. Yusof, N. D. Zakaria, N. A. M. Maamor, A. H. Abdullah and Md. J. Haron, *Int. J. Mol. Sci.*, Vol. 14, p. 3993, 2013.
72. G. Ciardelli, C. Borrelli, D. Silvestri, C. Cristallini, N. Barbani, and P. Giusti, *Biosens. Bioelectron.*, Vol. 21, p. 2329, 2006.
73. Y.H. Shim, E. Yilmaz, S. Lavielle, and K. Haupt, *Analyst*, Vol. 129, p. 1211, 2004.
74. F. Trotta, C. Baggiani, M.P. Luda, E. Drioli, and T.Massaria, *J. Membr. Sci.*, Vol. 254, p. 13, 2005.
75. R. Weiss, A. Molinelli, M. Jakusch, and B. Mizaikoff, *Bioseparation*, Vol. 10, p. 379, 2001.
76. Y.Q. Xia, T.Y. Guo, M.D. Song, B.H. Zhang, and B.L. Zhang, *React. Funct. Polym.*, Vol. 66, p. 1734, 2006.
77. M. Walshe, E. Garcia, J.M. Howarth, R. Smyth, and M.T. Kelly, *Anal. Commun.*, Vol. 34, p. 119, 1997,
78. P. K. Owens, L. Karlsson, E.S.M. Lutz, and L.I. Andersson, *Trends Anal. Chem.*, Vol. 18, p. 146, 1999.
79. P.T. Vallano, and V.T. Remcho, *J. Chromatogr. A*, Vol. 887, p.125, 2000.
80. L. Schweitz, P. Spe´gel, S. Nilsson, *Electrophoresis*, Vol. 22, p. 4053, 2001.
81. P. Spe´gel, L.Schweitz, and S. Nilsson, *Electrophoresis*, Vol. 24, p. 3892, 2003.
82. C. Liu, and C. Lin, *Electrophoresis*, Vol. 25, p. 3997, 2004.
83. J. Nilsson, P. Spe´gel, and S. Nilsson, *J. Chromatogr. B*, Vol. 804, p. 3, 2004.

84. Z. Liu, C. Zheng, C. Yan, and R. Gao, *Electrophoresis*, Vol. 28, p.127, 2007.

85. Y. Huang, Z. Liu, C. Zheng, and R. Gao, *Electrophoresis*, Vol. 30, p. 155, 2009.

86. M. L"ammerhofer, and A. Gargano, *J. Pharm. Biomed. Anal.*, Vol. 53, p. 1091, 2010.

87. C. Zheng, Y. Huang, and Z. Liu, *J. Sep. Sci.*, Vol. 34, p.1988, 2011.

88. N. M. Maier, and W. Lindner, *Anal. Bioanal. Chem.*, Vol. 389, p. 377, 2007.

89. S. Wei, and B. Mizaikoff, *J. Sep. Sci.*, Vol. 30, p. 1794, 2007.

90. J. Haginaka, *J. Chromatogr. B*, Vol. 866, p. 3, 2008.

91. R. Y. Wang, Y. Wang, C. Xue, T. T. Wen, J. H. Wu, J. L. Hong, and X. M. Zhou, *J. Sep. Sci.,* Vol. 36, p. 1015, 2013.

92. J. H. Kong, Y. Z. Wang, C. Nie, D. Ran, and X. P. Jia, *Anal. Methods*, Vol. 4, p. 1005, 2012.

93. C. D. Zhao, X. M. Guan, X. Y. Liu, and H. X. Zhang, *J. Chromatogr. A*, Vol. 1229, p. 72, 2012.

94. T. Muhammad, L. Cui, J. D. Wang, E. V. Piletska, A. R. Guerreiro, and S. A. Piletsky, *Anal. Chim. Acta*, Vol. 709, p. 98, 2012.

95. D. L. Deng, J. Y. Zhang, C. Chen, X. L. Hou, Y. Y. Su, and L. Wu, *J. Chromatogr. A*, Vol. 1219, p. 195, 2012.

96. M. Abdouss, S. Azodi-Deilami, E. Asadi, and Z. J. Shariatinia, *Mater. Sci. Mater. Med.*, Vol. 23, p. 1543, 2012.

97. B. Y. Huang, Y. C. Chen, G. R. Wang, and C. Y. Liu, *J. Chromatogr. A*, Vol. 1218, p. 849, 2011.

98. J. J. Hsu, I. B. Huang, C. C. Hwang, and M. C. Wu, *Indian J. Chem. Technol.*, Vol. 18, p. 7, 2011.

99. B. B. Prasad, M. P. Tiwari, R. Madhuri, and P. S. Sharma, *J. Chromatogr. A*, Vol. 1217, p. 4255, 2010.

100. M. Javanbakht, A. M. Attaran, M. H. Namjumanesh, M. Esfandyari-Manesh, and B. Akbari-Adergani, *J. Chromatogr. B*, Vol. 878, p. 1700, 2010.

101. F.L. Dickert, M. Tortschanoff, H. Basenböck, *Adv. Mat.*, Vol. 10, 145, 1998.

102. K. Haupt, *Anal.Chem.*, Vol. 377A, 2003.

103. R. Schirhagl, E. W. Hall, I. Fuereder and R. N. Zare, *Analyst*, Vol. 137, p. 1495, 2012.

104. E.V. Piletska, G. Stavroulakis, K. Karim, M.J. Whitcombe, I. Chianella, A. Sharma, K.E. Eboigbodin, G.K. Robinson, and S. A. Piletsky, *Biomacromolecules*, Vol. 11, p. 975, 2010.

105. E. Piletska, G. Stavroulakis, L. Larcombe, M. J. Whitcombe, A. Sharma, S. Primrose, G. Robinson, S. Piletsky, *Biomacromolecules*, Vol. 12, p. 1067, 2011.

14

The Role of Immunoassays in Urine Drug Screening

Niina J. Ronkainen, Ph.D.[1*] **and Stanley L. Okon, M.D.**[2]

[1]*Benedictine University, Department of Chemistry and Biochemistry,
Lisle IL, U.S.A*
[2]*Department of Psychiatry, 8S, Advocate Lutheran General Hospital,
Park Ridge, IL, U.S.A*

Abstract

Drug screening is routinely performed on urine for workplace drug testing, medical purposes at clinical and rehabilitation settings, criminal justice systems, schools and testing of professional athletes. In addition, federal, military, and corporate employees undergo periodic screening for illegal drugs such as opiates, amphetamines, cocaine, and cannabinoids as indication of drug abuse. Urine is the most common sample matrix for drug screening due to ease of sampling. Other advantages include the presence of higher drug and metabolite concentrations as well as longer lived chemical species. Furthermore, routine drug screening immunoassays have become largely automated to match the increasing demand and workload which in turn has led to rapid turnaround time for results. In addition, many simple and easy-to-use point of care devices based on solid state immunoassays are commercially available for the most commonly abused drugs. The simplicity, ease of use, low cost, and rapid results have made immunoassays the most popular initial screening test that can be performed on-site, at home, or in the clinical laboratory. The main limitation of immunoassays is the probability of cross-reactivity between other irrelevant drugs or metabolites and specific immunoassays, which can lead to false-positive test results.

Keywords: Addiction psychiatry, adulteration of urine samples, amphetamines, biological specimens, cannabinoids, chromatography, cocaine, confirmatory testing, crossreactivity of antibodies, cutoff limits, drugs and their metabolites, false negative results, false positive results, immunoassays, interferences, opiates, phencyclidine, point of care testing, substance abuse, synthetic cannabinoids, urine drug testing

---

*Corresponding author: NRonkainen@ben.edu

Ashutosh Tiwari (ed.) Advanced Healthcare Materials, (493–524) 2014 © Scrivener Publishing LLC

14.1 Introduction

Drug screening is routinely performed on urine for workplace drug testing (preemployment and random testing), medical purposes at clinical and rehabilitation settings (identification of substances associated with drug overdoses or substance abuse, compliance and/or abstinence monitoring in substance abuse treatment centers), forensics and the criminal justice system (cause of death, at the time of arrest, rehabilitation testing of convicts and persons on parole), schools and testing of professional athletes (eligibility to compete). Urine drug testing (UDT) is also an essential part of pain management as physicians seek to identify the use of illicit or unauthorized licit drugs in addition to verifying adherence to prescribed opioid analgesic regimens. In pain management, the expected testing result is often positive as compliance with the prescribed opioid analgesic is confirmed. Other hospital units that utilize UDT include intensive care, psychiatry, as well as labor and delivery. Drug testing in the criminal justice system provides judges with relevant information for bail-setting and sentencing, helps identify drug users in need of treatment, and provides indication on whether an individual is complying with rules or conditions of their probation, parole, work release, etc. In addition, federal, military, and corporate employees get screened periodically for illegal drugs such as opiates, amphetamines, cocaine, and cannabinoids as indication of drug abuse. For example, the U.S. Department of Transportation (DOT) requires alcohol and drug testing of all safety-sensitive transportation employees in aviation, railroads, mass transit, trucking, etc. As can be expected, the laws regarding random drug testing of employees as well as what drugs can be tested vary by state. Many federal agencies like the DOT limit drug testing to amphetamines, cocaine, opiates, cannabinoids, and phencyclidine (PCP). However, such screenings still generate a large number of samples, which must be analyzed in a timely and economically efficient manner that ultimately requires the use of high throughput analytical methods. In addition, it is critical that the analysis of the screening results be interpreted correctly which in turn requires a thorough understanding of how each assay/analytical method functions, which drugs and/or metabolites can be detected, knowing the limits of detection/cutoff values used to generate the result for the drug of interest, understanding the specificity of each assay, being aware of the possibility of obtaining false-negative or false-positive results, and also the possible need for confirmatory testing. Furthermore, one would also need to know how long each drug remains detectable in the body after ingestion as well as the effects any intentional and unintentional interferents would have on the assay results. Finally, it should be noted that drug screening methods are generally unable to distinguish between acute and chronic exposure to a substance.

This chapter will focus on the discussion of drug screening in a medical setting with some emphasis on drug screening as it pertains to substance abuse. There are very few published guidelines on drug screening for clinical purposes. In a hospital setting, a UDT for adults is usually ordered as a diagnostic test to guide medical treatment without a specific informed consent under the general "consent to treatment." UDT services at clinical laboratories are typically tailored with the needs of emergency care departments in mind and generally have short turnaround times. The results of the UDT are used to make decisions regarding further treatment in cases of substance abuse, drug overdose or drug/intoxicant related motor vehicle injuries among others. Therefore, providing accurate and precise UDT results in a timely and cost-effective manner is of great importance.

There are two primary analytical methods for drug testing: immunoassays and chromatography. Immunoassays are the most popular initial screening test and can be performed on-site, at home, or in the laboratory due to their simplicity, ease of use, low cost, and rapid results. The specificities of immunoassays vary depending on the particular assay and presence of chemically similar compounds in the sample, which can lead to false positive results. Most immunoassays designed for drug screening provide qualitative information about the presence or absence of a drug. The results from the sample specimen are compared to results from a calibrator or a standard, which contains a known quantity of the drug of interest. If the signal response from the biological specimen is equal or higher than that of the standard, then the test is considered positive. If the sample gives a lower response than the standard, then the test is interpreted as negative. Each test has different cutoff values, which usually correspond to federal workplace cutoff values established by the U.S. Department of Health and Human Services (DHHS) for the immunoassay and the chromatography method [1]. These cutoff values (which are relatively high) are chosen such that false positive results from cross reactivity due to consumption of legal drugs and food items with similar chemical structures are minimized. As a result, the cutoff values used by some health care facilities and substance abuse programs may not be appropriate because it assumes that "one size fits all" and increases the potential for false-negative results. Furthermore, most cutoff levels are developed for adults and may not be applicable to children whose urine is usually more dilute than that of adults leading to false-negative results. The specific cutoffs used for each test in the drug screening should be included with results on a drug screen report. The Substance Abuse and Mental Health Services Administration (SAMHSA), an operational division of the DHHS, has defined a panel of five drug classes (amphetamines, cocaine, opiates, PCP, and marijuana) called NIDA 5 for use in workplace drug testing. Many clinical laboratories offer more than one type of drug screening panel and also most tests are available on automated platforms that are capable of simultaneously and rapidly screening for multiple drugs on a single urine specimen.

Chromatographic analysis usually done by instrumental analysis is used as a confirmatory test in UDT. It is more specific, sensitive, accurate, and reliable than immunoassays in detection of most illicit drugs. However, it is more time-consuming to operate, expensive, often requires multiple sample preparation steps and must be performed by trained personnel. Also, instrumentation-based chromatographic methods may not be available in all hospital-based clinical laboratories due to relatively large capital investment required to acquire the instrumentation and are generally more common at laboratories that specialize in toxicology. Some easy to use, noninstrumental chromatography options like thin-layer chromatography drug screening methods such as Toxi-Lab (MP Products, Amersfoort, Netherlands) do exist. However, these assays may require 3 to 4 hours before the results can be obtained. Chromatographic methods are also used to quantify specific drugs and their metabolites in the biological specimen. More specifically, chromatographic analysis is coupled with a mass spectrometry detector to confirm positive test results obtained from drug screen immunoassays. For example, a positive immunoassay drug screen for opiates would be confirmed by a chromatographic method which would also quantify each opiate and their associated metabolites.

The confirmatory test is usually performed using tandem instrumental analysis such as gas chromatography-mass spectrometry (GC-MS), or liquid chromatography tandem mass spectrometry (LC-MS/MS). These analytical methods can also be used to identify drugs that are not routinely included in drug screen immunoassay panels or the so-called designer drugs for which screening immunoassays are currently not available. In addition, confirmatory tests are utilized in forensic toxicology laboratories to provide accurate and precise drug testing results, which can be used as evidence in the criminal justice system and in other legal settings. In most hospital laboratories, confirmatory testing is not routinely performed due to cost, longer turnaround times to obtain results and limited resources. Therefore, it is critical that clinicians have adequate knowledge regarding the limitations of analytical methods such as immunoassay based UDT, interpretation of the results and when to order confirmatory testing. Even though chromatographic methods are generally more specific and less prone to interference from sample matrix than immunoassays, false positive and false negative results can occur. However, when a confirmatory analytical method is properly designed and optimized, the chances of obtaining false positive results are virtually eliminated.

14.2   Urine and Other Biological Specimens

Biological specimens used as sample matrices for drug testing include urine, blood, oral fluid (saliva), hair, nails, meconium and sweat. Each sample matrix has different levels of sensitivity, specificity, and accuracy. Urine tends to be the preferred sample matrix because its collection is convenient, often rapid and noninvasive. Also, urine does not typically require preanalytical sample preparation and a relatively large quantity of urine is usually available for analysis. Depending on the analytical method, the volume of urine required for UDT varies from a few drops to 30 mL [2]. In addition, the concentrations of drugs and their metabolites tend to be higher and the chemical species longer lived in urine samples as opposed to blood and other sample matrices. Of note, there is no direct correlation between the concentration of drugs present in urine or blood. Furthermore, interference from other components in the sample matrix is less likely in immunoassays that utilize urine because it contains less proteins and lipids than blood. However, depending on the analytical method, urine that is turbid or contains sediments may need to be centrifuged prior to analysis in order to avoid false-negative results [2].

Like urine, obtaining saliva is noninvasive and convenient. In addition, saliva is less likely than urine to be adulterated and the sample collection does not require a person of the same gender witnessing the urine sample collection. Drug testing using saliva is also a better indication of current impairment and contains the parent drugs rather than their active metabolites, which may be detectable earlier compared to urine [2]. This makes saliva a convenient specimen for roadside testing by law enforcement officers. However, saliva is potentially more infectious than urine [3]. In addition, saliva-based testing is challenging because the sample volumes are considerably less compared to urine and the analytes are present in lower concentrations due to dilution, etc. Therefore, the cutoff limits for drug present in saliva are significantly lower than in urine making their detection that more challenging. Also, some patients may not be able to provide adequately large saliva specimens. Furthermore, artificial stimulation of saliva flow alters the normal pH of about 6.5 to a pH of about 8, which affects the protonation state of drugs with pK_as near these values and ultimately, the ability of an immunoassay to accurately measure their concentration [3, 4]. To further complicate matters, drug concentrations in saliva can vary depending on the sample collection method [5]. Finally, oral contamination with various substances may also affect test results in saliva-based testing [2].

Hair, nail, and meconium specimens require lengthy analyte extraction and sample preparation procedures prior to analysis, making them more difficult and less practical sample matrices to analyze. Concentrations of drugs in meconium reflect in utero drug exposure on the fetus over the last 20 weeks of gestation. As a specimen, sweat is cumbersome to utilize because elimination of drugs through the skin may take days, drug concentrations may vary depending on the sample collection site, and the possibility of the sample being contaminated during collection is high [2]. For these reasons, urine remains the specimen of choice for drug testing.

Drug concentrations in urine are affected by acute vs. chronic drug use, elapsed time since last use, pharmacokinetics, metabolism and renal clearance, body mass, fat distribution, and fluid consumption [1, 6, 7]. In addition, there are genetic variations in how drugs are metabolized by individuals [8]. Typically, when utilizing an immunoassay for drug screening, no pretreatment or predilution steps are performed on the urine sample. A positive urine drug screen indicates the presence of a drug, but does not provide any information about the dosage, how or when the drug was administered, or degree of physical impairment. A negative urine drug screen result does not guarantee that the person has not consumed the drugs of interest because the drug concentration may not be high enough to exceed the assay's cutoff level for detection.

Timing of urine sample collection and proper labeling of the samples are also important in all urine drug screening procedures. In addition, chain of custody procedures, such as maintaining a direct line of sight between the observer and the specimen bottle, must be strictly adhered to when a urine sample is to be used for workplace screening or legal situations. This ensures that the sample integrity is preserved. The guidelines for Federal Workplace Drug Testing Programs also include specimen integrity assessment and the use of split specimens. Detection time is defined as how long a drug or its metabolites can be detected in a biological sample after it was last used. In general, the detection time for drugs of abuse is longest in hair, followed by urine, sweat, saliva, and blood [7]. Blood and saliva have the shortest detection times, but provide the best correlation between current intoxication/impairment level and the concentration of the drug in the specimen. Although varying from drug to drug, typical detection time for a single dose of a drug is 1.5 to 4 days in urine and in chronic users up to one week after last use [7]. In chronic users, certain drugs may remain at detectable levels in the urine for up to 3 months in the case of cannabis and 22 days in the case of cocaine [7].

Also, possible substitution, dilution, and chemical adulteration of urine samples have to be considered and ruled out when utilizing urine samples for drug screening. Dilution, substitution and adulteration of urine samples by synthetic urine such as Sub-Solution or Dr. John's Concentrated Urine are often detected by comparing the specimen from the patient to the known characteristics of human urine. Validity of the urine sample can be verified by testing temperature (90^0-100^0 F within 4 minutes of collection), pH (4.0–9.0), specific gravity (1.003–1.030), and creatine (≥ 20 mg/dl) of the sample [8, 9]. Validity testing using pH and creatine concentration and well as testing for potential adulteration is required in federal workplace drug screening, but not at clinical laboratory settings. Easy-to-use, commercially available urine sample validity dipstick tests and devices such as TesTcup (Roche Diagnostics, Indianapolis, IN) can be used in non-regulated settings. There is a growing and profitable online industry from which one can obtain a plethora of chemical adulterants that are purported to and can invalidate urine drug screening results. Examples of such chemical adulterants include "Instant Clean Add-it-ive," "Stealth," and "Urine Luck," which contain pyridium chlorochromate and other assay interferents [8, 10, 11]. In addition, various household chemicals such as vinegar, lemon juice, vitamin C, table salt, baking soda, soap, bleach, and drain cleaner are also used to alter sample integrity tests and drug screening results [12, 13]. Ingestion of diuretics, salicylates, and sodium bicarbonate has also been done with the intention to alter the specific gravity or chemical composition of urine and to invalidate the test results [8]. There are also commercially sold "cleansers" such as Green Clean and XXTra Clean that may be ingested [8]. Adulterants that interfere with the immunoassay itself and are not detectable by routine urine integrity tests include halogens, chromates, pyridine, and glutaraldehyde. As an aside, nitrites may not affect immunoassay screening tests but have been shown to interfere with GC-MS confirmatory testing for drugs such as cannabinoids [3]. In short, if the sample does not have the expected characteristics of human urine, it should be rejected and if possible recollected before further testing for drugs of abuse.

14.3   Immunoassays

The first immunoassay was described by Solomon Berson, a physician, and Rosalyn Yalow, a biophysicist, in 1959 [14]. Yalow went on to receive the

Nobel Prize in Medicine in 1977 for their work on measurement of plasma insulin by a radioimmunoassay. The advantages of radioimmunoassay (RIA) include high sensitivity, excellent precision, the ability to detect the signal even without optimization of the assay conditions, and lack of interference from the sample or assay environment [15]. However, individuals performing the assay had to protect themselves against radiation exposure from the radioactive labels. RIAs are not commonly used for urine drug screenings.

The use of immunoassays for various biomedical applications has increased in popularity since 1970s. This popularity may in part be due to relatively simple assay procedures that require minimal to no sample preparation, availability of antibodies for a plethora of analyte molecules, and widely available sensitive analytical instrumentation for detection [16]. Improvements in labels used in generating the signal in the assays have also allowed for more sensitive detection, lower cost and safer testing. Many of the steps for quantitative immunoassays have also been automated for assays performed in clinical laboratory settings.

Immunoassays are highly sensitive and specific analytical chemistry methods, which allow the detection and/or measurement of trace levels of biologically relevant molecules in urine, blood and other sample matrices. Immunoassays provide qualitative (positive or negative) or quantitative (a number expressing amount or concentration with appropriate units) results for the analyte which are then compared to established ranges for the assay. Even qualitative immunoassays have a quantitative component as each test has a characteristic cutoff concentration above which the assay result is reported as positive and below which the results are considered negative. Therefore, the assays express whether the concentration of the drug analyte in the biological specimen is below or above a specific cutoff concentration. Qualitative immunoassays are more frequently utilized in drug screening than quantitative assays. The U.S. Food and Drug Administration (FDA) approves immunoassays for specific analytes and applications. Accuracy and precision studies must be performed to validate even FDA approved qualitative immunoassays for drugs of abuse at each lab [17]. For these studies, a large set of urine specimens are analyzed by utilizing both the new immunoassay tests and an instrumental analysis method with an established procedure such as GC-MS or LC-MS.

Immunoassays are often used to screen a large number of samples for a specific drug or a class of drugs. However, drugs that can be detected using immunoassays are limited by the availability of specific antibodies for an analyte and its cutoff concentration. The cutoff is defined as the concentration of an analyte drug in a specimen at or above which the test result is reported as positive and below which it is determined to be negative [9]. In drug screening, the cutoffs are based either on the analytical performance of the method or are determined by governmental or regulatory agencies [3]. Sensitivity of a drug screening test measures the proportion of drug users which are correctly identified as such (that is, the percentage of known drug abusers who are correctly identified as drug abusers). If the urine drug screening test is negative, generally no further testing is performed. If the screening test gives positive results, a confirmatory secondary test involving chromatography may be ordered.

The performance of the immunoassay screening tests varies between manufacturers as well as between several different lots from the same manufacturer [3]. Some of this variation is due to the quality of antibodies used in the assay. Therefore, quality control of assays must be performed at each testing site. Furthermore, many immunoassays respond to a class of drugs rather than a specific drug. For example, the immunoassay for opiate screening may give a positive result in the presence of hydromorphone or morphine making it difficult to determine which specific drug the person may be using and also rendering the assay results harder to interpret [17].

14.3.1 Assay Design

Immunoassays involve a highly specific, noncovalent interaction between the antibody (Ab) molecule, a large Y-shaped glycoprotein, and the antigen (Ag, i.e. the drug in this case). Each antibody molecule has two identical binding sites and are hence considered to be bivalent. Most antibodies used in urine drug screening methods are monoclonal antibodies and recognize a specific structural feature of the drug analyte or its metabolite [18]. This often irreversible binding interaction between the Ab's binding site and the Ag may be monitored using a colored, a fluorescent or a radioactive label. Several well studied conjugation methods exist for coupling the labels to the antigens. The development of new immunoassays to detect specific antigens can be time consuming, labor intensive, and costly, especially if new immunogenic reagents must be produced and tested. However, once developed, the assays can function rapidly, are reliable, inexpensive and easy to use. The most common immunoassay formats used in routine drug screening are enzyme multiplied immunoassay technique (EMIT), fluorescence polarization immunoassay (FPIA), cloned enzyme donor immunoassay (CEDIA) and immunoturbidimetric assay which will be discussed in this chapter.

Most immunoassays used in screening urine for the presence of drugs of abuse are based on the competitive immunoassay design [17]. Competitive immunoassays, also known as limited reagent assays, are often used for applications where the antigen molecule is small (with a molecular weight of a few hundred grams per mole) and has only one epitope [19]. In a competitive assay, a limited amount of Ab is used, which is usually not enough to capture all the Ag molecules in the sample. A fixed, predetermined amount of labeled Ag is mixed with the urine specimen and allowed to incubate. The label attached to the drug reagent, the labeled Ag, can either contain a chemiluminescent compound, a fluorophore, an electrochemiluminescent portion or an enzyme that catalyzes the formation of a detectable product molecule. Unlabeled Ag from the specimen and the labeled Ag compete for binding to a limited number of binding sites on the capture Abs. If quantitative results are desired from a competitive immunoassay, it is critical that the ratio of limited Ab reagent to the labeled Ag must remain constant between samples.

Rinses are required in some immunoassay formats to separate the unbound Ag from the bound Ag prior to the detection of a signal, which is proportional to the concentration of Ag in the sample. However, most immunoassays used for urine drug testing are homogeneous assays which do not require a separation or a washing step because they are qualitative and the signal generated by bound vs. unbound Ag is sufficiently different from each other to be reliably measured [17]. In addition, homogeneous assays are rapid and easier to use but lack the sensitivity of heterogeneous assays which include a physical separation step to isolate the Ab-Ag complex from the unbound constituents followed by a washing step to remove any unbound reagents and sample components [16]. On the other hand, homogeneous assays are more susceptible to interference from other chemical species in the sample. An advantage of using a heterogeneous immunoassay format for quantitative analysis is its ability to detect very low concentrations of Ag.

14.3.2   Antibody–Antigen Interactions

Drugs are relatively small antigen molecules compared to viruses, bacteria, and protein molecules that are detected by some immunoassays. Therefore, it is not unusual for the drug molecule or its metabolite that is being analyzed to have only one site for Ab binding. Many illicit drugs also have limited antigenic diversity due to their small size [8]. This is the main reason for limited specificity of some Abs for individual drugs and their broad cross-reactivity toward other closely related molecules. The binding interactions between Ab and Ag are noncovalent and cumulative [16]. The noncovalent intermolecular forces operate over short distances and include ionic bonding, hydrophobic interactions, hydrogen bonding, and van der Waals forces. The unique antigen binding region of the antibody is called the paratope which accepts and binds with high affinity to a complementary site on the antigen called the epitope. A high degree of complementarity is required between the paratope and the epitope in order for the noncovalent interactions to result in the formation of a stable Ab-Ag complex.

The analyte drug (Ag) must possess a certain degree of chemical complexity (unique functional groups) such that it can be selectively recognized by the antibody molecules. Although most Ab-Ag interactions are very specific, some antibodies can cross-react with other structurally similar but unrelated molecules [16]. Cross-reactivity occurs if two different molecules share an identical or very similar region that acts as the epitope in the binding interaction. Even though the antibody typically has lower affinity for the structurally similar molecules than the target Ag for which it was produced, cross-reactivity can still produce false positive results. As such the specificity of the immunoassay varies depending on the nature and quality of the antibodies used in the immunoassay reagent.

Most immunoassays incorporate monoclonal antibodies (MAb) as the reagent [18]. MAbs, which are identical in the primary structure, are more specific for their Ag than polyclonal antibodies [16]. MAbs are generated by a single Ab-producing cultured cell line (containing clones of a single parent cell) in a bioreactor or produced in microbial systems or transgenic mice [20, 21]. MAbs are known for being highly specific and having high affinity for a given Ag, the immunoassay target molecule. Thus, MAbs have a built-in specificity toward a single chemically unique epitope on the Ag molecule that allows the detection of small differences on the surface of the Ag. Therefore, they are used as the primary or capture Ab in most drug screening immunoassays.

Polyclonal Abs have a heterogeneous mixture of immunoglobulin molecules secreted against a specific antigen, each recognizing a slightly different epitope [19]. Polyclonal Abs have varying affinities for the Ag and are often used as the secondary Ab in ELISAs (Enzyme-Linked Immunosorbent Assays) and also for binding with the Ag in immunoassays where classes of compounds (such as opioids) are being analyzed instead of a specific drug. For some assays, antibodies are intentionally raised against a class of drug compounds rather than a specific parent drug because, urine for example, contains primarily drug metabolites instead of the parent drug that was consumed [17]. So, for a particular class of compounds, the same antibody assay reagent can have different affinities for the different drug metabolites in the urine and thus different concentrations of Ab-Ag complexes with which to trigger a positive result for each individual molecule or metabolite within that drug class.

For many drugs, the parent compound may not be excreted in the urine at detectable levels and the chemical structures of their metabolites may be significantly different from that of the parent drug that was consumed. These differences can be significant for Ab-Ag selectivity and binding in drug screening assays. For example, Δ^9-tetrahydrocannabinol (THC), the active constituent in cannabis, and other related cannabinoids undergo microsomal hydroxylation allylic to the Δ^9-THC double bond with two outcomes: the major product resulting in formation of an 11-CH$_2$OH moiety and the minor hydroxylation occurring on the C-8 carbon [22]. The major terminal metabolic products found in urine are generated after nonmicrosomal oxidation of the resultant 11-OH-Δ^9-THC to 11-nor-Δ^9-THC-9-carboxylic acid and to other more polar acids [22]. Also, the proportions of the two metabolites in the urine specimen vary depending on the route of cannabinoid administration. For example, after oral administration, approximately equal quantities of THC and its highly active 11-hydroxy metabolite are detected, whereas the latter metabolite is only a minor component after smoking or intravenous administration of THC [22]. Most commercially available THC immunoassays are not very specific and react with a variety of THC metabolites, whereas most confirmatory methods are designed to detect only the 11-nor-Δ^9-THC-9-carboxylic acid (Figure 14.1.), a major terminal THC metabolite in urine [8].

Of note, the use of synthetic cannabinoids has increased in popularity since 2008 due to their relative ease of acquisition via the Internet, in some gas stations, certain convenience stores, and smoke shops. Synthetic cannabinoids are laboratory synthesized chemicals sprayed on herbs and are purported to mimic the effects of natural THC. Some of these synthetic cannabinoids have similar chemical structures, which allow binding to the cannabinoid receptors in the human body. The inability of current standard drug screening immunoassays to detect synthetic cannabinoids has also added to their overall appeal among users. However, synthetic cannabinoids are known to be metabolized in the liver via hydroxylation

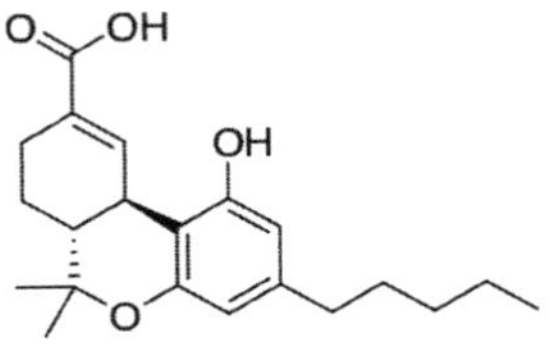

Figure 14.1 11-nor-Δ^9-THC-9-carboxylic acid or 11-*nor*-9-carboxy-delta-9-tetrahydrocannabinol (11-*nor*-9-Carboxy-THC). The main metabolite of THC which is formed after cannabis is consumed. This is the primary marker for THC use detected by UDTs.

and oxidation reactions and novel enzyme linked immunosorbent assays have been developed for the two most common synthetic cannabinoids JWH-018 and JWH-250 [23]. The immunoassays were calibrated with the hydroxylated metabolites of each synthetic cannabinoid, validated with urine samples containing their metabolites and confirmed with LC-MS resulting in greater than 95% sensitivity and specifity for both assays [23].

Like enzymes and other biological molecules with specific binding interactions, Abs (large glycoproteins) and their binding sites are sensitive to their environmental conditions. Special care has to be taken to properly immobilize the Abs on the assay. Improper orientation of antibodies on the solid surface can lead to loss of their biological binding activity because the two Ag binding sites on the Y-shaped arms of the antibodies have to be exposed in a specific orientation to the Ags. Also, the density of the Abs on the surface cannot be too high in order to minimize crowding and the resulting steric hindrance between Abs and Ags [24, 25].

14.3.3   Common Immunoassay Formats for Drug Screening

14.3.3.1   Enzyme Immunoassays

Quantitative enzyme immunoassays (EIAs) were introduced as an alternative to radioimmunoassays by Engvall, Perlmann, Van Weemen, and Schuurs in 1971 [26]. The previously used radioactive labels such as the isotopes iodine-125 and phosphorus-32, which would indicate the formation of an Ab-Ag complex in the sample, were replaced by safer and cheaper bioselective enzyme labels. Chosen enzymes have high selectivity for specific substrates, and can provide large signal amplification due to their high turnover rate, which ultimately results in the detection of very low concentrations of Ag. However, the catalytic activity of enzyme labels is often affected by reaction conditions such as temperature, pH, and ionic strength which may have to be controlled during the detection step.

14.3.3.1.1   Enzyme-Multiplied Immunoassay Technique (EMIT)
EMIT is a commonly used homogeneous enzyme immunoassay method for detection of relatively small molecules such as drugs. In this competitive assay, the drug molecules of interest (antigen, Ag) and enzyme-labeled drug molecules compete for a fixed, limited quantity of antibody (Ab) [27]. The Ag is immobilized onto the enzyme in such a way that the presence of an Ab restricts access of the enzyme substrate to the active site where the formation of the product is usually catalyzed, thereby inhibiting the enzyme activity. The Y-shaped Ab prevents the enzyme from undergoing the necessary conformational changes that would lead to induced fit and subsequent

substrate binding. The enzyme-labeled drug that is unbound, and thus free in solution, exhibits high enzyme activity, which can be detected as a change in absorbance or other measurable signal (Figure 14.2A). If there is no drug of abuse in the urine sample, the antibody binds to the enzyme-labeled drug and the enzyme activity decreases dramatically (Figure 14.2B). This may be due to structural changes at the active site of the enzyme from the Ab binding or due to increased steric hindrance. Ultimately, the enzyme activity increases as the unlabeled free drug concentration increases. The commonly used enzyme labels for EMIT assays are lysozyme, horseradish peroxidase, β-galactosidase, and glucose-6-phosphate dehydrogenase.

14.3.3.1.2   Cloned Enzyme Donor Immunoassay (CEDIA)

CEDIA is a homogenous, competitive enzyme immunoassay with an unusual assay approach, which utilizes recombinant DNA technology. The assay originally developed by Microgenics (now Thermo Scientific, Fremont, CA) depends on genetically engineered enzyme β- galactosidase having two types of individually inactive units: the Enzyme Acceptor (EA) unit and the Enzyme Donor (ED) unit. The ED unit has been conjugated with drug molecules for binding the antidrug antibody during the assay. During the assay, the drugs in the sample and the ED units with drug compete for available Ab binding sites. The free EA and ED units spontaneously reassemble into complete monomers, and ultimately, an active tetrameric enzyme in the presence of competing drug molecules from the urine sample (Figure 14.3A). The active enzyme reacts with its specific substrate catalyzing the formation

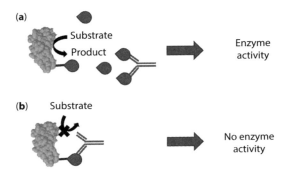

Figure 14.2 Enzyme-multiplied immunoassay technique (EMIT). A) Enzyme conjugated antigen exhibits high enzyme activity when it is free in solution. The capture antibody (Ab) ≻ is bound to free drug antigen (Ag) ● from the specimen. B) At low Ag concentration, the Ab binds to the labeled Ag and the enzyme activity is significantly diminished or absent due to steric hindrance or structural changes at the active site upon binding of the large Ab molecule.

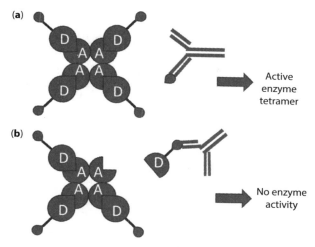

Figure 14.3 Cloned Enzyme Donor Immunoassay (CEDIA). A) When antigen (Ag), ●, from the specimen is available to bind with the antibody (Ab), ⤙, all Enzyme Acceptor (A) and Enzyme Donor (D) fragments are free to form complete monomers and subsequently the active tetrameric enzyme complex. B) In the absence of drug antigen, the Enzyme Donor fragment bound antigen is captured by the Y-shaped antidrug antibody and steric effects prevent the formation of a complete Donor-Acceptor monomer that is a vital part of the active enzyme tetramer resulting in an inactive enzyme in the sample.

of a product which efficiently absorbs visible light at 570 nm. Therefore, the absorbance of light increases over time as the active enzyme generates more product. This increase in absorbance is proportional to the quantity of drug found in the urine sample. In the absence of a drug of abuse in the urine sample, the ED-drug conjugate binds to antidrug antibodies thereby preventing the formation of an active tetrameric enzyme complex and resulting in a constant absorbance signal vs. time (Figure 14.3B).

14.3.3.1.3   Enzyme-Linked Immunosorbent Assays (ELISA) or Sandwich
              Immunoassays

Sandwich immunoassays are heterogeneous assays which use solid phase immobilized Abs to capture an Ag. Sandwich immunoassays are often referred to as enzyme-linked immunosorbent assays (ELISA) because the Ab or the Ag is immobilized on a solid surface such as a polystyrene well, a bead, or a membrane. Having the Abs of the ELISA physically immobilized makes it easy to separate bound from unbound sample components during the washing steps of the heterogeneous assay format [26].

ELISAs are typically semiquantitative or quantitative analytical assays that can be automated relatively easily. They have much lower limits of detection than competitive EIAs because, in the sandwich format, Ag capture can be driven to near-completion under optimized reaction conditions [28]. However, use of the ELISA format is limited to relatively large Ags (like opiates) because a sufficiently large surface area is required on the Ag molecule to simultaneously accommodate two bound Abs (the primary and secondary Abs). Sandwich EIAs are also relatively labor intensive and time consuming compared to other assay types because of the multiple incubation and washing steps required.

Figure 14.4 below details the main steps involved in utilizing a sandwich enzyme immunoassay. A nonspecific binding blocker is added and incubated to coat all exposed solid surfaces and minimize background signal that would result from immunoassay reagents binding to exposed areas. Addition of the biological (in this case urine) sample which contains the drug Ag is followed by an incubation period to allow time for the primary Ab to capture the drug Ag. Incubation times and temperatures are optimized and fixed for each enzyme immunoassay procedure, so that the Ag capture is complete in each standard and sample. Next, a washing step using a buffer solution removes all materials not captured by the Abs. An enzyme-labeled Ab probe (Ab*) that is directed toward a second binding site on the Ag molecule is then added, followed by an incubation period to allow binding to occur. Then, excess unbound Ab* is washed off. At this point, captured Ag molecules are "sandwiched" between the primary and the secondary antibodies (Ab and Ab*), hence the name sandwich immunoassay. The enzyme on the secondary Ab probe (Ab*) functions as a label for each captured Ag. Finally, the enzyme substrate is added and then incubated for a few minutes prior to detecting the enzymatically formed product. Each enzyme label rapidly converts several substrate molecules into detectable product thereby

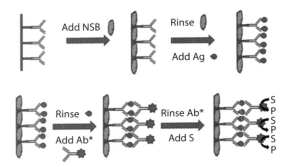

Figure 14.4 Enzyme-Linked Immunosorbent Assay (ELISA). ❘, nonspecific binding blocker (NSB); ≺ primary Ab; ●, drug Ag; ⤳ Ab*, enzyme-labeled secondary antibody; S, substrate for enzyme; and P, product of the enzymatic reaction that is detected.

resulting in significant signal amplification and allowing the detection of low Ag concentrations. The main disadvantage here is the occurrence of nonspecific binding (NSB) of the labeled secondary Ab* probe to any exposed solid surfaces of the assay which leads to increased background signal and therefore a diminished ability to detect low Ag concentrations [16, 29].

Quantification in ELISA relies on the proportionality that exists among the following factors which include: the signal response detected, the concentration of product generated by the enzyme-catalyzed reaction, the amount of enzyme label bound, the amount of Ag captured by the immobilized primary Abs, and ultimately the concentration of Ag in the urine sample. Calibration with standards of known drug concentrations are used to relate the signal response to the Ag concentration in the urine sample.

The selectivity of the assay is based on the immobilized primary Ab being able to selectively capture drug Ag from the biological specimen and on the washing steps to remove all other sample and reagent materials that might interfere with detection of the product. The secondary antibody, Ab* in ELISAs, provides a second level of selectivity prior to performing the detection step by recognizing a different region of the same Ag molecule. Incorporating these two selective binding reactions in the assay design, in addition to the washing steps, minimizes the need for significant sample preparation prior to analysis of the specimen.

14.3.4   Fluorescent Immunoassays

Fluorophore labels are used in some immunoassays to increase their sensitivity. Fluorescence is the process of a molecule absorbing energy in the form of light and then releasing the light energy at a longer, lower energy wavelength. The absorption of light excites the electrons in the fluorophore to higher energy states. Some energy is lost as heat, but a large amount is released as photons of visible light at a specific longer wavelength. Molecules that fluoresce well typically have somewhat rigid and often planar structures with aromatic groups. Fluorescein, a synthetic organic molecule, is a commonly used fluorophore label in immunoassays. Fluorescein in aqueous solution has its maximum absorbance at 494 nm and maximum emission at 521 nm. The most common fluorescent immunoassays used in drug screening are Fluorescence Polarization Immunoassays (FPIA).

14.3.4.1   *Fluorescence Polarization Immunoassay (FPIA)*

The FPIA, which was first described in 1970 by Dandliker, is a competitive homogeneous immunoassay where the antigen of the fluorophore labeled reagent and the Ag from the urine sample compete for access to an insufficient number of available Ab binding sites [30, 31]. FPIA relies on differences in the speed of rotation or tumbling of molecules due to the differences in their molecular weight. Fluorescein is a commonly used fluorophore label molecule in FPIAs [32, 33]. A polarizing lens or a prism is used in FPIA to resolve excitation light into rays in a single plane [15]. The plane polarized light is then used to excite the fluorophore molecules in the assay and their emission of partially polarized fluorescence is detected with a detector positioned at a 90° angle relative to the excitation beam originating from the light source. Polarized rays of light from the light source may excite the electrons in the fluorophores regardless of whether the labeled Ag molecules are bound to the Ab in the solution or not [15].

$$Ab + Ag\text{—}F^* \to Ab{:}Ag\text{—}F^* + Ag\text{—}F^*$$

When a drug Ag is not present in the sample, fluorophore-labeled Ag (Ag—F*) binds with the antidrug Abs. The fluorophore-labeled drug molecules bound to the higher molecular weight antibodies (Ab:Ag—F*) revolve slowly and maintain the polarization of the absorbed light. Smaller, unbound fluorophore-labeled Ag molecules (Ag—F*) that are free in solution when a drug Ag in present in the sample, rotate more freely and fast enough in the solution to depolarize the absorbed radiation resulting in low polarization (Figure 14.5A). For example, in a phenobarbitone FPIA that uses a fluorescein label, the free fluorescein labeled Ag has a molecular weight of about 500 g/mol which increases to over 150,000 g/mol upon antibody binding. The movement of the fluorophore-drug conjugate is severely restricted upon binding the antibody (Figure 14.5B). Therefore, the resulting signal detected from the sample is inversely proportional to the concentration of the drug in the urine specimen, i.e., the signal is high in the absence of drugs in the biological specimen and low when they are present. Abbott Laboratories (Abbott Park, IL) produces several types of analyzers for the detection of drugs of abuse as well as larger analytes of interest.

14.3.5   Immunoturbidimetric Assay

The particle agglutination immunoassay uses inert particles that help produce the detectable change [33]. Artificial particles made of latex or metal react with the drug molecules in the urine sample. The agglutination patterns are often so prominent as to be easily detected by the naked eye.

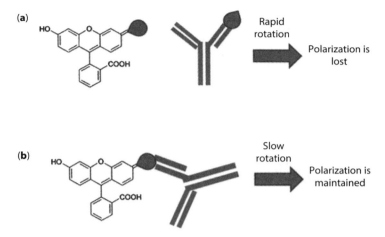

Figure 14.5 Fluorescence Polarization Immunoassay (FPIA). A) When drug Ag is present in the sample, it binds with the Ab binding sites and the fluorescein-labeled drug conjugate rotates fast enough to lose the previously absorbed polarized light. B) When no drug is present in the urine sample, the antidrug Ab binds with the fluorescein-conjugated drug resulting in the formation of a large, heavy complex that severely restricts its movement in solution. As a result, the polarization of absorbed light is maintained.

14.3.5.1   Kinetic Interaction of Microparticles in Solution Immunoassays (KIMS)

Kinetic interaction of microparticles in solution (KIMS) are homogeneous immunoassays that are essentially agglutination assays. In these assays, the particulate matter such as latex microparticles that have been conjugated with drug molecules become cross-linked by antidrug Abs in the absence of the drug in the specimen resulting in large complexes that visibly increase the turbidity of the solution (Figure 14.6A). The turbidity can be detected as increased absorbance, indicating that a binding reaction has taken place between the anti-drug Abs and the microparticle-bound Ags. When free drug molecules or their metabolites are present in the urine sample, they are captured by the antidrug Abs instead of the drug-conjugated microparticles and minimal/no aggregation or turbidity is observed in the solution, and the absorbance is much lower (Figure 14.6B). Therefore, the turbidity or the absorbance of the assay mixture is inversely proportional to the endogenous concentration of the drug analyte in the specimen. These FDA approved immunoassays that utilize this assay design in UDT are produced by Roche Diagnostics (Indianapolis, IN).

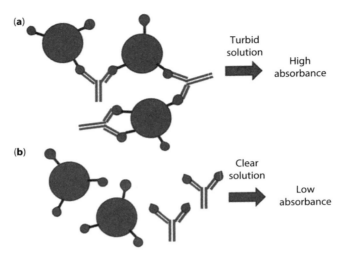

Figure 14.6 Kinetic Interaction of Microparticles in Solution (KIMS). A) The urine sample does not contain any drug Ag and microparticle-conjugated Ag become crosslinked by the capture Abs resulting in the formation of large complexes that significantly increase the absorbance of the solution. B) When the urine sample contains drug Ag, it binds and saturates the Ab binding sites. Therefore, the capture Abs will not crosslink the Ag-microparticle conjugates in the assay resulting in a relatively clear solution with low absorbance.

14.3.6   Lateral Flow Immunoassay

Beginning in the late 1970s, point-of-care testing with simplified immuno-assay formats became commonplace for workplace UDT protocols. These assays provide rapid results, are simple to use (consisting typically of only one step where a sample is applied onto the device), and yield reproduc-ible results especially when used according to the instructions. Most of these tests are solid-phase competitive immunoassays, which consist of porous, absorbent membranes with detection zones onto which the capture Abs are immobilized. The urine sample flows spontaneously through the porous membrane. The main advantage of the assay is its ability to be utilized at the point of care urine testing. Only samples with positive responses are gener-ally forwarded to the laboratory for secondary testing and confirmation.

14.4   Drug Screening with Immunoassays

14.4.1   On-Site Drug Testing

On-site drug testing with immunoassays refers to assays utilized outside of a laboratory setting, usually at the time of specimen collection, for example at homes, workplaces, or schools [9]. There are many commercially available immunoassay kits for testing urine for commonly abused drugs. Such easy to use tests include iCup Drug Screen (BioScan Screening Systems Smyrna, TN), iCassette Drug Screen Screen (BioScan Screening Systems, Smyrna, TN), Rapid K2 Synthetic Marijuana Urine Drug Test (Rapid Detect Inc., Poteau, OK), Rapid Detect Dip Drug Test 6 Panel (Rapid Detect Inc., Poteau, OK), and Rapid TOX Cup® II (Amedica Biotech Inc., Hayward, CA). FDA-approved, qualitative drug screening devices using saliva as the sample are also gaining popularity and are available for analytes such as opiates. Cozart® RapiScan Opiate Oral Fluid Drug Testing System (Opiate Cozart® RapiScan System or Opiate CRS, now Concateno, Abingdon, UK) is an example of such device. These kits yield rapid results, are easy to use, and require no prior training to perform. However, the interpretation of these tests is somewhat subjective. The cutoffs for these commercially sold test kits are not standardized and may differ significantly from those suggested by the SAMHSA. In addition, these tests are generally more expensive than those performed in laboratory settings.

14.4.2   Point of Care Drug Testing

Point of care (POC) testing of urine for the most common illicit substances is a multimillion dollar business. It is an attractive alternative to the collection, storage, transport, and laboratory analysis of a urine specimen needed to produce rapid results. In clinical settings, POC testing can be performed at various sites such as emergency departments, pain-management centers, drug treatment clinics, and detoxification clinics. Non-laboratory personnel or persons with minimal or no training on the assays used in POC are often involved in the interpretation, quality control, and documentation of the test results because the process is easy to perform and results easy to interpret.

Most POC devices are based on competitive immunoassays and are available in formats ranging from dipsticks to cards, cups, and plastic cassettes [3]. Most of these devices for are based on the immunoturbidimetric assay design or include antibodies or drugs conjugated to a chromophore or a fluorophore. In addition to urine, POC testing devices have been developed to screen for drugs of abuse in saliva. In addition, POC devices are available for the detection of a specific drug such as methadone or class of drugs such as opiates. POC test can also be used to detect the presence of several common drugs simultaneously from a single specimen. However, it is important to keep in mind that some POC testing devices may not only detect the specific drug they were designed for but others as well. For example, the POC testing device RapiTest MOP (One Step Morphine Test,

Morwell Diagnostics, Zurich, Switzerland) is designed to detect morphine but it actually also detects other opioids such as codeine [34]. As is the case with immunoassay drug screening tests performed in laboratory settings, most POC devices detect the presence of drug metabolites rather than the parent drug that was ingested.

The complexity and length of each POC testing device varies. One such device, the Triage Panel for Drugs of Abuse (ASCEND Multi Immunoassay, Biosite Diagnostics, San Diego, CA), which tests for seven drugs (PCP, benzodiazepines, cocaine, amphetamines, marijuana, opiates, and barbiturates) has three steps: 1) introduction of the urine sample to the reaction cup and incubating it for 10 minutes, 2) transferring the solution from the reaction cup to the detection area and 3) washing the detection area with three drops of the washing solution and reading the test results. The incubation period results in a total assay time of about 10 minutes. Although some devices require 15–30 minutes for analysis, in most cases results can be obtained in less than 15 minutes [3]. The interpretation of the results obtained from most devices used in POC drug testing is different from similar devices, like pregnancy tests, (which test for the hormone beta-hCG) as they do not develop a line or undergo a color change indicating a positive result. However, with certain devices such as Triage, a color line appears in the detection zone next to the name(s) of the drug(s) that is (are) present in the urine sample making interpretation of results somewhat easier.

Following manufacturer instructions closely in order to perform the test and interpret results correctly is critical when using POC testing devices as most steps involve operator input. Important steps include: introduction of the sample, possible transfer of the sample to another portion of the device, adhering to the recommended time needed to complete the test, and recording the results [2, 3]. For example, the presence of a faint line in the observation window should be interpreted as a negative result in most devices. Certain POC devices are more difficult to interpret as indicated by an increase in false-positive results from screening tests as shown by confirmatory methods [4, 35]. Also, the device will typically only show the proper response for the sample in the window for 5–10 minutes. If the person performing the test is distracted or engaged in other tasks, they may forget to read the result at the recommended time for that test, leading to false results. Other POC testing devices such as the LifeSign Status DS (LifeSign, Skillman, NJ), the American Bio Medica Rapid Drug Screen (American Bio Medica, Kinderhook, NY), and the Triage TOX Drug Screens (Biosite Diagnostics, San Diego, CA) are also commercially available for purchase.

14.5   Immunoassay Specificity: False Negative and False Positive Test Results

Major limitations of immunoassays include: false negative, false positive, and pseudo-false positive test results. False negatives can result from improper sample storage or treatment, reagent deterioration, or improper washing techniques when using the assay. In addition, the co-ingestion of prescription or over-the-counter drugs may also result in false negatives in certain UDTs [8].

False positives may be caused by other molecules present in the urine sample that cross-react or bind to the capture Ab in the immunoassay. Some of these interfering molecules include relatively common and legal substances (such as poppy seeds or coca leaf tea), nutritional supplements (such as hemp seed oil), over the counter medications (such as cough suppressants, antihistamines, and pain relievers) and prescription medications (such as antipsychotics, antibiotics, antidepressants, and antiepileptics). Table 14.1 lists common interfering agents contributing to false positive results in immunoassay and GC-MS analysis of urine (and/or plasma) [1, 38–42]. Abs with poor specificity may bind with molecules that possess similar chemical structures to the analyte of interest resulting in a false positive. For example, common over-the-counter (OTC) drugs such as decongestants (pseudoephedrine, Vicks inhaler), antihistamines, and appetite suppressants (phentermine) can cause false-positive results in urine immunoassays for the detection of amphetamines. The small amounts of morphine and codeine naturally present in poppy seeds may also cause false positive results if the cutoff for the assay is set low. In order to avoid obtaining too many false positive results that require time-consuming and costly confirmatory testing, the detection cutoffs/thresholds for amphetamine and opiate assays are set relatively high at 1,000 and 2,000 ng/ml respectively by the workplace drug testing programs, even though the actual detection limits of the immunoassay used are significantly lower (Table 14.1.) [8]. Interestingly, with some drug analytes, the chromatographic confirmatory test has higher limits of detection than the immunoassay screening test [8]. This can be problematic, as it may potentially create numerous, unconfirmed positive drug screening results which would ultimately erode the credibility of the drug screening process.

Immunoassays are also used to monitor therapeutic levels of certain drugs such as tricyclic antidepressants (TCAs) in order to avoid toxicity. The therapeutic ranges and toxicity levels for four commonly used TCAs are listed in Table 14.1 [38]. Even though UDTs are not routinely performed for TCAs, rapid screening for TCAs in urine is of great importance in

Table 14.1 Summary of common agents potentially contributing to positive and false positive test results in urine (and/or plasma) immunoassays. The standard cutoff values used for the Federal Workplace and other immunoassay screening and GC/MS confirmatory testing are also listed in (ng/ml).

Drug/drug class	Immunoassay screen cutoff (ng/ml)	GC/MS confirmation (ng/ml)	Common interferents
Alcohol	–	*	Short-chain alcohols (isopropyl alcohol, *etc.*)
Amphetamines/ methamphetamine	1000	500	Bupropion, *l*-Deprenyl, Desipramine, Dextroamphetamine Ephedrine, Fluoxetine, Isometheptene, Isoxsuprine, Labetalol, MDMA[a], *l*-Methamphetamine (Vick's inhaler) [b], Methylphenidate, Phentermine, Phenylephrine, Phenylpropanolamine, Promethazine, Pseudoephedrine, Ranitidine, Ritodrine, Selegiline, Thioridazine, Trazodine, Trimethobenzamide, Trimipramine
Barbiturates	300	200	Ibuprofen, Naproxen
Benzodiazepines	300	200	Oxaproxin, Setraline
Cannabinoids	50	15	Dronabinol, Efavirenz, Hemp-containing foods, Ibuprofen, Ketoprofen, Naproxen, Proton pump inhibitors (Pantoprazole, *etc.*)
Cocaine	300	150	Amoxicillin, coca leaf tea
Lysergic acid diethylamide (LSD)	0.5	0.5	Amitriptyline, Dicyclomine, Ergotamine, Promethazine, Sumatriptan
Methadone	300	200	Chlopromazine, Clomipramine, Diphenhydramine, Doxylamine, Ibuprofen, Quetiapine, Thioridazine, Verapamil

Drug/drug class	Immunoassay screen cutoff (ng/ml)	GC/MS confirmation (ng/ml)	Common interferents
Opioids, opiates, and heroin	2000	2000	Dextromethorphan, Diphenhydramine, Fluoroquinolones (Ciprofloxacin, etc.), Poppy seeds and oil, Quinine, Rifampin, Verapamil and metabolites
Phencyclidine	25	25	Dextroamphetamine, Dextromethorphan, Doxylamine, Ibuprofen, Imipramine, Ketamine, Meperidine, Thioridazine, Tramadol, Venlafaxine, O-Desmethylvenlafaxine
Tricyclic antidepressants (TCA)			Carbamazepine[c], Cyclobenzaprine, Cyproheptadine[c], Dipehhydramine[c]
Amitriptyline	120–150 (toxic at >500)[d]	[e]	Hydroxyzine[c], Quetiapine
Desipramine	150–300 (toxic at >500)[d]	[e]	
Imipramine	150–300 (toxic at >500)[d]	[e]	
Nortriptyline	50–150 (toxic at >500)[d]	[e]	

*Alcohol is not considered illegal. In most states, the Blood Alcohol Content (BAC) that is the legally accepted presumptive concentration for impairment is 100 mg/dl or 0.1%.

a Methylenedioxymethylamphetamine

b Interference has not been seen in newer immunoassays

c Interference was reported to occur in serum only

d The values represent therapeutic ranges for the specific TCA

e Concentrations of 50 to 20,000 ng/mL can be measured simultaneously in urine following enzymatic hydrolysis using GC-MS References 1, 38–42.

emergency medical situations such as intentional overdose. However, many over the counter as well as prescribed medications can lead to false positive TCA assay results (see Table 14.1 for commonly interfering agents). It is thought that the cause of the cross-reactivity between the interferents listed in Table 14.1. and TCAs is due to the fused 3-ring system (two of which are usually aromatic) which results in limited unique structural features for Ab production and ultimately decreasing the selectivity of the assay.

Pseudo-false positive tests occur when screening or confirmatory tests detect an illicit drug or its metabolite in urine when the drug was not actually ingested, but rather, may have originated from the use of a legal substance that contained detectable amounts of an illicit or licit substance or the metabolic conversion of an ingested licit or prescribed drug into a different yet structurally-related species.

The specificity of capture Abs has been reported to vary between different manufacturers of immunoassay reagents and can also affect assay results. Therefore, positive results are always considered presumptive until the urine sample undergoes confirmatory testing via gas chromatography-mass spectrometry (GC-MS) or high performance liquid chromatography (HPLC) coupled with mass spectrometry (LC-MS). Commonly used confirmatory testing methods will be discussed in the next section.

To improve detection of drugs of abuse, drug screening immunoassays are often calibrated with a single representative drug within a class of drugs. A description of a barbiturate assay with a cutoff value of 200 ng/ml may give the false impressions that the assay can be used in urine drug screening to detect any barbiturate at that concentration when that cutoff is actually only accurate for secobarbital. Other related drugs such as amobarbital, pentobarbital, and phenobarbital may vary widely in their ability to bind to capture Abs in the same immunoassay resulting in differing cutoff values.

New drugs that may interfere and cause false-positive results with specific immunoassays continue to be identified. Fenderson et. al. recently reported the first case of atomoxetine (a selective norepinephrine reuptake inhibitor commonly prescribed as a nonstimulant treatment for ADHD) and/or its metabolites resulting in a false-positive urine toxicological drug screen in a young female patient with CEDIA developed for amphetamine detection.[36]

14.6   Confirmatory Secondary Testing Using Chromatography Instruments

Gas chromatography–mass spectrometry (GC-MS) analysis of urine can be used to confirm the identity of a specific drug of abuse and/or its relevant metabolites. Liquid chromatography–mass spectrometry (LC-MS) UDT is typically used to detect for specific categories of illicit drugs rather than an individual drug or its common metabolite. A typical drug screen confirmatory testing procedure begins with sample preparation by hydrolysis (where applicable), organic solvent extraction of the drug metabolites, and derivatization (when necessary) [43]. Confirmatory testing of drugs of abuse such as cannabis, cocaine, amphetamine, and phencyclidine by GC-MS requires only 0.2–0.5 mL of urine sample [43]. The concentrations of major metabolites are confirmed by comparing the abundance of major ions (i.e., y-axis peak heights that are indicative of the species concentration) and the observed retention times of drug metabolites in the urine sample (i.e. x-axis values for major peaks used to confirm the identity of the chemical species) with those of the established analytical standards. Libraries of mass spectra for common illicit and licit drugs can also help confirm the identity of the drug of abuse as each chemical species has a unique mass spectrum with series of sharp lines that can act as a "chemical fingerprint." All newly developed GC-MS methods are tested against urine samples already screened to be positive or negative by well-validated immunoassays to establish the accuracy of the new methods.

14.6.1   Gas Chromatography–Mass Spectrometry (GC-MS)

Gas chromatography (GC) is an instrumental chromatographic method that can be used to separate volatile organic compounds. A GC-MS consists of a flowing mobile phase (MP), a heated injection port, a long column containing the stationary phase (SP), which aids in separation of the different sample components, and a detector (in this case a mass spectrometer (MS)). The MP is generally an inert gas such as helium, argon, or nitrogen. The injection port for sample introduction is maintained at a higher temperature than the boiling point of the least volatile component in the sample. The GC column, the heart of the instrument, is housed in a temperature-controlled oven. Long (5–100 m) capillary GC columns with small (0.05–0.5 mm) inner diameter are usually used and only require microliter volumes of the sample. The drugs and/or their metabolites contained in the urine sample are separated due to differences in their partitioning behavior when in the mobile gas phase or the stationary phase (contained in the GC columns) and thus, arrive at the detector at different times.

Mass spectrometers (MS) are extremely sensitive detectors commonly used in various tandem instrumental analysis approaches such as in GC-MS. A mass spectrometer utilizes the difference in mass-to-charge ratio (m/z) of ionized gas-phase sample molecules and their fragments to separate them, in space or time, inside a mass analyzer resulting in sharp peaks or lines on a signal vs. m/z spectrum. Mass spectrometers can be used to obtain chemical and structural information such as the molar mass of the sample molecule via analysis of a molecular ion peak, which helps identify the drug molecule or its metabolite. The molecular ion peak (M^+) is often the peak with the highest molecular weight and lowest intensity. M^+ represents an ion, which consists of the sample molecule minus an electron. Different molecules have distinctive fragmentation patterns inside a MS, akin to a chemical fingerprint, that allows the determination of their structural components such as functional groups, molecular fragments and ultimately the molecular identity. The magnitude of the MS signal at each m/z peak may also be used to quantify the molecules in the specimen.

The series of events that occur inside a typical MS begin with the ionization of the gas-phase sample, thereby generating gas-phase ions. During the separation step, a mass analyzer uses electric and magnetic fields to change the speed and direction of the ions according to their mass-to-charge ratios. There are various different types of mass analyzers such as quadrupole and time-of-flight to name a few. Eventually the ions terminate their journey by striking the sensor of the detector, which measures the electrical charge induced or the current produced by the ions. The entire MS is in a high-vacuum system, which is required to avoid unnecessary collisions between the sample ions and air molecules prior to reaching the detector.

When the combination of an immunoassay screening test and a GC-MS confirmatory test both give a positive test result for the presence of an illicit drug, the proof to a legal standard of certainty has been achieved. Other uses for GC-MS in a clinical or forensic setting include the accurate determination of blood alcohol level and the identification of unknown substances.

14.6.2 Liquid chromatography–Mass Spectrometry/Mass Spectrometry (LC-MS/MS)

Liquid chromatography–mass spectrometry/mass spectrometry is a hybrid instrumental analysis method that combines the physical separation capabilities of high performance liquid chromatography (HPLC) with the structural, chemical and mass analysis capabilities of mass spectrometry detection that were discussed above. HPLC is a popular automated chromatographic technique in which the liquid mobile phase flows through an analytical column packed with solid stationary phase (SP) particles. Unlike GC, HPLC can also be used to analyze nonvolatile sample components that have higher molecular weights. The physical separation and the time of elution are determined by the intermolecular interactions of the sample components with the SP particles inside the column vs. the mobile phase (MP).

Validated LC-MS/MS procedures exist as confirmatory tests for many drugs of abuse. However, these procedures can be very time-consuming as significant amount of sample preparation including extractions and derivatization steps are required prior to analyzing the sample. For example, a method to detect metabolites of five synthetic cannabinoids (JWH-018, JWH-019, JWH-073, JWH-250 and AM-2201) consists of adding two internal standards to aliquots of urine sample that are first enzymatically hydrolyzed at pH 5.5, then made basic with the addition of sodium hydroxide and extracted into methyl t-butyl ether [23]. Following evaporation, the samples were reconstituted using a mixture of water and mobile phase for chromatography prior to analysis by LC-MS/MS. Confirmation of positive results from analysis was based on signal responses that exceeded those of the cutoff calibrators as well as correlating retention times and transition ion ratios within defined limits [23].

Conclusion

Urine drug screening plays an important role in modern health care, athletics, the military, legal and criminal justice system. For instance, it can be used to detect drug abuse at the workplace, drug overdose, monitoring therapeutic drug levels as well as adherence to a prescribed drug regimen at a pain management clinic and certain other medical settings. In addition, urine drug testing with immunoassays is an easy, noninvasive method, which provides access to rapid results. Also, easy-to-use point of care urine drug screening devices that can be used in homes, the workplace or in clinical settings are available from numerous manufacturers. However, urine specimen validity must be considered by personnel handling the samples prior to any analytical testing. Due to certain inherent limitations such as cross-reactivity between assay Abs and molecules that are structurally similar to the target analyte (a drug), it is recommended that validation of routinely used immunoassays be performed for each analyte. When positive results are obtained from immunoassay screening test, it is important to note that they are presumptive, and further confirmatory testing based on different separation and detection principles (i.e., chromatography-mass spectrometry) may be required. In legal settings, a combination of the urine drug screening by a well validated immunoassay with a confirmatory chromatographic test by a GC-MS or an LC-MS are used to provide reliable test results. However, the cost of and time involved in performing confirmatory testing limits their use in some hospitals.

References

1. K.E. Moeller, K.C. Lee, J.C. Kissack, "Urine Drug Screening: Practical Guide for Clinicians," *Mayo Clin. Proc.*, Vol. 83 (1), pp. 66–76, 2008.

2. S.E.F. Melanson, "Drug-of-Abuse Testing at the Point of Care," *Clin. Lab Med.*, Vol. 29, pp. 503–509, 2009.

3. I.D. Watson, R. Bertholf, C. Hammett-Stabler, B, Nicholes, B. Smith, S. George, S. Welch, A. Verstraete, B. Goldberger, "Drugs and ethanol," In: *Laboratory medicine practice guidelines: evidence-based practice for point-of-care testing.* Washington (DC): National Academy of Clinical Biochemistry (NACB), pp. 63–75, 2006.

4. S. George, R.A. Braithwaite, "Use of on-site testing for drugs of abuse," *Clin. Chem.*, Vol. 48, pp. 1639–1646, 2002.

5. C.L. O'Neal, D.J. Crouch, D.E. Rollins, A.A. Fatah, "The effects of collection methods on oral fluid codeine concentrations," *J. Anal. Toxicol.*, Vol. 24, pp. 526–542, 2000.

6. W.B. Jaffee, E. Trucco, S. Levy, R.D. Weiss, "Is This Urine Really Negative? A Systematic Review of Tampering Methods in Urine Drug Screening and Testing," *J. Subst Abuse Treat.*, Vol. 33, pp. 33–42, 2007.

7. A.G. Verstraete, "Detection Times of Drugs of Abuse in Blood, Urine, and Oral Fluid," *Ther. Drug Monit.*, Vol. 26 (2), pp. 200–205, 2004.

8. G.M. Reisfield, E. Salazar, R.L. Bertholf, "Rational use and interpretation of urine drug testing in chronic opioid therapy," *Ann. Clin. Lab. Sci.*, Vol. 37, pp. 301–314, 2007.

9. C.A. Cavacuiti, *Principles of Addiction Medicine The Essentials*, Wolters Kluwer/Lippincott Williams & Wilkins Health, Philadelphia, 2011.

10. J.T. Cody, S. Valtier, "Effects of Stealth $^{(TM)}$ adulterant on immunoassay testing for drugs of abuse," *J. Anal. Toxicol.*, Vol. 25 (6), pp. 466–470, 2001.

11. A.H.B. Wu, B. Bristol, K. Sexton, G. Cassella-McLane, V. Holtman, D.W. Hill, "Adulteration of urine by 'Urine Luck.'" *Clinical Chemistry*, Vol. 45 (7) pp. 1051–1057, 1999.

12. H.M. Phan, K. Yoshizuka, D.J. Murry, P.J. Perry, "Drug Testing in the Workplace," *Pharmacotherapy*, Vol. 32 (7), pp. 649–656, 2012.

13. R. Wong, "The Effect of Adulterants on Urine Screen for Drugs of Abuse: Detection by an On-site Dipstick Device," *Am. Clin. Lab.*, Vol. 21 (3), pp. 14–18, 2002.

14. S.A. Berson, R.S. Yalow, "Quantitative aspects of the reaction between insulin and insulin-binding antibody," *J. Clin. Invest*, Vol. 38, pp. 1996–2016, 1959.

15. R.A. McPherson, and M.R. Pincus, Eds. *Immunoassay and Immunochemistry*, authors Y. Ashihara, Y. Kasahara, and R.M. Nakamura, In Henry's Clinical Diagnosis and Management by Laboratory Methods, 21st edition, W.B. Saunders, Philadelphia, PA, 851–876.

16. N.J. Ronkainen, H.B. Halsall, W.R. Heineman, "Electrochemical Biosensors," *Chem. Soc. Rev.*, Vol. 39, pp. 1747–1763, 2010.

17. S.E.F. Melanson, "The Utility of Immunoassays for Urine Drug Testing," *Clin. Lab. Med.*, Vol. 32, pp. 429–447, 2012.

18. R.L. Bertholf, M.A. Bowman, "Microbeads, magnets, and magic: the enchanting science of immunochemistry," *Ann. Clin. Lab. Sci.*, Vol. 26, pp. 377–388, 1996.

19. R.A. Goldsby, T.J. Kindtm, B.A. Osborne, J. and Kuby, *Immunology*, W.H. Freeman and Company, New York, New York, USA, 2003.

20. K. Graumann and A. Premstaller, *Biotechnol. J.,*, Vol. 1 (2), p. 111, 2006.

21. S.A. Kellermann and L.L. Green, *Curr. Opin. Biotechnol.*, Vol. 13 (6), p. 593, 2002.

22. M.E. Wall and M. Perez-Reyes, "The Metabolism of Δ^9-Tetrahydrocannabinol and Related Cannabinoids in Man," *Journal of Clinical Pharma*, Vol. 21, pp. 178S–189S, 1981.

23. A. Arntson, B. Ofsa, D. Lancaster, J.R. Simon, M. McMullin, B. Logan, "Validation of a Novel Immunoassay for the Detection of Synthetic Cannabinoids and Metabolites in Urine Specimens," *J. Anal. Toxicol.*, Vol. 37, pp. 284–290, 2013.

24. B. Lu, M.R. Smyth, R. and Kennedy, *Analyst*, Vol. 121 (3), p. R29, 1996.

25. D. Grieshaber, R. MacKenzie, J. Vörös, and E. Reimhult, *Sensors*, Vol. 8, p. 1400, 2008.

26. N.R. Ronkainen-Matsuno, H.B. Halsall, W.R. Heineman, In *Immunoassay and Other Bioanalytical Methods*; J.M Van Emon, Ed.; CRC Press: Boca Raton, FL, 2007; pp. 385–402.

27. S.R. Mikkelsen, E. Cortón, *Bioanalytical chemistry*; Wiley-Interscience: Hoboken, NJ, Vol. 1, pp. 1–346, 2004.

28. V.C. Yang, T.T. and Ngo, *Biosensors and Their Applications*, Kluwer Academic/Plenum Publishers, New York, 2000.

29. B. Eggins, *Biosensors—An Introduction*, Wiley Teubner, 1–8, 149–151, 1999.

30. Dandliker, W.B., de Saussure, V.A. (1970). Fluorescence polarization in immunochemistry. *Immunochemistry* 7:799–828.

31. W.B. Dandliker, R.J. Kelly, J. Dandliker, J. Farquhar, J. Leirn, *Immunochemistry*, Vol. 10, pp. 219–227, 1973.

32. D. Pérez-Bendito, A. Gómez-Hens, A. Gaikwad, "Direct stopped-flow fluorescence polarization immunoassay of abused drugs and their metabolites in urine," *Clin. Chem.* Vol. 40 (8), pp. 1489–1493, 1994.

33. Y. Kasahara, "Homogeneous enzyme immunoassays," In R.M. Nakamura, Y. Kasahara, G.A. Rechnitz, eds. *Immunochemical Assays and Biosensor Technology for the 1990s*. Washington, DC: American Society for Microbiology; pp. 169–182, 1992.

34. T. Korte, J. Pykäläinen, P. Lillsunde, T. Seppälä, "Comparison of RapiTest with Emit d.a.u. and GC-MS for the analysis of drugs in urine," *J. Anal. Toxicol.*, Vol. 24, pp. 49–53, 1997.

35. C.A. Burtonwood, A. Marsh, S.P. Halloran, B.L. Smith, *Sixteen devices for the detection of drugs of abuse in urine*. Norwich UK, Medicines and Healthcare Regulatory Agency, 2005.

36. J.L. Fenderson, A.N. Stratton, J.S. Domingo, G.O. Matthews, C.D. Tan, "Amphetamine Positive Urine Toxicology Screen Secondary to Atomoxetine," *Case Reports in Psychiatry*, article ID 381261, 3 pages, 2013.

37. S. Rana, V.P. Uralets, W. Ross, "A New Method for Simultaneous Determination of Cyclic Antidepressants and Their Metabolites in Urine Using Enzymatic Hydrolysis and Fast GC-MS," *J. Anal. Toxicol.*, Vol. 32 (5), pp. 355–63, 2008.

38. Medline Plus. National Institutes of Health. Therapeutic Drug Levels. http://www.nlm.nih.gov/medlineplus/ency/article/003430.htm. Accessed June 28, 2013.

39. R.I. Hawks, C.N. Chiang, eds. *Urine Testing for Drugs of Abuse.* Rockville, MD: Department of Health and Human Services, National Institute on Drug Abuse, 1986. National Institute on Drug Abuse Research Monograph Series, No. 73. http://archives.drugabuse.gov/pdf/monographs/73.pdf. Accessed June 28, 2013.

40. N.C. Brahm, L.L. Yeager, M.D. Fox, K.C. Farmer, T.A. Palmer, "Commonly Prescribed Medications and Potential False-Positive Urine Drug Screens," *Am. J. Health-Syst. Pharm.*, Vol. 67 (16), pp. 1344–1350, 2010.

41. E.C. Vincent, A. Zebelman, C. Goodwin "What Common Substances Can Cause False Positives on Urine Drug Screens for Drugs of Abuse?" *J. Fam. Pract.*, Vol. 55 (10), pp. 893–897, 2006.

42. S. Schneider, P. Kuffer, R. Wennig, "Determination of lysergide (LSD) and phencyclidine in biosamples," *J. Chromatogr. B Biomed. Sci. Appl.*, Vol. 713 (1), pp. 189–200, 1998.

43. S.J. Mulé and G.A. Casella, "Confirmation of Marijuana, Cocaine, Morphine, Codeine, Amphetamine, Methamphetamine, Phencyclidine by GC/MS in Urine Following Immunoassay Screening," *J. Anal. Toxicol.*, Vol. 12 (2), pp. 102–107, 1988.

Index

Also of Interest

Check out these published and forthcoming related titles from Scrivener Publishing

Books marked with an * denote a title in the Advanced Materials Series

***Advanced Materials for Agriculture, Food and Environmental Safety**
Edited by Ashutosh Tiwari and Mikael Syväjärvi
Forthcoming July 2014. ISBN: 978-1-118-77343-7

***Advanced Biomaterials and Biodevices**
Edited by Ashutosh Tiwari and Anis N. Nordin
Forthcoming June 2014. ISBN 978-1-118-77363-5

***Biosensors Nanotechnology**
Edited by Ashutosh Tiwari and Anthony P. F. Turner
Forthcoming June 2014 ISBN 978-1-118-77351-2

***Advanced Sensor and Detection Materials**
Edited by Ashutosh Tiwari and Mustafa M. Demir
Forthcoming May 2014. ISBN 978-1-118-77348-2

***Advanced Healthcare Materials**
Edited by Ashutosh Tiwari
Published 2014 ISBN 978-1-118-77359-8

***Advanced Energy Materials**
Edited by Ashutosh Tiwari and Sergiy Valyukh
Published 2014. ISBN 978-1-118-68629-4

***Advanced Carbon Materials and Technology**
Edited by Ashutosh Tiwari and S.K. Shukla
Published 2014. ISBN 978-1-118-68623-2

Introduction to Surface Engineering and Functionally Engineered Materials
By Peter Martin
Published 2011 ISBN 978-0-470-63927-6

Encapsulation Nanotechnologies
Edited by Vikas Mittal
Published 2013. ISBN 978-1-118-34455-2

Introduction to Industrial Polypropylene: Properties, Catalysts, Processes
by Dennis P. Malpass and Elliot Band.
Published 2012. ISBN 978-1-118-06276-0

Introduction to Industrial Polyethylene: Properties, Catalysts, Processes
by Dennis P. Malpass.
Published 2010. ISBN 978-0-470-62598-9

Handbook of Bioplastics and Biocomposites Engineering Applications
Edited by Srikanth Pilla
Published 2011. ISBN 978-0-470-62607-8

Biopolymers
Biomedical and Environmental Applications
Edited by Susheel Kalia and Luc Avérous
Published 2011. ISBN 978-0-470-63923-8

Renewable Polymers: Synthesis, Processing, and Technology
Edited by Vikas Mittal
Published 2011. ISBN 978-0-470-93877-5

Polymer Nanotube Nanocomposites
Synthesis, Properties, and Applications
Edited by Vikas Mittal.
Published 2010. ISBN 978-0-470-62592-7

Incredibly **Disgusting** Drugs™

Steroids

Doretta Lau

rosen publishing's
rosen central®

New York

For Mom and Dad

Published in 2008 by The Rosen Publishing Group, Inc.
29 East 21st Street, New York, NY 10010

First Edition

Library of Congress Cataloging-in-Publication Data

Lau, Doretta.
Steroids / Doretta Lau.—1st ed.
    p.cm.—(Incredibly disgusting drugs)
Includes bibliographical references.
ISBN-13: 978-1-4042-1376-0 (library binding)
1. Anabolic steroids—Toxicology—Popular works. 2. Doping in sports. I. Title.
RC1230.L38 2007
362.29—dc22

2007029657

Manufactured in the United States of America

Contents

Introduction

teroids should not be used without a doctor's permission. If abused, they will wreak havoc, ravaging the mind and body the way locusts devour a field of crops. Like any other drug, steroids will make you sick if used improperly. Many of the side effects are irreversible, so if you start abusing the drug, your health will decline and your body will suffer permanent damage.

Anabolic steroids are a man-made substance similar to the male sex hormones known as androgens. Slang words for steroids include "'roids," "gear," "juice," "gym candy," "pumpers," "Arnolds," and "weight trainers." There are more than 100 different types of anabolic steroids, which can be taken orally, injected, or absorbed through the skin via gels and creams. According to the National Institute on Drug Abuse (NIDA), the most commonly abused forms of the drug are Anadrol (oxymetholone), Oxandrin (oxandrolone), Dianabol (methandrostenolone), Winstrol (stanozolol), Deca-Durabolin (nandrolone

decanoate), Durabolin (nandrolone phenylpropionate), Depo-Testosterone (testosterone cypionate), and Equipoise (boldenone undecylenate).

Medical professionals first used steroids in the late 1930s to treat a condition in boys called hypogonadism, in which the testicles produce too little testosterone, delaying puberty and preventing normal growth. During the Second World War, German scientists tested the effects of steroids on prisoners and soldiers. In 1945, doctors used steroids to help World War II concentration camp survivors regain lost weight and rebuild muscle mass. By the 1950s, athletes from Soviet and Eastern European countries were using steroids to enhance their performance at sporting events like the Olympic Games. The success of the Soviet and Eastern European athletes prompted competitors from other countries to follow suit, creating a legacy of doping that has affected amateur and professional sports to this day.

Some athletes mistakenly think that taking anabolic steroids will make them swifter and stronger, and help them in their quest to win gold medals and championships. But the truth is that long-term usage of the drug makes a person slower and weaker. Simply put, the misuse of anabolic steroids can lead to illness and possible death. What good is a gold medal if you're in poor health or, worse yet, dead?

In sports, the body is the most important piece of equipment, much more crucial than a pair or shoes or a helmet. Would you rip the soles off your shoes or continuously bash your helmet against a brick wall? It would be ridiculous to do something so clearly detrimental to your ability to compete. By the same token, you should care for your body as if it is

an expensive and crucial piece of sporting equipment. In fact, shoes and helmets can be replaced, whereas no amount of money can buy you a new body or good health. In order to be at your best and achieve your personal goals, you need to stay healthy. Since steroids can destroy your mental and physical health, it is best to stay far, far away from the drug.

Though sporting glory may be tempting to some, it is clear that no championship is worth the risk of taking steroids. Most steroid abusers take a dosage ten to 100 times higher than the levels prescribed for legitimate medical usage, making the drug dangerous rather than beneficial. Think about eating ice cream. One scoop is perfectly delicious, but 100 scoops in one sitting will make you sick.

The list of side effects for steroids is as long as a summer's day: stunted growth in adolescents, heart attack, stroke, cancer, severe depression leading to possible suicide, aggressive behavior (popularly known as 'roid rage), acne, liver damage, rapid weight gain, rashes, elevated blood pressure, elevated cholesterol levels, hives, cysts, and oily hair. In men, there is a possibility of prostate enlargement, shriveled testicles, breast development, reduced sperm production, and impotence. Female-specific side effects include disrupted menstrual cycles, facial hair growth, enlarged clitoris, and a deepened voice. Clearly, abusing steroids can have monstrous effects. In physical terms, men become more like women, while women become more like men.

Anabolic steroid usage has been banned by a number of athletic governing bodies, including the International Amateur Athletic Federation, International Federation of Bodybuilders, International Olympic Committee,

Sporting fans at a major league baseball all-star game show their displeasure at the alleged prevalence of anabolic steroid usage by top baseball players.

National Collegiate Athletic Association, and National Football League. Athletes who use steroids are deemed cheaters and may be permanently banned from competition. Can you imagine if you could no longer do the sport you loved dearly because you decided to cheat? A lifetime's hard work goes down the drain if one abuses steroids.

Not only are large doses of anabolic steroids terrible for your health and unethical, possession without a prescription is a criminal offense. In the United States, illegal possession is punishable by fines and a prison

sentence of up to five years. In order to obtain the massive dosages required for doping, one would have to buy the drugs off the black market. Drugs obtained in this manner are dangerous because they are not regulated by national and international governing bodies. Black market dealers will not necessarily care if the correct doses and chemicals are present in the drugs they are pushing. Profit comes before health and safety for illegal drug manufacturers and distributors, many of whom are not trained pharmacists.

Without a doubt, taking steroids without medical supervision is a terrible life choice. Ask yourself this: would you get in an elevator that you knew was going to free-fall, causing a serious injury or fatality? Of course you wouldn't! Staying away from steroids is simply common sense. It's just like making sure that you don't get into a faulty elevator. It's the smart thing to do.

1
What Are **Steroids,** and Why Are They **Banned** from Sports?

Anabolic steroids are a synthetic form of the male hormone testosterone. That means that the drug doesn't occur naturally. It must be made in a laboratory. Hormones control the body's growth, development, and reproductive capabilities in both men and women. This means that adding extra testosterone to the body affects everything related to these three things, which can have unwelcome consequences. A man's body naturally makes 2.5 to 11 milligrams of testosterone a day, whereas an anabolic steroid abuser takes doses that may exceed 100 milligrams. Think of it this way: abusing anabolic steroids is like pouring a giant tank of water into an already full glass. The large volume of water cannot possibly fit into the small glass. Water flows everywhere, resulting in a giant mess that could have been easily prevented. The same idea can be applied to anabolic steroids. If you don't abuse steroids, your body won't

be flooded with an excess of testosterone, and therefore you won't end up with a giant mess on your hands.

According to the National Institute on Drug Abuse (NIDA), anabolic-androgenic steroids are "man-made substances related to male sex hormones." "Anabolic" refers to muscle building, and "androgenic" refers to increased masculine characteristics. "Steroids" refers to the class of drugs. There are often newspaper, radio, and television reports about sporting scandals involving anabolic steroids because some athletes abuse the muscle-building element of the drug as a shortcut to success. Rather than training hard, eating right, and sleeping well, some athletes decide to cheat, tarnishing the outcome of sporting events for both competitors and spectators. No one really wins when anabolic steroids come into play.

Medical Uses

Due to the numerous sporting scandals, anabolic steroids have become synonymous with cheating. There are, however, legitimate medical uses for small doses of the drug. Anabolic steroids help the body synthesize protein, build muscle mass, and increase strength, appetite, and bone growth. These characteristics make the drug useful in treating many ailments and diseases.

During the late 1930s, European researchers developed steroids to treat hypogonadism, a male-specific condition in which the testes do not produce an adequate amount of testosterone, resulting in insufficient development and growth. Medical professionals also discovered that anabolic steroids could rebuild disease-ravaged tissue in the human body.

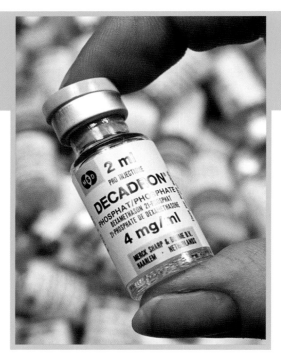

A vial of steroids intended for intravenous use. This means that this form of the drug is meant to be injected into a vein with a needle and syringe.

When the Second War World ended, doctors gave steroids to starving concentration camp survivors to repair tissue damage in their bodies and help them gain weight.

Between the 1960s and 1980s, pediatric endocrinologists prescribed anabolic steroids to children suffering growth failure. However, due to the social stigmatization surrounding steroids and the development of synthetic growth hormone, doctors stopped using the drug to treat growth failure. Anabolic steroids were also used to stimulate the bone marrow in patients with hypoplastic anemia, but medical professionals now prefer to use synthetic protein hormones as treatment for this ailment.

Today, doctors prescribe steroids for patients suffering from chronic wasting conditions such as AIDS and various forms of cancer to help stimulate appetite and preserve existing muscle mass. Medical professionals also use steroids in cases of gender dysmorphia to help increase red blood

cell levels, to help deepen the voice, and to increase bone, muscle mass, and facial hair. Testosterone is also given to boys with delayed puberty, helping increase height and muscle mass.

Use in Sports

When scientists discovered that anabolic steroids helped increase the growth of skeletal muscle, weight lifters and bodybuilders added the drug to their training regiments, unaware of the long-term side effects and adverse health risks that accompanied the positive aspects of the drug. By the 1950s, Soviet and Eastern European athletes in other sports began using anabolic steroids to improve their results, not questioning the ethics of their choice. During the mid-1970s, the International Olympic Committee acknowledged that anabolic steroids were providing an unfair and risky advantage and officially banned athletes from using the drug.

At the 1984 Olympic Games, officials administered random drug tests, and half the athletes registered a positive result for steroid usage. Yet, it wasn't until the 1988 Summer Olympics in Seoul, Korea, that steroid abuse in amateur sports received serious worldwide attention. Canadian superstar sprinter Ben Johnson, who had won the 100-meter race in a world-record time of 9.79 seconds on September 24 against longtime rival Carl Lewis, tested positive for the anabolic steroid stanozolol. On September 27, officials disqualified Johnson. He had previously been the 1987 100-meter world champion and had won two bronze medals at the 1984 Olympic Games. But after proof of

Canadian sprinter Ben Johnson celebrates his world record–breaking 9.79-second
100-meter race win at the 1988 Summer Olympics in Seoul, Korea. Three days
later, officials disqualified Johnson because he tested positive for the anabolic
steroid stanozolol.

his steroid use was revealed, he was stripped of his gold medal and suspended from competition until 1991. The third-place finisher in the race, Linford Christie, later tested positive for banned substances and was also barred from competition.

Throughout the media storm at the 1988 Seoul Olympics, Johnson denied doping. In response to the fallout from Johnson's disgrace on the world stage, in January 1989, the Canadian federal government launched the Commission of Inquiry Into the Use of Drugs and Banned Practices Intended to Increase Athletic Performance. Ontario Appeals Court justice Charles Dubin presided and the proceedings became known as the Dubin Inquiry. The inquiry lasted ninety-one days and featured hundreds of hours of testimony by 122 witnesses. Coaches, athletes, sporting officials, doctors, and members of government testified that there was a widespread performance-enhancing drug problem among athletes. During this time, Johnson admitted to abusing anabolic steroids. His coach, Charlie Francis, reported that Johnson had been a steroid user since 1981.

Johnson attempted a comeback in 1991, but he didn't come close to matching his former glory. At the 1992 Barcelona Olympics, he wasn't able to make the finals in the 100m. In 1993, he tested for steroids once again at a race in Montreal. This time, he received a lifetime ban from competition. At one time, Johnson commanded hundreds of thousands of dollars a month in endorsements. Those days are now long gone.

The History of Doping and International Anti-Doping Agencies

When an athlete abuses a banned substance or method, it is called doping. Athletes have tried all sorts of methods since the early days of competitive sport to gain an edge. Ancient Greek athletes used stimulants, while nineteenth-century cyclists and endurance athletes used strychnine, caffeine, cocaine, and alcohol to improve their results. During the 1904 Saint Louis Olympic marathon, Thomas Hicks won with the help of strychnine injections and doses of brandy, both administered during the race.

The term "doping" first came into use during the twentieth century to refer to the illegal drugging of racehorses. In 1928, prior to the development of anabolic steroids, the International Amateur Athletic Federation became the first international sporting federation to ban the use of stimulating substances. Other sporting federations elected to do the same. However, no one was testing for banned substances, so none of the anti-doping agencies could enforce the restrictions they had set out. This meant that some athletes continued to use banned substances in a quest to win.

The pressure on sporting federations to mandate drug tests increased during the 1960 Rome Olympic Games, when Danish cyclist Knud Enemark Jensen died during competition from amphetamine abuse. In 1966, cycling and football international sporting federations introduced drug tests during their world championships to help keep their competitions fair and safe. The International Olympic Committee launched a medical commission in

English cyclist Tom Simpson took a toxic cocktail of amphetamines, alcohol, and diuretics during the 1967 Tour de France. He died of exhaustion while attempting to climb Mont Ventoux during the thirteenth stage of the race.

1967, creating a list of banned substances. That same year, cyclist Tom Simpson died during the Tour de France from doping complications, which prompted sporting federations to tackle the doping problem. By 1968, there were drug tests at the Grenoble Winter Games and the Mexico City Summer Games.

By the 1970s, most international sporting federations had mandated drug testing. At first, however, there was no proper test to detect anabolic

steroid abuse, so athletes were widely abusing the drug for personal gain. By 1974, researchers had developed a reliable testing method, prompting sporting federations to officially ban anabolic steroids. Drug-related disqualifications increased after the implementation.

Despite the work of international sporting federations to regulate and prohibit doping, many people suspected that some countries, such as the German Democratic Republic, or East Germany, were sponsoring doping programs. This made it difficult for the anti-doping agencies to police the use of stimulants. The world's attention on the problem of doping in sports became intensely focused after the Ben Johnson scandal at the 1988 Seoul Olympics.

Despite ongoing efforts to ban illicit substances and impose sanctions upon athletes for violating anti-doping rules, there was no one single international agency setting the rules for all sports, making it difficult for many rules to be properly enforced. In 1998, a police raid during the Tour de France revealed that many athletes were still using banned substances to get an edge. As a result, in February 1999, the International Olympic Committee headed the World Conference on Doping in Sport in Lausanne, Switzerland. The World Anti-Doping Agency (WADA), a private law foundation, was formed on November 10, 1999, in response to the conference findings. The goal of the independent agency, according to its Web site, is to "promote, coordinate, and monitor the fight against doping in sport in all its forms." The agency drafted the World Anti-Doping Code and currently works with governments and sporting agencies to keep sports drug free.

World Anti-Doping Agency's 2007 Prohibited Anabolic Steroids List

The following anabolic steroids are banned from use in competition:

1. Exogenous Anabolic Androgenic Steroids (AAS) including:

T1-androstendiol; 1-androstendione; bolandiol; bolasterone; boldenone; boldione; calusterone; clostebol; danazol; dehydrochlormethyltestosterone; desoxymethyltestosterone; drostanolone; ethylestrenol; fluoxymesterone; formebolone; furazabol; gestrinone; 4-hydroxytestosterone; mestanolone; mesterolone; metenolone; methandienone; methandriol; methasterone; methyldienolone; methyl-1-testosterone; methylnortestosterone; methyl-trienolone; methyltestosterone; mibolerone; nandrolone; 19-norandrostenedione; norboletone; norclostebol; norethandrolone; oxabolone; oxandrolone; oxymesterone; oxymetholone; prostanozol; quinbolone; stanozolol; stenbolone; 1-testosterone; tetrahydrogestrinone; trenbolone; and other substances with a similar chemical structure or similar biological effect(s).

2. Endogenous Anabolic Androgenic Steroids (AAS) including:

androstenediol; androstenedione; dihydrotestosterone; prasterone; testosterone; and the following metabolites and isomers:

5α-androstane-$3\alpha,17\alpha$-diol; 5α-androstane-$3\alpha,17\beta$-diol; 5α-androstane-$3\beta,17\alpha$-diol; 5α-androstane-$3\beta,17\beta$-diol; androst-4-ene-$3\alpha,17\alpha$-diol; androst-4-ene-$3\alpha,17\beta$-diol; androst-4-ene-$3\beta,17\alpha$-diol; androst-5-ene-$3\alpha,17\alpha$-diol; androst-5-ene-$3\alpha,17\beta$-diol; androst-5-ene-$3\beta,17\alpha$-diol; 4-androstenediol; 5-androstenedione; epi-dihydrotestosterone; 3α-hydroxy-5α-androstan-17-one; 3β-hydroxy-5α-androstan-17-one; 19-norandrosterone; 19-noretiocholanolone.

Steroid Users

Though we often hear about high-level athletes abusing anabolic steroids, not all users are sporting professionals or even recreational athletes. Some users are regular men who take the drug for cosmetic reasons. Some steroid abusers just want to look muscular and lean at the expense of good health.

Medical researchers and technicians work hard to develop new testing and detection methods for various anabolic steroids and other drugs used in doping.

The 1999 Monitoring the Future Study, which surveyed adolescent drug abuse across the United States, reported that as many as 2.7 percent of high school students use steroids. This means that in a school of 1,000 students, as many as thirty can be abusing steroids. Your brother, best friend, teammate, or lab partner might be risking a lot for very little. No teenager should abuse steroids. The personal and health risks are far too great.

2

Health
Risks

t first, a steroid abuser thinks that the drug will help him or her become faster and stronger. Or perhaps the abuser doesn't even have performance goals in mind. Maybe he or she just wants to look lean and muscular. But beneath the skin, the deadly doses required for noticeable cosmetic results—nearly 100 times greater a dosage than a doctor would prescribe for a real health problem—are hard at work destroying the body.

A person on steroids will be blind to the terrible things happening to his or her body because he or she will only see the cosmetic gains. Many steroid abusers only have the short term in mind, such as winning a race, a meet, or a tournament. But an anabolic steroid abuser is forgetting that any gains made in sports while on steroids are unethical and carry a huge health risk. An anabolic steroid abuser becomes a shadow of him- or herself. Friends and family, coaches, and officials will

Steroids

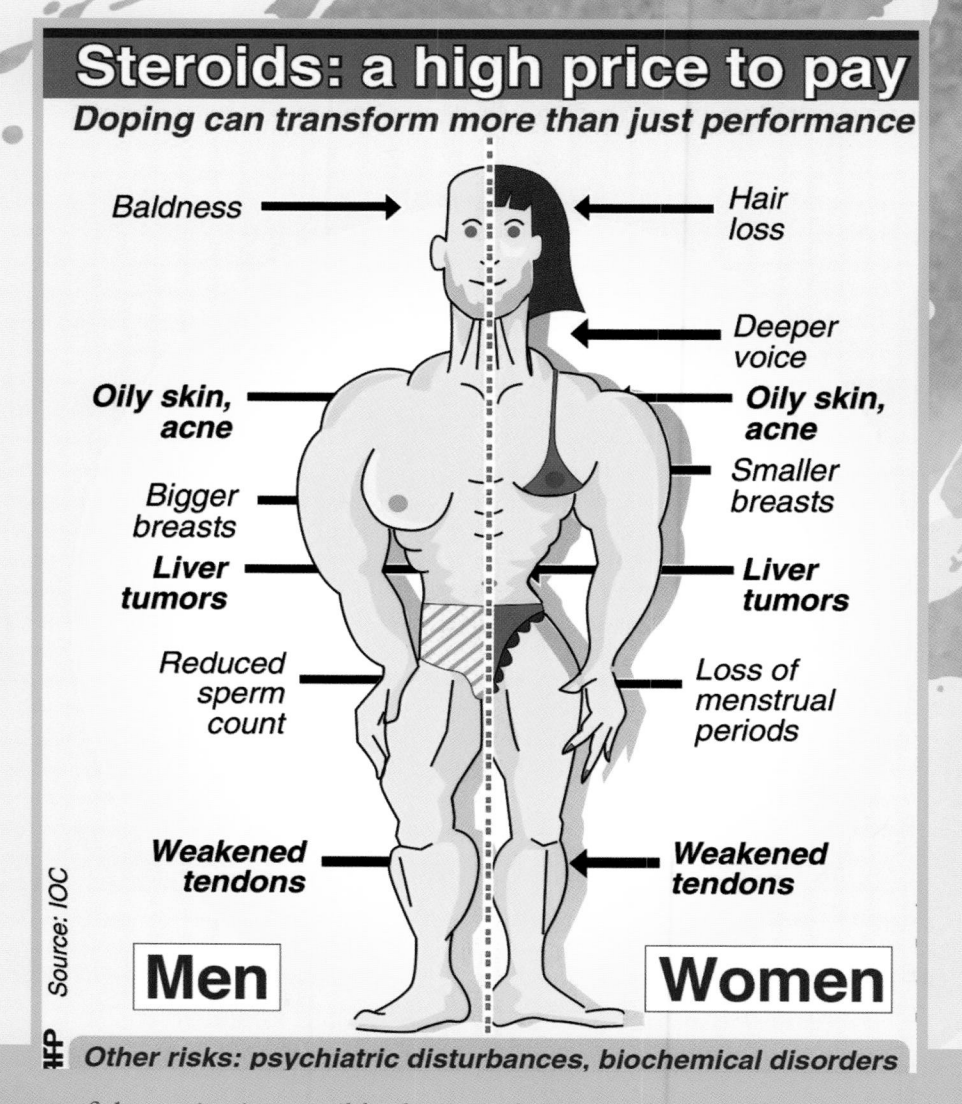

Steroids: a high price to pay
Doping can transform more than just performance

Baldness

Hair loss

Deeper voice

Oily skin, acne

Oily skin, acne

Bigger breasts

Smaller breasts

Liver tumors

Liver tumors

Reduced sperm count

Loss of menstrual periods

Weakened tendons

Weakened tendons

Source: IOC

Men

Women

Other risks: psychiatric disturbances, biochemical disorders

On top of the major irreversible damage that steroids can inflict, there are also disgusting changes to the appearance that can occur in both men and women.

be able to tell that something is amiss. Telling physical changes merge when a person abuses anabolic steroids.

Stunted Growth

Imagine taking a drug that prevents your bones from growing. This seems like a silly thing to do, doesn't it? But this is what anabolic steroids can do. Testosterone and other sex hormones tell the body when to grow and when to stop growing. Puberty begins when the body experiences a surge in hormones. If you flood your body with large doses of anabolic steroids, your brain will think that puberty is complete and will send a signal to the bones to stop lengthening. When your bones stop lengthening, you won't grow taller. Abusing anabolic steroids will stunt your growth, preventing you from becoming the strongest and fastest that you can be.

Less of a Man

Even though anabolic steroids increase muscle mass and decrease fat, which is a masculine ideal sought after by bodybuilders and other body-conscious men, the drug has a surprising side effect. Anabolic steroids have a feminizing effect on men. This means that men on steroids develop female characteristics!

Men who take large quantities of steroids sometimes develop a condition called gynecomastia, in which they grow breasts. The disruption of the normal hormone balance, which is like pouring water into an already full glass, triggers these effects. The result of gynecomastia is not like a

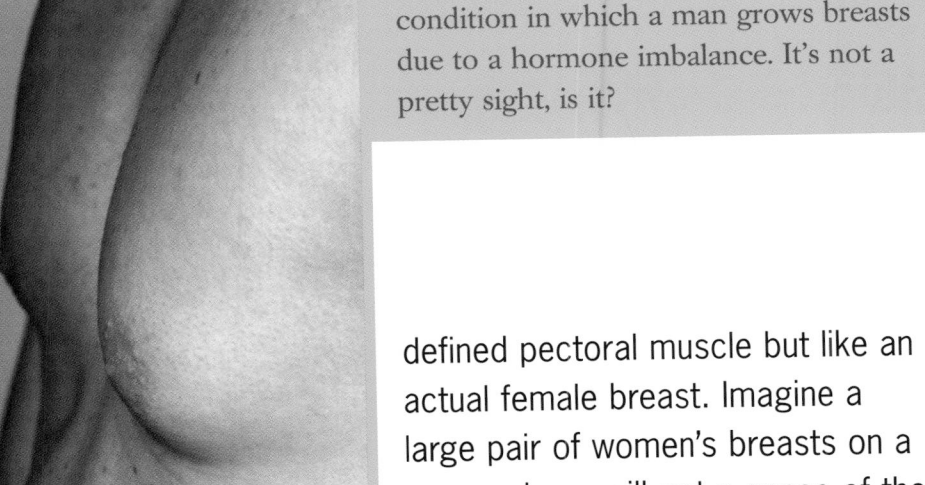

Here is an example of gynecomastia, a condition in which a man grows breasts due to a hormone imbalance. It's not a pretty sight, is it?

defined pectoral muscle but like an actual female breast. Imagine a large pair of women's breasts on a man and you will get a sense of the monstrous effect that steroids can have on a man's body.

Another feminizing effect that anabolic steroid users suffer from is shrinking testicles and a lower sperm count. The testicles and sperm count may not return to normal size and levels, even after the man stops taking the drug. This can lead to impotence, as well as infertility, meaning the man will not be able to father a child.

Less of a Woman

Women who abuse steroids have the opposite problem than men do. Instead of becoming more feminine, women who abuse steroids become more masculine.

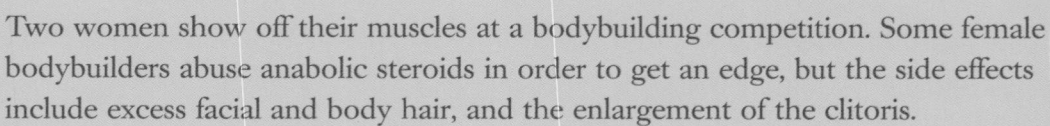

Two women show off their muscles at a bodybuilding competition. Some female bodybuilders abuse anabolic steroids in order to get an edge, but the side effects include excess facial and body hair, and the enlargement of the clitoris.

A woman on steroids will experience an irregular menstrual cycle, which affects fertility. As well, a female steroid abuser will start to grow excess facial and body hair: mustache, beard, chest hair, and super-hairy arms and legs. A woman on steroids will start looking like a freak-show spectacle: the Bearded Woman or the Hairiest Woman Alive!

Another effect that a female steroid abuser experiences is the deepening of the voice, like a boy undergoing puberty. As well, steroids cause a

reduction in body fats, which causes the breasts to shrink, giving the woman a masculine look. And since steroids are a sex hormone, the effect on the female sex organs is the enlargement of the clitoris, making the body part look like a small penis.

Bad Hair

At this moment, you probably have a full and healthy head of hair. Even with healthy hair, some days you still wake up and discover that you just can't get your hair right. No matter how hard you try, applying gel and mousse or using a hair dryer and curling iron, you're just having a bad hair day.

An anabolic steroid abuser experiences endless bad hair days. One of the major side effects of the drug is that it causes the body to secrete excess oil. This means the scalp and hair become oily, giving the steroid abuser a greasy appearance that no amount of shampoo can fix. The result is limp and dirty-looking hair.

As well, steroids cause premature hair loss in both men and women. Imagine going bald at sixteen! Male pattern baldness happens to many steroid abusers. The hair does not grow back, even after the steroid abuser stops taking the drug. Steroids create permanent damage to the hair. For women, even as an excess of facial and body hair appears, the hair on the head falls out.

You don't want to suffer from a series of endless bad hair days, do you? No product, no matter how expensive it is, can help recover a healthy head of hair.

Bad Skin

Another major side effect of abusing anabolic steroids is acne. Steroids increase the secretion of excess oils in the skin, making it a breeding ground for bacteria. The presence of bacteria leads to pore blockage. This leads to pimples. Think about having a bad pimple or painful zit. Multiply that feeling by fifty and you can begin to understand what it feels like to have anabolic steroid-induced acne. In addition to severe acne, steroid abusers may develop cysts, closed sacs with membranes that may contain air or fluids. If you think that the acne and skin problems that accompany puberty are bad, the acne that steroid abusers suffer is far worse. And it can be prevented!

Insomnia

You have probably heard people call sleep "beauty rest." During sleep, the body has time to recover and heal. Think about those nights when you just can't get to sleep. You toss and turn, count sheep, and drink warm milk, yet you can't achieve deep sleep. The state of being unable to fall asleep is called insomnia. Some anabolic steroid abusers have trouble sleeping. This means that they don't have adequate time to recover and heal. The hours spent lying in bed, unable to sleep, feel very long. An anabolic steroid abuser who suffers from insomnia can't be at his or her best for a competition, which defeats the purpose of training and competing in a sport.

Scary Consequences

3

n addition to a physical price, anabolic steroid abuse also has a mental and social price. Mental health side effects of the drug include depression, mania, delusions, and homicidal rage.

In the United States, possession of steroids without a prescription is a felony offense. You could be fined or even spend time in prison. Prison is a very scary place! You lose your freedom and are trapped in a tiny cell. You would have no privacy at all.

Stacking and Pyramiding

Many athletes who abuse anabolic steroids usually use multiple types of drugs at the same time. This usage is called stacking. Stacking adds to myriad steroid abuse dangers because it increases drug dosages, thus increasing side effects. Different drugs have different side effects. Using concurrent drugs means that the

chemicals found in each individual drug are at work, telling the body to do different things.

Another form of steroid abuse is called pyramiding. Users who follow this method escalate the number of drugs and the dosage in a cycle in order to peak for competitions. The body is tricked over and over again into thinking that it needs to increase musculature and decrease fat. Some anabolic steroid abusers also take other drugs in order to minimize the side effects. These drugs include tamoxifen, diuretics, probenecid, epitestosterone, and human chorionic gonadotropin. This cocktail of drugs, which sends conflicting signals throughout the body, can be very damaging.

Rage

One of the possible side effects of steroid usage is aggression. Some people handle their emotions and impulses well, while others do not. It's impossible to tell whether or not a steroid abuser will suddenly erupt and become

A young woman backs away in fear from a loved one who is abusing anabolic steroids.

full of rage. Some researchers have found that anabolic steroids can trigger feelings of anger or hostility, which can manifest as violent behavior. Imagine feeling angry and frustrated but not being able to control your emotions. Steroid abuse may lead to this bad place, where it seems perfectly accept-able to fight or commit armed robbery or beat up someone you love.

There is some controversy about the existence of 'roid rage.' The NIDA reports that some scientists believe that steroid abusers may act aggres-sively because popular media links steroid usage and aggression. Other scientists have conducted studies that show that test subjects on steroids behaved more aggressively than those on a placebo.

Mania

One possible psychiatric side effect of steroid abuse is mania, which is characterized by elevated energy and mood, as well as abnormal thought patterns. People in a manic state exhibit such symptoms as hypersexuality, euphoria, rapid speech, racing thoughts, decreased need for sleep, irritability, grandiosity, and a drive to participate in goal-directed activities. Individual reactions to drugs differ, but in some cases of mania, patients experience hallucinations and grandiose delusions so intensely that a hospital stay is necessary in order to protect the patient from harming him- or herself, as well as others.

Depression

Every high is followed by a low. Many steroid abusers experience depression while taking or withdrawing from the drug. Depression is an intense sadness

A young man, suffering from depression due to steroid withdrawal, sits on the street, unable to overcome the intense feeling that life is meaningless.

that takes over your life. Some depressed people cannot even get out of bed to face the day and neglect their loved ones. What's the use of taking steroids to improve in sports if you end up not wanting to participate in the sport because you're so depressed?

Sometimes, the depression becomes so great and life seems so meaningless that some people commit suicide. They have allowed steroids to take over their lives to the point where the drugs destroy them.

Delusions

Another mental or neurological illness that anabolic steroid abuse can trigger is delusions. A delusion is a false belief. In psychiatric terms, the term "delusion" implies that the belief is pathological in nature, which means that it is the result of an illness.

Some steroid abusers begin to experience religious delusions and delusions of grandeur, absolutely certain that their beliefs are real no matter how impossible or bizarre they appear to others.

Crime

Doctors do not prescribe the levels of anabolic steroids that are used for doping. High dosages of the drug can only be obtained through illegal channels. Many anabolic steroids enter the United States through drug smuggling operations. Some of the drugs may be counterfeit, as the black market is not regulated by any national or international governing bodies. The dangers of abusing anabolic steroids are very high because some of the drugs are meant for veterinary, not human, use. This means that the dosages could be for a horse or cow, rather than for a person. Imagine giving a pill meant for a human to a mouse. The pill might have very minimal effects on the human, but to the mouse the pill could be fatal. The same logic applies to drugs that are meant for horses but abused by humans. The effects may be lethal.

According to the U.S. Drug Enforcement Administration, the most common black market steroids are testosterone, nandrolone, methenolone, stanozolol, and methandrostenolone. As well, boldenone, fluoxymesterone, methandriol, methyltestosterone, oxandrolone, oxymetholone, and trenbolone are widely available in illicit trade.

Buying steroids off the black market is a crime. In the United States, it is a third-degree felony, which carries a prison sentence of up to five

shot putters C. J. Hunter and Kevin Toth; and Oakland Raiders members middle-distance runner Regina Jacobs; hammer thrower John McEwen; and client lists, which named sprinters Marion Jones and Tim Montgomery; mones. During a search of Greg Anderson's home, authorities found cash

Federal agencies raided BALCO and found steroids and growth hor-

had supplied the drug.

athletes were abusing the drug. He told the authorities that Victor Conte steroid that was not detectable during doping tests, saying that many the U.S. Anti-Doping Agency and delivered a syringe of a newly developed that the authorities caught on. A sprint coach in the United States contacted Olympic athletes and major league baseball stars. It wasn't until June 2003 that also sold food supplements. The company began supplying steroids to

BALCO was founded in 1984 as a blood and urine analysis business

track coach Remi Korchemny.

BALCO vice president James Valente, weight trainer Greg Anderson, and four main culprits named in the scandal were BALCO owner Victor Conte, hormones to high-profile athletes from the United States and Europe. The Co-Operative was supplying the steroid tetrahydrogestrinone and growth huge sports scandal. An American company called the Bay Area Laboratory In 2003, journalists Lance Williams and Mark Fainaru-Wada uncovered a

Bay Area Laboratory Co-Operative

worth expensive legal fees and jail time.

years, as well as the possibility of fines. No sporting event victory is

Chris Cooper, Bill Romanowski, Barrett Robbins, Dana Stubblefield, and Tyrone Wheatley. Marion Jones lost lucrative endorsements after this scandal came to light.

After pleading guilty to money laundering and illegal steroid distribution, Victor Conte and Greg Anderson received four months of prison time and four months of probation. Just like that, their criminal activities caused a lifetime's work in business and sports to become meaningless.

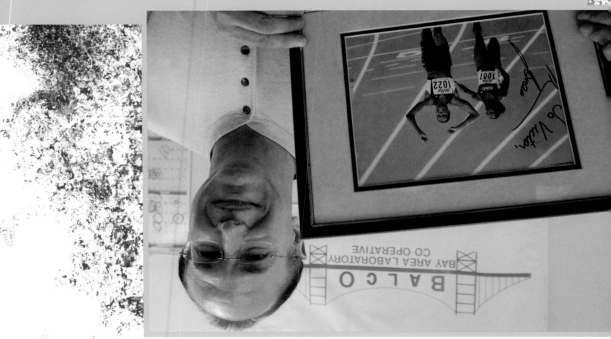

Bay Area Laboratory Co-Operative owner Victor Conte holds up a signed photograph of track star Marion Jones. Later, federal agents charged him with money laundering and illegal steroid distribution.

Dying to Win

n the surface, a steroid abuser seems to be faster and stronger than ever before. But in reality, his or her body is suffering from being poisoned with dosages that can be up to 100 times the amount a doctor would prescribe. It's just like the example of eating one scoop of ice cream, rather than 100 scoops of ice cream. When you eat one scoop of ice cream, your body has an easier time processing and turning the sugar into energy. But 100 scoops of ice cream will be stored as fat. A medical dose of steroids may be beneficial to the body, but a dosage 100 times greater becomes a poison.

Drug abuse slowly wears on the body the way the rain, wind, and sunlight will erode even the tallest mountain. Long-term use of steroids takes a toll greater than cosmetic side effects like acne and oily hair. Anabolic steroid abuse gives only short-term rewards. In the long term, taking steroids can be detrimental to the body.

Taking steroids is like giving your body a death sentence. There are many chronic and terminal diseases associated with steroid abuse, like kidney cancer, hyper- tension, prostate enlargement and cancer in men, and hepatitis B and C.

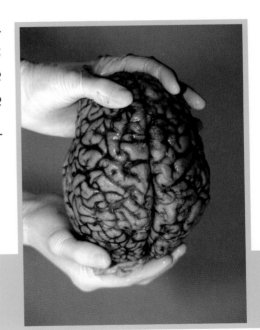

The brain is a delicate organ. Abusing anabolic steroids will destroy it.

Dangerous Infections

Steroid abusers who inject their dosages can get diseases from sharing needles or by using non-sterile injection techniques. This dangerous behavior increases the risk of contracting diseases like HIV, which can cause AIDS, bacterial endocarditis, hepatitis B and C, and other diseases that are transmitted through the blood.

AIDS makes for a long and slow death because HIV affects every organ and system in the body. Patients with full-blown AIDS suffer from severe weight loss, fever, chills, and swollen glands. AIDS damages the immune

system, making it impossible for someone who has it to fight off simple infections.

HIV makes a patient more susceptible to cervical cancer, lymphomas, and Kaposi's sarcoma. Kaposi's sarcoma is characterized by lesions on the skin, inside the mouth, and on the gastrointestinal and respiratory tracts.

Cardiovascular Diseases

Anabolic steroid abuse wreaks havoc on the cardiovascular system, causing permanent damage. Users may develop hypertension and high cholesterol, even at a young age.

Blood clots may form, leading to heart damage. The risk of atherosclerosis, in which fatty substances build up inside the arteries, increases with steroid usage. This slows the flow of blood in the body. If blood does not reach the heart, there is a chance of a heart attack. If blood is cut off from the brain, there is a chance of a stroke. Teenage athletes who abuse anabolic steroids are at risk for heart attacks and strokes.

The Liver

Doctors have linked steroid usage to a number of major illnesses. The body has trouble breaking down orally administered steroids, which ends up putting stress on the liver that can be damaging.

Steroid users who inject the drug can become infected with hepatitis B or C, which is the inflammation of the liver tissue. People ill with either disease can suffer flu-like illness, nausea, vomiting, diarrhea, loss of appetite,

weight loss, itchy skin, and jaundice. Jaundice is the condition where a person's skin and whites of the eyes become an unnatural shade of yellow, while his or her urine becomes a dark yellow and the feces become a pale color.

If someone doesn't get treatment for hepatitis B or C, he or she can develop chronic hepatitis, liver cirrhosis (the hardening of the liver tissue), or liver cancer.

Other liver problems include the condition peliosis hepatitis, where blood-filled cysts form in the liver. Both the blood-filled cysts and cancer tumors can rupture, causing internal bleeding.

Anabolic steroid users only end up suffering. Winning at a sport or developing larger muscles seems very dull when compared to the risks that are involved. Your looks suffer. Your body breaks down. You leave yourself susceptible to invasion by other diseases. The body that you have is the only one that you get. Do you really want to sacrifice it?

The liver acts as a filter in your body, processing nutrients and breaking down various substances. Drug abuse takes a toll on the liver, preventing it from per-forming its important role in the body.

Glossary

anabolic steroid Synthetic substance related to the male sex hormone testosterone.

black market An underground economy in which merchants evade applicable taxes; it can also refer to illegal activities such as drug distribution and prostitution.

cardiovascular system The heart and blood vessels.

cyst A closed sac with a membrane that may contain air or fluids.

doping The use of performance-enhancing drugs, like anabolic steroids, in sports.

hormone A chemical substance made by the glands in the body and circulated to the organs and tissues through the blood; it affects structure, function, and behavior.

juice A slang word for anabolic steroids.

placebo A substance with no chemical effects, used in experiments to distinguish between real drug effects and the drug effects expected by experiment subjects.

synthetic Made by humans, rather than occurring in nature.

testosterone A male sex hormone.

For More Information

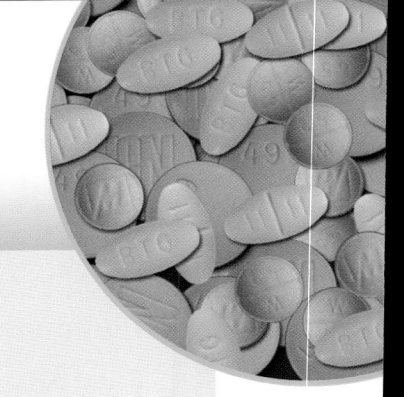

Alateen
P.O. Box 862
Midtown Station
New York, NY 10018-0862
(212) 302-7240
Web site: http://www.al-anon.org

American Council for Drug Education
136 East 64th Street
New York, NY 10021
(212) 758-8060
E-mail: acde@phoenixhouse.org

Athletics Canada
National Office
Suite B1-110 2445 St-Laurent Boulevard
Ottawa, ON K1G 6C3
Canada
(613) 260-5580
E-mail: athcan@athletics.ca
Web site: http://www.athletics.ca

Canadian Centre for Ethics in Sport
350-955 Green Valley Crescent
Ottawa, ON K2C 3V4
Canada
(613) 521-3340
E-mail: info@cces.ca
Web site: http://www.cces.ca

Drug Enforcement Administration
Demand Reduction Section
Washington, DC 20537
(202) 307-7936
Web site: http://www.usdoj.gov/dea

National Clearinghouse for Alcohol and Drug Information
P.O. Box 2354
Rockville, MD 20852
(800) 729-6686

National Families in Action
Century Plaza II
2957 Clairmont Road, Suite 150
Atlanta, GA 30329

(404) 248-9676
Web site: http://www.emory.edu/NFIA

World Anti-Doping Agency (WADA)
Maison du Sport International
Av. de Rhodanie 54
1007 Lausanne
Switzerland
Web site: http://www.wada-ama.org

Web Sites

Due to the changing nature of Internet links, Rosen Publishing has developed an online list of Web sites related to the subject of this book. This site is updated regularly. Please use this link to access the list:

http://www.rosenlinks.com/idd/ster

For Further Reading

Aretha, David. *Steroids and Other Performance-Enhancing Drugs.* Berkeley Heights, NJ: Enslow Publishers, Inc., 2005.
This book contains a number of Web site links helpful for research and writing reports.

Clayton, Lawrence. *Steroids.* New York, NY: Rosen Publishing Group, Inc., 1996.
This book covers steroid usage in sports history and details the health risks in abusing the drug.

Dolan, Edward F. *Drugs in Sports.* New York, NY: Franklin Watts, 1986.
Drugs in Sports provides an in-depth look at a number of drugs, including steroids, that athletes abuse in sports.

Jendrick, Nathan. *Dunks, Doubles, Doping: How Steroids Are Killing American Athletes.* Guildford, CT: The Lyons Press, 2006.
This book focuses on the health risks, scandals, and legal issues pertaining to steroid abuse and athletes.

McCloskey, John, and Julian Bailes. *When Winning Costs Too Much: Steroids, Supplements, and Scandal in Today's Sports.* New York, NY: Taylor Trade Publishing, 2005.
This is a study of steroid use in sports, the health risks, and attempts by sporting governing bodies to control the use of steroids.

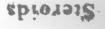

Monroe, Judy. *Steroid Drug Dangers*. Berkeley Heights, NJ: Enslow Publishers, Inc., 1999.
Charts, photographs, and endnotes in this book illustrate the dangers of steroids and the places one can seek help for a steroid abuse problem.

Yesalis, Charles E., and Virginia S. Cowart. *The Steroids Game: An Expert's Inside Look at Anabolic Steroid Use in Sports*. Champaign, IL: Human Kinetics, 1998.
This is an expert's investigation of anabolic steroid abuse among athletes.

Bibliography

Aretha, David. *Steroids and Other Performance-Enhancing Drugs*. Berkeley Heights, NJ: Enslow Publishers, Inc., 2005.

AVERT—Averting HIV and AIDS. "Hepatitis A, B, and C: Symptoms, Treatment, and Facts." Retrieved June 14, 2007 (http://www.avert.org/hepatitis.htm).

Brainpop. "Steroids." Retrieved April 23, 2007 (http://www.brainpop.com/health/personalhealth/steroids).

Monroe, Judy. *Steroid Drug Dangers*. Berkeley Heights, NJ: Enslow Publishers, Inc., 1999.

National Institute on Drug Abuse. "Info Facts: Steroids (Anabolic-Androgenic)." Retrieved May 2, 2007 (http://www.nida.nih.gov/Infofacts/Steroids.html).

National Institute on Drug Abuse. "Research Report Series: Anabolic Steroid Abuse." NIH Publication Number 00-3721. Printed 1991, reprinted 1994, 1996. Revised April 2000.

Schott, Ben. *Schott's Almanac*. New York, NY: Bloomsbury USA, 2006.

Taubert, Kathryn A., and Adnan S. Dajani. "Preventing Bacterial Endocarditis: American Heart Association Guidelines." American Academy of Family Physicians. Retrieved June 2, 2007 (http://www.aafp.org/afp/980201ap/taubert.html).

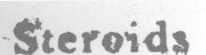

U.S. Department of Health and Human Services and SAMHSA's National Clearinghouse for Drug & Alcohol Information. Retrieved June 3, 2007 (http://ncadi.samhsa.gov).

U.S. Drug Enforcement Administration. "Chapter 10: Steroids." Retrieved May 8, 2007 (http://www.usdoj.gov/dea/pubs/abuse/10-steroids.htm).

You and AIDS. "Symptoms." Retrieved June 3, 2007 (http://www.youandaids.org/About%20HIVAIDS/Symptoms/index.asp).

Index

About the Author

Doretta Lau earned her MFA in writing at Columbia University. As a former competitive karate practitioner, she is interested in fair play in sports at both the amateur and professional levels.

Photo Credits

Cover, p. 1 © www.istockphoto.com/csaba fikker; p. 7 © Reuters/Corbis; p. 11 © Ton Kinsbergen/Photo Researchers, Inc.; p. 13 © Romeo Gacad/AFP/Getty Images; p. 16 © Keyston/Hulton Archive/Getty Images; p. 19 © Ian Waldie/Getty Images; p. 22 © AFP Graphics/Newscom; p. 24 © Dr. P. Marazzi/Science Photo Library/ Photo Researchers, Inc.; p. 25 © Tiziana Fabi/AFP/Getty Images; p. 29 © www. istockphoto.com/Stockphoto4u; p. 31 © www.istockphoto.com/Ronald Bloom; p. 34 © AP Images; p. 36 © Jeff Greenberg/PhotoEdit; p. 38 © CMSP.

Editor: Bethany Bryan; **Photo Researcher:** Amy Feinberg